Computer Systems

An Embedded Approach

Computer Systems

An Embedded Approach

Professor Ian Vince McLoughlin

School of Computing
Medway Campus
University of Kent
Chatham, Kent
United Kingdom

Mc
Graw
Hill
Education

New York   Chicago   San Francisco   Athens
London   Madrid   Mexico City   Milan
New Delhi   Singapore   Sydney   Toronto

Library of Congress Control Number: 2018944194

Computer Systems: An Embedded Approach

1 2 3 4 5 6 7 8 9   QVS   23 22 21 20 19 18

ISBN 978-1-260-11760-8
MHID 1-260-11760-X

Sponsoring Editor
Lauren Poplawski

Project Manager
Radhika Jolly,
Cenveo® Publisher Services

Indexer
Robert Swanson

Editorial Supervisor
Stephen M. Smith

Copy Editor
Surendra Shivam,
Cenveo Publisher Services

Art Director, Cover
Jeff Weeks

Production Supervisor
Pamela A. Pelton

Proofreader
Manish Kumar,
Cenveo Publisher Services

Composition
Cenveo Publisher Services

Acquisitions Coordinator
Elizabeth M. Houde

About the Author

Ian Vince McLoughlin is a professor of computing and is currently head of the School of Computing on the Medway Campus of the University of Kent, in Chatham, United Kingdom. Over a career spanning more than 30 years (so far), he has worked for industry, government, and academia on three continents, with an emphasis on research and innovation. At heart he is a computer engineer, having designed or worked on computing systems that can be found in space, flying in the troposphere, empowering the global telecommunications network, being used underwater, in daily use by emergency services, embedded within consumer devices, and helping patients speak following larynx surgery. Professor McLoughlin is proud to be a Fellow of the IET, a Senior Member of the IEEE, a Chartered Engineer (UK), and an Ingenieur European (EU).

About the Author

Ian Vince McLoughlin is a professor of computing and is currently head of the School of Computing on the Medway Campus of the University of Kent, in Chatham, United Kingdom. Over a career spanning more than 30 years (so far), he has worked for industry, government, and academia on three continents, with an emphasis on research and innovation. At heart he is a computer engineer, having designed or worked on computing systems that can be found in space, flying in the troposphere, empowering the global telecommunications network, being used underwater, in daily use by emergency services, embedded within consumer devices, and helping patients speak following larynx surgery. Professor McLoughlin is proud to be a Fellow of the IET, a Senior Member of the IEEE, a Chartered Engineer (UK), and an Ingenieur European (EU).

Contents

Preface xvii
Acknowledgments xxi
List of Boxes xxiii

1 Introduction 1
   1.1 The Evolution of Computers 1
   1.2 Forward Progress 3
   1.3 Computer Generations 5
        1.3.1 First Generation 6
        1.3.2 Second Generation 7
        1.3.3 Third Generation 8
        1.3.4 Fourth Generation 9
        1.3.5 Fifth Generation 10
   1.4 Cloud, Pervasive, Grid, and Massively Parallel Computers 12
   1.5 Where To from Here? 13
   1.6 Summary 16

2 Foundations 17
   2.1 Computer Organization 17
        2.1.1 Flynn's Taxonomy 18
        2.1.2 Connection Arrangements 18
        2.1.3 Layered View of Computer Organization 20
   2.2 Computer Fundamentals 21
   2.3 Number Formats 25
        2.3.1 Unsigned Binary 26
        2.3.2 Sign Magnitude 26
        2.3.3 One's Complement 27
        2.3.4 Two's Complement 27
        2.3.5 Excess-n 28
        2.3.6 Binary-Coded Decimal 29
        2.3.7 Fractional Notation 29
        2.3.8 Sign Extension 30
   2.4 Arithmetic 31
        2.4.1 Addition 32
        2.4.2 The Parallel Carry-Propagate Adder 32
        2.4.3 Carry Look-Ahead 34
        2.4.4 Subtraction 35
   2.5 Multiplication 37
        2.5.1 Repeated Addition 38
        2.5.2 Partial Products 38

	2.5.3	Shift-Add Method	42
	2.5.4	Booth's and Robertson's Methods	42
2.6	Division		44
	2.6.1	Repeated Subtraction	44
2.7	Working with Fractional Number Formats		46
	2.7.1	Arithmetic with Fractional Numbers	47
	2.7.2	Multiplication and Division of Fractional Numbers	48
2.8	Floating Point		49
	2.8.1	Generalized Floating Point	49
	2.8.2	IEEE754 Floating Point	50
	2.8.3	IEEE754 Modes	51
	2.8.4	IEEE754 Number Ranges	55
2.9	Floating Point Processing		58
	2.9.1	Addition and Subtraction of IEEE754 Numbers	59
	2.9.2	Multiplication and Division of IEEE754 Numbers	62
	2.9.3	IEEE754 Intermediate Formats	62
	2.9.4	Rounding	63
2.10	Summary		64
2.11	Problems		64

3	**CPU Basics**		**69**
3.1	What Is a Computer?		69
3.2	Making the Computer Work for You		70
	3.2.1	Program Storage	70
	3.2.2	Memory Hierarchy	71
	3.2.3	Program Transfer	73
	3.2.4	Control Unit	74
	3.2.5	Microcode	79
	3.2.6	RISC versus CISC Approaches	81
	3.2.7	Example Processor—the ARM	83
	3.2.8	More about the ARM	85
3.3	Instruction Handling		86
	3.3.1	The Instruction Set	86
	3.3.2	Instruction Fetch and Decode	90
	3.3.3	Compressed Instruction Sets	95
	3.3.4	Addressing Modes	97
	3.3.5	Stack Machines and Reverse Polish Notation	101
3.4	Data Handling		102
	3.4.1	Data Formats and Representations	103
	3.4.2	Data Flows	107
	3.4.3	Data Storage	107
	3.4.4	Internal Data	108
	3.4.5	Data Processing	109
3.5	A Top-Down View		113
	3.5.1	Computer Capabilities	113
	3.5.2	Performance Measures, Statistics, and Lies	114
	3.5.3	Assessing Performance	116

3.6   Summary   118
3.7   Problems   119

4   **Processor Internals** **123**
4.1   Internal Bus Architecture   123
       4.1.1   A Programmer's Perspective   123
       4.1.2   Split Interconnection Arrangements   124
       4.1.3   ADSP21xx Bus Arrangement   126
       4.1.4   Simultaneous Data and Program Memory Access   127
       4.1.5   Dual-Bus Architectures   129
       4.1.6   Single-Bus Architectures   131
4.2   Arithmetic Logic Unit   132
       4.2.1   ALU Functionality   132
       4.2.2   ALU Design   133
4.3   Memory Management Unit   136
       4.3.1   The Need for Virtual Memory   136
       4.3.2   MMU Operation   136
       4.3.3   Retirement Algorithms   139
       4.3.4   Internal Fragmentation and Segmentation   140
       4.3.5   External Fragmentation   140
       4.3.6   Advanced MMUs   142
       4.3.7   Memory Protection   143
4.4   Cache   144
       4.4.1   Direct Cache   146
       4.4.2   Set-Associative Cache   148
       4.4.3   Full-Associative Caches   149
       4.4.4   Locality Principles   149
       4.4.5   Cache Replacement Algorithms   150
       4.4.6   Cache Performance   154
       4.4.7   Cache Coherency   156
4.5   Coprocessors   158
4.6   Floating Point Unit   159
       4.6.1   Floating Point Emulation   160
4.7   Streaming SIMD Extensions and Multimedia Extensions   162
       4.7.1   Multimedia Extensions   162
       4.7.2   MMX Implementation   163
       4.7.3   Use of MMX   164
       4.7.4   Streaming SIMD Extensions   164
       4.7.5   Using SSE and MMX   165
4.8   Coprocessing in Embedded Systems   165
4.9   Summary   167
4.10   Problems   167

5   **Enhancing CPU Performance** **173**
5.1   Speedups   174
5.2   Pipelining   174

	5.2.1	Multifunction Pipelines	176
	5.2.2	Dynamic Pipelines	177
	5.2.3	Changing Mode in a Pipeline	178
	5.2.4	Data Dependency Hazard	180
	5.2.5	Conditional Hazards	181
	5.2.6	Conditional Branches	183
	5.2.7	Compile-Time Pipeline Remedies	185
	5.2.8	Relative Branching	187
	5.2.9	Instruction Set Pipeline Remedies	188
	5.2.10	Run-Time Pipeline Remedies	189
5.3		Complex and Reduced Instruction Set Computers	193
5.4		Superscalar Architectures	194
	5.4.1	Simple Superscalar	194
	5.4.2	Multiple-Issue Superscalar	197
	5.4.3	Superscalar Performance	197
5.5		Instructions per Cycle	198
	5.5.1	IPC of Difference Architectures	198
	5.5.2	Measuring IPC	200
5.6		Hardware Acceleration	201
	5.6.1	Zero-Overhead Loops	201
	5.6.2	Address Handling Hardware	204
	5.6.3	Shadow Registers	207
5.7		Branch Prediction	208
	5.7.1	The Need for Branch Prediction	209
	5.7.2	Single T-Bit Predictor	211
	5.7.3	Two-Bit Predictor	212
	5.7.4	The Counter and Shift Registers as Predictors	214
	5.7.5	Local Branch Predictor	214
	5.7.6	Global Branch Predictor	217
	5.7.7	The Gselect Predictor	219
	5.7.8	The Gshare Predictor	221
	5.7.9	Hybrid Predictors	222
	5.7.10	Branch Target Buffer	224
	5.7.11	Basic Blocks	225
	5.7.12	Branch Prediction Summary	227
5.8		Parallel and Massively Parallel Machines	227
	5.8.1	Evolution of SISD to MIMD	230
	5.8.2	Parallelism for Raw Performance	232
	5.8.3	More on Parallel Processing	234
5.9		Tomasulo's Algorithm	239
	5.9.1	The Rationale behind Tomasulo's Algorithm	239
	5.9.2	An Example Tomasulo System	240
	5.9.3	Tomasulo in Embedded Systems	245
5.10		Very Long Instruction Word Architectures	246
	5.10.1	What Is VLIW?	246
	5.10.2	The VLIW Rationale	248
	5.10.3	Difficulties with VLIW	249
	5.10.4	Comparison with Superscalar	250

5.11   Summary   250
5.12   Problems   251

6   **Externals** **255**
6.1   Interfacing Using a Bus   255
    6.1.1   Bus Control Signals   256
    6.1.2   Direct Memory Access   257
6.2   Parallel Bus Specifications   258
6.3   Standard Interfaces   260
    6.3.1   System Control Interfaces   260
    6.3.2   System Data Buses   260
    6.3.3   I/O Buses   267
    6.3.4   Peripheral Device Buses   267
    6.3.5   Interface to Networking Devices   268
6.4   Real-Time Issues   269
    6.4.1   External Stimuli   269
    6.4.2   Interrupts   269
    6.4.3   Real-Time Definitions   270
    6.4.4   Temporal Scope   270
    6.4.5   Hardware Architecture Support for Real Time   272
6.5   Interrupts and Interrupt Handling   273
    6.5.1   The Importance of Interrupts   274
    6.5.2   The Interrupt Process   274
    6.5.3   Advanced Interrupt Handling   280
    6.5.4   Sharing Interrupts   280
    6.5.5   Reentrant Code   281
    6.5.6   Software Interrupts   281
6.6   Embedded Wireless Connectivity   282
    6.6.1   Wireless Technology   282
    6.6.2   Wireless Interfacing   284
    6.6.3   Issues Relating to Wireless   284
6.7   Summary   285
6.8   Problems   285

7   **Practical Embedded CPUs** **291**
7.1   Introduction   291
7.2   Microprocessors Are Core Plus More   291
7.3   Required Functionality   295
7.4   Clocking   298
    7.4.1   Clock Generation   300
7.5   Clocks and Power   301
    7.5.1   Propagation Delay   303
    7.5.2   The Trouble with Current   303
    7.5.3   Solutions for Clock Issues   304
    7.5.4   Low-Power Design   304
7.6   Memory   306
    7.6.1   Early Computer Memory   306

7.6.2   ROM: Read-Only Memory 307
7.6.3   RAM: Random-Access Memory 314
7.7   Pages and Overlays 321
7.8   Memory in Embedded Systems 323
7.8.1   Booting from Non-Volatile Memory 325
7.8.2   Other Memory 327
7.9   Test and Verification 328
7.9.1   IC Design and Manufacture Problems 328
7.9.2   Built-In Self-Test 331
7.9.3   JTAG 333
7.10   Error Detection and Correction 336
7.11   Watchdog Timers and Reset Supervision 340
7.11.1   Reset Supervisors and Brownout Detectors 341
7.12   Reverse Engineering 343
7.12.1   The Reverse Engineering Process 344
7.12.2   Detailed Physical Layout 348
7.13   Preventing Reverse Engineering 353
7.13.1   Passive Obfuscation of Stored Programs 355
7.13.2   Programmable Logic Families 356
7.13.3   Active RE Mitigation 357
7.13.4   Active RE Mitigation Classification 357
7.14   Soft Core Processors 358
7.14.1   Microprocessors Are More Than Cores 359
7.14.2   The Advantages of Soft Core Processors 360
7.15   Hardware Software Codesign 363
7.16   Off-the-Shelf Cores 365
7.17   Summary 367
7.18   Problems 368

8   Programming 371
8.1   Running a Program 372
8.1.1   What Does Executing Mean? 372
8.1.2   Other Things to Note 375
8.2   Writing a Program 376
8.2.1   Compiled Languages 377
8.2.2   Interpreted Languages 381
8.3   The UNIX Programming Model 383
8.3.1   The Shell 384
8.3.2   Redirections and Data Flow 385
8.3.3   Utility Software 387
8.4   Summary 388
8.5   Problems 388

9   Operating Systems 391
9.1   What Is an Operating System? 391
9.2   Why Do We Need an Operating System? 392
9.2.1   Operating System Characteristics 393
9.2.2   Types of Operating Systems 394

9.3   The Role of an Operating System 396
    9.3.1   Resource Management 396
    9.3.2   Virtual Machine 396
    9.3.3   CPU Time 397
    9.3.4   Memory Management 398
    9.3.5   Storage and Filing 400
    9.3.6   Protection and Error Handling 401
9.4   OS Structure 402
    9.4.1   Layered Operating Systems 403
    9.4.2   Client-Server Operating Systems 404
9.5   Booting 405
    9.5.1   Booting from Parallel Flash 406
    9.5.2   Booting from HDD/SSD 408
    9.5.3   What Happens Next 409
9.6   Processes 410
    9.6.1   Processes, Processors, and Concurrency 411
9.7   Scheduling 413
    9.7.1   The Scheduler 414
9.8   Storage and File Systems 417
    9.8.1   Secondary Storage 417
    9.8.2   Need for File Systems 421
    9.8.3   What Are File Systems? 424
    9.8.4   Backup 431
9.9   Summary 433
9.10   Problems 434

10   **Connectivity** **437**
10.1   Why Connect, How to Connect 437
    10.1.1   One-to-One Communications 438
    10.1.2   One-to-Many Communications 439
    10.1.3   Packet Switching 440
    10.1.4   Simple Communications Topologies 441
10.2   System Requirements 443
    10.2.1   Packetization 443
    10.2.2   Encoding and Decoding 445
    10.2.3   Transmission 445
    10.2.4   Receiving 445
    10.2.5   Error Handling 446
    10.2.6   Connection Management 451
10.3   Scalability, Efficiency, and Reuse 453
10.4   OSI Layers 454
10.5   Topology and Architecture 455
    10.5.1   Hierarchical Network 455
    10.5.2   Client-Server Architecture 456
    10.5.3   Peer-to-Peer Architecture 456
    10.5.4   Ad Hoc Connection 457
    10.5.5   Mobility and Handoff 458

10.6   Summary   458
10.7   Problems   459

11   Networking   461
11.1   The Internet   461
       11.1.1   Internet History   462
       11.1.2   Internet Governance   462
11.2   TCP/IP and the IP Layer Model   464
       11.2.1   Encapsulation   465
11.3   Ethernet Overview   469
       11.3.1   Ethernet Data Format   470
       11.3.2   Ethernet Encapsulation   471
       11.3.3   Ethernet Carrier Sense   473
11.4   The Internet Layer   474
       11.4.1   IP Address   474
       11.4.2   Internet Packet Format   476
       11.4.3   Routing   477
       11.4.4   Unicasting and Multicasting   478
       11.4.5   Anycasting   478
       11.4.6   Naming   478
       11.4.7   Domain Name Servers   479
11.5   The Transport Layer   482
       11.5.1   Port Number   482
       11.5.2   User Datagram Protocol   483
       11.5.3   Transmission Control Protocol   483
       11.5.4   UDP versus TCP   484
11.6   Other Messages   485
       11.6.1   Address Resolution Protocol   485
       11.6.2   Control Messages   486
11.7   Wireless Connectivity   486
       11.7.1   WiFi   487
       11.7.2   WiMax   487
       11.7.3   Bluetooth   488
       11.7.4   ZigBee   488
       11.7.5   Near-Field Communications   489
11.8   Network Scales   490
11.9   Summary   490
11.10 Problems   490

12   The Future   493
12.1   Single-Bit Architectures   494
       12.1.1   Bit-Serial Addition   494
       12.1.2   Bit-Serial Subtraction   495
       12.1.3   Bit-Serial Logic and Processing   496
12.2   More-Parallel Machines   497
       12.2.1   Clusters of Small CPUs   497
       12.2.2   Parallel and Cluster Processing Considerations   501
       12.2.3   Interconnection Strategies   502

12.3 Asynchronous Processors 505
    12.3.1 Data Flow Control 507
    12.3.2 Avoiding Pipeline Hazards 508
12.4 Alternative Number Format Systems 509
    12.4.1 Multiple-Valued Logic 509
    12.4.2 Signed Digit Number Representation 510
12.5 Optical Computation 514
    12.5.1 The Electro-Optical Full Adder 514
    12.5.2 The Electro-Optic Backplane 515
12.6 Science Fiction or Future Reality? 517
    12.6.1 Distributed Computing 518
    12.6.2 Wetware 518
12.7 Summary 519

A **Standard Memory Size Notation** **521**

B **Standard Logic Gates** **523**

Index **525**

Preface

Computers in their widest sense—including smartphones, portable gaming systems, and so on—surround us and increasingly underpin our daily lives. This book is dedicated to peeling back the layers from those systems to examine and understand what "makes them tick." That is the motivation behind the emphasis on embedded systems—that and the fact that embedded systems contain truly fascinating technology. Looking inside them, to a technically minded person, is like unwrapping a Christmas gift of knowledge and understanding.

Bookshops (particularly in university towns) seem to overflow with textbooks on topics like computer architecture, computer system design, networking, operating systems, and even embedded systems. Many famous technical authors have tried their hands at writing in this area, yet computers constitute a fluid and ever-advancing area of technology that is difficult to describe adequately with a static textbook that may rapidly become out-of-date. In particular, the rise of embedded computing systems over the past decade or so seems to have surprised some of the traditional authors: Some textbooks persist in regarding computers as being the room-sized machines of the 1950s and 1960s. Other textbooks regard computers as being primarily the desktop and server machines of the 1980s and 1990s. Only a handful of authors have truly acknowledged that the vast majority of computers in modern use are embedded within everyday objects. Few textbooks acknowledge that the future of computing is embedded, connected, and pervasive: There will come a time when the concept of a desktop or even notebook computer seems as anachronistic as the punched card machines of 50 years ago.

In *Computer Systems: An Embedded Approach*, as mentioned, we point our discussion squarely toward this embedded future wherever possible, and use examples from the embedded world, for all three subareas of computer architecture, operating systems, and connectivity. Some topics naturally relate better to embedded processors and are described in that way, but others are handled alongside the more traditional topics that dominate other texts. Wherever possible, examples are given from the embedded world, and related material introduced to describe the relevance of the topics to the readers of today.

Book Structure

The 12 chapters that constitute this textbook can, apart from the introduction and conclusion, be divided roughly into three parts that explain, in turn, the following questions:

- What hardware is inside a modern computer system (embedded or otherwise), how does it work, and how does it fit together?

- What is needed to program a computer to "do" things? How does that software get written, loaded, and executed; how is it organized and presented; and how are the systems inside a modern computer managed?

- How do computers connect together to exchange information and provide distributed servers for users?

In general, following the introduction and foundations chapter, the discussions about hardware are confined to Chapters 3 to 7, those concerning software programming and operating systems are in Chapters 8 and 9, while the networking and connectivity discussions involve Chapters 10 and 11. There is significant internal referencing between parts, but there is no reason that readers must progress through the book sequentially—the three parts contain material that can be read and understood independently of the other parts.

Target Audience

The target audience of readers includes undergraduate students of computing, computer science, computer engineering, computer systems engineering, electronic and computer engineering, electronic and electrical engineering, and variants. The book will equally appeal to those working in other technical disciplines who wish to study the foundations and fundamentals of computers and computing. The introductory sections and the foundations chapter are designed to be suitable for undergraduates in their first 2 years of study, but sufficient depth and pointers to further topics are provided to make this a suitable text for final-year students undertaking courses on computer architecture or computer systems design.

Computer programmers and engineers who are working in the embedded systems industry will find this textbook to be a useful reference, and the emphasis on the ARM processor—which is widely used across industry—will be welcome for many of those for whom this technology is now a livelihood.

Book Design

The material in this book was written from the bottom up, without being based on existing textbooks, apart from some sections in the hardware or computer architecture part of the book, which make use of sections from the author's previous work *Computer Architecture: An Embedded Approach* as a foundation. By taking a fresh approach, and planning the book without being constrained by traditional structures, the text avoids many of the historical blind alleys and irrelevant sideshows that have occurred in computer evolution. This leads to a more precisely defined flow in the writing structure that maintains the sharp focus on embedded systems (although it does not totally ignore the larger machines—many of them contained fascinating examples of ideas that morphed over the years into more important and better technology that has since became ubiquitous).

In creating this book, the author has aimed to write easy-access and readable text that builds reader interest and tries to maintain relevance with the popular forefront of technology as it impacts daily life. However, there are tricky concepts in any field of study, and where those have been encountered, care has been taken to write clear explanatory text, and in many cases provide an informative and intuitive illustration to aid understanding. In addition, there are many explanatory boxes provided throughout, which contain material such as extra worked examples, interesting snippets of information, and additional explanations. All of those features aim to augment the main text and assist the reader in absorbing the information.

SI (System International) units are used throughout the book, including the unusual-sounding "kibibyte" and "mebibyte" capacity measures for computer memory (which are explained in Appendix A). Each of the main chapters in the book is followed with end-of-chapter problems, which have answers available to instructors online. An accompanying website, www.mcloughlin.eu/computer, provides students and other readers additional reference material, links, and opportunities to find out more about many of the topics presented here.

Book Preparation

This book has been prepared and typeset with LaTeX using **TeXShop**. The author wrote the content using **TeXstudio** on Linux Ubuntu– and Apple OS-X–based computers. Line diagrams were all drawn using the OpenOffice/LibreOffice drawing tools, and all graphics conversions have made use of the extensive graphics processing tools that are freely available on GNU/Linux systems. Code examples and most of the embedded system descriptions were sourced from the author's own hardware and software designs. Thanks are gratefully expressed to the GNU Project for the excellent and invaluable GCC ARM compiler, as well as to the Busybox (the Swiss Army knife of embedded systems coding) and ARM/Linux projects.

Several of the images in this book were supplied by online resources such as Wikimedia Commons, as credited in the figure captions. All such figures are under Creative Commons (CC) by attribution (BY) or share alike (SA) licenses. The author would like to acknowledge in particular the excellent resource provided by Wikimedia Commons, as well as the protective licenses from Creative Commons.[1]

Before You Begin

Please take a moment to remember the generations of computer engineers who have worked hard for decades to bring us the mobile, smart, computer, and embedded technologies that modern society thrives on. While we can rightly applaud those great efforts of the past, it is the author's hope that readers will enjoy unwrapping the gifts of understanding and knowledge of embedded computer systems, and will work just as hard to build a better technology future for us all.

Ian Vince McLoughlin

---

[1] The full license text for all CC images used in this book can be viewed at https://creativecommons.org/licenses.

Acknowledgments

Thanks are due most of all to my patient wife Kwai Yoke and children Wesley and Vanessa, all of whose patience and encouragement have contributed to this book. More practically I would like to thank all of the editorial and production staff at McGraw-Hill Education in New York, whose encouragement and enthusiasm have been much appreciated, and whose professionalism has been key in getting this published. But I also acknowledge a debt of gratitude to Gerald Bok and others at McGraw-Hill Asia in Singapore, in particular Gerald's support of my writing career from 2006 to 2013.

I have many friends who I could (and probably should) thank here for their support, encouragement, and influence; however, I would simply like to dedicate this work to my mother. I acknowledge her constant encouragement, not just for writing this and other books, but throughout my entire lifetime. Her high expectations led to my entering academia, and she was always enthusiastic about anything related to education. I can truly say that her memory lives on in all that I do and accomplish. But above all I give glory to the God who made me, guided me, refined me, gave His son to save me, and will eventually welcome me into His presence. All that I am, accomplish, obtain, and achieve, I ultimately dedicate to Him. Except the errors (in this book or elsewhere); those are all mine.

List of Boxes

2.1   Worked Endiness Example 1   23

2.2   Worked Endiness Example 2   23

2.3   Worked Endiness Example 3   24

2.4   Worked Endiness Example 4   24

2.5   What Is a Number Format?   25

2.6   Negative Two's Complement Numbers   27

2.7   Worked Examples of Number Conversion   27

2.8   Is Binary a Fractional Number Format?   29

2.9   Fractional Format Worked Example   30

2.10 Sign Extension Worked Example   31

2.11 Exercise for the Reader   33

2.12 Worked Example   33

2.13 Exercise for the Reader   34

2.14 Worked Examples of Two's Complement Multiplication   40

2.15 Exercise for the Reader   43

2.16 Booth's Method Worked Example   43

2.17 Long Division Worked Example   45

2.18 Worked Examples of Fractional Representation   47

2.19 Worked Example of Fractional Division   49

2.20 IEEE754 Normalized Mode Worked Example 1   52

2.21 IEEE754 Normalized Mode Worked Example 2   52

2.22 Exercise for the Reader   53

2.23 IEEE754 Denormalized Mode Worked Example   54

2.24 IEEE754 Infinity and Other "Numbers"   54

2.25 Worked Example Converting from Decimal to Floating Point   57

2.26 Floating Point Arithmetic Worked Example   60

2.27 IEEE754 Arithmetic Worked Example   61

3.1   How the ARM Was Designed   83

3.2   Illustrating Conditionals and the S Bit in the ARM   89

3.3   Condition Codes in the ARM Processor   91

3.4   Understanding the MOV Instruction in the ARM   94

3.5   A Huffman Coding Illustration   95

3.6   Recoding RPN Instructions to Minimize Stack Space   102

3.7   Data Types in Embedded Systems   104

3.8   Standardized Performance   115

4.1   Exploring ALU Propagation Delays   134

4.2   MMU Worked Example   137

4.3   Trapping Software Errors in the C Programming Language   143

4.4   Cache Example: The Intel Pentium Pro   145

4.5   Direct Cache Example   147

4.6   Set-Associative Cache Example   148

4.7   Cache Replacement Algorithm Worked Example 1   152

4.8   Cache Replacement Algorithm Worked Example 2   153

4.9   Access Efficiency Example   155

4.10 MESI Protocol Worked Example   158

4.11 An Alternative Approach—FPU on the Early ARM Processors   160

5.1   Pipeline Speedup   176

5.2   WAW Hazard   181

5.3   Conditional Flags   182

5.4   Branch Prediction   185

5.5   Speculative Execution   186

5.6   Relative Branching   187

5.7   Scoreboarding   196

5.8   ZOL Worked Examples   205

5.9   Address Generation in the ARM   207

5.10 Aliasing in Local Prediction   217

6.1   DMA in a Commercial Processor   257

6.2   Bus Settings for Peripheral Connectivity   259

6.3   The Trouble with ISA   262

6.4   Scheduling Priorities   272

6.5   ARM Interrupt Timing Calculation   277

6.6   Memory Remapping during Boot   279

7.1 Configurable I/O Pins on the MSP430   295

7.2 Pin Control on the MSP430   296

7.3 NAND and NOR Flash Memory   310

7.4 Memory Map in the MSP430   326

7.5 Using JTAG for Finding a Soldering Fault   333

7.6 Using JTAG for Booting a CPU   335

7.7 Hamming (7, 4) Encoding Example   339

7.8 Hamming (7, 4) Encoding Example Using Matrices   340

7.9 Bus Line Pin Swapping   352

9.1 More about the Kernel   402

9.2 Finding Out about Processes   411

9.3 Practical File Systems—ext4   422

9.4 File System Configuration—ext4   423

10.1 Everyday Errors   448

11.1 RFCs   463

11.2 IPv4 Address Allocation   475

12.1 Some Examples of Large-Scale Cluster Computers   504

12.2 Example CSD Number   513

Computer Systems

An Embedded Approach

Introduction

1.1 The Evolution of Computers

A glance at the picture in Figure 1.1 of Charles Babbage's Analytical Difference Engine of 1834 reveals that computers have come a long way since then. Their story has been one of ever-increasing processing power, complexity, and miniaturization, but also of increasing usefulness, which has driven greater and greater levels of adoption and ubiquity.

Despite looking nothing like a modern smartphone, quite a few of the techniques used in Babbage's machine, as well as the early electrical computers of the 1940s, can still be found in today's computer systems. This stands testament to the amazing foresight of those early pioneers, but also demonstrates that certain basic operations and structures are common to computers of almost all types, over all ages. With the considerable benefit of hindsight, we have the opportunity to look back through computing history and identify the discovery of those truths, or emergence of those techniques. We will also see many short-lived evolutionary branches that seemed, at the time, to be promising paths to future progress, but that quickly disappeared.

What seems likely then is that the computers of tomorrow will build on many techniques found in those of today. A snapshot of current techniques (as any computing textbook has to be) needs to recognize this fact, rather than presenting the technology as being set in stone.

This book will loosely follow the evolutionary trend. Early chapters will focus on computer fundamentals. Mastery of these will allow a student to understand the basic workings of a computer on paper, however slow and inefficient this might be. These early chapters will be followed by consideration of architectural speedups and advanced techniques in use today. These are separated from the fundamentals because some of them may turn out to be the current "evolutionary blind alleys," but nevertheless they are among the many techniques currently driving Moore's law[1] so quickly forward.

Every now and then something completely revolutionary happens in computer architecture—these break the evolutionary trend and consign many past techniques that gave incremental performance increases to oblivion. Without a crystal ball, this book will not attempt to identify future disruptive technology (although that will not prevent us from making an informed guess about some advanced technologies; see Chapter 12),

---

[1] Gordon Moore of Intel remarked, in 1965, that transistor density (and hence computer complexity) was increasing at an exponential rate. He predicted that this rate of improvement would continue in the future, and the idea of this year-on-year improvement subsequently became known as Moore's law.

POR.TION OF BABBAGE'S DIFFERENCE ENGINE.

FIGURE 1.1 A portion of Babbage's Analytical Difference Engine, as drawn in *Harper's New Monthly Magazine*, Vol. 30, Issue 175, p. 34, 1864. The original engine documents, and a working reconstruction, can be seen today at the Science Museum, London.

but we are free to examine the performance accelerators of past and present computer systems. Above all, we can learn to understand and appreciate the driving factors and limitations within which computer architects, computer system designers, software developers, and networking engineers have to work.

CHAPTER 1

1.2 Forward Progress

Computers have long followed an evolutionary path of improvement. Despite occasionally rare, but welcome, disruptive breakthroughs, computing history is full of many small incremental improvements over the years.

Of course, something as complex as a computer requires an intelligent engineer to have designed it, and we can often identify them by name, especially those who have made significant improvements (a few of them are still alive today to tell us about it). Furthermore, the design and history of the pioneering machines, often constructed at great expense, was often very well documented.

Given this fact, one would expect the history of development in computing to be very definite: There should be little confusion and controversy regarding the pioneering machines from a few decades ago. Unfortunately that is not the case, and there exists a very wide range of opinions, with little agreements upon exact dates, contributions, and "firsts." Just pick up any two books on computer architecture or computer history and compare them. For our present purposes, we will begin the modern era of computing with the invisible giant, Colossus.

The Colossus, constructed by engineer Tommy Flowers in 1943, and programmed by mathematician Alan Turing and colleagues in Bletchley Park, England, is now generally credited with being the world's first programmable electronic computer. A picture of one of the Colossus machines in use is shown in Figure 1.2. Colossus was built during the Second World War as part of the ultimately successful U.K. code-breaking effort. In particular, it was designed to crack the German Enigma code, which was fiendishly difficult to decipher. Unfortunately for the British computer industry, Colossus was classed *top secret*, and it remained hidden from the world for 50 years. With a typically grandiose—although secret—pronouncement, Prime Minister Winston Churchill ordered the machines to be "broken into pieces no larger than a man's hand." All documents and papers relating to it were ordered to be destroyed after the war. The plans and schematics were burned by the designers, and its codebreaker operators were sworn to secrecy under peril of imprisonment (or execution) for treason.

The action to hide this machine was successful. Despite the occasional unverified rumor over the years, the existence of Colossus was only revealed publicly when the few remaining documents were declassified in the year 2000, and a government report containing the information was released. For this reason, Colossus is not even mentioned in older descriptions of computer history: An entire generation of computer architects had never even heard about it.

However, there were other very well-known and reported machines of similar vintage to Colossus that began operation in the following years. One of the most famous, Electronic Numerical Integrator and Computer (ENIAC) was commissioned and built in the United States. While Colossus remained totally hidden, ENIAC, operational by 1944, is said to have snapped up worldwide patents to digital computing devices. Many textbook authors, not knowing anything about Colossus, hailed ENIAC as the first modern computer. In fact apart from being operational earlier, Colossus was more like today's computers than ENIAC. That is because Colossus was binary, while ENIAC was decimal. However, neither were easily reprogrammable, requiring adjustments to switch settings (Colossus) and changing wire plug positions (ENIAC) to make changes to its operating code.

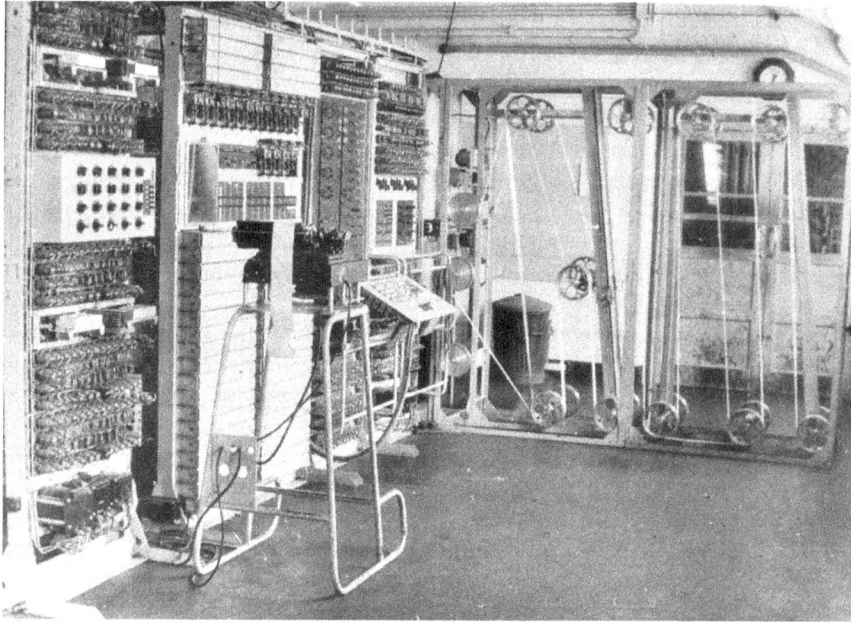

FIGURE 1.2   A 1945 photograph of Colossus 10 in Block H at Bletchley Park. The room now contains the Tunny galley of The National Museum of Computing. (Photo courtesy of the U.K. Public Record Office.)

Amazingly, Charles Babbage's Analytical Difference Engine of over a century earlier, being both digital rather than analog and fully programmable, was in some ways more advanced than either of these first electronic computers. Babbage even designed a printer peripheral that could literally "write out" the results of numerical computations. Babbage's machine also had a full programming language that could handle loops and conditional branching. This led Babbage's friend who worked on the machine, Ada Byron, Countess of Lovelace (the child of famous poet Lord Byron), to write the world's first computer program. Possibly the first and last time in history that poetry and programming have mixed.

Between the Analytical Difference Engine and Colossus, the computing field was not totally deserted: German Konrad Zuse had an electrical computer working around 1940–1941, based on relays (therefore, classified as electrical rather than electronic). Another creditable early attempt at an electronic computer was the Atanasoff-Berry machine at Iowa State College in 1941. Although not programmable, and plagued by unreliability, this demonstrated several early concepts that helped to advance the state of the art in computing.

The advent of the transistorized computer has similarly contentious dates. The transistor, invented at Bell Labs in 1948, was low power and small—ideal characteristics for building a computer, though the early transistors were somewhat less reliable than valves.[2] The first transistor-based machine was probably the transistor computer at Manchester University running in 1953, although several texts again afford pride of place to the TX-0 at Massachusetts Institute of Technology, which became operational in 1956 and was a fine machine that advanced the state of knowledge significantly.

Most of the machines mentioned up to this point were "programmed" by flipping switches or manually plugging and unplugging wires. In that sense they were programmable, but very slowly. Modern computers, by contrast, have programs stored in memory, which is potentially much more convenient to program and debug. The first stored-program computer was again at Manchester University and was called the SSEM (known affectionately as the "Baby"). This was first demonstrated working in 1948. Meanwhile, another famous stored-program computer was being built by Maurice Wilkes of Cambridge University. This was EDSAC (Electronic Delay Storage Automatic Calculator), which began operation in May 1949. The equally famous U.S. Army machine with a similar name, EDVAC (Electronic Discrete Variable Automatic Computer), was also a stored-program binary device, reported operational in 1951–1952 (despite construction starting as early as 1944).

Clearly, early computers were either university-built or related to military calculation in some way (e.g., Colossus and EDVAC). Among the former, Manchester University made several important contributions to the history of computing, although its role has been somewhat overlooked in some computer textbooks. Interestingly, it was Manchester that produced the world's first *commercial* computer, the Ferranti Mark 1, in 1951. The Ferranti Mark 1, very closely followed by the EDSAC-derived LEO, was a commercial computer that ran accounting programs for the ubiquitous Lyons Tea Houses from Spring 1951 onward. Despite the U.K.-centric nature of the early computer business, the largest and most important commercial computer ventures for the following half century were indisputably located in the United States. That remained true until the rise of embedded systems, but the astonishingly rapid growth in the adoption of the ARM processor, developed in Cambridge, has swung the pendulum back to the United Kingdom. At least for a while; ARM was sold to Japanese company SoftBank in September 2016.

Table 1.1 identifies a handful of the computing world firsts, along with the year when they were reported to have become operational. A glance at the progression of steps in this table goes a long way toward explaining how today's computer is very much evolutionary rather than revolutionary, although one wonders what happened to the 1960s.

1.3 Computer Generations

Sometimes computers, just like humans, were historically described in terms of their generation. This was a classification that built up over the years, based mostly around the construction method, computing logic devices, and usage of computers.

---

[2] Glass thermionic tubes containing tiny filament electrodes in a partial vacuum were the basic logic switches used in most early computers. Valves are usually better known as "vacuum tubes" or simply "tubes" in North America. Interestingly, although they are long defunct in computing, they have become sought-after items for very high-end audio amplification equipment.

Year	Location	Name	Description (first used as)
1834	Cambridge	Analytical Difference Engine	Programmable computer
1943	Bletchley	Colossus	Electronic computer
1948	Manchester	SSEM (Baby)	Stored-program computer
1951	Boston	MIT Whirlwind 1	Real-time I/O computer
1953	Manchester	The transistor computer	Transistorized computer
1971	California	Intel 4004 CPU	Mass-market CPU and IC
1979	Cambridge	Sinclair ZX-79	Mass-market home computer
1981	New York	IBM PC	Personal computer
1987	Cambridge	ARM-based Acorn A400	Home computer with RISC CPU
1990	New York	IBM RS6000 CPU	Superscalar RISC processor
1998	California	Sun picoJAVA CPU	Computer based on a language

TABLE 1.1   A short list of some of the more prominent landmark machines in the evolution of computer technology.

Anyone who saw computer magazine advertisements in the 1980s may remember how manufacturers cashed in on these generations to repeatedly advertise new products as being the fifth generation (even though they evidently were not). Those were the days when virtually every computer came with a different operating system—in fact there was an enormous variation in both hardware and operating software, whereas today's world is dominated by just a handful of operating systems (mainly Linux, Microsoft Windows, Mac-OS, Android) and computing devices tend to sell on their hardware features. Although the term "computer generation" is seldom used these days, we will briefly review the five generations, their characteristics, and give some examples.

1.3.1   First Generation

- Based on vacuum tubes—usually occupying an entire room.
- Short MTBF (mean time before failure)—only a few minutes between failures.
- Base-10 arithmetic.
- Programming may use switches, cables with plugs, or be hard wired. A few were stored-program machines.
- No programming languages above basic machine code.
- Introduction of von Neumann architecture.

The best-known example, the ENIAC, consumed over 100 kW of power yet could only deliver around 500 additions per second. This monster used 1800 valves, constructed in a 30-ton, 1300-square-meter machine. The user interface (typical for machines of this generation) is shown in Figure 1.3. ENIAC was designed by the U.S. Army for solving ballistic equations as a means of calculating artillery firing tables.

FIGURE 1.3   Two women operating the ENIAC's main control panel. (U.S. Army photo.)

The Colossus (also shown during wartime; see Figure 1.2) was equally vast and dedicated—at least in its early years—to code breaking: number crunching that broke the powerful and secret Enigma code, contributing to Allied victory in the Second World War. It is reported that, having finally broken the Enigma code, one of the first decoded German messages said something like "we're going to bomb Coventry." Not wanting to alert the enemy that the code had been cracked, the government decided not to warn the inhabitants of that major Cathedral city, many of whom were later killed or injured as the bombs rained down.

1.3.2   Second Generation

- Transistor based, but still heavy and large
- Much better reliability
- Generally using binary logic
- Punched card or tape used for program entry
- Support for early high-level languages
- These machines were often bus based

The CDC6000 of the time was renowned for its intelligent peripherals. But it is another example, the PDP-1 with 4 Ki words of RAM running at up to 0.2 MHz, that is perhaps the best known. This remarkable machine led the now sadly defunct Digital Equipment Corporation (DEC) to prominence. The PDP-1 was available at a price of around

FIGURE 1.4   Photograph of the Oslo PDP-7 taken by Tore Sinding Bekkedal.

US$100,000, but it had an impressive array of peripherals: light pen, EYEBALL digital camera, quadraphonic sound output, telephone interface, several disc storage devices, a printer, keyboard interface, and a console display. The PDP-1 was first in an innovative range of 16 PDP computers from DEC. A later example is shown in Figure 1.4, the PDP-7 plus several of its peripherals.

1.3.3   Third Generation

- Computation using integrated circuits (ICs)
- Good reliability
- Emulation possible (microprograms)
- Multiprogramming, multitasking, and time sharing
- High-level languages common, some attempts at a user interface
- Use of virtual memory and operating systems

The very popular and versatile IBM System/360 boasted up to 512 kibibytes of 8-bit memory and ran at 4 MHz. It was a register-based computer with a pipelined central processing unit (CPU) architecture and memory access scheme that would probably appear familiar to programmers today. IBM constructed many variants of the basic machine for different users and most importantly opted for a microcode design that could

FIGURE 1.5 IBM System/360. (Photograph by Ben Franske, courtesy of the Wikipedia IBM System/360 page.)

easily emulate other instruction sets: This guaranteed backward compatibility for users of the second-generation computers who had invested very significant sums of money in their now-obsolete machines (so being able to run their legacy programs on the new System/360 made great business sense—an approach to computing that IBM followed into the 1990s and beyond). Modified and miniaturized, five of these computers performed the number crunching in the NASA space shuttles.

Although not quite room-sized, the basic S/360 was still a physically large device as Figure 1.5 amply illustrates.

1.3.4 Fourth Generation

- Using VLSI (very large-scale integration) ICs.
- Highly reliable and fast.
- Possible to integrate the entire CPU on a single chip.
- DOS and CP/M operating systems and beyond.
- These are today's computers.

Examples are profuse, including all desktop and notebook PCs. Figure 1.6 shows an example of a customized RiscPC. Acorn's RiscPC (1994–2003) was one of the last computer model ranges produced by that company, most of which were based around their own RISC-based processor design called the ARM (which we will meet again many times while progressing through this book). Despite creating an innovative processor architecture and one of the most advanced windowing operating systems of the time,

FIGURE 1.6   An Acorn RiscPC computer, customized by OmniBus Systems Ltd., and running their OUI software.

Acorn ceased trading before it could release RiscPC-2. Apple, by contrast, displayed substantially better marketing skills when they released their original 333 MHz iMac with a choice of five flavors (colors). They subsequently reverted to a premium all-aluminum design, illustrated in Figure 1.7.

1.3.5   Fifth Generation

- Natural interaction between humans and computers
- Very high-level programming languages—maybe even programming in English
- May appear intelligent to the user

There are no confirmed examples at time of this writing. When such examples arrive, it is quite possible that there will be nothing worth photographing: a cloud service located in a data center hundreds of miles away, or scattered clusters of tiny embedded computers distributed around us—probably not a beige (or aluminium) box in sight, but perhaps a loudspeaker and microphone.

FIGURE 1.7   The Intel CPU-based Apple iMac is a stylish and user-friendly machine running a reliable UNIX-based operating system. This machine has a 27-inch screen size and dates from 2007. (Photograph courtesy of Matthieu Riegler, Wikimedia Commons.)

In fact, maybe something like the Amazon Echo device, shown in Figure 1.8. Or perhaps it is Apples' smaller but equally desirable iPhones, which are reputed to contain at least eight separate ARM processor cores, *excluding* the main CPU.

FIGURE 1.8   The Amazon Echo device, providing speech-based interfacing to Alexa, a nascent virtual digital assistant. (Courtesy Frmorrison, Wikimedia Commons.)

1.4 Cloud, Pervasive, Grid, and Massively Parallel Computers

Consider the history of computers. In the beginning these were room-sized machines, whether mechanical or electrical, serviced by a dedicated staff of technicians. Relentless technological progress allowed electrical valve-based hardware to be replaced with smaller transistors. The room-sized computer started to shrink. Integrated circuits were then invented to carry multiple transistors, starting with hundreds, then thousands, millions, billions, and beyond. The 8-bit MOS Technology Inc./Rockwell 6502 processor released in 1975 contained around 4000 transistors in a 40-pin dual in-line package (DIP). By 2008 Intel had reached 2 billion transistors on a single chip.

So the story so far has been room-sized computers shrinking, first to several refrigerator-sized units, then to a single unit. Further shrinkage into a desktop box heralded the era of the personal computer (PC). PCs in turn became smaller; "luggables" appeared in the early 1980s, and then portables, laptops, notebooks, and palm computers appeared. Today, it is possible to purchase a fully embedded computer with sensors and digital video camera within a capsule that can be swallowed to aid in medical diagnosis.

So is this a story of one-way miniaturization? Well the answer has to be "no" because computers have also become larger in some respects. The benefits of networking, such as the Internet, allow computers to easily interlink and potentially to share computation resource between themselves. What were once single computing jobs can now be parallelized across multiple computing elements, or computer clusters, even in geographically diverse configurations. We will consider all of these types of parallelism and interconnection later, in particular looking at how it is achieved—but at this point we simply recognize that systems are becoming more distributed and yet more interlinked, and this statement is true whether we are considering computers, cellphone technology, or industrial computing.

Given that the tasks we need to perform can either be executed on a single small box, or spread around and shared among several machines surrounding us (including embedded ones), the question becomes how do we define "a computer"—is it the box itself, or is it the "thing" (or "things") that execute the tasks?

Fifty years ago the question was easy to answer because "the computer" was in "the computer room." Today, a single beige box resting on my desk may well contain two or more CPUs, each of which may contain several computing cores, and yet we refer to that box in the singular as "a computer." When we perform a Web search, the query will probably be sent to Google where it is processed by a "server farm" containing upward of 10,000 computer elements (each one similar to a desktop PC). When server farms like this cooperate to perform processing, they can be classed as a supercomputer, again in the singular ("a supercomputer" rather than "a bunch of supercomputers").

What this means is that the computer has become large again, and yet it consists of many small individual computing elements. The collective computers are becoming bigger and bigger, but the elements they are made of are becoming smaller and smaller.

One beautiful example of a large collection of computers working together is the Barcelona Supercomputer Center. Installed in the Torre Girona chapel in Barcelona, MareNostrum 4 began operation in June 2017 (Figure 1.9).

FIGURE 1.9 The beautiful MareNostrum installation developed by the Barcelona Supercomputing Center in the Torre Girona chapel (www.bsc.es). (Photograph courtesy Wikimedia Commons.)

1.5 Where To from Here?

The process of miniaturization is set to continue. More and more products, devices, and systems contain embedded computers, and there is no sign that this trend will die out. Computer speeds also will continue to increase. After all, there is a pretty amazing track record to this: Consider the graph in Figure 1.10, showing how computers have progressed in speed since the earliest days. The y-axis shows the log of the computer peak performance in floating point operations (FLOP) per second (FLOP/s), because the figures are impossible to plot meaningfully *without* using a log scale. For example:

log(FLOP/s)	FLOP/s	
1	1	
3	1000	1 kFLOP/s
6	1,000,000	1 MFLOP/s
9	1,000,000,000	1 GFLOP/s
12	1,000,000,000,000	1 TFLOP/s
15	1,000,000,000,000,000	1 PFLOP/s
18	1,000,000,000,000,000,000	1 EFLOP/s

Peak computer performance, floating point operations (FLOP)/second

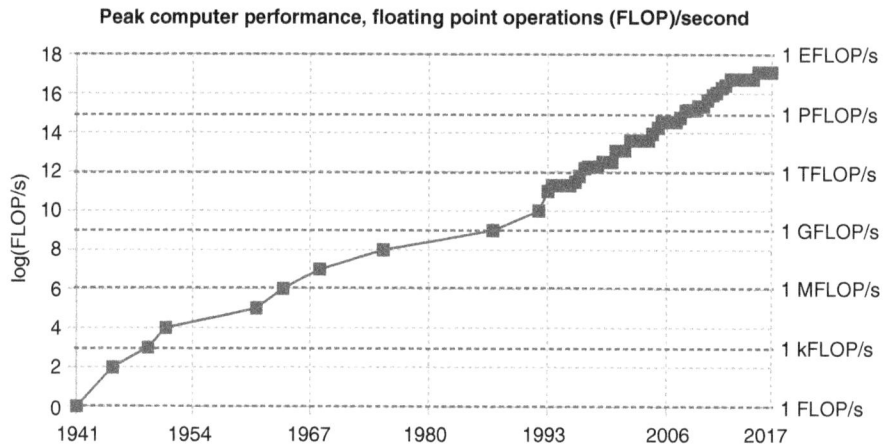

FIGURE 1.10   The amazing progression of computer calculating speeds from the earliest days. The 1941–1993 data was provided to the author by Professor Jack Dongarra of the University of Tennessee, one of the founders of http://www.top500.org, from which the remaining 1993–2017 data was obtained.

At the time of this writing, the largest supercomputers are just on the verge of reaching exaFLOP (EFLOP) per second performance, but if Figure 1.10 shows us anything about the relentless increase in performance, it is that 1 EFLOP/s machines will be with us soon. Although, with all these figures, we should remember that the various definitions of the word "computer" have changed several times throughout their brief history captured here.

At this point, it is also worth pausing for a moment to consider the sheer magnitude of the progress that has been achieved. I can think of no other sphere of life in which such incredible sustained performance improvement year on year has been achieved. Much of the driving force has been there, thanks to the miniaturization techniques and ingenuity of major industry players such as ARM, Intel, and AMD, but also the communications industry and the success of the global Internet. Parallel computing has emerged as the main technique of choice in building the world's fastest computers, and these parallel computers are physically large. In fact, the days of a central computer facility, the main-frame, or computer room, could well be returning. The difference is that the modern "mainframe" may now be located in a different country to its users (instead of down the corridor), and users connect through a mixture of wireless and wired Internet. These mainframes are also energy hungry. Some server farms are said to consume as much energy as an entire town. With that in mind, perhaps the most sensible approach is to locate all such mainframes in cold countries where excess heat can go toward warming nearby communities?

The technology to separate bulk computing resources from the point at which that computer power is needed mostly exists today—namely wired or wireless connectivity—but the services and software able to make use of such a computing model have been slower to appear.

However, the fact that the computing technology we need mainly exists already does not mean that it is time to abandon the advance and improvement of computers and

FIGURE 1.11   Samsung Galaxy S8+, Galaxy S7 Edge, and Galaxy S8 (left to right). (Photograph by Vernon Chan from Kuala Lumpur, Malaysia, courtesy of Wikimedia Commons.)

their architecture (which would mean you can stop reading here), but it does mean that the focus must change. The focus needs to shift from big and powerful to small and low power, from large-scale number crunching to embedded and application specific, and from isolated computing in a fixed location to networked computing everywhere.

Returning to the educational aims of this book for a moment, those designing embedded or even desktop computer systems have traditionally asked questions such as "What processor shall I use in my system?" and "How do I get this processor to work in my system?" This book does provide the kind of background necessary to enable answering both of those questions, but also allows new questions to be asked, and answered, such as "Should I use a lightweight CPU and connect to a remote server, or do all processing internally in a more powerful CPU?" or "Is it better to have a single fast CPU in my application or multiple slower ones?"

The fact that computing is now overwhelmingly about embedded and mobile systems, despite the existence of huge supercomputers like the MareNostrum 4, is due to the pervasiveness of computer technology that impacts our daily lives more and more. Consider the case of modern smartphones, shown in Figure 1.11, which reportedly contains something like nine separate microprocessors, almost all of which are ARM based. If we are to predict the computing future at this point, the smart money would probably be on two converging trends: *fewer but bigger* for large-scale number-crunching (specifically, huge clusters of powerful computers in remote server farms), and *more but smaller*, referring to personalized computing devices (pervasive computing; meaning having computers everywhere).

We finish this discussion by noting that the operating software for both extremes of computing is primarily, and increasingly, Linux based. Meanwhile the processor of choice for embedded systems is the ARM (and perhaps for future server farms too). Both technologies will be well explored in the remainder of this book.

1.6   Summary

You, the reader, may not build the world's fastest supercomputer (or maybe you will, who knows?), but hopefully you will be designing or programming some amazing embedded systems in future.

This chapter has presented a historical perspective of computing: relentless forward progress, many huge leaps in technology and understanding, but millions of small incremental improvements. Isaac Newton is famously said to have written:

If I have seen further it is by standing on the shoulders of Giants.

And this could not be more true of most computer designers. You cannot really get closer to standing on the shoulders of giants than when you use an existing computer to design the next one.

With this perspective behind you and confidence in ongoing future progress in this field, it is now time to learn the techniques (and some secrets) from the designers of computing systems over the past few decades. The following chapters will begin this process by covering basic and foundational techniques, before considering speedups and performance-enhancing techniques—whether mainframe computers, desktop machines, or embedded systems. Later we will spend more time investigating embedded systems themselves, before we move on to programming, operating systems, and exploring how to interconnect computers with networking and the Internet. At the end of this book, we will do some more crystal-ball gazing to try and identify some of the promising, but unusual, techniques on the horizon of the computing world.

CHAPTER 2
Foundations

In this chapter, we will introduce much of the background knowledge that is needed to appreciate the design of a modern central processing unit (CPU). We will consider how computers are organized and classified, and define many of the terms used to describe computer systems throughout this book. In addition, we will discuss computer arithmetic and data representations, and look at a few of the structural building blocks that we are going to encounter later when analyzing computer systems.

2.1 Computer Organization

What is a computer? Although we asked this question in the previous chapter, we did so by looking externally: We had defined a computer as a device able to perform computing tasks—either big or small, single CPU or multiple CPU, distributed or localized.

In this chapter, we now look internally and ask what is inside something to make it be called a computer. We ask what are those internal elements, and how are they connected. But to answer such questions, we need to first recognize that there exists a vast range of possibilities inherent in the structure of different computers. Looking at some of today's desktop computers, many of the peripheral elements traditionally connected around a CPU are subsumed within a single integrated circuit (IC). This would not be recognizable as a computer to the early pioneers. However, the main, historic computer elements are usually still present—even when they are not at first immediately identifiable. In embedded systems the trend toward integration is even more apparent—system on chip (SoC) processors that integrate almost all required functions on a single chip are now predominant.

Furthermore, not all computers are organized in the same way, or have the same requirements. After all they could range in size from a warehouse-sized supercomputer, down to a tablet computer, a wrist-worn fitness tracker, or even tiny computers embedded in a pill that can be swallowed or injected under the skin. It would be a little unusual if such different-sized systems all had identical design or identical software, so we should expect a large range of possibilities.

Despite this vast range of sizes and uses, most computer systems comprise functional blocks that are broadly similar. The placement of these blocks as being inside or outside an integrated CPU is a design or cost consideration, as are the speed and type of interconnections between them. The size, complexity and performance of the blocks will vary widely, but they are essentially doing the same type of job (although at different speeds). Internal interconnections (buses) are generally parallel, and again the "width" and clocking speed are design considerations, but serial buses are becoming more popular. Lower cost and complexity serial buses also now dominate external connectivity, and they use similar principles but speeds vary widely.

Finally, as well as there being big and small, fast and slow functional blocks inside a computer, another design consideration is to have multiple copies of each functional block present, or have multiple interconnections between blocks.

With such bewildering variety, there is a need to classify the range of architectural possibilities in some way. We will attempt to do so, starting with Flynn's taxonomy.

2.1.1 Flynn's Taxonomy

This widely used scheme was first devised by Michael Flynn in 1966 as a comprehensive classification scheme for describing computer systems and their multiple functional units. It categorizes computers based on the number of instruction streams and the number of data streams that are present.

An instruction stream can be thought of as the flow of commands to a data processing unit to modify data (in a data stream) passing through the unit. This is represented diagrammatically in Figure 2.1, which shows four examples of different connection arrangements. These are namely:

- *Single instruction stream, single data* (SISD): A traditional computer containing a single CPU receiving its instructions from a stored program in memory, and acting on a single data stream (shown in this case as one instruction acting upon one item of data).

- *Single instruction stream, multiple data* (SIMD): A single instruction stream acting on more than one item of data. For example, given the numbers 4, 5 and 3, 2, a single SIMD instruction could perform two separate additions of $4+5$ and $3+2$. An example of this arrangement is an array or vector processing system which can perform identical operations on different data items in parallel.

- *Multiple instruction stream, single data* (MISD): A rare combination of overspecified multiple instructions acting on a single data stream. This redundancy could possibly be useful in fault-tolerant systems.

- *Multiple instruction stream, multiple data* (MIMD): These systems are arranged similarly to multiple SISD systems. In fact, a common example of an MIMD system is a multiprocessor computer.

Although Flynn originally designed his taxonomy to describe processor-level arrangements, the same considerations can equally be applied to units within a processor. For example, Intel's Multimedia Extension (MMX) which was introduced in Pentium processors and later expanded into Intel's streaming SIMD extensions (SSE), is an example of an SIMD arrangement—it allows a single instruction to be issued which can cause an operation on multiple data items (such as eight simultaneous additions on different pairs of data). We will cover MMX along with SSE later in Section 4.7.

2.1.2 Connection Arrangements

Another common description of processor architectures is based on whether the program instructions and data items are handled together or separately.

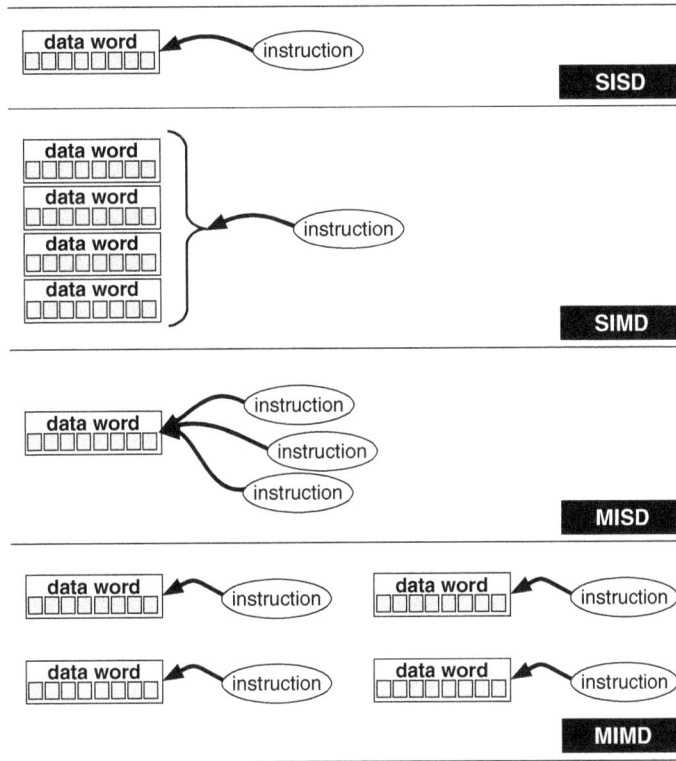

FIGURE 2.1   Illustration of Flynn's taxonomy of SISD, SIMD, MISD, and MIMD processing showing the relationship between instructions and data being acted upon. The taxonomy actually refers to streams of data and instructions rather than individual items, so consider this image as representing a snapshot in time.

- *Von Neumann* systems are those that share resources for storage and transfer of data and instructions. Many modern computers fall into this category by virtue of storing programs and data in a single block of shared random-access memory (RAM), and using a single bus to transfer items from memory to CPU. However, buses have a maximum bandwidth (meaning you can only "squirt" through a certain amount of data per second, usually measured in Mbits/s or Gbits/s), and instructions inside a CPU usually operate on data items. So both the instructions and the data items have to be transferred to the CPU during normal operation. If both are located in shared memory, and share a bus, then they share the bandwidth too. This means that performance can be limited by having to share the bus bandwidth.

- *Harvard architecture* systems have separate data and instruction storage and transfer. This allows instructions and data items to be transferred simultaneously so that both can have the full bandwidth of their respective memory buses. Such systems can be high performance; however, this comes at a price, because fast memory and fast buses are both costly in terms of silicon real estate.

- Other architectures include systems with multiple dedicated buses (such as the ADSP2181 internal buses), shared data/instruction address bus but separate data buses or similar. Chapter 4 will introduce and explain internal bus arrangements further.

Some CPUs such as the Intel StrongARM were advertised as being Harvard architecture, although they interfaced to shared memory via a single bus. In the case of the StrongARM, it had a Harvard architecture internally since it contained separate blocks of internal data and instruction cache memory, although it had an external von Neumann connection arrangement (i.e., single shared external bus).

2.1.3 Layered View of Computer Organization

It is sometimes useful to consider a computer system as a number of interlinked layers. This is illustrated in Figure 2.2 in which the operation of connecting between layers is described, also as a hierarchy of operations.

From the bottom up, a computer (or CPU) can be viewed as being simply a collection of electronic gates performing logical operations on binary. These logical operations are switched, or controlled, to perform the required tasks by an internal control unit. The control unit either uses microprograms or a state machine to operate as directed by a sequence of instructions that are issued from a program (which is just a structured list of instructions—and all of this terminology will be described more fully later). A basic input output stream (BIOS) can be thought of as a small built-in program that runs when the computer first starts up, and this too issues instructions to the hardware. User

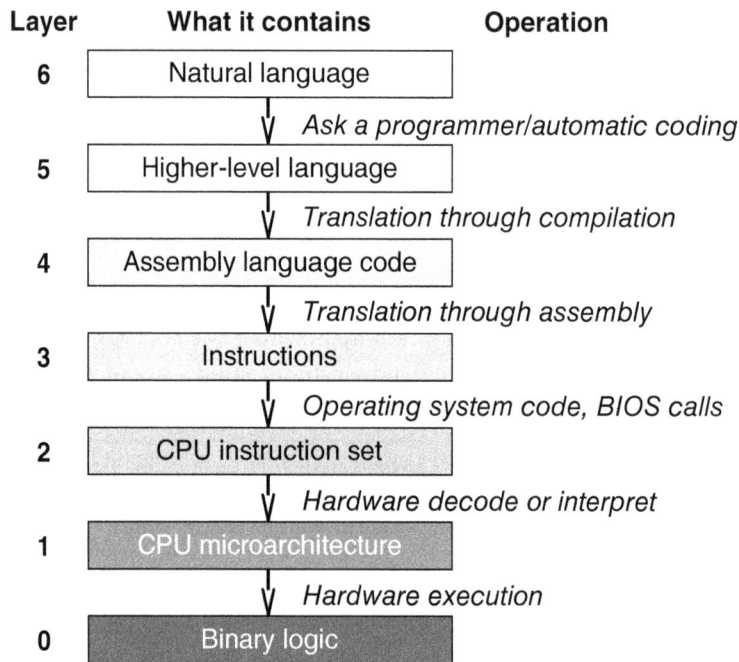

Layer	What it contains	Operation
6	Natural language	
		Ask a programmer/automatic coding
5	Higher-level language	
		Translation through compilation
4	Assembly language code	
		Translation through assembly
3	Instructions	
		Operating system code, BIOS calls
2	CPU instruction set	
		Hardware decode or interpret
1	CPU microarchitecture	
		Hardware execution
0	Binary logic	

FIGURE 2.2 A layered view of computer organization and structure.

programs are sequences of instructions to the hardware, but may additionally ask the BIOS or the operating system (OS) to perform functions. Like the BIOS, OS functions are small programs (usually; although some of them can be quite extensive) that perform predefined tasks. Being programs, these functions are also implemented as a sequence of instructions that are issued to the hardware.

Interestingly, this layer-like model is a reflection of the Open Systems Interconnection (OSI) model, applied to computer hardware and software (we will discuss the OSI model more fully later when we consider computer networking in Section 10.4).

2.2   Computer Fundamentals

The computer systems described in this book, such as the SISD machine shown in Section 2.1.1, generally comprise a number of discrete functional units interconnected by buses. Some of these units are now briefly introduced, before being covered in more detail in subsequent sections and chapters:

- *Central processing unit* (CPU): The part of a computer that controls operation through interpretation of instructions and through built-in behavior. It handles input/output functions and performs arithmetical and logical operations on data (in other words, it contains an ALU). Originally the "CPU" was a collection of hardware, but eventually this became miniaturized so that it could be implemented in a single IC. In recent years, the term "CPU" tends to refer to a physical IC which now contains much more than just a CPU; it may contain memory, networking hardware, a number of peripherals, and even contain power conditioning circuitry (especially for single-chip computers).[1]

- *Arithmetic logic unit* (ALU): It can perform simple arithmetic and logic operations such as add, subtract, AND, OR. It is an asynchronous unit which takes two data inputs from parallel connected registers or bus(es), and outputs either direct to a register or is connected through a tristate buffer to a bus. In addition, it has a control input to select which function to perform, and interfaces to a status register. It handles fixed point binary (and occasionally BCD) numbers only, and is located on-chip in modern processors.

- *Floating point unit* (FPU): Either an on-chip, or an external coprocessor, it performs arithmetic on floating-point numbers. The particular floating point format supported in most modern FPUs is called IEEE754 (explained further in Section 2.8.2). FPU calculations are usually slower than integer operations in a CPU (they can take tens or hundreds of instruction cycles to perform a calculation). The FPU often interfaces to the main CPU through special floating point registers.

- *Memory management unit* (MMU): It provides a layer of abstraction between how the processor addresses memory and how that memory is physically arranged. This abstraction is termed "virtual memory." The MMU translates a *virtual address* that the processor needs to access into a real *physical address* in

---

[1] Since the term "CPU" is very loosely defined, we will specify separately in the text whenever necessary, if we are referring to a central processing unit or to a single-chip computer.

memory. The processor typically sees a large linear continuous address space in memory, with the MMU hiding a physical memory organization which may be of different size (larger or smaller), noncontinuous, or comprising partly of RAM and partly of hard disc storage.

In addition, there are a number of items that we will include in our discussion that are useful to define now, prior to being covered in more detail later:

- *Register*: On-chip[2] storage locations that are directly wired to internal CPU buses to allow extremely fast access (often in one instruction cycle). The distinction blurs with on-chip memory for some CPUs, and with the stack in the picoJava II processor.

- *Tristate buffer*: A device to enable or disable driving a bus. Usually placed between a register and a bus to control when the bus will be driven by that register. The first two states are when the tristate drives the bus voltage to be either logic high or logic low; the third (tri-) state is high impedance, meaning that the device does *not* drive the bus at all.

- *Complex instruction set computer* (CISC): Think of any useful operation and directly insert this into the CPU hardware. Do not be too concerned about how big, power hungry, or slow this will make the CPU; then you will end up with a CISC machine. Early VAX machines were extreme examples, reputedly including an instruction that required more than 2000 clock cycles to execute.

- *Reduced instruction set computer* (RISC): CPUs are limited by their slowest internal components, and by silicon size. Based on the premise that 80 percent of instructions use only 20 percent execution time, and the remaining 20 percent use up 80 percent of the chip area, CPUs were reduced to contain the 80 percent most useful instructions. This left them compact and fast, and able to execute those common instructions very quickly. Sometimes a working definition of RISC meant "supporting a set of less than 100 instructions." Interestingly, some of the more modern CISC processors actually contain a RISC CPU core that can execute their complex instructions using small RISC programs.

- *Instruction cycle*: The time taken to fetch an instruction, decode it, process it, and return the result. This may be one or more periods of the main clock cycle (derived from an external oscillator). For RISC processors, instructions typically execute in a single clock cycle. For CISC processors, many instructions require multiple cycles for execution (some need very many cycles).

- *Big or little endian*: Big endian means that the most significant byte is presented first, and was traditionally used in processors such as the 68000 and SPARC. Little endian means that the least significant byte is presented first and was the traditional choice of the Intel x86 family. Some processors (such as the ARM7) allow for switchable "endiness."

    Unfortunately, endiness is complicated by the variable memory-width of modern computers. When memory storage was all "byte wide" it was easier,

---

[2] Originally these were separate hardware devices, but are now exclusively incorporated on-chip so that their access speed is as fast as possible.

but now there is an added dimension to defining big or little endian, which is bus width. To classify whether an unknown computer is big or little endian, it is easier to check first whether it is little endian, and if not, class it as big endian, rather than working the other way around. Boxes 2.1 to 2.4 explore this issue in more detail.

Box 2.1   Worked Endiness Example 1

Question: Given a 32-bit word stored in 16-bit wide memory (e.g., that used by a 16-bit architecture processor) as shown below, and given that the stored word is made up of least significant byte (LSB), second byte (B1), third byte (B2), and most significant byte (MSB), is the following a little- or big-endian representation?

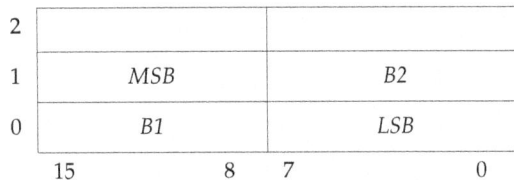

2		
1	*MSB*	*B2*
0	*B1*	*LSB*
	15 8	7 0

In the diagram, the memory address, counting in 16-bit words, is given on the left, and the bit positions (from 0 to 15) are shown below the memory cells.

Answer: Checking for little endian first, we identify the lowest byte-wise memory address, and count upward. In this case, the lowest address line is 0 and the lowest byte starts at bit 0. The next byte up in memory starts at bit 8 and is still at address line 0. This is followed by address line 1 bit 0 and finally address line 1 bit 8. Counting the contents from lowest byte address upward, we get {*LSB, B1, B2, MSB*}. Since this does follow least to most ordering, it must be in little endian format.

Box 2.2   Worked Endiness Example 2

Question: A 32-bit word is stored as shown below. Is this a little or big-endian representation?

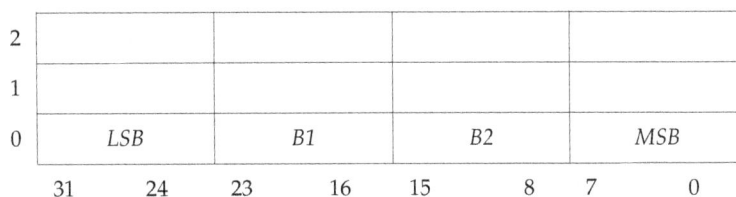

2				
1				
0	*LSB*	*B1*	*B2*	*MSB*
	31 24	23 16	15 8	7 0

Answer: First identify the lowest byte-wise memory address. This is clearly address line 0, starting at bit 0. Next is address line 0, bit 8, and so on. Counting from least significant to most significant and writing out the contents of the memory cells gives us {*MSB, B2, B1, LSB*}. This does *not* follow strictly from least to most, so it is not little endian. Therefore, it must be classed as big endian format.

Box 2.3 Worked Endiness Example 3

Question: Given the memory map shown below, write in the 32-bit number represented by MSB, B1, B2, and LSB bytes using a little-endian representation.

28				
24				
20				

| 0 | 7 | 8 | 15 | 16 | 23 | 24 | 31 |

Answer: Little endian is always easier: Its LSB is at the lowest byte address, and then we count upward in memory to the MSB. First, we need to identify the location of the lowest byte address in memory. In this case note the bit positions written along the bottom—they start from left and increment toward the right. Lowest address of those shown is therefore address 20 and bit 0. Next byte will be address 20, bit 8 onward, and so on. The end result should then be:

28				
24				
20	LSB	B1	B2	MSB

| 0 | 7 | 8 | 15 | 16 | 23 | 24 | 31 |

Note: Look also at the addresses. Instead of being consecutive locations, incrementing by one each line (as the other examples showed), these addresses jump by 4 bytes each line. This indicates that memory is byte addressed instead of word addressed. This is typical of ARM processors which, despite having 32-bit wide memory, address each byte in memory separately.

Box 2.4 Worked Endiness Example 4

Question: Given the memory map shown below, write in the 16-bit number represented by MSB and LSB bytes using a big-endian representation.

50	
51	
52	

| 7 | 0 |

Answer: In this case, we again need to identify which is the lowest byte address in the memory pictured, and then place the MSB there since we are big endian. In this case, the memory map is written from top down—a common format from some processor manufacturers. The top position is the lowest address, and we count downward. Since memory is byte wide, this is relatively easy. The answer is thus:

50	*MSB*
51	*LSB*
52	

7   0

2.3   Number Formats

Modern computers are remarkably homogeneous in their approach to arithmetic and logical data processing: most utilize the same number format, and can be classified as 8-, 16-, 32, or 64-bit architectures (known as their *data width*), which means the number of bits of binary data that are operated on by typical instructions, or held in typical data registers. There are also a relatively small number of techniques for number handling. This was not the case in early computers where a profusion of nonstandard data widths and formats abounded, most of which are purely of historical significance today.

Only about seven binary number formats remain in common use in computers today. Box 2.5 discusses exactly what constitutes a number format, but in order for us to consider processing hardware later in this chapter, it is useful to now review the main formats that we will be encountering as we progress onward.

Box 2.5   What Is a Number Format?

We are all aware of decimal format, either integer as in the number 123, or fractional as in 1.23, which are both examples of base 10. This is the natural format we use for handling numbers in mathematics, when counting money, computing GPA, and so on.

However, there are actually infinite numbers of ways to represent any number (i.e., an infinite number of different bases), although only very few of those bases are in common use. Apart from decimal, hexadecimal format (base 16) is used frequently in software, and binary (base 2) is frequently used in hardware.

We will use either binary, hexadecimal (hex) or decimal in all examples, and will display hex numbers with the prefix 0x in places where they might be confused, for example, 0x00FF0010 (which can be written 0x00FF,0010, or 0x00FF 0010 to break up the long strings of digits).

For binary, we will sometimes use a trailing b to denote a number as binary in places where it might be misinterpreted, for example, 00001101b (and again we will often break up long strings of digits into groups of four for clarity, for example, 0000 1101b).

2.3.1   Unsigned Binary

In unsigned binary, each bit in a data word is weighted by the appropriate power of two corresponding to its position. This sounds complicated at first, but in fact is really easy to use in practice. For example, the 8-bit binary word 00110101b (which is equivalent to 53 in decimal) is interpreted by reading from right to left as the summation

$$1(2^0) + 0(2^1) + 1(2^2) + 0(2^3) + 1(2^4) + 1(2^5) + 0(2^6) + 0(2^7)$$

In general, the value, v of a n-bit binary number x, where $x[i]$ is the ith bit reading from the right to the left, starting from bit 0, is

$$v = \sum_{i=0}^{n} x[i].2^i$$

Unsigned binary format is easy for humans to read after a little practice, and is handled efficiently by computer. To practice conversion, the best way is to draw out a grid such as the following.

2^7	2^6	2^5	2^4	2^3	2^2	2^1	2^0
128	64	32	16	8	4	2	1

Then to convert from binary to decimal, write the binary number that needs converting and read off the sum. To illustrate this, using the example above of 00110101b, that would be

128	64	32	16	8	4	2	1
0	0	1	1	0	1	0	1

From this, we can read off the sum $1 + 4 + 16 + 32$ and compute the answer 53 in decimal.

Conversion in the other direction (from decimal to binary), the fastest method is to start from left and proceed to right. Starting with 128, determine whether the number being converted is larger than 128. If so, write a 1 in the first column and subtract 128 from the number. If not, write a 0 and continue. Moving to the second column, determine whether the number being converted is larger than 64. If so, write a 1 in the first column and subtract 64 from the number. If not, write a 0 and continue, in turn, until reaching the final column. Interestingly, the final column with a bit weight of 1 will show whether the number is odd (1) or even (0).

Some other worked examples of converting numbers for different formats are given in Box 2.7.

2.3.2   Sign Magnitude

This format uses the most significant bit (MSB) to convey polarity (the MSB is often called the "sign bit"). The remaining less significant bits then use unsigned binary notation to convey magnitude. By convention, an MSB of 0 indicates a positive number while an MSB of 1 indicates a negative number.

For example, the 4-bit sign-magnitude number 1001 is -1, and the 8-bit number 10001111b is equivalent to $8 + 4 + 2 + 1 = -15$ decimal. A value of $+15$ decimal will be the same apart from the sign bit: 00001111b.

2.3.3   One's Complement

This format has largely been replaced by two's complement but can still occasionally be found, particularly in specialized hardware implementations. Again, the MSB conveys polarity while the remaining bits indicate magnitude. The difference with sign-magnitude is that, when the number is negative, the polarity of the magnitude bits is reversed (i.e., when the sign bit is 1, the remaining bits are all flipped in meaning, so a "0" becomes "1" and a "1" becomes "0" when compared to sign-magnitude).

For example, the 8-bit one's complement number 11110111b is equal to −8 decimal, whereas a value of +8 decimal would be written identically to the sign-magnitude representation: 00001000b.

2.3.4   Two's Complement

This is undoubtedly the most common signed number format in modern computers. It has achieved predominance for efficiency reasons: identical digital hardware for arithmetic handling of unsigned numbers can be used for two's complement numbers. Again, the MSB conveys polarity, and positive numbers are similar in form to unsigned binary. However, a negative two's complement number has magnitude bits that are formed by taking the one's complement and adding one (Box 2.6 provides a binary example of this method which is literally "taking the two's complement of a number").

For example, the 4-bit two's complement number represented by binary digits 1011 is equal to $-8 + 2 + 1 = -5$ decimal, and the 8-bit two's complement number 10001010 is equal to $-128 + 8 + 2 = -118$ decimal.

Box 2.6   Negative Two's Complement Numbers

Negative two's complement numbers can be easily formed in practice by taking the one's complement of the binary magnitude then adding 1. As an example, suppose we wish to write −44 in 8-bit two's complement:

Start by writing +44 in 7-bit binary:	010 1100
Next, flip all bits (take the 1's complement):	101 0011
Add 1 to the least significant bit position:	101 0100
Finally, insert the sign bit (1 for negative):	1101 0100

As an aside, if you are not used to writing binary numbers, try to write them in groups of 4. That way it is easier to line up the columns, and aids in the conversion to hexadecimal (since a group of 4 bits corresponds to a single hex digit).

It is undoubtedly harder for humans to read negative two's complement numbers than some of the other formats mentioned above, but this is a small price to pay for reduced hardware complexity. Box 2.7 provides some examples of two's complement number formation, for both positive and negative values.

Box 2.7   Worked Examples of Number Conversion

Question 1: Write the decimal value 23 as a two's complement 8-bit binary number.

Answer: We can start by drawing the bit weightings of an 8-bit two's complement number. Starting from the left, we begin with the sign bit.

-128	64	32	16	8	4	2	1

The sign bit is only set if the number we want to write is negative. In this case it is positive so write a zero there. Next we look at 64. If our number is greater than 64 we would write a 1 here, but it is not so we write 0. The same for 32, so now we have

0	0	0	16	8	4	2	1

Moving on to 16, we find that our number (23) *is* bigger than 16, and so we subtract 16 from the number to leave $23 - 16 = 7$. A "1" goes in the 16 box.

Next we compare our remainder with 8, and find the remainder smaller so a "0" goes in the 8 box. Moving on to 4, our remainder is bigger than this so we subtract 4 to make a new remainder $7 - 4 = 3$ and write a 1 in the 4 box. Continuing with 2 and 1, both get 1's in their boxes. The final answer is thus

0	0	0	1	0	1	1	1

Question 2: Write the decimal value -100 as a two's complement 8-bit binary number.

Answer: Again looking at the number line above, we realize that, as a negative number, we need a "1" in the -128 box. Doing the sum $-100 - (-128)$ or $-100 + 128$ leaves a remainder of 28. The rest of the numbers act as normal – a "0" in the 64 box, a "0" in the 32 box, then a "1" in the 16 box. The remainder will then be $28 - 16 = 12$. Continuing, there will be "1" in the 8 box, remainder 4, then a "1" in the 4 box and 0's beyond that

1	0	0	1	1	1	0	0

Note: The only really easy things to see, at a glance, about two's complements numbers is whether they are negative or not (a 1 in the most significant position), and whether they are odd or not (a 1 in the least significant position).

2.3.5   Excess-*n*

This representation crops up later when we discuss floating point. In excess-*n*, a number v is stored as the unsigned binary value $v + n$. The example we will see later is excess-127 representation in 8 bits, which can store any number between -127 and $+128$ (stored in binary bit-patterns that look like the unsigned values 0 and 255, respectively).

This format requires a little thought to decode correctly, but looking at the extreme (largest and smallest, respectively) values helps to clarify. Consider the smallest value; 8-bit excess-127 binary number `00000000b`, which is equivalent to -127 decimal. We get -127 by working out the unsigned binary value for the digits, in this case zero, and then subtracting 127. By contrast, the largest value, `11111111b`, would equal $+255$ if it were in unsigned format. But since we know it is in excess-127 format, we subtract 127 from 255 to convert it, giving a resulting value of $255 - 127 = 128$. Another example is

11000010 which in unsigned binary would be $128 + 64 + 2 = 194$, but since we are told it is an excess-127 number, we need to subtract 127 from the unsigned binary amount to give a result of $194 - 127 = 67$ decimal.

To convert from decimal to excess-127, we follow the reverse process. First add 127 to the decimal number, and then write in unsigned binary. For example, converting decimal 101 to excess-127, we add $101 + 127 = 228$ and then write the result as unsigned binary: 11100100 ($4 + 32 + 64 + 128$).

There are other excess-n formats in use in different specialized computing fields, and those follow the same general principle—but as mentioned, we will only be encountering excess-127 in this book.

2.3.6   Binary-Coded Decimal

Binary-coded decimal (BCD) was used extensively in early computers. It is fairly easy for humans to read in practice, because each decimal digit (0 to 9) of a number to be stored in BCD is encoded using a group of four binary digits. Thus decimal value 73 is stored as separate digits 7 and 3, namely 0111 0011 in BCD. Since four binary digits can store a value from 0 to 15, but BCD only stores decimal values 0 to 9 in each group of four, there are also some binary patterns that are never used in BCD. Ultimately, BCD has been superseded in mainstream use because it is neither efficient in storage nor easy to design hardware for. There is, however, one area in which BCD is common, and that is in driving the LED and LCD digits on a seven-segment display (i.e., a digital number display). The driver ICs for such displays usually require the data to be output in BCD format.

2.3.7   Fractional Notation

Both binary and decimal formats can be used to store fractional numbers (see Box 2.8), but we will consider just unsigned and two's complement numbers, as these are most common within the computer architecture field. Fractional notation is strictly a *conceptual* interpretation of the numbers. The written numbers look the same as any other binary, but our interpretation of the bit weightings changes.

The usual bit weightings of 2^0 for the LSB, 2^1 for the next bit, 2^2 for the third bit, and so on are replaced by a scaled weighting pattern. In some digital signal processing (DSP) circles, fractional notation is described as *Q-format*. Otherwise, fractional notation binary is typically described as $(m.n)$ format, where m is the number of digits before the imaginary radix[3] and n is the number of digits after the radix. So the bit width in this case would be $m + n$.

Box 2.8   Is Binary a Fractional Number Format?

Remember that there is nothing special about binary—it is simply a way of writing a number in base 2 instead of base 10 (decimal) that we are familiar with.

Just as we can write fractional numbers in decimal (such as 2.34) as well as integers (such as 19), we can also write any other base number in fractional as well as integer format. So far we have only considered integer binary format; however,

---

[3]In decimal, the radix is known as the decimal point, but when dealing with another number base we cannot refer to it as a "decimal" point, so we call it the radix.

it is also important to realize that fractional binary format is used extensively in areas such as DSP.

An example of two 8-bit binary number arrangements in unsigned and (6.2) format are shown below:

unsigned	2^7	2^6	2^5	2^4	2^3	2^2	2^1	2^0
(6.2) format	2^5	2^4	2^3	2^2	2^1	2^0	2^{-1}	2^{-2}

Refer to Box 2.9 for some further examples of fractional format numbers in binary.

The beauty of fractional notation applied to unsigned or two's complement numbers is that the values are handled in hardware exactly the same way as the nonfractional equivalents: it is simply a "programming abstraction." This means that any binary adder, multiplier, or shifter can add, multiply, or shift fractional numbers the same as unsigned numbers (in fact the hardware will never "know" that the numbers are fractional—only the programmer needs to keep track of which numbers are fractional and which are not).

Box 2.9   Fractional Format Worked Example

Question: Write the decimal value 12.625 as a (7.9) fractional format two's complement binary number.

Answer: First start by looking at the bit weightings of the (7.9) format:

−64	32	16	8	4	2	1	1/2	1/4	1/8				

where the weightings below 1/8 have been removed for space reasons. Next we realize that the number is positive, so there is a "0" in the −64 box. We then scan from left to right in exactly the same way as for a standard two's complement representation (or unsigned binary for that matter), using the weights shown above.

It turns out that $12.625 = 8 + 4 + 0.5 + 0.125$ and so the result will be

0	0	0	1	1	0	0	1	0	1	0	0	0	0	0	0

2.3.8   Sign Extension

This is the name given to the process by which a signed two's complement number of a particular width is extended in width to a larger number of bits. For example, converting an 8-bit number to a 16-bit number. While this is done occasionally as an explicit operation specified by a programmer, it is more commonly performed as part of operations such as addition and multiplication.

Sign extension can be illustrated in the case of moving from a 4-bit to an 8-bit two's complement binary number. First write the 4-bit two's complement number 1010 in 8-bit two's complement.

If we are considering signed numbers, we know that the 4-bit number involves bit weightings of [−8, 4, 2, 1] while the 8-bit weightings are [−128, 64, 32, 16, 8, 4, 2, 1]. For the 4-bit number, the value 1010 is clearly

$$-8 + 2 = -6$$

If we were to simply write the 8-bit value as a 4-bit number padded with zeros as in 00001010 then, referring to the 8-bit weightings the value that this represents would be

$$8 + 2 = 10$$

Which is clearly incorrect. If we were then to note that a negative number requires the sign bit set and responded by simply toggling the sign bit to give 10001010 then the value would become

$$-128 + 8 + 2 = -118$$

Which is again incorrect. In fact, in order to achieve the extension from 4 to 8 bits correctly, it is necessary that not only the original MSB must be set correctly, but every additional bit that we have added (every bit to the left of the original MSB) must also be set to the same value as the original MSB. The sign bit has thus been extended to give **1111**1010 with a value of

$$-128 + 64 + 32 + 16 + 8 + 2 = -6$$

Finally achieving a correct result. Another example of sign extension is given in Box 2.10.

Box 2.10   Sign Extension Worked Example

Question: Write the value -4 in 4-bit two's complement notation. Copy the MSB four times to the left. Read off the result as an 8-bit two's complement number.
Answer: 1100 $(-8 + 4 + 0 + 0)$
MSB is 1, so copying this to the left four times gives

11111100

Reading off in 8-bit signed binary $(-128 + 64 + 32 + 16 + 8 + 4) = -4$.
    For further thought: Repeat the exercise with a positive number such as 3. Does the method still apply equally for positive numbers?

There is evidently no difficulty with positive two's complement numbers, but the sign extension rule can still be applied (it has no effect, but makes a hardware design easier if it applies to *all* numbers rather than just *some*).

2.4   Arithmetic

This section discusses the hardware used to perform addition or subtracting of two binary numbers. This functionality is used within the ALU in almost all processors, which also handle basic logic functions such as AND, OR, NOT, and so on. The ALU itself—from the perspective of being a CPU functional unit—will be discussed later in Section 4.2.

FIGURE 2.3 A full adder, showing two bits being added, together with a carry in, and a single output bit with carry.

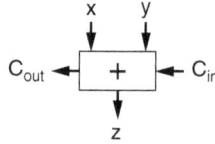

2.4.1 Addition

Binary arithmetic is accomplished bit wise with a possible carry from the adjacent less significant bit calculation. In hardware, a *full adder* calculates the addition of two bits and a carry in, and generates a result with an additional carry output.

A full adder is shown symbolically in Figure 2.3, where each arrow represents a single logic bit. A half adder is similar, but does not have any provision for the carry in.

2.4.2 The Parallel Carry-Propagate Adder

To create an 8-bit parallel adder, the full adder hardware would typically be repeated eight times for each of the input bits—although the least significant bit position could use a slightly smaller half adder, as shown in Figure 2.4.

In Figure 2.4, $x[7:0]$ and $y[7:0]$ are the two input bytes and $z[7:0]$ is the output byte. C_{out} is the final carry output. For the case of adding unsigned numbers, when C_{out} is set it indicates that the calculation has resulted in a number that is too large to be represented in 8 bits. For example, we know that the largest magnitude unsigned number that can be represented in 8 bits is $2^8 - 1 = 255$. If two large numbers such as 200 and 100 are added together, the result (300) cannot fit into 8 bits. In this case, the carry would be set on the adder and the result (z) would hold the remainder $300 - 256 = 44$.

The topmost C_{out} therefore doubles as an overflow indicator when adding unsigned numbers: If it is set following a calculation, this is indicating that the result cannot be represented using the number of bits present in the adder. Some further thoughts on this are explored in Box 2.11.

This behavior, and the add mechanism, is common to almost any binary adder. Although the parallel adder appears to be a relatively efficient structure, and even works in a similar way to a human calculating binary addition by hand (or perhaps using an abacus), it suffers from a major speed limitation that bars its use in the fastest microprocessor ALU implementations: carry propagation.

Since the input numbers are presented to the adder simultaneously, one measure of adder speed is the length of time it takes after that until the answer is ready at the output.

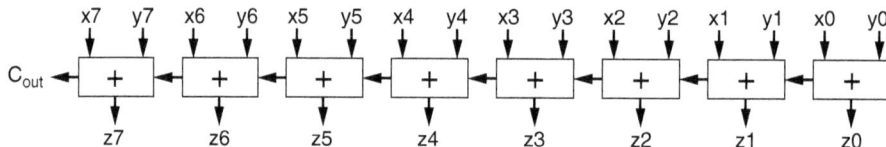

FIGURE 2.4 The carry-propagate or ripple-carry adder constructed from a sequence of full adders plus one-half adder.

Box 2.11   Exercise for the Reader

How does the topmost C_{out} signal from an adder behave when adding signed two's complement numbers?

1. Try working by hand using a 4-bit adder. With 4-bit two's complement numbers the representable range is -8 to $+7$.

2. Try adding some values such as

   - $2 + 8 =?$
   - $2 + (-8) =?$
   - $7 + 7 =?$
   - $(-8) + (-8) =?$

3. What do you conclude about the C_{out} signal: does it mean the same for signed two's complement numbers as it does when adding unsigned numbers?

Each full or half adder in the chain is relatively quick: both the carry out and the result from a single adder could be available just a few nanoseconds after the carry in and input bits are presented (for modern silicon devices). The problem is that the least significant half adder (adder 0) must finish calculating before the next bit calculation (adder 1) can start. This is because adder 1 needs to get the carry results from adder 0 *before* it can even start its calculation, and that carry signal is not valid until adder 0 *finishes* its own calculation. When adder 1 finishes, it supplies its carry output to adder 2, and so on. So they all must wait their turn. Further up the chain, by the time the signal has propagated up to adder 7, a significant length of time will have elapsed after the input words were first presented to the adder.

A worked example of calculating an entire ripple-carry adder propagation delay is presented in Box 2.12. It is important because, if such an adder were present in a computer, this propagation delay might well be the part of the system that limits the overall maximum system clock speed.

Box 2.12   Worked Example

Question: The adders and half adders used in a 4-bit parallel carry-propagate adder are specified as follows:

Time from last input bit (x or y) or carry in to result z: 15 ns

Time from last input bit (x or y) or carry in to carry out: 12 ns

If input words $x[3:0]$ and $y[3:0]$ are presented and stable at time 0, how long will it be before the 4-bit output of the adder is guaranteed stable and correct?

Answer: Starting from the least significant end of the chain, adder 0 receives stable inputs at time 0. Its result z is then ready at 15 ns and its carry is ready at 12 ns. Adder 1 requires this carry in order to begin its own calculation, so this only starts at 12 ns. It takes until 24 ns until it can provide a correct carry result to adder 2, and this will not provide a carry to adder 3 until 36 ns. Adder 3 then begins its calculation. Its output z is then ready at 51 ns, and its carry out is ready at 48 ns.

So even though the adders themselves are fairly quick, when chained, they require 51 ns to calculate the result.

Note: The phrase "begins its calculation" when applied to the full or half adders may be misleading. They are actually combinational logic blocks. A change of state at the input will take some time (up to 15 ns in this case) to propagate through to the output. Since they are combinational logic, they are always "processing" input data and their outputs are always active. However, from the specification, we know that the outputs are only guaranteed correct 15 ns or 12 ns after the inputs are correctly presented (for result z and carry out, respectively).

2.4.3 Carry Look-Ahead

In order to speed up the parallel adder described above, a method is required for obtaining the carry inputs to adders as early as possible.

This is achieved with a carry predictor, which is a piece of combinational logic that calculates the carry values directly. In fact, it can supply carry values to each adder in the chain at the same time, with approximately the same propagation delay as a single half adder. A carry predictor is shown in Figure 2.5 for a 3-bit adder. It is interesting to note the formation of the logic equations describing the carry look-ahead units (see Box 2.13).

Box 2.13 Exercise for the Reader

1. Write the logic equation of a single bit full adder.

2. Extend this to a three bit adder as shown above.

3. Rearrange the equations to give C_0 and C_1 in terms of the input (rather than any carry ins). Note that the number of basic calculations required to give C_1 is small and thus the propagation delay through gates required to do this calculation is also small.

4. Now extend the equations to derive C_2? How many calculation steps are needed for this? Is it more than for C_1? Can you deduce anything about the scaling of this method to longer adder chains (thinking in terms of propagation delay—and also logic complexity)?

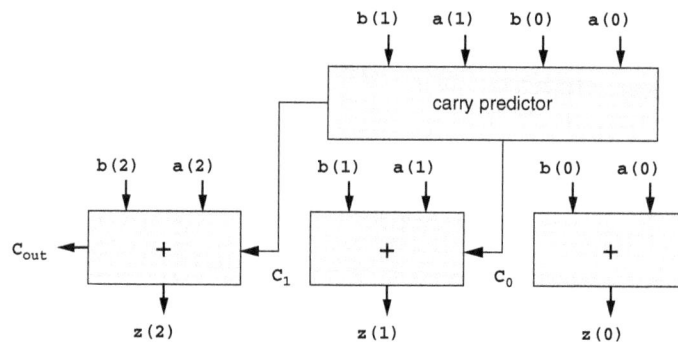

FIGURE 2.5 The carry look-ahead adder constructed from several full adders and carry predict logic.

2.4.4 Subtraction

Similar to addition, subtraction is performed bit-wise. But when performing subtraction, do we need to consider the result from neighboring bits? The answer is yes, but these are now linked through "borrows" from higher bits, rather than "carries" from lower bits. Unfortunately, the "borrows" are just as problematic, in terms of timing delay, as the "carries" in addition.

In terms of computational hardware, a specialized subtractor would be required if it were not for the fact that addition and subtraction can be interchanged in many number formats. As an example, consider the decimal calculation $99 - 23 = 76$ which can be written in an alternative arrangement as $99 + (-23)$ giving an identical result.

Although the result is identical, it is achieved by performing an addition rather than a subtraction, and changing the sign or the second operand (which is easy in hardware). Many commercial ALUs work in a similar fashion: they contain only adding circuitry and a mechanism to change the sign of one operand. As we saw in Section 2.3.4, changing the sign of a two's complement number is relatively easy: first change the sign of every bit, and then add 1 to the least significant bit position. In fact, adding 1 to the LSB is the same as setting the carry input for that adder to 1.

Needless to say, this is easily achieved in hardware with a circuit such as the subtract logic shown in Figure 2.6. In this circuit, the exclusive-OR gate at the bottom, acting on input operand y, is used to change the sign of each bit (an exclusive-OR acts as a switched inverter in that if one input is held high, every bit present on the other input will be inverted, otherwise it will be unchanged; see Appendix B). If the circuit is performing a subtraction, the add/subtract line is held high, one operand is negated and C_{in} is also set high—having the effect of adding 1 to the least significant bit.

There is one further area of subtraction that needs to be explored, and that is overflow: when performing an addition, you will recall that the topmost C_{out} can be used to indicate an overflow condition. This is no longer true when performing subtractions as some examples on 4-bit two's complement numbers will reveal:

```
0010 + 1110 = ?                    2 + (−2) = ?
0010 + 1110 = 0000 + Cout
```

Clearly the result should be an easily represented zero, and yet the C_{out} signal is set. Consider another example where we would normally expect an overflow:

```
0111 + 0110 = ?                    7 + 6 = ?
0111 + 0110 = 1101                 Answer = −3 ?
```

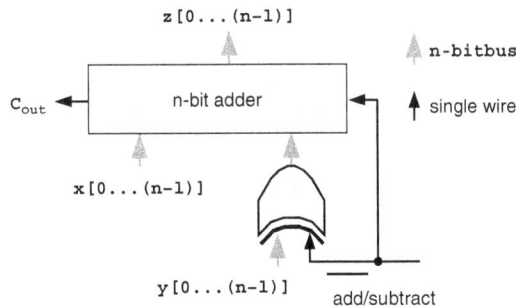

FIGURE 2.6 Subtraction logic consisting basically of an adder with external exclusive-OR gates.

Again, the result should not be -3, it should be 13. Evidently the circuitry shown is not sufficient alone, and some account needs to be taken of the values being processed. The answer is that the sign bits must be examined prior to adding and the result checked based on this. This is not computationally hard—a simple look-up table will suffice:

positive + positive = positive
positive + negative = unknown
negative + positive = unknown
negative + negative = negative

For the mixed calculation (one positive and one negative number), the sign of the answer is unknown, but is not problematic since by definition it can never result in an overflow (think of it this way: the negative number will be reducing the size of the positive number, but the most it can do would be if the positive number is zero, in which case the answer is the same as the negative input, and since the inputs themselves do not include carry flags, the outputs will not).

Let us consider some examples of 4-bit two's complement signed number calculations and examine the role of the carry (**C**) and overflow (**V**) flags.

Example 1:
`0111 + 1000`
In this example, we are adding a positive and a negative number, which will never cause an overflow according to our criteria. Performing the binary addition, we get

$$0111 + 1000 = 1111$$

which we see from the sign bits is *positive + negative = negative*. Converting to decimal we find that the decimal version of the calculation is $7 - 8 = -1$, which is obviously correct.

Example 2:
`0111 + 0100`
This time we are adding two positive numbers, and therefore need to carefully check the sign bit of the result. Let us perform the addition:

$$0111 + 0100 = 1011$$

from which we see that the result is negative (the MSB or "sign bit" is set). This does not seem to be correct—adding one positive value to another positive value should give us a positive answer (i.e., should give us a sign bit which is 0), but in this case it gives us a *negative* result and no carry. Converting to decimal, we find that the calculation is $7 + 4 = -5$, which is certainly not what we would expect.

When doing unsigned arithmetic, the programmer relies upon the C_{out} signal (or **C** flag) after a calculation to find out whether there was a "problem" with the calculation—to find out when the result was too large to fit into the number of bits available. Clearly the programmer needs to rely upon a different mechanism for signed arithmetic.

It turns out that CPU designers have a solution in the overflow or **V** flag. This flag does precisely what we had identified above, and indicates when the sign bit of the output result was unexpected. For the case of two positive numbers being added, the result sign bit should be 0. If not, then an overflow has occurred and the **V** flag will be set.

For the case of two negative numbers, the result sign bit should be 1, and if not, an overflow has occurred, so the **V** flag will also be set. The following table provides some further examples of 4-bit calculations, showing the **V** and **C** flags that are set as a result:

Operand A	Operand B	Result A + B	C flag	V flag
0000	0000	0000	0	0
0010	0010	0100	0	0
0100	0011	0111	0	0
1111	1111	1110	1	0
0111	0111	1110	0	1
0100	1000	1100	0	0
1111	0001	0000	1	0

2.4.4.1 A Programming Perspective

The binary computations in the above table are all correct; all of them are pure binary that could be directly and simply calculated by any computer's ALU. The ALU does not "know" whether the operands are unsigned or signed; only the programmer knows that. It is therefore the programmers' responsibility to write a program that checks either the **V** or the **C** flag after a computation. The **V** flag should be checked if the operands were two's complement, whereas the **C** flag should be checked if they were unsigned binary.

To illustrate the point further, consider the fourth operation shown in the table, 1111 + 1111 = C+1110. If the operands are unsigned, the programmer would need to look at the **C** flag after the calculation. In this case it is set, indicating that the result of the addition is too large to represent in 4 bits. However, if the operands are signed two's complement, then the programmer would need to look at the **V** flag after the addition. In this case it is not set, indicating that the result *can* be represented in four bits. Is this correct? Let's see.

Using unsigned representation, 1111b=15 in decimal, and then $15 + 15 = 30$, which is too large to be held in four bits, as indicated by the carry. But using two's complement representation, 1111b=-1 in decimal, and then $-1 + -1 = -2$, which is easily held in 4 bits. The signed binary result 1110b=-2, which is correct. In each case, the ALU, the binary input and the binary output are unchanged. It is all a matter of what those binary numbers represent—and whether we, as programmers, examine the **V** (because we used signed operands) or the **C** (because we used unsigned operands).

2.5 Multiplication

In the early days of microprocessors, multiplication was too complex to be performed in logic within the CPU and hence required an external unit. Even when it was finally squeezed onto the same piece of silicon, it could be a tight fit: the multiplication hardware in early ARM processors occupied more silicon area than the entire ARM CPU core.

In more recent times, however, manufacturers have tuned multipliers to the target application. For fast real-time embedded processors (perhaps a GSM cellphone handling video calling), there is a need to perform fast multiplications and hence a high-performance multiplier will be used. This kind of hardware will evidently occupy a large silicon area compared to a slower multi-cycle multiplier used on a non-real-time processor such as those in small and simple embedded systems.

There are many methods of performing the multiplication $m \times n$ at various rates (and with various complexities). Some of the more typical methods are listed here:

1. Repeated addition (add m to itself n times).
2. Add shifted partial products.
3. Split n into a sequence of adds and left shifts applied to m.
4. Booth's and Robertson's methods.

Each of these will be considered in the following subsections in turn. There are, of course, other more esoteric methods as this is an active research area. Interestingly, some methods may perform estimation rather than calculation, or involve loss of precision in the result. These would include converting operands to the logarithmic domain and then adding them, or using an alternative or redundant number format.

Alternative number formats are briefly described in Section 12.4, but when it comes to hardware for performing binary calculations, there are so many alternatives that imagination alone seems to be the limiting factor.

2.5.1   Repeated Addition

The simplest method of performing a multiplication is one of the smallest in implementation complexity and silicon area but at the cost of being slow. When multiplying integers $m \times n$, the pseudo-code looks like:

```
set register A ←  m
set register B ←  0
loop while (A ←  A − 1)  ≥  0
        B ←  B + n
```

Since this involves a loop that repeats n times, the execution time is dependent on the value of n. If n is small, the result, B, is formed early. However, a 32-bit number can represent an integer value of 4 billion or more, so many iterations of the loop might be necessary to perform the calculation. This could imply rather a long execution time in practice.

2.5.2   Partial Products

Instead of iterating based on the magnitude of n (as in the repeated addition method above), the partial products method iterates based on the number of bits in number n.

Each bit in the number n is examined in turn, from least to most significant. If a bit is set, then a partial product derived from number m shifted left to line up with the bit being examined is accumulated. In multiplier terminology, the two numbers are termed "multiplier" and "multiplicand" although we also know for decimal numbers that it does not matter which way around the multiplication is performed since $(m \times n) \equiv (n \times m)$.

Here is a partial products example:

```
 ‾1001     multiplicand 9
  1011     multiplier 11
  1001     (since multiplier bit 0 = 1, write 9 shifted left 0 bits)
 1001      (since multiplier bit 1 = 1, write 9 shifted left 1 bits)
0000       (since multiplier bit 2 = 0, write 0 shifted left 2 bits)
1001       (since multiplier bit 3 = 1, write 9 shifted left 3 bits)
01100011   result = 99 (sum of the partial products)
```

The situation is complicated slightly when it comes to working with two's complement signed numbers—first in that the MSB of the multiplier represents sign, and second in that sign extension must be used (see Section 2.3.4).

For the signed case, all partial products have to be sign extended to the length of the result (which by default would be the sum of the lengths of the input representations minus 1 to account for the sign bit, such that a 6-bit signed number plus a 7-bit signed number would require 12 bits to represent the result).

Since each partial product corresponds to 1 bit of the multiplier and is shifted to account for the multiplier bit weighting, the partial product corresponding to the MSB is a special case: the bit weighting is negative and this partial product must therefore be subtracted from the accumulator rather than added. This is shown in the flowchart of Figure 2.7, where it is assumed that the gray-colored two's complement accumulate blocks are able to take account of sign extension.

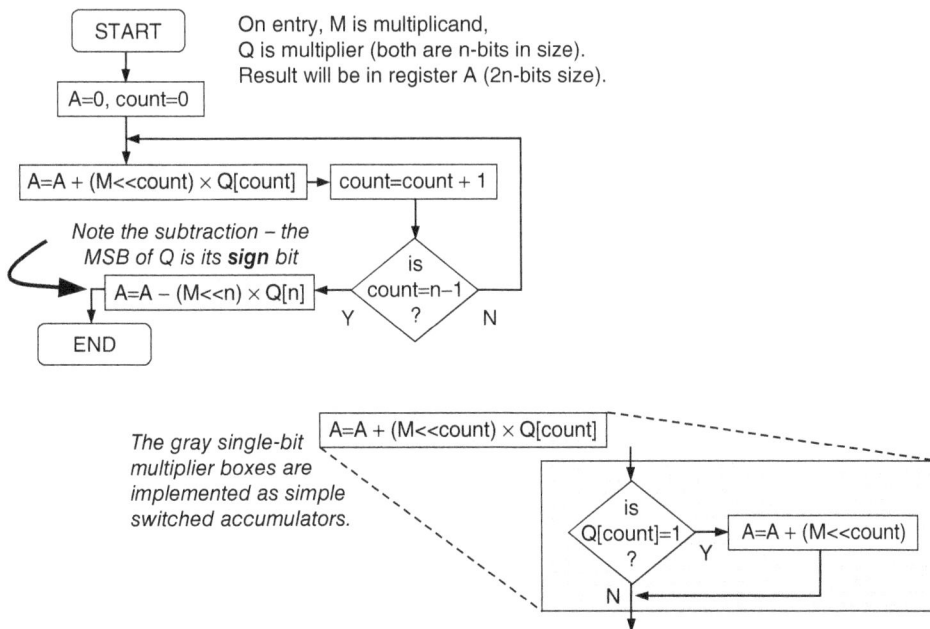

FIGURE 2.7 Flowchart for performing partial product multiplication.

As an aid to understanding the process, it is useful to attempt some simple binary multiplication by hand using those methods, and the reader can follow some examples in Box 2.14.

Box 2.14   Worked Examples of Two's Complement Multiplication

Look at −5 × 4 (signed):

```
      1011    multiplicand −5
      0100    multiplier 4
  00000000    (since multiplier bit 0 = 0, write 0 shifted left 0 bits and sign extend)
 +0000000     (since multiplier bit 1 = 1, write 0 shifted left 1 bit and sign extend)
 +111011      (since multiplier bit 2 = 0, write −5 shifted left 1 bit and sign extend)
 +00000       (since multiplier bit 3 = 0, write 0 shifted left 1 bit and sign extend)
 =11101100    result = −128 + 64 + 32 + 8 + 4 = −20
```

Similarly, let us look at 4 × −5 (signed):

```
      0100    multiplicand 4
      1011    multiplier −5
  00000100    (since multiplier bit 0 = 1, write 4 shifted left 0 bits and sign extend)
 +0000100     (since multiplier bit 1 = 1, write 4 shifted left 1 bit and sign extend)
 +000000      (since multiplier bit 2 = 0, write 0 shifted left 2 bits and sign extend)
 -00100       (since multiplier bit 3 = 1, write 0 shifted left 3 bits and sign extend)
 =11101100    result = −128 + 64 + 32 + 8 + 4 = −20
```

But the last term needs to be subtracted. What we will do is change the sign by flipping all the bits and adding 1 ($00100000 \rightarrow$ flip $\rightarrow 11011111 \rightarrow +1 \rightarrow 11100000$) then we can simply add it to the other partial products. This gives:

```
  00000100
 +0000100
 +000000
 +11100
 =11101100   result = −20
```

As we can see the result is the same. We have illustrated the cases of needing sign extension and of handling a negative multiplier causing the final partial product to be subtracted instead of being added.

In reality, the accumulation of partial products may be more efficiently performed in the reverse direction. In the best case this would also remove the need to treat the partial product of the multiplier sign bit differently (since this is not accumulated, it is merely the value in the accumulator before additions begin, thus allowing its sign to be negated during the load-in process).

A block diagram of an alternative partial product multiplication method, for unsigned numbers only, is shown in Figure 2.8 (although extending this method to two's complement is a relatively simple task). The figure shows the sequence of operations to be taken once the setup (operand loading) is complete.

The setup phase resets the accumulator Q to zero and loads both multiplier and multiplicand into the correct locations. In step 1 the least significant bit of the multiplier is tested. If this is a 1 (step 2), then the multiplicand is added to the accumulator (step 3).

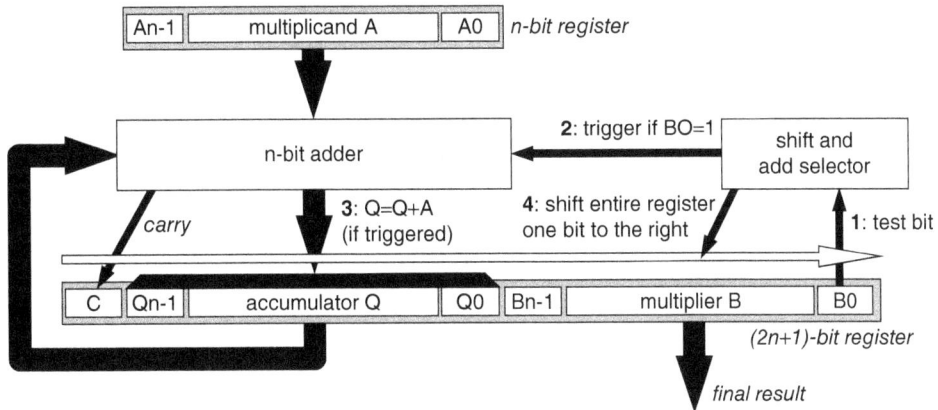

FIGURE 2.8 Bit-level block diagram of signed partial-product multiplication using an accumulator.

Step 4 occurs regardless of the two previous conditional steps, and shifts the entire accumulator 1 bit to the right. The system loops n times (using control logic which is not shown) before terminating with the answer in the long register.

Consider the differences between this and the original flowchart of Figure 2.7 in terms of the number of registers needed, bus wires, connections, switches, adder size, and control logic involved.

Interestingly, this method of multiplication, including the right shift (as a divide by 2), was reportedly used by Russian peasants for hundreds of years, allowing them to perform quite complex decimal multiplications with ease. The algorithm starts with the two numbers to be multiplied, A and B, written at the head of two columns, respectively. We will give as an example, 31 multiplied by 17:

$$B = 17 \qquad A = 31$$

Working downward, divide the B column by two each line, discarding the fractional part, until 1 is reached. Fill the A column similarly, but double the number on each successive line:

$$
\begin{array}{rr}
B = 17 & A = 31 \\
8 & 62 \\
4 & 124 \\
2 & 248 \\
1 & 496 \\
\end{array}
$$

Next simply add up all of the numbers in the A column that correspond to *odd* numbers in the B column. In this example, only 17 and 1 are odd in the B column; therefore, the final answer will be $31 + 496 = 527$, which is of course correct.

Note that the alternatives given in this section are by no means the only partial product hardware designs available, and far from being the only multiplication methods available (even among Russian peasants).

X_i	X_{i-1}	Rule
0	0	No action
0	1	Add shifted multiplicand
1	0	Subtract shifted multiplicand
1	1	No action

TABLE 2.1 Predefined rules for bit-pair scanning in Booth's method.

2.5.3   Shift-Add Method

The shift-add method relies on the fact that, for binary numbers, a shift left by 1 bit is equivalent to multiplying by 2. A shift left by 2 bits is equivalent to multiplying by 4, and so on.

Using this property to perform a multiply will still suffer from the same issue as the repeated addition method in that the number of operations depends on the *value* of the multiplier rather than the number of bits in the multiplier word. For that reason, this method is not normally found as a general multiplier in commercial processors; however, it can be very efficient where the multiplier is fixed and close to a power of 2. For this reason it is often used in digital filters (devices that perform a sequence of multiplications) with predetermined multiplier values.

This method is also easy to implement as a fixed filter in FPGA[4]-based designs since in this case moving from one adder to the next is simply wiring up two logic elements (logic cells) and a right shift can be accomplished simply by wiring output bits 0, 1, 2 . . . of one cell to input bits 1, 2, 3 . . . on the next.

2.5.4   Booth's and Robertson's Methods

Booth's method is similar to partial products in that the multiplier bits are scanned from right to left and a shifted version of the multiplicand added or subtracted depending on these. The difference is that the multiplier bits are examined in pairs rather than singly. An extension of this examines 4 bits in parallel, and in Robertson's method, an entire byte in parallel.

The advantage of these methods is that they are extremely fast. However, the logic required becomes complex as the number of bits considered in parallel increases.

The trick in Booth's method is to define a rule by which the multiplicand is subtracted or added depending on the values of each pair of bits in the multiplier. If two consecutive bits from the multiplier are designated as X_i and X_{i-1}, when the multiplier is scanned from $i = 0$, then the action taken upon detecting each possible combination of 2 bits is shown in Table 2.1.

When a multiplicand is added or subtracted to/from an accumulator, it is first shifted left by i bit positions, just as it is for partial products. This process can be examined in detail by following the example in Boxes 2.15 and 2.16.

---

[4]FPGA: field programmable logic array—a programmable binary calculation device.

Box 2.15 Exercise for the Reader

Consider 9×10 (unsigned):

```
      1001    multiplicand 9
  00001010    multiplier 10
```

```
      0000    (i = 0, no action since bit pair = 0 and a hidden zero)
     -1001    (i = 1, subtract multiplicand since bit pair = 10)
     +1001    (i = 2, add multiplicand << 2 since bit pair = 01)
     -1001    (i = 3, subtract multiplicand << 3 since bit pair = 10)
     +1001    (i = 4, add multiplicand << 4 since bit pair = 01)
              (i = 5 and onward, no action since all bit pairs = 00)
```

The result is therefore obtained as the summation of

```
  10010000
  -1001000
  +100100
   -10010
```

Or by converting the subtractions into additions (see Section 2.4.4)

```
  10010000
 +10111000
  +100100
 +11101110
 =01011010
```

With the result being

$1011010 = 64 + 16 + 8 + 2 = 90$ (correct)

It is important to note that when $i = 0$, the bits considered are the least significant bit of the multiplier and a hidden zero. Thus when the least significant bit of the multiplier is a 1, the multiplicand must be subtracted (i.e., treated as a 10 instead). This can be seen in the second worked example (Box 2.16).

Box 2.16 Booth's Method Worked Example

Consider -9×11 (signed):

```
  11110111    multiplicand −9
  00001010    multiplier 11
 -11110111    (i = 0, subtract multiplicand since bit pair = 10)
  0000000     (i = 1, no action since bit pair = 11)
   +110111    (i = 2, add multiplicand << 2 since bit pair = 01)
   -10111     (i = 3, subtract multiplicand << 3 since bit pair = 10)
   +0111      (i = 4, add multiplicand << 4 since bit pair = 01)
   000        (i = 5 and onward, no action since all bit pairs = 00)
```

The result is therefore obtained as the summation of

```
 -11110111
 +11011100
 -10111000
 +01110000
```

Or by converting the subtractions into additions (see Section 2.4.4):

```
  00001001
+11011100
+01001000
+01110000
─────────
=10011101    + Carry
```

With the result being:

$10011101 = -128 + 16 + 8 + 4 + 1 = -99$ (correct)

There are two points worth mentioning here. The first is that, when dealing with two's complement signed operands, the partial products must be sign extended in the same way as the full partial-product multiplier. The second point to mention is that, when scanning from right to left, the hidden bit at the right-hand side means that the first pair of non-equal bits that is encountered will always be a 10, indicating a subtraction. This regularity may be useful in when designing a hardware implementation.

As an aside, it is worth mentioning to the reader that, even for someone who has been doing this for many years, the preparation of this book highlighted how easy it can be to make very trivial binary addition mistakes—if you are required to do this as part of an examination, you would be well advised to always double-check your binary arithmetic. Getting this right first time is not as simple as it may seem.

As mentioned previously, Booth extended his method into examination of 4 bits at a time, using a look-up table-type approach, and Robertson took this one step further into an 8-bit look-up table. These methods are in fact common in various modern processors, although they require considerable resources in silicon.

2.6   Division

For many years, commodity CPUs and even DSPs did not implement hardware division due to the complexity of silicon required to implement it. Analog Devices DSPs and several others did include a DIV instruction, but this was generally only a hardware assistance for the very basic primary-school method of repeated subtraction.

2.6.1   Repeated Subtraction

Since division is the process of deciding how many times a divisor M goes into a dividend Q (where the answer is the quotient Q/M), it is possible to simply count how many times M can be subtracted from Q until the remainder is less than M.

For example, in performing 13/4, we could illustrate this loop:

iteration $i = 1$, remainder $r = 13 - 4 = 9$
iteration $i = 2$, remainder $r = 9 - 4 = 5$
iteration $i = 3$, remainder $r = 5 - 4 = 1$; Remainder 1 is less than divisor 4 so the answer is 3 with remainder 1.

When working in binary the process is identical, and perhaps best performed as a long division as in the worked example in Box 2.17.

Box 2.17 Long Division Worked Example

Consider $23 \div 5$ (unsigned).

First step would be to write the values down formatted as a long division:

```
101 | 010111
```

       divisor   dividend

Then, starting from most significant end (left) and working toward the least significant end (right), scan at each bit position in the dividend to see if the divisor can be found in the dividend. In each case if it is not found, writing a 0 in the corresponding position above the dividend, and look at the next bit. After three iterations, we would have

```
        000   (quotient)
101 | 010111
```

But now, at the current bit position in the dividend, `101` can be found. We thus write `101` below the dividend and a "1" above the dividend at the correct bit position. Then subtract the divisor (at that bit position) from the dividend to form a new dividend:

```
        0001
101 | 010111
    -    101

      000011
```

Next we continue working from left to right but this time looking in the new dividend for the divisor. In this case it is not found, so that after scanning all bit positions we are left with:

```
        000100
101 | 010111
    -    101

      000011
```

The answer is seen above: The quotient is `000100` with a remainder of `000011`. Since we were dividing 23 by 5, we expect an answer of 4 (correct) and a remainder of 3 (also correct).

So now the question is, how to handle signed integer division? Well it turns out that the most efficient method is probably to note the signs of both operands, convert both to unsigned integers, perform the division, and then apply the correct sign afterward: division uses the same sign rules as multiplication in that the answer is only negative if the signs of the operands differ.

The division process for one popular microprocessor can be seen in the flowchart of Figure 2.9. A close examination of this may prompt some questions, such as why shift *both* A and Q left each iteration? Also, why perform an addition of $Q = Q + M$ inside the

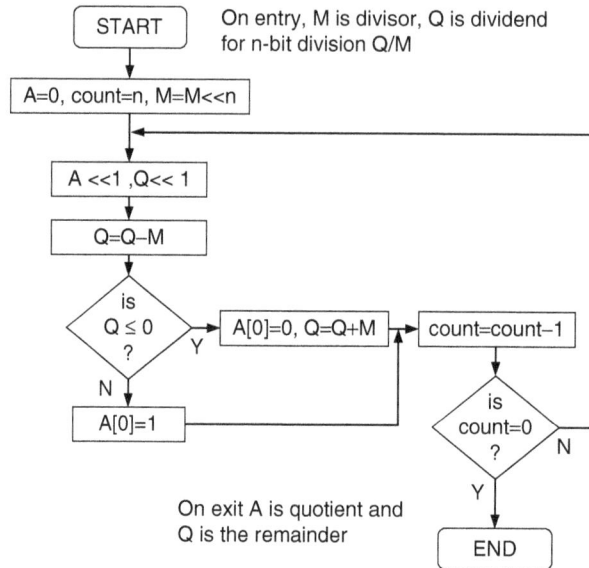

FIGURE 2.9　Flowchart for division algorithm.

loop? The answers to these questions may be answered by considering how the operations are performed using registers within a CPU. It is left as a pencil-and-paper exercise for the reader to follow the operation of the algorithm for one example division, perhaps of two 6-bit numbers: making the effort to do so will really help to clarify how this system works.

Just note that at the completion of the algorithm, register A holds the answer, with any remainder being in register Q. The algorithm will have iterated for n cycles, where n is the number of bits in the input words. As always, it is entirely possible to derive other flowcharts that work differently; for example, some will even iterate and scan through the bits in the opposite direction.

2.7　Working with Fractional Number Formats

Section 2.3.7 introduced the representation of fractional numbers using Q-format notation. Although there are many reasons for requiring fractional notation, one major reason is in DSP, where a long digital filter[5] may require hundreds or thousands of multiply-accumulate operations before a result is determined.

Imagine if some of the filter "weights" (the fixed values in the filter that the input numbers are multiplied by) are very small. In this case, after multiplying by these small values many times, the result could be tiny, rounded down to zero by the number format

---

[5] A digital filter uses a series of linked multiply-add operations to change the characteristics of a signal (presented as a stream of data). For example, an audio signal might be filtered to block out background noise.

used. On the other hand, if some filter weights are large, the result of many multiplications could be huge, resulting in an overflow. It can be difficult to design a filter to avoid either extreme.

Fortunately, there is a reasonable and efficient solution by ensuring that the operands used are in fractional format, and are less than, but as close to, 1.0 as possible. The rationale being that anything multiplied by a number that is less than or equal to 1.0 cannot be greater than itself—so we can be sure that the result of multiplying two of those numbers together will never result in an overflow. Similarly, anything multiplied by a value slightly less than 1.0 does not become much smaller than it was, so results are less likely to quickly round down to zero.

This is possible numerically because we are only multiplying and adding in a filter and these are linear processes: the calculation $(a + b)$ has the same result as $(10a + 10b)/10$. Remember also that the actual fractional format used is not relevant to the hardware used to perform the calculations. The hardware does not change, the only difference is that the software engineer must keep in mind what format is used to store the numbers. This will be illustrated with various examples as we progress below.

2.7.1 Arithmetic with Fractional Numbers

When adding two fractional format numbers, the hardware adder will blindly add whatever binary data it is presented with. However, the answer will only be correct if the formats of each operand prior to the addition are identical. The format of the answer will be the same as that of the operands. If the programmer tries to add operands that have different formats, the answer will make no sense. Assuming the programmer got this right;

$$(m.n) + (m.n) = (m.n)$$

$$(m.n) - (m.n) = (m.n)$$

The arithmetic of calculating with such fractional format numbers is illustrated using two examples in Box 2.18.

Box 2.18   Worked Examples of Fractional Representation

Question 1: Represent 1.75 and 1.25 in (2.2) format fractional notation, perform an addition between the two and determine the result.

Answer: First calculate the bit weightings for (2.2) format notation: We need two digits to the right and two digits to the left of the radix point. Digits to the left of the radix are integer powers of 2 starting with 1 (which is 2^0) and doubling with every digit to the left. Digits to the right of the radix are fractional, starting with ½ (which is 2^{-1}) and halving with every digit to the right:

2^1	2^0	2^{-1}	2^{-2}
2	1	1/2	1/4
2	1	0.5	0.25

We then represent 1.75 as $1 + 0.5 + 0.25$ and represent 1.25 as $1 + 0.25$. Writing them in (2.2) binary format, we get `0111` and `0101`.

Performing the binary addition of these operands results in 1100. Is this correct? 1100 in (2.2) format equals $2 + 1 = 3$. Of course $1.75 + 1.25 = 3$; so yes, the answer is correct.

Next we will illustrate what happens when something goes wrong with the number format.

Question 2: Represent 1.75 in (2.2) format fractional notation, represent 0.625 in (1.3) format fractional notation, perform an addition between the two and determine the result.

Answer: 1.75 was represented in question 1 and is 0111. For 0.625, we note that (1.3) format has bit weightings 1, 0.5, 0.25, 0.125, and we decompose 0.625 into $0.5 + 0.125$ to get (1.3) binary number 0101.

Next we perform the addition 0111 + 0101 to give 1100. However, we do not know the fractional format of the result. Let us speculate whether this is (2.2) or (1.3) by working out the decimal value in each case.

In (2.2) format the result is $2 + 1 = 3$, in (1.3) the result is $1 + 0.5 = 1.5$; however, the answer should be $1.75 + 0.625 = 2.375$. Clearly this does not match with either of the potential answers.

What we should have done was change one operand so that both in the same format before we performed the addition.

Note: Did you notice that the binary patterns of both examples were identical? It was only our *interpretation* of the format of those bit-patterns that changed between examples. By using different interpretations in this way, the same bit-pattern can have multiple meanings to the programmer—but the hardware used to perform the calculation does not need to change.

2.7.2 Multiplication and Division of Fractional Numbers

In the case of multiplication, there is more flexibility in that the operands can have different fractional formats, and the fractional format of the answer is derived from those of the operands. For example, if we multiply an $(m.n)$ format number by a $(p.q)$ format number:

$$(m.n) \times (p.q) = (m + (p \times n) + q)$$

It is evident that the number of bits in the answer of the multiplication will be the sum of the number of bits in the two operands, and this is what we saw when we looked at examples of multiplier hardware in Section 2.5.

Division is rather more complex. Sometimes the best way to perform division is first to remove the radix points of both numbers by shifting both radix positions, in step, digit by digit to the right until they are below the least significant digit of the largest operand, extending the smaller operand where appropriate. The division then occurs in standard binary fashion. In decimal, this is like doing the computation $5.4 \div 0.9$ by first multiplying each operand by 10 so that the computation becomes $54 \div 9$. Most people would find the second division easier than the first, even though they give exactly the same answer. In binary, which is base-2 rather than base-10, we need to double the numbers each time we shift the radix point instead of multiplying by 10. This is illustrated with a worked example in Box 2.19.

Box 2.19 Worked Example of Fractional Division

Consider `11.000 ÷ 01.00` (unsigned).

This has the trivial meaning of $3 \div 1$. The first step in performing this operation is to shift the radix point in step to the right by one position:

`110.00 ÷  010.0`

This is insufficient since the numbers still contain the radix so we repeat one step:

`1100.0 ÷  0100.`

This is still insufficient so we repeat again, extending the `0100` as a side-effect of removing the radix from the `1100.0` as follows:

`11000. ÷  01000.`

The division then occurs as a standard binary division:

```
01000 | 11000
```

Continuing in the long-hand binary division:

```
                  00011
01000          | 11000
  –              1000
                 01000
        –        01000

                 00000
```

The answer is seen to be `11`, which as decimal value 3, is correct.

Looking at the worked examples in Box 2.19, it is clear that the actual division is no more complex than standard binary arithmetic; however, the consideration of radix position can be confusing. This requires careful coding on the part of the programmer (although in practice most division calculations are probably done using compiled languages, in which a mathematics library does all of the hard work, so the programmer does not have to).

2.8 Floating Point

Floating point numbers are similar to fractional format binary, with the additional flexibility that the position of the radix point is variable (and is stored as part of the number itself). It is this flexibility that allows floating point numbers to encode an enormous range of values with relatively few bits.

2.8.1 Generalized Floating Point

A floating point number is one which has a mantissa S (or fractional part) and an exponent E (or power). There is probably also a sign (σ) such that the value of the number represented in base B is given as

$$n = \sigma \times S \times B^E$$

Or more correctly, since the sign digit is binary (with 1 indicating negative and 0 indicating positive), the floating point number is

$$n = (-1)^\sigma \times S \times B^E$$

An example in base-10 decimal would be 23×10^8 which we know is just a shorthand method of writing the value 2300000000. This actually illustrates one of the main benefits of floating point: That floating point numbers generally require less writing (and in binary they require fewer digits) to represent a value compared to the decimal (or binary) values they represent.

In binary the difference is that $B = 2$ instead of $B = 10$ and thus the example will typically be something like 01001111×2^6 which, if the mantissa (01001111) is unsigned, becomes: $01001111 \times 2^6 = 79_{10} \times 64_{10} = 5056_{10} = 1001111000000$, where the subscript $_{10}$ is used to indicate a decimal value.

The final value represented is the same as the mantissa shifted by the exponent (in the same way that we added eight zeros to the base-10 example above).

Normally, all the bits that constitute a floating point number (σ, S, and E) are stored in a single binary word whose length is a convenient number of bits such as 16, 32, 64, or 128. Some bit-level manipulation is therefore required when processing them to separate out the three different parts from the full stored number.

2.8.2   IEEE754 Floating Point

Although there are many possible floating point formats and various implemented examples scattered through computing history; IEEE standard 754 that emerged in 1985, has become by far the most popular, adopted by all major CPU manufacturers. Much of its popularity is because IEEE754 is generally seen to be a well-designed and relatively efficient standardized floating point format.

It is not the intention of this text to describe the entire IEEE754 standard, but we will look at some of the more common features from the 2008 update of the original 1985 version. In particular, we will consider the **binary32** and **binary64** formats (which used to be known simply as **single** and **double** precision, respectively). These are designed to fit into 32-bit and 64-bit storage locations, and are accompanied in the standard by **binary16** half-precision, **binary128** quadruple precision, and **binary256** octuple precision. In the C programming language, **binary32** and **binary64** would normally correspond to *float* and *double* data types (although the precise mapping to hardware is compiler dependent):

Name	Bits	Sign, σ	Exponent, E	Mantissa, S
binary32	32	1	8	23
binary64	64	1	11	52

In addition, further bits can be added to the representation during intermediate calculation stages (in particular, within a hardware floating point unit) to ensure that overall accuracy is maintained during calculations, as will be described later in Section 2.9.3.

Despite all 32- or 64-bits in the table above being used for sign, mantissa, and exponent, the IEEE754 format cleverly manages to signify four alternative modes using unique bit-patterns that do not occur for normal numbers. These states are shown in the following table, beginning with the default state, called *normalized* and ending with *denormalized* (also known as *subnormal*) numbers—which fit between the smallest magnitude normalized number and zero:

Name of Mode	σ	E	S
Normalized	1 or 0	Not all zero, not all one	Any
Zero	1 or 0	All zero	All zero
Infinity	1 or 0	All one	All zero
Not a number (NaN)	1 or 0	All one	Nonzero
Denormalized	1 or 0	All zero	Nonzero

When an IEEE754 number is written down, we typically write the bits from left to right in the order (σ, E, S) as shown below:

σ	E	S

This box could contain a 32-bit or 64-bit binary IEEE754 format value with a single bit representing sign (σ) and the remaining bits conveying exponent and mantissa. All examples given in this book will use only the binary32 format to save paper.

2.8.3   IEEE754 Modes

In the discussion that follows, we will use S to represent the mantissa bit-pattern in unsigned fractional (0.23) format, E to represent the 8-bit binary exponent bit-pattern and σ to represent the sign. Remember that those bitfields have different meanings in the five IEEE754 modes (i.e., E may not be an exponent, and S may not be the mantissa in some modes – indicated by them having the special bit-patterns).

An example IEEE754 binary32 number, while it would be stored in a computer as a single block of numbers such as 01011001011100000000000000000000, we will write it in a box that conveniently separates the three parts of the number like this:

0	10110010	11100000000000000000000000

which would be said to have $\sigma = 0$ and therefore a positive sign, and then

$$E = 128 + 32 + 16 + 2 = 178 \text{ and}$$

$$S = 0.5 + 0.25 + 0.125 = 0.875$$

We will maintain this naming convention for E and S throughout. So from this point on the words "mantissa" and "exponent" are used to indicate the *meaning* of the written bit-pattern, whereas E and S are the actual values written down in binary.

For example, an S of 11100...0 = 0.875 in decimal might mean the mantissa is 0.875 or is 1.875 or is irrelevant (not a number, *NaN*) depending on the number mode. In other words, the meaning of the bit-patterns in E and S change with mode, as we shall see below.

2.8.3.1 *Normalized Mode*

It is the number format that most nonzero numbers will be represented in. It is the one mode where the number format can truly be called *floating point*. In this mode, the number represented by the bit-patterns (σ, E, S) is given by

$$n = (-1)^{\sigma} \times (1 + S) \times 2^{E-127}$$

From this it can be seen firstly that the exponent is in an excess-127 notation (introduced in Section 2.3.5) and secondly that the mantissa always has a 1 added to it. In other word the mantissa value is equal to $S+1$, and we know that S was written in (0.23) format, so the mantissa value has to lie somewhere between 1 and 2.

All this may be very confusing, so we will return to the IEEE754 number format above as a worked example in Box 2.20, and give a second example in Box 2.21.

Box 2.20 IEEE754 Normalized Mode Worked Example 1

Given the following binary value representing an IEEE754 binary32 number, determine its decimal value.

0	10110010	11100000000000000000000000

First of all, we note that here $\sigma = 0$ and therefore the value has positive sign, plus we also note that the number is in normalized mode. Therefore:

$$E = 128 + 32 + 16 + 2 = 178 \text{ and}$$

$$S = 0.5 + 0.25 + 0.125 = 0.875$$

Using the formula for normalized mode numbers, we can calculate the value that this conveys:

$$n = (-1)^0 \times (1 + 0.875) \times 2^{178-127}$$

$$= 1.875 \times 2^{51}$$

$$= 4.222 \times 10^{15}$$

As we can see, the result of the worked example is a fairly large number, amply illustrating the ability of floating point formats to represent some quite big values.

Box 2.21 IEEE754 Normalized Mode Worked Example 2

Given the following binary value representing an IEEE754 binary32 number, determine its decimal value.

1	00001100	01010000000000000000000000

In this case, $\sigma = 1$ and therefore has negative sign, and remaining bit-patterns give

$$E = 8 + 4 = 12 \text{ and}$$

$$S = 1/4 + 1/16 = 0.3125$$

Using the formula for normalized mode numbers, we can calculate the value that this conveys:

$$n = (-1)^1 \times (1 + 0.3125) \times 2^{12-127}$$
$$= -1.3125 \times 2^{-115}$$
$$= -3.1597 \times 10^{-35}$$

This time the result is a very small magnitude number. This illustrates the enormous range of numbers possible with floating point, and also the fact that all through the represented number range (explored further in Section 2.8.4), precision is maintained.

Many of our example numbers have long tails of zeros. We can obtain an idea about the basic precision of IEEE754 by considering what difference would result if the least significant bit at the end of one of those tails is flipped from a 0 to a 1. Box 2.22 provides a guide as to how we can investigate the effect.

Box 2.22   Exercise for the Reader

Notice in the worked examples (Boxes 2.20 and 2.21) that our 23-bit long mantissa values began with a few 1's but tailed off to a long string of 0's at the end. This was done to reduce the difficulty in calculating the value of the example. As a (0.23) fractional format number, the weightings of S at the left-hand end are easier to deal with, having values 0.5, 0.25, 0.125, and so on. Moving to the right, the bit weights quickly become quite difficult to write down.

The exercise in this case is to repeat one of the worked examples, but with the least significant bit of the mantissa set to 1. If the weighting for the most significant mantissa bit, bit 23, is 2^{-1} (0.5) and for the next bit, bit 22, is 2^{-2} (0.25), what will be the weighting for bit 0?

When this is added into the answer, what difference, if any, does it make to the written result?

The real question now is whether this indicates anything about the precision of IEEE754 numbers?

2.8.3.2   *Denormalized Mode*

Some numbers have such small magnitude that IEEE754 cannot represent them. Generalized floating point would round these values down to zero, but IEEE754 has a special denormalized or subnormal number mode that is able to extend the represented numbers downward in magnitude toward zero—at the cost of gradually decreasing numerical precision, until zero is reached.

Denormalized mode is not truly floating point, because the exponent (which is the part of the number that specifies the radix position) is set to all zeros and thus no longer "floats." However, this mode is an important advantage of the IEEE754 floating point standard; it allows useful range extension (at the cost, as we have noted, of reduced precision when in that mode).

In "denorm" mode, the number represented by the bit-patterns (σ, E, S) is given by

$$n = (-1)^\sigma \times S \times 2^{-126}$$

It can be seen first that the exponent is fixed as we had noted above, and second that the value 1 is no longer added to S to form the mantissa. The reason for this will be apparent when we explore number ranges a little later in Section 2.8.4.

Since the exponent is fixed, the bit-pattern is always all-zero and the mantissa nonzero. A worked example will, as always, help to illustrate this, in Box 2.23.

Box 2.23   IEEE754 Denormalized Mode Worked Example

Given the following binary value representing an IEEE754 binary32 number, determine its decimal value.

0	00000000	11010000000000000000000000000000

First, we note that since $\sigma = 0$, the number represented by these bit-patterns therefore has positive sign.

$E = 0$ so we look at the mode table in Section 2.8.3 to see that we are either dealing with a zero or a denormalized number. We need to examine the mantissa to decide which mode the number is (a zero must have all zero S, otherwise it is a denormalized number).

Looking at the mantissa we see it's nonzero and therefore a denormalized mode number:

$$S = 0.5 + 0.25 + 0.0625 = 0.8125$$

Using the formula for denormalized mode numbers, we can calculate the value is being conveyed:

$$n = (-1)^0 \times 0.8125 \times 2^{-126}$$
$$= 9.5509 \times 10^{-39}$$

Since denormalized numbers extend the range of IEEE754 downward, one thing we know about them is that they will always have very small magnitude.

2.8.3.3   Other Mode Numbers

Zero, infinity and *NaN* (not a number) are identified by their special bit-patterns. These can all be positive as well as negative, and require special handling in hardware (see Box 2.24).

Box 2.24   IEEE754 Infinity and Other "Numbers"

Infinity is most commonly generated by a divide-by-zero or by a normalized mode overflow. Infinity can be positive or negative to indicate which direction the overflow occurred in.

NaN, indicating Not-a-Number, is generated by an undefined mathematical operation such as infinity multiplied by zero or zero divided by zero.

Zero itself may indicate an operation that really did result in zero, for example
$(2-2)$, or it could result from an underflow, when the result is too small to be
represented even by denormalized mode, in which case the meaning of $+/-$ zero
indicates whether the unrepresentable number was slightly above or slightly
below zero.

2.8.4 IEEE754 Number Ranges

One excellent way of understanding IEEE754 is through construction of a number line
that represents the ranges possible in the format. The following number line, representing an unsigned 8-bit number, will illustrate what this involves:

Minimum magnitude = 0		Maximum magnitude = $2^8 - 1 = 255$
<————————————————————————>		
Accuracy (distance between number steps) = 1.0		

Three parameters are indicated which describe the format. The first is the smallest
magnitude number (`0000 0000`), the second is the largest magnitude number (`1111
1111`), and the final is the accuracy. Accuracy is defined as the distance between steps in
the format. In this case, the numbers count upward as integers: $1, 2, 3, 4, 5, \ldots, 255$, and
so the step size is simply 1.

Now we will undertake to define a number line for IEEE754 format in the same way.
To simplify matters we will only consider positive numbers, but we will look at both
normalized and denormalized modes—but only for single precision (binary32) numbers.

2.8.4.1 Normalized Mode

This requires that E is neither all-zero nor all-one, but S can have any bit-pattern, while
the actual value represented is

$$n = (-1)^\sigma \times (1 + S) \times 2^{E-127}$$

The smallest magnitude value is conveyed by the bit-pattern having the smallest S
and the smallest allowable E. The smallest S is simply 0, but the smallest E cannot
be 0 (because that would make it either denormalized or zero mode); hence it has to be
`00000001` instead. The smallest magnitude value is thus

0	00000001	00000000000000000000000

Inserting these values into the formula and assuming a positive sign gives us

$$\text{Min. norm} = (1 + 0) \times 2^{1-127} = 1 \times 2^{-126} = 1.175 \times 10^{-38}$$

Next, looking for the largest magnitude number, we remember that S can be anything, but E cannot be all ones `11111111` (because that would put it into infinity or *NaN*
modes). The largest E is therefore `11111110` and the largest S is all-ones.

Considering E first, that bit-pattern is simply 254. However, S is now slightly harder to evaluate:

```
111 1111 1111 1111 1111 1111
```

But realizing that this is (0.23) format and is slightly less than 1.0 in value, we can see that if we added a binary 1 to the least significant digit then all the binary 1's in the word would ripple-carry to zero as the carry is passed up the chain and we would get a value like this:

```
  111 1111 1111 1111 1111 1111
 +000 0000 0000 0000 0000 0001
 =1000 0000 0000 0000 0000 0000
```

We can use this fact, knowing that there are 23 bits, the bit weight of the first MSB is 2^{-1} and the weight of the second MSB is 2^{-2} then the 23rd MSB (which is actually the least significant bit) must have a weight of 2^{-23}.

Therefore, the value of S has to be $(1.0 - 2^{-23})$ since we saw that adding 2^{-23} to it would make it exactly equal 1.0

0	11111110	11111111111111111111111

Putting all that into the formula, we have

$$\text{Max. norm} = (1 + 1 - 2^{-23}) \times 2^{254-127} = (2 - 2^{-23}) \times 2^{127} = 3.403 \times 10^{38}$$

What about number accuracy? Well it turns out if we look at the numbers we have decoded so far, it is not constant. However, we can see that the smallest bit is always 2^{-23} times the exponent, across the entire range.

Finally starting a number line for normalized mode, we get

min. 1.175×10^{-38}		max. 3.403×10^{38}
<-->		
Accuracy (distance between number steps) = 2^{-23} of the exponent		

Since the sign bit affects only the sign and does not affect the magnitude, the range line for negative numbers must be a mirror image.

2.8.4.2   Denormalized Mode

This can be handled in a similar way, although by definition the exponent is always zero and the value represented is

$$n = (-1)^\sigma \times S \times 2^{-126}$$

Remembering that a mantissa of zero is disallowed, the smallest denormalized number has just the least significant mantissa bit set:

0	00000000	00000000000000000000001

And therefore a value of 2^{-23} following the argument for normalized mode maximum number. The formula becomes

$$\text{Min. denorm} = 2^{-23} \times 2^{-126} = 2^{-149} = 1.401 \times 10^{-45}$$

As for the largest denormalized number, this is simply the number where S is a maximum. Looking at the mode table in Section 2.8.3 reveals that an S which is all-ones is allowable in denormalized mode:

0	00000000	11111111111111111111111

Again using the same argument as the normalized maximum value case, this has a meaning of $(1 - 2^{-23})$, giving a value represented of

$$\text{Max. denorm} = (1 - 2^{-23}) \times 2^{-126} = 2^{-149} = 1.175 \times 10^{-38}$$

Now work out the number accuracy. In this case, since the exponent is fixed the accuracy is simply given by the value of the least significant bit of the mantissa, multiplied by the exponent:

$2^{-23} \times 2^{-126} = 1.401 \times 10^{-45}$ (i.e., the same as the smallest value representable).

min. 1.401×10^{-45}	max. 1.175×10^{-38}
<—————————————————————————————————————>	
Accuracy (distance between number steps) $= 2^{-23} \times 2^{-126}$	

Putting the number lines together, we see the huge range spanned by IEEE754 single precision numbers. Remembering too that this is actually only half of the full number line that would show negative values positive as well as positive ones:

Zero	Denormalized		Normalized	
0	1.401×10^{-45}	1.175×10^{-38}	1.175×10^{-38}	3.403×10^{38}
0	<—————————————>		<———————————>	
0	Accuracy $2^{-23} \times 2^{-126}$		Accuracy $2^{-23} \times 2^{E-127}$	

The number line becomes useful when we want to convert decimal numbers to IEEE754 floating point since it identifies which mode: zero, denormalized, normalized, or infinity, should be used to represent a particular decimal number. To illustrate this, worked examples of conversion from decimal to floating point are given in Box 2.25. There will also be several more examples of conversions in Sections 2.9.1 and 2.9.2.

Box 2.25   *Worked Example Converting from Decimal to Floating Point*

Question: Write decimal value 11 in IEEE754 single precision format.
Answer: Looking at our number line in Section 2.8.4, we can see that this value lies squarely in the normalized number range, so we are looking for a normalized

number of the form:

$$n = (-1)^{\sigma} \times (1 + S) \times 2^{E-127}$$

To obtain this, it is first necessary to write $N = 11$ decimal as $A \times 2^B$, where A is equivalent to $(1 + S)$ and knowing that $0 \geq S < 1$ then it follows that $1 \geq A < 2$. Probably the easiest way is to take the number N and repeatedly halve it until we get a value A between 1 and 2:

This gives 11 followed by 5.5 followed by 2.75 and then finally 1.375.

So we have settled on $A = 1.375$ and therefore $N = 1.375 \times 2^B$, and it is evident that B equals the number of divide by 2 we did on number N: in this case, 3. Therefore our number is: $n = (-1)^0 \times (1.375) \times 2^3$

Examining the formula for normalized numbers, we see that this requires:

$\sigma = 0$
$E = 130$ (so that $E - 127 = 3$)
$S = 0.375$ (so that $1 + S = 1.375$)

Now finding a binary bit-pattern for E, gives $128 + 2$ or `10000010` and since 0.375 is easily represented as $0.25 + 0.125$ then the full number is

0	10000010	01100000000000000000000

2.9 Floating Point Processing

Up to now we have considered only the representation of floating-point numbers, in particular the IEEE754 standard. Such a representation is only useful if it is possible to process the numbers to perform tasks, which is considered further here.

In many computer systems, floating point processing is accomplished through the use of special-purpose hardware called a floating point coprocessor, or floating point unit (FPU). In fact, even though this is often included on-chip in commercial CPUs, it is often still accessed as if it were a separate coprocessor rather than as part of the main processor.

For computers that do not have hardware floating point support, software emulation is widely available, and apart from longer execution times (refer to Section 4.6.1), the user may be unaware whether their float computations are being done in hardware or software. Most floating point support (whether hardware or software) is based on the IEEE754 standard although there are occasional software options to increase calculation speed at the expense of having less than full IEEE754 accuracy.

IEEE754 number processing involves the following steps:

1. Receive operands.

2. Check for number format modes of operands. If result value is fixed (e.g., anything multiplied by zero), immediately generate answer from a look-up table.

3. Convert exponents and mantissas, if necessary, to expand numbers into binary.

4. Perform operation on binary numbers.

5. Convert back to valid IEEE754 number format—in particular ensuring that the most significant 1 of the mantissa is as close to the left as possible, since this maintains maximum numerical precision.

2.9.1 Addition and Subtraction of IEEE754 Numbers

In generalized floating point, the exponents of the numbers must all be the same before addition or subtraction can occur. This is similar to ensuring fractional format $(n.m) + (r.s)$ has $n = r$ and $m = s$ before adding, as we saw in Section 2.7.1.

For example, consider the decimal numbers $0.824 \times 10^2 + 0.992 \times 10^4$. In order to do this addition easily, we must have both exponents equal—then we simply add the mantissas. But do we convert both exponents to be 10^2 or do we convert both to be 10^4, or even choose something in between such as 10^3?

In answering this question, first, let us consider how to convert an exponent downward. We know that 10^3 is the same as 10×10^2 and 10^4 is the same as 100×10^2. Since we are talking decimal, we multiply the mantissa by the base value for every time we decrement the exponent. Performing this in our calculation would give us the sum:

$$0.824 \times 10^2 + 99.2 \times 10^2$$

Converting up is the converse since 10^2 is the same as 0.01×10^4, this would result in the sum being

$$0.00824 \times 10^4 + 0.992 \times 10^4$$

But the question remains—which action do we take in practice: Do we convert the smaller exponent to be bigger, or the bigger exponent to be smaller, or compromise with an exponent that is between the two?

The answer is, first, that we do not want to compromise and convert both numbers because that is introducing extra work (two conversions, instead of one). Second, when we consider the bit-fields of binary numbers, and knowing that by making an exponent smaller the mantissa has to get bigger, it means that there is a danger of the mantissa overflowing if it has to become too big. We therefore opt not to increase the magnitude of a mantissa. Therefore, we have to decrease a mantissa, which means increasing the smaller exponent instead:

$$0.00824 \times 10^4 + 0.992 \times 10^4$$

This is termed equalizing the exponents or normalizing the operands. Later we will see that methods exist to help prevent the mantissa from disappearing entirely if it is in danger of being rounded down to zero during this process.

Once the exponents are equal, we can perform an addition on the mantissas:

$$0.00824 \times 10^4 + 0.992 \times 10^4 = (0.00824 + 0.992) \times 10^4$$

IEEE754 addition/subtraction is similar to the decimal case except that since the base is 2, the action of increasing one exponent to be the same value as the other causes the mantissa of that number to be reduced by a factor of two for each integer increase in exponent. The reduction by a factor of two, in binary, is a right shift.

There is also one other factor we must consider and that is the format of the resulting number. Remember that in normalized mode the mantissa bit-pattern cannot be greater than 1, and if the result of a calculation on the mantissa becomes too big then we must right shift the mantissa and consequently increment the exponent.

Similarly, if the mantissa becomes small, it must be shifted left and the exponent decremented. These factors will be explored through a worked example in Box 2.26.

Box 2.26 Floating Point Arithmetic Worked Example

Question: Convert decimal values 20 and 120 to IEEE754 format, add them, and convert the result back to decimal.

Answer: Looking at our number line from Section 2.8.4, we realize that both values lie in the normalized number range of IEEE754, but initially will simply consider a generic $A \times 2^B$ format. Furthermore, we will not look at the exact IEEE754 bit-patterns here, simply remember that $A = (1 + S)$ and $B = (E - 127)$.

Starting with 20, we divide repeatedly by 2 until we get a remainder between 1 and 2: $10, 5, 2.5, 1.25$, and so $A = 1.25$. We had to divide four times so $B = 4$.

120 similarly divides down to $60, 30, 15, 7.5, 3.75, 1.875$ so $A = 1.875$. Since we divided six times, $B = 6$.

The information is inserted into the following table. We do not actually need to derive the E and S bit-patterns at this stage, we are more concerned with their interpretation:

σ	B	A	Binary value	Decimal value
0	4	1.25	1.25×2^4	20
0	6	1.875	1.875×2^6	120

Next step is to equalize the exponents. As discussed in the text, we have to make both equal the largest exponent value, reducing the mantissa of the smaller number as appropriate.

1.25×2^4 thus becomes 0.625×2^5 and then 0.3125×2^6 to reform the operands into the following table:

σ	B	A	Binary value	Decimal value
0	6	0.3125	0.3125×2^6	20
0	6	1.875	1.875×2^6	120

Since both exponents are identical, it is now possible to proceed and add the mantissas $(0.3125 + 1.875)$ to form a result:

σ	B	A	Binary value	Decimal value
0	6	2.1875	2.1875×2^6	?

However, this is not a valid representation for IEEE754 because the mantissa value is too large. Remember the $(1 + S)$ in the formula? Well $A = (1 + S) \leq 2$ is our constraint. If both operands were IEEE754 compliant then we should be able to guarantee that no more than 1 shift is needed to put this correct, so we shift the A value right by one binary digit and then increment B:

σ	B	A	Binary value	Decimal value
0	7	1.09375	1.09375×2^7	?

A check on a calculator will reveal that 1.09375×2^7 is indeed correct, giving us a decimal answer of 140. Number conversion into IEEE754 format will proceed in Box 2.27.

We can now take the process further—having determined how to equate the exponents prior to performing arithmetic, we can tie that in with our knowledge of IEEE754 format, and perform these operations directly on IEEE754 format numbers themselves.

Referring to the previous worked example in Box 2.26, we can now write the IEEE754 bit-patterns of the numbers we have been dealing with and perform the conversion in Box 2.27.

Box 2.27   IEEE754 Arithmetic Worked Example

First we begin with the normalized mode formula:

$$n = (-1)^\sigma \times (1 + S) \times 2^{E-127}$$

Beginning with the value of 20 decimal. In the worked example, it was determined to be 1.25×10^4. Slotting this into the formula reveals that $(1 + S) = 1.25$ and so $S = 0.25$, $(E - 127) = 4$ and thus $E = 131$. This is represented below:

0	10000011	01000000000000000000000

120 decimal was 1.875×2^6 which gives us $S = 0.875$ and $E = 133$:

0	10000101	11100000000000000000000

The result of the addition was 1.09375×2^7 such that $S = 0.09375$ and $E = 134$.

Since the 0.09375 is not an obvious fraction of 2, we can use a longhand method to determine the bit-patterns. In this, we repeatedly multiply the value by 2, subtracting 1 whenever the result is equal to or more than 1, and ending when the remainder is zero:

0 : 0.09375
1 : 0.0187
2 : 0.375
3 : 0.75
4 : 1.5 − 1 = 0.5
5 : 1 − 1 = 0

We can see that we subtracted 1 on iterations 4 and 5. We make use of this by setting the fourth and fifth bits from the left to 1. In fact, we could have used this method for the first two numbers, but they were too easy.

0	10000110	00011000000000000000000

Subtraction is similar to addition—all steps remain the same except the mantissas are subtracted as appropriate. Of course we still have to consider overflow on the result mantissa because we could be subtracting a negative number, such that the result is larger than the operands.

2.9.2 Multiplication and Division of IEEE754 Numbers

For multiplication and division we do not need to normalize the operands first, but we do need to perform two calculations on the numbers, one for the mantissas and one for the exponents. The following relationships hold for these operations on base B numbers:

$$(A \times B^C) \times (D \times B^E) = (A \times D) \times B^{(C+E)}$$

$$(A \times B^C)/(D \times B^E) = (A/D) \times B^{(C-E)}$$

Another decimal example will illustrate the point, for multiplication:

$$(0.824 \times 10^2) \times (0.992 \times 10^4)$$

$$= (0.824 \times 0.992) \times 10^{(2+4)}$$

$$= 0.817408 \times 10^6$$

Once again, in the case of IEEE754 format numbers the result must be converted to a correct representation, and special results (zero, infinity, *NaN*) checked for.

2.9.3 IEEE754 Intermediate Formats

Although a particular IEEE754 calculation may have IEEE754 operands as input and as output, there are cases where the output will be numerically incorrect unless there is greater precision used *within* the calculation. This will be illustrated using a short example subtraction on 9-bit numbers:

$$1.0000\ 0000 \quad \times 2^1 \quad A$$
$$-1.1111\ 1111 \quad \times 2^0 \quad B$$

Before we can proceed with the subtraction, it will of course be necessary to normalize the numbers to the same exponent. We do this by increasing the smaller one as we have seen in Section 2.9.1:

$$1.0000\ 0000 \quad \times 2^1 \quad A$$
$$-0.1111\ 1111 \quad \times 2^1 \quad B$$

Now we can proceed with the calculation to give the following result:

$$0.0000\ 0001 \quad \times 2^1 \quad C$$

Then shift the mantissa left as far as possible:

$$1.0000\ 0000 \quad \times 2^{-7}$$

Let's look at the actual numbers we used. Operand A has value 2.0 and operand B has value $(2.0 - 2^{-8})$ which in decimal is 1.99609375. So the result should be

$$2.0 - 1.99609375 = 0.00390625$$

however, the result from our calculation is 1×2^{-7} or 0.0078125. There is obviously a problem somewhere.

Now let's repeat the calculation but this time including something called a *guard bit* during the intermediate stages. The guard bit acts to extend the length of the mantissa by introducing another digit at the least significant end. We start at the point where the numbers have been normalized. Note the extra digit:

$$
\begin{array}{llll}
1.0000\ 0000\ \mathbf{0} & \times\ 2^1 & \text{A} \\
-\ 1.1111\ 1111\ \mathbf{0} & \times\ 2^0 & \text{B}
\end{array}
$$

Next shifting to normalize the exponents, the LSB of B shifts into the guard bit when we shift the number right by 1 bit:

$$
\begin{array}{llll}
1.0000\ 0000\ \mathbf{0} & \times\ 2^1 & \text{A} \\
-\ 0.1111\ 1111\ \mathbf{1} & \times\ 2^1 & \text{B}
\end{array}
$$

subtraction to give a new result:

$$
0.0000\ 0000\ \mathbf{1} \quad \times\ 2^1 \quad \text{C}
$$

Then shift the mantissa left as far as possible:

$$
1.0000\ 0000\ \mathbf{0} \quad \times\ 2^{-8}
$$

Notice that in line C this time the most-significant (only) 1 occurred in the guard bit whereas previously it was located at the bit above that. The normalized value is 1×2^{-8} or 0.00039065, a correct answer this time.

Although this example showed generalized 8-bit floating point numbers, the principle is the same for IEEE754 numbers.

The example above showed a loss of precision error causing an incorrect result during a subtraction. Of course the same error could occur during an addition since $A - B$ is the same as $A + (-B)$. But can it also occur during multiplication and division? It is left as an exercise for the reader to try and find a simple example that demonstrates this.

In IEEE754 terminology more than one guard bit is used, and the method is called extended intermediate format. It is standardized with the following bit widths:

Name	Bits	σ	Exponent E	Mantissa S
Extended single precision	43	1	11	31
Extended double precision	79	1	15	63

Obviously it becomes awkward to handle 43-bit and 79-bit numbers in computers that are based around 8-bit binary number sizes, but this should not normally be an issue because extended intermediate format is designed for use within a hardware floating point unit during a calculation. The input numbers and output numbers will be 32 bit or 64 bit only.

2.9.4 Rounding

Sometimes an extended intermediate value needs to be rounded in order to represent it in a desired output format. At other times, a format conversion from double to single precision may require rounding. The requirement for rounding is true for both fixed and floating point number formats.

There is more than one method of performing numeric rounding, and many computer systems will support one or more of these methods under operating system control:

- Round to nearest (most common): round to the nearest representable value, and if two values are equally near, default to the one with LSB = 0, for example, 1.1 to 1, 1.9 to 2, and 1.5 to 2.

- Round toward +ve: round toward the most positive number, for example, −1.2 to −1 and 2.2 to 3.

- Round toward −ve: round toward the most negative number, for example, −1.2 to −2 and 2.2 to 2.

- Round toward 0: equivalent to always truncating the number, for example, −1.2 to −1 and 2.2 to 2.

For very-high-precision computation, it is possible to perform each calculation twice, rounding toward negative and rounding toward positive, respectively, during each iteration. The average of the two results found could be the answer (at least in a linear system). Even if a high-precision answer is not obtained using this method, the difference between the two answers obtained will give a good indication of the numerical accuracy involved in the calculations.

2.10   Summary

This chapter, entitled "Foundations," has really begun our journey inside the computer—whether that is a room-sized mainframe, a gray desktop box, or a tiny embedded system. It is foundational too, since almost all computers, whatever their size, are based upon similar principles. They use the same number formats and perform the same type of calculations such as logical, arithmetic, multiplication, and division. The main differences that we have seen are that there exist some faster methods that can be chosen for any of these operations, at the expense of increased complexity, size, and usually power consumption.

We began the chapter by considering the definition of what a computer is and what it contains. We introduced the useful classification of computer types (or CPUs) by Flynn, viewed them in terms of their connectivity and the layers of functionality that they contain. We then refreshed our knowledge of number formats, and the basic operations, before looking in a little more detail at how these calculations are achieved.

With the foundations having been covered here, Chapter 3 will focus on how to achieve the connectivity and calculations that we know are required—how to fit these functional units together, write and store a program, and control the internal operation required in a working CPU.

2.11   Problems

2.1   A programmer wrote a C language program to store 4 bytes (b0, b1, b2, b3) to consecutive memory locations, and ran this on a little-endian computer with 32-bit wide memory. If she examined the memory in her computer after running the program, would she see something like A or B in the diagram below?

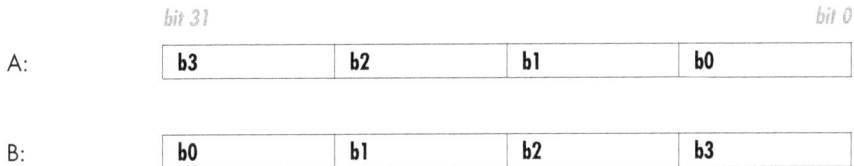

bit 31				bit 0

A:

b3	b2	b1	b0

B:

b0	b1	b2	b3

2.2 Complete the following table (for 8-bit binary numbers), indicating any instances where conversion is impossible for the given value:

Value	Unsigned	Two's complement	Sign-magnitude	Excess-127
123				
−15				
193				
−127				

2.3 With a two's complement (2.30) format number, how can the value 0.783203125 be represented? Can it be represented exactly within (1) 32 bits, (2) 16 bits, and (3) 8 bits?

2.4 One BCD digit consists of 4 bits. Starting with a 4-bit ripple-carry adder, introduce a modification using extra single-bit adders and logic gates to create an adder that can add two BCD digits, and produce a BCD sum. Extend the design so that it can add two 4-digit BCD numbers.

2.5 Using partial products (long multiplication), manually multiply the two 4-bit binary numbers $X = 1011$ and $Y = 1101$ assuming they are unsigned numbers.

2.6 Repeat the previous multiplication using Booth's algorithm.

2.7 If ADD, SHIFT, and compare operations each require a single CPU cycle to complete, how many CPU cycles are needed to perform the calculation in Problem 2.5? Compare this with the steps of Booth's method in Problem 2.6. Also state whether Booth's algorithm would become more efficient for a larger word width?

2.8 Consider an RISC CPU that has an instruction named "MUL" that can multiply the contents of two registers and store the result into a third register. The registers are 32 bits wide, and the stored result is the top 32 bits of the 64-bit logical result (remember that 32 bits × 32 bits should give 64 bits). However, the programmer wants to determine the full 64-bit result. How can she obtain this? (*Hint: You will need to do more than one multiply, and also a few ANDs and adds to get the result.*) Verify your method and determine how many instructions are needed.

2.9 If we multiply the two (2.6) format unsigned numbers $X = 11010000$ and $Y = 01110000$ then we should get a (4.12) format result. We can shift the result two digits left, giving (2.14) [i.e., effectively removing the top two bits] and then truncate it to (2.6) [by discarding the lower 8 bits]. Will this cause an overflow, and will the truncation lose any bits?

2.10   Consider the IEEE754 single-precision floating point standard.

   a) Express the value of the stored number N in terms of its storage bits (σ, E, S) for the following cases:

      i. $E = 255$, $S \neq 0$
     ii. $E = 255$, $S = 0$ successfully
    iii. $0 < E < 255$
    iv. $E = 0$, $S \neq 0$
     v. $E = 0$, $S = 0$

   b) Express the following values in IEEE754 single-precision normalized format:

      i. -1.25
     ii. $1/32$

2.11   Can a standard exponent/mantissa floating point number format represent zero in more than one way? Can IEEE754 represent zero in more than one way? If so, explain any differences.

2.12   Use the division flowchart of Figure 2.9 to obtain the quotient and remainder values for the unsigned 5-bit binary division Q/M, where Q = 10101b and M = 00011b.

2.13   Use the multiplication flowchart from Figure 2.7 to perform partial-product multiplication of two 5-bit unsigned binary numbers 00110 and 00101. Determine the number of registers used, their sizes, and their content during each iteration for the two methods.

2.14   Repeat the previous question using the multiplication block diagram of Figure 2.8, to compare and contrast the two approaches in terms of efficiency, number of steps, and so on.

2.15   Consider the following calculation in the C programming language:

$$0.25 + (\text{float})(9 \times 43)$$

Assuming that integers are represented by 16-bit binary numbers and the floats are represented by 32-bit IEEE754 single precision representation, follow the numerical steps involved in performing this calculation to yield a result in IEEE754 format.

2.16   How would Michael Flynn classify a processor that had an instruction able to simultaneously right shift by 1 bit position every byte stored in a group of five internal registers?

2.17   Justify whether self-modifying code (i.e., software that can modify its own instructions by rewriting part of its code) would fit better in a von-Neumann or Harvard architecture system.

2.18   Using a 16-bit processor and only a single-result register, follow the process to add the (2.14) format unsigned number X = 01.11000000000000 and the (1.15) format unsigned number Y = 0.110000000000000. What format would the result need to be in to avoid any overflow? Is there any loss of precision caused by the calculation in this case?

2.19 Identify the IEEE754 modes of the following numbers:

1	10100010	10100000000000000000000000000
0	00000000	10100000000000000000000000000
0	11111111	00000000000000000000000000000

2.20 What would be the mantissa and the exponent of the result of the following base 7 calculation, expressed in base 7:

$$(3 \times 7^8)/(6 \times 7^4)$$

Hint: You should not need to use a calculator to obtain the answer.

2.21 Using partial products (long multiplication), manually multiply the two 6-bit binary numbers $X = 100100$ and $Y = 101010$ assuming they are signed.

2.22 Repeat the previous multiplication by swapping the multiplier and multiplicand (i.e., multiply the two 6-bit signed binary numbers $X = 101010$ and $Y = 100100$). Compare the number of additions that are required to perform the partial product summation. Is it possible to simplify the process by swapping multiplier and multiplicand, and if so, why?

2.23 Repeat the previous two multiplications using Booth's method. Is there any difference in the number of partial product additions when the multiplier and multiplicand are swapped?

2.24 Referring to Section 2.9, determine the number of basic integer addition, shift and multiplication operations required to perform a single precision IEEE754 floating point normalized mode multiply, and compare this with the basic operations required to perform a (2.30) × (2.30) multiply. Ignore extended intermediate mode, overflow and saturation effects, and assume the floating point numbers had different exponent values.

2.25 How many computational operations are required to perform an 8-bit division using repeated subtraction?

CHAPTER 3

CPU Basics

I n Chapter 1, we viewed computers mainly in terms of their external characteristics (size, scale, processing ability, general usefulness) from old to new, and from large to small and back to large again. Chapter 2 then delved inside the box to see what happens inside so that the devices can be called computers; we considered number formats, binary arithmetic, a brief mention of types of computation and CPU control, and then introduced the idea of functional units such as the arithmetic logic unit (ALU) and multiplier. In this chapter, we now look in more detail at the brains of a computer—the central processing unit (CPU) and how it works to perform calculation, process instructions and, more importantly, how it controls everything within a computer. This more traditional view of computer architecture will not consider state-of-the-art extensions and speedups (which Chapter 5 will cover), and neither look too deeply at the individual functional units within a computer (since these will be presented more thoroughly in Chapter 4). Instead, this chapter will concentrate on describing what a computer is comprised of, how it is organized and controlled and how it is programmed.

3.1   What Is a Computer?

When the general public think of a CPU, they often envisage a beige colored box with monitor, keyboard, and mouse. While the box they imagine does contain a CPU, we know there is a whole lot more in there besides. Many people may not also realize that there is a similar CPU inside their smartphone, as well as inside their digital watch and even inside a modern 'fridge or vacuum cleaner (although those CPUs are likely to be far less complex)!

The "computer" part of the system comprises the CPU, memory subsystem, and any required buses to link them—in fact those items that allow it to function as a stored-program digital computer. A computer does not require any graphics capability, wireless connectivity, hard disc, or sound system in order to act as a computer and execute stored programs. In essence, a stored-program digital computer is basically just a very flexible, but generally quite basic, calculator with memory that is programmable to perform the required functions.

These days most people in the developed world will be surrounded by tens, if not hundreds, of computing devices. As mentioned, these may be inside vacuum cleaners, microwave ovens, toasters, music players, doorbells, or electronic door locks (all of which are examples of embedded systems). In 2010, it was estimated that luxury cars contained well over 100 processors, and even an entry model contained over 40 separate devices. However, buyers of even the most basic vehicle today are likely to be purchasing well over 100 processors—even the individual bulbs in a modern car each contain built-in CPUs capable of signaling to the vehicle control unit over a controller area network (CAN bus) when a bulb fails. Airbags, steering, gas and temperature sensors, type pressure monitors, airflow monitors, and wiper controls are all examples of devices containing tiny processors.

This scale of usage is not confined to cars—smart homes, industrial systems, modern healthcare, finance, education, power generation, power distribution, and even agriculture have similarly embraced the embedded systems revolution. It is probably becoming quite easy to foresee that computing's "future is embedded." The contents of this chapter—which apply equally to computers that are room sized, as well as those the size of an ant—will be supplemented by examples that particularly demonstrate the relevance to embedded systems. One of the main ways of accomplishing that aim is to use the ARM processor for many examples throughout this book. The ARM, as we will see in this chapter, did not begin life as an embedded CPU (in fact the term "embedded systems" was not in popular use at the time), but over the intervening years it's architectural advantages, regularity, and power efficiency have cemented it as the prime choice of processor for new embedded systems, across every application area.

3.2   Making the Computer Work for You

We have read that, at its most basic level, a computer is simply a unit able to perform logical operations on data. All higher-level computational functions are a sequence or combination of basic data moves and logic operations on the data. Various units inside a computer are dedicated to performing different tasks, but all are made up from operational building blocks that either transfer or operate on data. For example, the ALU performs arithmetic, while a bus transfers data from one point to another. Similarly, memory storage involves lodging an item of data to a particular location, until it is *retrieved* again. A graphics display involves taking an item of data (such as an integer representing a single pixel), lodging it in the particular location that maps to a position on the screen, and the display interpreting its value to change the color and intensity of that pixel. All of these examples show that the essence of a computer is simply to be a device that moves or transforms data.

Obviously, some method is needed to *direct* the computer, that is, decide when and where to move the data, and which logic operations to perform on it. The computer (comprising its internal units and buses) must be *programmed* to perform the work that we wish it to undertake.

The work required needs to be divided into a sequence of available operations. Such a sequence is a *program*, and each operation is commanded through an *instruction* plus operands. The list of supported operations in a computer defines its *instruction set*.

We will discuss various ways to program a computer in Chapter 8, but for now we will concentrate on what is inside a computer that allows a program to operate.

3.2.1   Program Storage

Instructions clustered into a program need to be stored in a way accessible to the computer. The very first electronic computers were programmed by plugging wires into different holes. Later, manual switches were used, and then automated with punched card readers. Punched and then magnetic tape were invented, but whatever the storage format, a new program was entered by hand each time after power up.

Modern computers store programs on magnetic disc, ROM, EEPROM, flash memory, or similar media. Programs are often read from their storage device into RAM before execution for performance reasons: RAM is faster than most mass-storage devices.

Items stored in memory need to have a location that is accessible, and their storage place also needs to be identified in order to be accessed. Early computer designers termed the storage location a "memory address" (today we tend to shorten that to simply *address*), since this allows the CPU to select and access any particular item of information or program code which reside at unique addresses. The most common way for a computer to store data to memory (using parallel-connected RAM) is for the CPU to notify the memory storage device (we refer to it simply as *memory*) which address it should use, present the data to the memory and wait for it to be lodged within the device. When the CPU wishes to read from memory, it tells the memory which address to look up, then waits for the memory to retrieve the content and provide it to the CPU.[1]

CPUs are programmed, at the lowest level, in machine code instructions which have fixed [in most reduced instruction set computer (RISC) devices such as ARM, PIC, or MIPS], or variable numbers of bytes per instruction [as in several complex instruction set computer (CISC) devices such as ADSP and Motorola 68000]. A bunch of these instructions, in some particular sequence, instructs a computer to perform required tasks. This bunch of instructions is called a program.

For the sequence of instructions in a program to be able to do something useful, they probably require access to some data which requires processing. This historically encouraged a separation between program and data storage spaces, particularly since the two types of information have different characteristics: programs are typically sequential and read-only (since they are providing a prewritten sequence of program instructions, which does not change during program execution usually), whereas data may require read/write access and may be either sequential or random access in nature.

3.2.2 Memory Hierarchy

Storage locations within a computer can all be defined as "memory," because once written to, they remember the value that was written while power is applied (and some even remember the value after the power is off). However, we usually reserve the term "memory" to refer to solid-state RAM and ROM rather than registers, CDs, and so on. Whatever the naming, storage is defined by various trade-offs and technology choices that include the following characteristics:

- Cost
- Density (bytes per cm^3)
- Power efficiency

---

[1] Readers might be wondering why the CPU needs to store and retrieve items using external memory in the first place—the answer is that while computers have a bank of built-in registers which are very fast to access, these relatively expensive parts of the CPU almost never have enough storage capacity to be useful in real programs (the ARM has just 16 registers, each of which can hold a single integer). Even single programs may, in reality, require a million or more times as much storage space. Consider also that modern CPUs might be multitasking between a hundred and a thousand separate programs at any one time. . . additional addressed memory is a necessity.

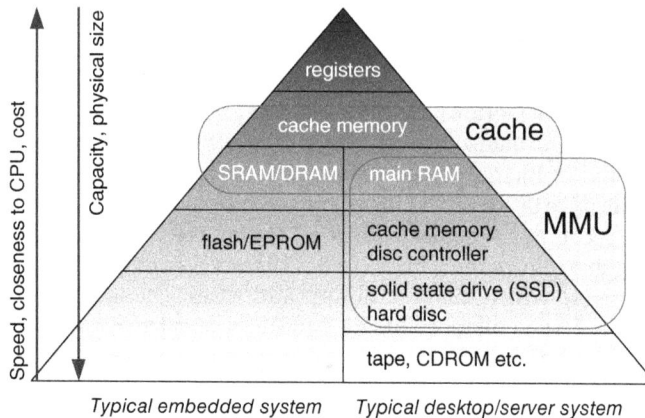

FIGURE 3.1 A pyramidal diagram illustrating the hierarchy of memory in terms of speed, size, cost, and so on for embedded systems (on the left) and traditional desktop computers (on the right).

- Access speed (including seek time, and average access time)
- Access size (byte, word, page, etc.)
- Volatility (i.e., is data lost when the device is unpowered)
- Reliability (does it have moving parts, does it age?)
- CPU overhead to manage it

Combining some of these factors for different types of technology leads to a hierarchy of memory as shown in the pyramid in Figure 3.1, for both a large desktop/server and a typical embedded system. Two items shown will be explored subsequently in Chapter 4: the memory management unit (MMU) and cache. For the present discussion, notice that registers—the temporary storage locations within the CPU located very close to the ALU—are the fastest, but most expensive resource (and are therefore generally few in number, ranging from 1, 2, or 3 in simple microcontrollers up to 128 or more in some large UNIX servers).

Moving down the pyramid, cost per byte decreases (and thus the amount provided tends to increase), but the penalty is that access speed also decreases. A computer, whether embedded, desktop, or supercomputer, almost always comprises several of the levels in the hierarchy:

Registers: Stores temporary program variables, counters, status information, return addresses, stack pointers, and so on.

RAM: Holds stack, variables, data to be processed and often a temporary store of program code itself.

Nonvolatile memory such as flash, EPROM, or hard disc: Stores programs to be executed, which is particularly important after initial power up (booting) when volatile RAM memory would start empty.

Other levels are there for convenience or speed reasons, and since there are so many levels in the hierarchy, there are several places capable of storing required items of information. Thus, a convenient means is required to transfer information between locations as and when required.

3.2.3 Program Transfer

For reading a program from external storage into RAM, an I/O interface such as IDE (integrated drive electronics—a very popular but now largely superseded interface for hard discs), SATA bus (serial advanced technology attachment—also for hard disc and peripheral attachment), or serial buses like USB are used. Such interfaces will be explained later in Sections 6.3.2 and 6.3.4. The connection between the RAM and CPU, and also between CPU and I/O devices is via a bus, and these transfer programs or data either a word (typically a byte, 16 bits, or 32 bits) at a time or serially (which means transferring a bit at a time). Data transfer is thus a stream of bits or a stream of words, depending upon the bus technology. RAM may be external or internal to the physical integrated circuit (IC) on which the CPU resides, but always connects to it over some kind of bus.

When an instruction from a program is read from RAM into a CPU, it needs to be decoded, and then executed. Since different units inside the CPU perform different tasks, data being processed needs to be sent to the correct unit, able to perform the required function on that data. To convey information (comprising instructions or data items) around the inner parts of a CPU, there needs to be an internal bus between an instruction fetch/decode unit and the various processing units, and perhaps a bus to collect the result from each processing unit, and place it somewhere else.

Data to be processed by an instruction is often already available in internal registers, usually having either been the output of the previous instruction or having previously been transferred from external memory into a register. In particular, many modern RISC CPUs have a *load-store* architecture, which constrains the data processing instructions in the instruction set so that the data to be processed by each instruction must come from a register. If the data is not currently in a register and is instead in memory, then a separate instruction is needed to fetch the data from memory to a register, before the operation can take place. This is in contrast to other CPUs (including most CISC devices) in which data processing instructions can operate on data residing either in a register or in memory. Load-store appears to be less flexible in operation, but this reduced flexibility is key to ensuring that the data processing instructions in a load-store architecture machine are fast and efficient. If additional flexibility were introduced to also operate on data in memory, the increased complexity would result in the instructions not being able to execute as fast, even for operands in registers.[2]

During data processing operations in a load-store machine, data is transported from the selected register to the appropriate processing unit over parallel buses. A calculation result will then be sent back to registers, again over parallel buses. It is often convenient to group all internal registers together into a bank, and in a *regular architecture* machine, every processing unit will be connected to this bank of registers—again, this is done over a bus.

In Chapter 4, we will look at computer buses in a different way, as we examine many of the functional blocks found in modern CPUs. We will also consider the effect of different bus arrangements on performance. Here though, we can be content with the knowledge that such things as internal buses are necessary, and do exist.

---

[2]Transfer of data between a register and the CPU is fastest—remember registers are very fast memory located close to the CPU. Transfer of data to and from memory is always slower. However, it is the instruction decode, control, and timing processes that becomes more complex when accessing data from memory.

Given a (possibly quite complex) bus interconnection network inside a CPU, plus multiple internal functional units and registers that connect to this, the question arises as to what arbitrates and controls the multitude of data transfers across and between the buses and functional units.

3.2.4   Control Unit

Multiple buses, registers, various functional units, memories, I/O ports, and so on need to be controlled. This is the job of the imaginatively named *control unit*. Its job is simplified by the fact that most operations require there to be a well-defined process flow within a CPU, such as

- Fetch instruction
- Decode instruction
- Execute instruction
- Save result of instruction (if any)

The control unit needs to ensure that these steps occur in the correct order, and that any data for operands or result is moved to/from the appropriate functional units (e.g., for an ADD, it needs to ensure that the two items of data being added are provided to the inputs of the ALU before the ALU is told to execute an addition). Every functional unit and bus has control ports, and so there needs to be a set of control wires and signals within a device that runs from the control unit to each of the on-chip elements that must be controlled.

In early processors, the control unit was a simple finite-state machine (FSM) which endlessly stepped through a set of predefined states (much like the list above). Control wires ran from the FSM to each of the devices requiring control, in a spiderweb of wires and interconnects. When CPUs became integrated into single chip devices, the spiderwebs of control wires, in many cases followed, and also became integrated on chip.

Control is not only needed to handle the fetching and distribution of instructions to functional units, it is also needed for carrying out the actions of single instructions (most of which have several substeps). Consider an example of a simple data transfer from register A to register B in a system that has a single 32-bit data bus as shown in Figure 3.2 (an example of the assembly language instruction to do this could be LDR B, A, where LDR means "load register"). The two triangles within the figure are tristate buffers—devices similar to a switch in that when the control signal is enabled, signals can pass through the buffer and "drive" whatever it is connected to. But when the control signal is disabled,

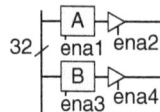

FIGURE 3.2   A block diagram of a very simple computer control unit showing two registers, each with selectable tristate buffers, and a single 32-bit bus connecting all ports.

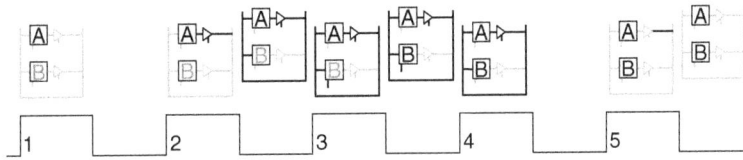

FIGURE 3.3 An illustration of the cycle-by-cycle timing of the simple control unit that was shown in Figure 3.2 as it transfers data from register A to register B. Darker lines indicate that the particular bus or signal is active at that time.

signals do not pass through the buffer or affect its output. This kind of device is used between registers and the buses they are connected to in order to control which register is allowed to drive the bus wires at any time.[3] Not only registers, all functional units that can output values to a shared bus need to have a tristate between their outputs and the bus.

Bearing this in mind, the actions that need to be taken for a data transfer are summarized as

1. Turn off any tristate buffers driving the bus (in this case de-assert *ena1–4*).

2. Assert *ena2* to turn on the 32-bit tristate, driving the content of register A onto the shared bus.

3. Assert *ena3* to feed the bus data into register B.

4. De-assert *ena3* to lock the bus data into register B.

5. De-assert *ena2* to free up the bus for other operations.

Perhaps the details of the process will differ from device to device (in particular we did not consider the exact timing, since enable signals are usually edge-triggered on different clock edges), but something like this process is needed—in the order given—and more importantly sufficient time is required between stages for

1 to 2: wait for the "off" signal to propagate along the control wires, hit the tristate buffers, and for them to act on it.
2 to 3: wait for the bus voltage to stabilize (i.e., the content of register A to be reflected by the bus voltage levels).
3 to 4: give the register sufficient time to capture the bus value.
4 to 5: wait for the control signal to hit the register, and the register to stop "looking at the bus" before the bus can be freed for another purpose.

Sometimes the waiting time is most important—in modern processors it is counted in system clock cycles, with each stage of the process being allocated at least one cycle.

Figure 3.3 illustrates cycle-by-cycle timing for the case of one clock between actions, showing the sequence of events at each stage in the process. It is evident that a

---

[3] Only a single tristate buffer can drive a bus at any one time, with the tristates for all other connected registers being turned off (or *tristated*, which means that they do not "drive" the bus). If two registers are allowed to drive a bus simultaneously, the bus value is corrupted—or the entire device may even malfunction and short out.

synchronous control system is needed to carry out this sequence of actions, which is required for even the most simple of CPU instructions.

Not all instructions need to step through the same states—some such as those that return no result—can be terminated early. Those instructions could either be supported by allowing a state machine to continue running through all states (but with dummy actions for the unused states), or could be supported by early termination or custom state transitions. In the example given, the instruction could potentially finish earlier, because there are states in which nothing (useful) is happening.

Some instructions are likely to need specialized handling that expands the state machine further (i.e., instructions that do not fit into the same sequence). CPU designers generally cater for those cases by increasing the complexity of the state machine to handle the exceptional instructions, or would use a second state machine. Over the years, more and more weird and wonderful instructions have been introduced. It does not require a genius to figure out where all this ended up—more and more complex state machines! In some cases, the CPU control unit became the most complex part of the CPU design, and required up to half of the on-chip area. In other cases, the state machine was so complex that it was itself implemented by another CPU—in effect a simpler processor was used just to handle the control needs of a larger and more complex one. In IC design terms (as in many other fields), complexity is known to lead to errors, and for these reasons simpler alternatives were researched.

However, we have only considered handling of different instructions within a processor and how the controller steps from state to state. If we now consider the actual task of distributing the control signals across larger and ever-growing IC sizes with increasing numbers of internal bus interconnects, larger register banks, more functional units, and a larger degree of clocking complexity and flexibility, we can imagine that a large portion of the internal processor routing logic (i.e., wires that traverse the device from one side to another) is going to need to be reserved for the controller. This presents difficulties beyond the complexity of instruction control. It turns out that in a silicon IC, interconnects that can reach across an entire chip are a scarce resource: these are normally reserved for fast data buses. The need to utilize more and more of these for dedicated control purposes provided another impetus to the research into alternative control strategies.

Three general methodologies resulted from this, namely *distributed control, self-timed control*, and *simplification* (through increased regularity). The main example of distributed control is in the use of microcode, explored in Section 3.2.5. An example of simplification is in the move to RISC processors—explored after that in Section 3.2.6. Let us briefly examine each control method.

Figure 3.4 now shows part of the internals of a very simple CPU. A bank of registers is shown (with four visible), and there are two ALUs all connected through two shared

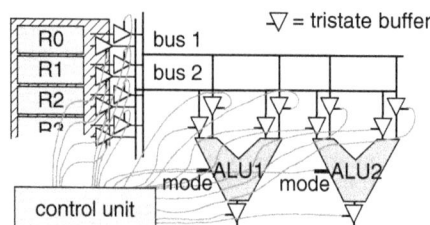

FIGURE 3.4 A block diagram of the centralized control wiring required for a very simple CPU.

data buses. At each point of bus entry/exit there is a tristate buffer. Each bus, tristate, register, and ALU port is several bits wide, that is, they would all be 16 bits wide if this were a 16-bit device. For clarity we draw just single, but slightly thicker, lines to denote the buses.

There are many thin control wires emanating from the control unit in Figure 3.4 and, even for such a simple system, the connectivity appears messy. These wires are used to enable or disable each of the tristate buffers, and set the mode of the ALUs (which can each select between several functions, such as ADD, AND, SUB, and so on). Some wires, such as those used to select which register is used, are omitted for clarity. In Chapter 4 and beyond, different bus arrangements will be discussed, but we will not display control signals such as these in subsequent chapters: if we did so, the diagrams would quickly become too complicated. It has to be understood, therefore, that when we draw a register, ALU, multiplier, bus output, bus input, and so on, there is always some control logic needed for managing them "behind the scenes."

One simplification that can be introduced is to use a control bus, or several control buses. Instead of two control signals needed for each register in Figure 3.4, the fact that each data bus can only carry a single value at a time can be exploited to need only a 2-bit selection bus to drive each data bus (i.e., 4 bits total control for the system shown). This is termed a "register select" bus. Such an approach may not seem particularly beneficial in a four-register system, but with 32 registers it would reduce the number of register select control wires from 64 to 6.

A small example is shown in Figure 3.5. The number of wires emanating from the control unit on the left of the register bank is four. In practice these are decoded *in the register bank itself* to select which register, or registers are active. This approach does not necessarily minimize logic, but it does halve the number of connections (wires) between the control unit and CPU.

To summarize, control is needed for arbitration of internal buses, for initiating the fetch, decoding and handling of instructions, for interactions with the outside world (such as I/O interfacing), and pretty much everything sequential in a CPU—which is a great deal. Control may even extend to handling external memory, and the next chapter carries an important example of this in the memory management unit.

Self-timed control is an alternative strategy that distributes control throughout a CPU, following from the observation that most instructions need to follow a common "control path" through a processor—fetch, decode, execute, and store. And during execution, the processes for different instructions are also fairly common—drive some registers onto buses, drive values from buses into one or more functional unit, then sometime later allow the result to be collected (again using one or more buses), and latched back into registers.

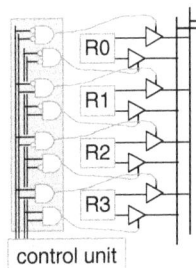

FIGURE 3.5 A small control unit is shown in this diagram wired to the input select logic for a bank of four registers.

Self-timed control in this instance does not imply an asynchronous system, since each block is synchronous; albeit to a faster clock (note that self-timing *is* used within some esoteric asynchronous systems which we will explore in Chapter 9, but in this case we are only dealing with synchronous logic).

A centralized control unit could specify in turn "fetch now" then "decode now" then "execute now" and finally "store now." This would require control connections from the IC areas responsible for each of these tasks, back to the central control unit. However, the self-timed strategy requires the control unit to simply start the process of instruction fetch. The signal "decode now" would be triggered from the fetch unit, when ready, rather than from a central location. Similarly, "execute now" would be a signal generated by the decode unit and passed to the execute unit. In this way a control interconnect is needed from each unit to the next unit, but it does not need to run all the way back to a single central location. In effect, the control signals in this method, are actually following the data paths—something that becomes even more effective in a pipelined machine (which will be covered in Chapter 5).

The two alternative approaches of centralized and self-timed control are shown in the flowcharts of Figure 3.6. In this figure, data buses are not shown, but they would originate from external memory, and traverse the path through fetch, decode, execute, and store (FDES). On the left of Figure 3.6 is a control unit with four control buses, each one linked to the enable inputs of the four separate units. At the relevant times, as specified in an internal state machine, the control unit will initiate operations in the FDES units.

Depending upon the instruction being processed, the control unit state machine may need to operate the FDES differently (perhaps a longer execution stage, or skip the store stage). This knowledge must be encoded within the control unit, which must remember every combination of operations for every unit connected to it. The state machine must firstly contain detailed information on the timings and requirements of each unit, but must keep track of potentially multiple instructions progressing simultaneously through these units.

On the right-hand side of Figure 3.6, a self-timed system is shown: the control unit still initiates the process, but in this case each subsequent unit is triggered from the previous unit as and when necessary. Since the individual units themselves initiate the next step, the data buses (not shown) are assumed to handle the correct information at the correct times.

Depending upon the instruction being processed, units may decide to skip the next step, and pass the request directly to the subsequent unit. Each unit must thus encode the knowledge of its own responsibilities and timings.

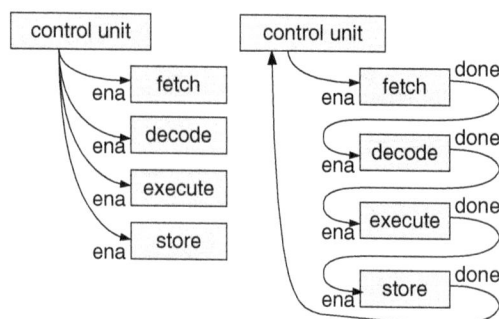

FIGURE 3.6 Control flowcharts of the alternative strategies of centralized control (left) and self-timed control (right).

Perhaps more problematic is the need to convey different information to the various units: for example, the execute unit needs to know what function is to be performed—is it an AND, OR, SUB, and so on. It does not need to know where to store the result from the execution—this information is needed by the store unit which in turn does not need to know what function was performed. In the self-timed case, either a full record of needed information is passed between units, with units only reading the items relevant to them, or there is a centralized store for such information. The choice of implementation strategy depends on complexity and performance requirements.

3.2.5 Microcode

As early CPUs grew and became more complex, they ended up being an amalgamation of basic required functionality, with assorted customized one-off instructions that their designers considered important. Rapid changes in the applications and use of computing, coupled with the slow design cycle of creating a new CPU, meant that some of those instructions were out of fashion by the time computers based on those CPUs saw widespread adoption (ignore anyone who claims that the history of computing does not have techniques that go into and out of fashion periodically). An example from the 1980s would be the binary-coded-decimal handling instructions of the 8086, 80386, and 80486. These instructions were required for backward compatibility with several decades-old legacy business software.[4] The commercial drive was for greater processing speed, and that was achieved partly through increasing clock rates, and partly through performing more functions with a single instruction.

One of the main factors leading to increasingly complex instructions came from the disparity between the speed and cost of external and internal memory. Fast internal (on-chip or next to the CPU core) memory was often very expensive—but could be 1000 times faster than inexpensive external memory (off-chip, further from the CPU core). A big bottleneck, in CPU performance terms, is how long it takes to drag an instruction from external memory and bring it into the processor. Programs are typically large and hence reside in external memory (slow), but the instructions within the programs are executed internal to the CPU (fast). It was therefore sensible to create single complex instructions that replaced a sequence of 100 separate smaller instructions. It meant waiting for just one instruction to be conveyed from slow memory instead of 100.

It is possible to think of the instructions as being tokens. A program, stored externally, was just a sequence of tokens that get fed slowly into the CPU. Each token caused the CPU to execute a fixed sequence of internal operations. In fact, when you think about this, it means that each token was really causing a small program to be run inside the CPU. We call the small internal programs microcode, and the external tokens are written using the CPU's advertised instruction set. Internal microcode often did not particularly resemble the external instructions, in fact there is no need for them to be similar. Every external instruction would be translated into a microcode program, or microprogram,

---

[4] Backward compatibility meant ensuring that the latest CPUs could correctly execute code written for older machines—this was highly important when companies had invested heavily in computer hardware and software. There were examples of the central operating software for a particular business being decades old, yet too expensive or disruptive to be replaced. Replacing computer hardware could be expensive, but if that also meant that software needed to be replaced too, then the costs could become prohibitive. Hence the need for new CPUs to maintain backward compatibility with older designs.

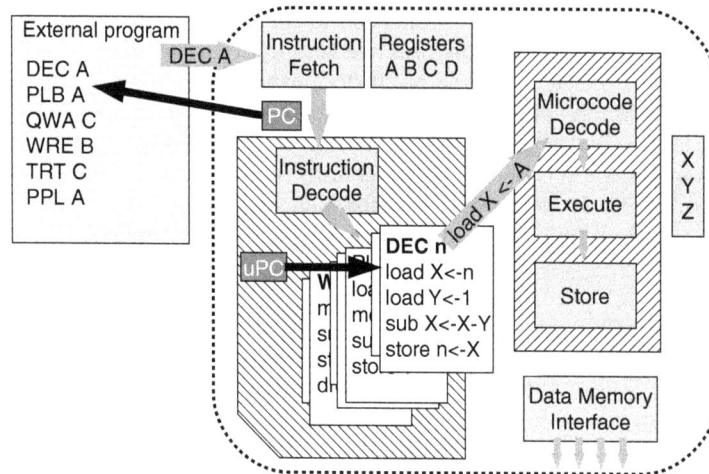

FIGURE 3.7   A block diagram of an instruction being fetched from slow external memory, being decoded inside a CPU, and executed as a sequence of much simpler microcode instructions.

upon entering the CPU. Internally, the CPU would support a set of operations that microprograms relied upon—and this was the microcode instruction set.

Microprogramming, as a technique, was actually invented by Maurice Wilkes in the early 1950s at Cambridge University, although the IBM System/360 range was probably the first commercial machine to implement this technology.

Some of the microcoding concepts are illustrated in Figure 3.7 where an external program in slow memory is being executed by the CPU. The current program counter (PC) is pointing at the instruction DEC A, presumably a command to decrement register A. This is fetched by the CPU and decoded into a sequence of microcode instructions to load register X from A, then load register Y with 1, then subtract Y from X and finally to store the result back in A.

The small 4-instruction microprogram that the DEC instruction launched is contained entirely inside the CPU, in fast, on-chip read-only memory (ROM), and requires an internal microprogram counter. None of this is visible from the "outside world" of the external program. In fact, the actual programmer may not even know that registers X, Y, and Z exist inside the CPU.

Extending this approach further led to a processor which used nanocode: instructions from an externally stored program would be converted to a microprogram of microcode instructions, each of which would in turn translate to a nanoprogram of nanocode instructions! Despite the elegance of this Cat-In-A-Hat technique, there were decreasing returns with the microcode approach. It relied upon the fact that external memory was a bottleneck. In the days when external random-access memory (RAM) was expensive and slow, but internal ROM was very fast, this was undoubtedly true. But then advances in RAM technology, including static RAM (SRAM), dynamic RAM (DRAM), and then synchronous dynamic RAM (SDRAM) all chipped away at the speed advantages of ROM until by the 1990s there was little speed difference between the technologies.

With minimal speed advantage, the popularity of microcode began to wane. A notable exception, however, was where the benefits of *instruction translation* were required. This feature is inherent in the microcode approach, and allows a CPU of one type to use the instruction set of another machine.

In the late 1990s processors were being developed that were internally RISC machines, but which could execute CISC instruction sets (see next section). Nowhere was this more advantageous than with the humble x86 series of processors. With a design heritage harking back to 1971, these CPUs had to not only guarantee backward code compatibility by executing an ancient and poorly optimized CISC instruction set, but had to do this faster than competing processors. The old-fashioned CISC instructions that entered some of these processors would be translated into sequences of much faster optimized RISC-style assembler. The RISC instructions thus took the place of modern-day microcode.

A further advantage of the microcode translation was the design of a processor that could mimic other devices. Such a device could execute an ARM program as if it were a native ARM processor, and then switch to executing Texas Instruments DSP code as if it were a TI DSP: the ultimate approach to being all CPUs to all programmers.

Despite such niche markets, the driving factors behind microcode had disappeared, and it became less popular in the 1980s. The trend was constantly toward doing more, and doing it faster: Moore's law in full swing.

3.2.6 RISC versus CISC Approaches

The ideas behind RISC (reduced instruction set computer) and CISC (complex instruction set computer) have been briefly mentioned in Section 2.2. In summary, CISC architectures encompass many complicated and powerful instructions, whereas RISC architectures concentrate on a smaller subset of common and useful instructions, which it processes extremely quickly. Even when complex operations are synthesized through multiple RISC instructions they will be as fast, or faster, than if encoded directly as a CISC instruction.

This concept is illustrated in Figure 3.8 showing two programs—one running on a RISC machine, with its fast one-cycle per instruction operation completing a program of 12 instructions (A-L) in 12 clock cycles. Below that is a CISC computer with its slower clock cycle (it is slower because the hardware is larger and more complicated, designed to process its more complex instructions) completing the same process in roughly the same

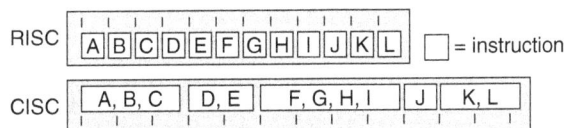

FIGURE 3.8 A diagram illustrating the difference in size, speed, and functionality of CISC and RISC instructions. RISC instructions (above) are uniformly small, each occupying a single CPU cycle, indicated by the vertical tick marks. By contrast the CISC instructions (below) require multiple cycles to execute and often accomplish more per instruction than in the RISC case. Both CPUs perform tasks A to L, but using 12 fast or 5 slower instructions, respectively.

number of clock cycles, but in this case using only five complex instructions instead of the 12 in the RISC machine. Since the CISC clock cycles are longer, it completes the task slower than the RISC machine. The example illustrated is typical for general software; however, there are some processing tasks for which programs can be written that would allow a CISC processor to complete its program faster.

This description above is useful, but it does not capture all of the important points relating to CISC and RISC. To consider further, let us take a historical perspective. We note that early computers were operated by the designers of the machines themselves. Designers knew what basic operations were required in their programs, and catered for these directly in hardware. As hardware became more capable, it became possible to add instructions to the computer that could perform functions that would otherwise require long and time-consuming strings of instructions.

As time progressed, computer programmers specialized on software, and computer architects specialized on hardware. Programmers would then approach architects asking for custom instructions to make their programs faster. Architects often complied, but sometimes took the initiative to add what they imagined were useful instructions (but on occasion could leave the programmers scratching their heads).

By the mid-1980s various design groups, most notably at Berkeley and then Stanford in the United States, began to question the prevailing design ethos. This academic thought was accompanied by some groundbreaking work performed at IBM, in which less complex machines that could clock much faster because of simple and regular design, were investigated. These machines demonstrated that simple instructions could be processed very quickly. Even though it meant that often it would require several RISC instructions to perform the same job as a single CISC instruction, a RISC program typically still executed significantly faster overall.

The name reduced instruction set computer (RISC) pays tribute to the simplicity of the original designs, although there was no actual reason to reduce the size of the instruction set, just to reduce the complexity of the instructions. Groups that popularized RISC technology produced, in turn the RISC I, RISC II, and MIPS processors. These evolved into commercial devices powering powerful workstations, where backward compatibility with x86 code was not required, particularly the commercial SPARC and MIPS CPUs.

In the meantime, over in Cambridge, United Kingdom, a tiny design group at Acorn Computers Ltd., the highly successful producer of the 6502-powered BBC microcomputer range (that contributed to giving the United Kingdom the highest rate of computer ownership in the world), had designed their own processor, inspired by the earliest Berkeley work. This *Acorn RISC Machine*, the ARM 1, was designed on a 2-MHz BBC microcomputer running BASIC—Acorn wrote their own silicon design tools for this—and was very soon followed by the ARM 2 processor which became the world's first commercial RISC processing chip. This powered the novel Acorn Archimedes range of computers. By 2002 ARM, now renamed *Advanced RISC Machine*, became the world's top selling 32-bit processor claiming 76.8 percent of the market. By mid-2005 over 2.5 billion ARM processor-powered products had been sold, and by the start of 2009 that had increased to be more than one sold for every person on the planet. The popularity of the ARM continues to increase today, and it is everywhere—in 2018 this includes powering almost every smartphone on the market. Box 3.1 briefly explores the background to the amazing ARM processor.

Box 3.1 How the ARM Was Designed

In the mid-1980s, groundbreaking British computer company Acorn, with a contract from the British Broadcasting Corporation (BBC) to design and market BBC Microcomputers, was looking for a way to move beyond their hugely successful 8-bit BBC Microcomputers. These were powered by lean and efficient Rockwell 6502 processors. The BBC initiatives had encouraged computer use in the United Kingdom so much so that there were reportedly far more computers per-capita in England than anywhere else in the world. Sir Clive Sinclair's ZX Spectrum, for example, had sold 4 million units by the time the IBM PC had reached sales of 1 million. Acorn is also reputed to have sold over 1 million BBC computers overall.

In the early explosion of the "computer revolution" it quickly became apparent to Acorn that 16-bit processors from companies such as Intel and Motorola were not powerful enough to meet their projected future needs—needs which included releasing the world's first multitasking graphical desktop operating system in the late 1980s (later some observers would conclude that this was copied by Microsoft as the basis for Windows 95, XP, and beyond).

In typical pioneering fashion, Acorn decided that, since nothing good enough was available commercially, they would create their own processor. They designed the ARM 1 and its support ICs (such as MEMC and VIDC) within 2 years, despite having never developed any silicon previously.

Acorn wanted a machine with a regular architecture—similar to the 6502, but vastly more powerful. They chose to use the RISC approach, but revisited their software needs by analyzing operating system code to determine most used instructions which they then optimized for. The same approach yielded an instruction set (see Section 3.3) and its coding. Later much-needed additions were the multiply and multiply-accumulate instructions.

This heritage leaves the globally successful ARM processor with a direct link back to the U.K. Government-funded BBC initiatives: the ARM software interrupt, supervisor modes, fast interrupt, no microcode, static pipeline, and load-store architecture are all derived either from the hardware or the software architectures adopted by Acorn.

Since Intel rode the wave of the desktop PC boom, ARM is riding the much larger wave of the embedded processor boom. CPUs are now inside almost every electronic product: and most of these are ARM based. Meanwhile, Acorn itself no longer exists, having self-destructed as a company in 1999.

3.2.7 Example Processor—the ARM

Over the years since the IBM research group published their initial results, the RISC approach has impacted almost every sphere of processor design. In particular, the ARM RISC processor family now dominates the world of embedded systems. This book, which aims to ensure examples are relevant to embedded computation, therefore uses ARM assembly language for almost all code examples. ARM-based software and hardware examples are provided wherever relevant. Readers need not become ARM assembler experts, but a little familiarity with the format of ARM instructions will be very useful. As an example, consider the following:

```
ADD R0, R1, R2
```

This instruction causes the contents of registers R1 and R2 to be added together, and then the result to be saved into register R0. It is thus of the format:

```
INSTRUCTION DEST_REG, SOURCE_REG1, SOURCE_REG2
```

Today, although it is easy to find examples of "pure" RISC processors such as the ARM and MIPS, even the modern devices from die-hard CISC CPU ranges (such as Motorola/Freescale 68000/Coldfire and the Intel x86 range) are often implemented with CISC-to-RISC hardware translators and internal RISC cores. Pure CISC processors are less popular these days and for this reason, when referring to CISC processors, we define a pseudo-ARM assembler format. In other words, we use ARM-style instructions (even though the ARM is a RISC CPU) rather than use the format from an actual CISC CPU. For example:

```
ADD A, B, C
```

Adds together registers B and C, placing the result in register A. Examples in this text are usually identified clearly as being RISC or CISC, but can otherwise be differentiated because the RISC examples use ARM-style registers R0 to R15 whereas the CISC examples use alphabetical registers A, B, C, and so on. Some special purpose registers are also mentioned in later sections. In the ARM, SP is the stack pointer and LR is the link register (these are actually the same as registers R13 and R14, respectively—whereas R15 is the program counter, usually called PC).

The main exception to the use of pseudo-ARM instructions in this book is in discussions relating to the Analog Devices ADSP21xx processor range, a single Texas Instruments TMS320 range example, and a few specialized machines (e.g., very long instruction word and Tomasulo architecture). Each of these examples is used to emphasis particular aspects of computer systems or architecture, and will be briefly introduced before being presented in context.

Although our examples and code snippets mainly assume an ARM architecture and contain ARM assembly language, it is important to acknowledge the vast array of different CPU architectures that are still in use. Many of these differ from the ARM substantially, not least in the types of instructions in their instruction sets, their addressing modes, but also in the syntax of their assembly language. One point to note right away is that assembly language instructions for some processors (including the 68000 which is still used by several textbooks) specifies the *destination* register last (e.g., it would write ADD A, B, C meaning A+B=C, whereas in ARM assembly language format the same instruction would mean B+C=A). In this book, the destination register is always specified ARM-style, and any comment is written following a semicolon (';'):

```
SUB R3, R2, R1   ; Comment text to explain, R3 = R2 − R1
```

Sometimes the destination and first source register will be the same:

```
ADD C, D   ; C = C + D
```

or perhaps there is only a single source register (single operand):

NOT E, F ; E= not F[5]

Some instructions have no source register but instead use an immediate value:

B −4 ; jump backward by 4 memory locations

This type of instruction is unusual in the ARM, but common with some CPUs. Generally, the instructions themselves are self-explanatory (ADD, AND, SUB, and so on), although the following section will provide more examples on the ARM instruction format, including a list of all instruction families in the basic architecture.

Beware, in the ARM, the destination register is specified first for all instructions *apart from* the store to memory instruction (and its variants):

STR R1, [R3]

This would store the content of register R1 to the memory address held in R3.

3.2.8   More about the ARM

Before moving on, we note that the ARM architecture, although stable and widely adopted, is continually undergoing improvements and enhancements. It is thus a dynamic, modern, and evolving design, rather than a static historical fixed example. The owning company (also named ARM) are continually developing new CPU design products based on the ARM, particularly for important application markets.

Apart from their earliest designs when users were numbered in thousands rather than billions, processors and system-on-chip devices (something we meet later in Chapter 7) based on the ARM processor are designed and manufactured by many IC design companies. Those companies primarily contract to use ARM Ltd CPU core (or cores) in their designs, and then cluster their own unique functionality around those cores. This allows their designs to make use of the very substantial ARM ecosystem of development tools, existing code-base, operating system support, and so on, while differentiating their products through on-chip peripherals, accelerators, and similar features. A few ARM-based designs have involved the IC design companies in making adjustments to the ARM core itself. A notable example that we will mention in several places is the DEC/Intel StrongARM—in which DEC (Digital Equipment Corporation) engineers worked with the ARM Ltd design team to introduce several very substantial performance improvements to the ARM core, including an extended pipeline, Harvard architecture data and instruction caches, fast multiplier, and so on. The DEC designs were subsequently purchased by Intel, and formed the basis of Intel's XScale range. However, the vast majority of ARM-based ICs, irrespective of the manufacturer they are sourced from, will include a standard ARM core.

In this book, we confine the discussion mainly to the common ARM cores (in old terminology, those are ARM7 and ARM9), and use standard ARM assembler examples that would be understood by almost all of those devices. Readers should understand, however, that many other processors have been created over the past few decades. There is

---

[5] While many ARM assemblers will handle an instruction NOT R0, R0 with ease, we should note that the ARM does not actually have NOT in its instruction set. In practice it uses an equivalent called move-and-negate or MVN. Hence the assembler would translate the NOT instruction to MVN R0, R0 during assembly. The programmer may not even notice this since there would probably be no error or warning given, and the code would simply work.

a great deal more variety in currently available, and historic, processors than could be described in a static textbook.

3.3 Instruction Handling

As mentioned in Section 3.2, computers are operated through sequences of instructions known as *programs*. The generic term for such programs is *software*. Various schemes exist for creating software through writing in high-level languages (HLL), where each HLL command is made up of a sequence of perhaps several tens of CPU instructions. In low-level languages, typically each language-level command invokes few, or perhaps only a single CPU instruction. All of this will be discussed more fully in Chapter 8, but for now, we will define programming as essentially making sure a CPU is fed with a sequence of instructions in order to make it undertake the wanted operation.

CPU operations themselves are data moves or logical transactions carried out inside the CPU. So instructions from a program are commands to the CPU to perform the wanted operation. As mentioned, a typical HLL command will result in one or more CPU instructions, and a stored program is a list of those commands, and thus a sequence of instructions.

In some computers, a single instruction can be used to invoke multiple CPU operations, which may be required for performance reasons: especially where the rate at which a program can be read from external memory is far slower than the speed at which the processor can execute the operations. It was this thinking that led in the past to the idea of microcode, as explored in Section 3.2.5.

Machine code is the name given to (usually) binary numerical identifiers that correspond to known actions in a CPU. This may mean, for example, that when examining program memory, hexadecimal byte 0x4E followed by byte 0xA8 might represent two instructions in an 8-bit processor, or a single instruction, 0x4EA8, in a 16-bit processor. In modern processors, programmers are very rarely required to deal with the underlying binary numerical identifiers that the processor understands, but instead handle them through a set of abbreviated mnemonics called *assembly language* or *assembly code*, which we met in the previous section.

The *instruction set* is a list of all of the available assembly language mnemonics—it is a list of all instructions supported by a particular CPU.

3.3.1 The Instruction Set

The instruction set of a particular CPU describes the set of operations that CPU is capable of performing—with each operation being encoded through an instruction which is part of the set. Some instructions require one or more operands (e.g., ADD A,B,C where A, B, and C are called the source and destination operands and may be immediate values, registers, memory locations or others depending on the addressing modes available—see Section 3.3.4. There are often restrictions with particular instructions—such as which registers they can be used with, the maximum values they can handle, or the range of numbers that the operands can contain. For example, a shift instruction may be limited by the maximum shift allowed by the shift hardware; issuing an instruction to shift by a value that the CPU cannot support will result in an assembler (or compiler) error.

The instruction set contains every instruction and thus describes the full capability of the processor hardware. Instruction sets are often split into groups based upon which functional processor unit they invoke, such as the following defined for the ADSP2181 processor:

Instruction group	Example operations within the group
ALU	add, subtract, AND, OR, etc.
MAC	multiply, multiply-accumulate, etc.
SHIFT	arithmetic/logical shift left/right, derive exponent, etc.
MOVE	register/register, memory/register, register/memory, I/O, etc.
PROGRAM FLOW	branch/jump, call, return, do loops, etc.
MISC	idle mode, NOP, stack control, configuration, etc.

Many processors would add an FPU or SSE group to those defined, but the ADSP2181 is a fixed-point only processor, with no multimedia extensions and hence supports the basic functional groups only.

The instruction set for the ARM processor, specifically the ARM7, is shown for reference in Figure 3.9 (note this shows the ARM mode instructions and does not include the 16-bit Thumb mode, and variants, that many ARM processors also support). Notations used in the instruction set table include the following:

S in bit 20 indicates instruction should update condition flags upon completion (see later).

S in bit 6/22 indicates transfer instruction should restore status register.

U signed/unsigned for multiplication and up/down for data transfer index modifications.

I an indicator bit used to select immediate addressing.

A accumulate/do not accumulate answer.

B unsigned byte/word.

W write back.

L load/store.

P post-/pre-increment/decrement.

R indicates one of the 16 registers.

CR indicates a coprocessor register (one of 8 coprocessors that can be identified).

Many of these modifiers are specific to the ARM processor and not considered further in the text; however, we shall look in more detail at the "S" bit, and the flexible addressing capabilities (see Section 3.3.4). The interested reader is also referred to the ARM Ltd website[6] where a large amount of additional documentation may be found. The instruction set varies slightly between different ARM processors, and it is the very common ARM7TDMI version that we reproduce.[7]

---

[6] The ARM Ltd website is at http://www.arm.com

[7] This information was extracted from ARM Ltd Open Access document DDI 0029E.

cond	27	26	25	24	23	22	21	20	19	18	17	16	15	14	13	12	11	10	9	8	7	6	5	4	3	2	1	0	
conditions	0	1	1	X	X	X	X	X	X	X	X	X	X	X	X	X	X	X	X	X	X	X	X	1	X	X	X	X	undefined
conditions	0	0	I	opcode				S	Rn				Rd				Second operand													data processing
conditions	1	0	1	L	address offset to destination																								branch
conditions	0	0	0	0	0	0	A	S	Rn				Rd				Rs				1	0	0	1	Rm				MUL
conditions	0	0	0	0	1	U	A	S	RdHi				RdLow				Rn				1	0	0	1	Rm				long multiply
conditions	0	1	I	P	U	B	W	L	Rn				Rd				address offset													LDR/STR
conditions	0	0	0	P	U	1	W	L	Rn				Rd				offset				1	S	H	1	offset				halfword transfer
conditions	0	0	0	P	U	0	W	L	Rn				Rd				0	0	0	0	1	S	H	1	Rm				halfword transfer
conditions	1	0	0	P	U	S	W	L	Rn				list of registers																block transfer
conditions	0	0	0	1	0	0	1	0	1	1	1	1	1	1	1	1	1	1	1	1	0	0	0	1	Rn				BX
conditions	0	0	0	1	0	B	0	0	Rn				Rd				0	0	0	0	1	0	0	1	Rm				single data swap
conditions	1	1	0	P	U	N	W	L	Rn				CRd				CP no.				offset								LDC
conditions	1	1	1	0	CP opcode				CRn				CRd				CP no.				CP			0	CRm				CDP
conditions	1	1	1	0	CP opcode			L	CRn				Rd				CP no.				CP			2	CRm				MCR

FIGURE 3.9 The ARM instruction set arranged in a table format, showing all 14 classes of instruction available in this particular version of the ARM instruction set (which does not include 16-bit Thumb instructions). Columns are aligned by the 32-bit fields in the instruction word.

After more than a decade of continuous development, ARM Ltd rebranded their processor ranges, so that their processor cores are now known as Cortex devices. The original ARM7, ARM9, and ARM11 devices, in as much as they are still used in legacy processors or referred to in documentation, are termed "classic." Most likely, this move has been part of an effort to counter the fragmentation of the huge ARM market in which one basic architecture (the ARM) was required to span a very wide and diverse set of needs, ranging from tiny and slow sensor systems to larger and faster handheld computers. At the time of writing, the new processors are classed into three ranges which better subdivide the traditional strength areas for different-sized ARM devices:

Cortex-A series processors are application-oriented. They have the in-built hardware support suited for running full-featured modern operating systems such as Linux, with graphically rich user interfaces such as Apple's iOS and Google's Android. The processing power of these cores runs from the efficient Cortex-A5, through the A8, A9, and up to the highest performance Cortex-A15 device. All support ARM, Thumb, and Thumb-2 instructions sets (Thumb-2 reportedly improves upon Thumb in terms of performance and compactness, but has similar design rationale).

Cortex-R series devices are targeted at real-time systems that also have significant performance requirements. These include smartphone handsets, media players, and cameras. ARM Ltd have also promoted Cortex-R cores for automotive control and medical systems; ones in which reliability and hard real-time response are often important. These generally do not require complex and rich operating systems, just small, hard and fast real-time arrangements.

Cortex-M family processors are at the lower end of the range for use in very cost-sensitive and low-power systems. It could be argued that these are for traditional microcontroller-type applications that often do not need advanced operating system support (and possibly do not even need an operating system in the first place). These cores are designed for applications that do not have rich user interface requirements, and for which the clock speed will be no more than several tens of MHz.

Although most variants of the classic ARM7 series were designed to support a 16-bit Thumb mode (see Section 3.3.3), all ARM7 devices support the standard fixed length

32-bit instructions shown above. It can be seen that, as in the ADSP21xx, there are various groups of instructions, such as data processing, multiply or branch. With 15 instruction groups, 4 bits are needed to represent the instruction group, and further bits are used within this to represent the exact instruction in each group.

Notice the fixed condition bits available for every instruction. No matter which instruction is being used, these bits are located at the same position in the instruction word. This regularity aids in instruction decoding within the processor. It is important to note that the consequence of this is that every instruction can operate conditionally. This is unusual, and among common modern processors is found only in the ARM: most other processors support conditionals for branching instructions only. In the ARM, the S bit within many instruction words controls whether that instruction can change condition codes on completion (see Box 3.2). These two features, when used in conjunction with each other, are very flexible and efficient.

Box 3.2   Illustrating Conditionals and the S Bit in the ARM

Consider the efficiency of the ARM processor compared to a mythical standard RISC processor that does not allow conditional operation for every instruction.

The instruction mnemonics used are similar to those of the ARM (but not completely realistic). First, we will examine the program on the standard RISC processor that adds the numbers in registers R0 and R1 and then, depending on the answer, either places a zero in register R2 (if the result was less than zero) or places a one in register R2 otherwise.

```
      ADDS R0, R0, R1
      BLT pos1    (branch if less than zero)
      MOV R2, #1
      B pos2
pos1 MOV R2, #0
pos2 .....
```

The program occupies five instructions and will always require a branch no matter what registers R0 and R1 contain on entry.

The following code segment reproduces the same behavior for the ARM processor, but uses conditional moves to replace the branch. In this case, R0 and R1 are added. The S after the ADD mnemonic indicates that the result of the addition should update the internal condition flags. Next, a value 1 is loaded into R2 if the result of the last condition-code-setting instruction was less than zero. A 0 is loaded into R2 if the result was greater than or equal to zero.

```
ADDS R0, R0, R1
MOVLT R2, #1
MOVGE R2, #0
 . . . . .
```

The ARM version is obviously shorter—only three instructions are required, and in this case no branches are needed. It is this mechanism that allows ARM programs to be efficient whereas RISC processors are traditionally known for less efficient code density. In high-level languages, the structure that leads to this code

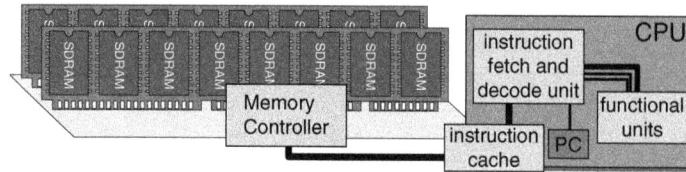

FIGURE 3.10 A diagram showing the connectivity of the memory controller in a typical CPU system.

arrangement is very common:

```
IF condition THEN
    action 1
ELSE
    action 2
```

Also, note that for every instruction, the destination register (if required) is in the same place in the instruction word.

3.3.2 Instruction Fetch and Decode

In a modern computer system, programs being executed normally reside in RAM (they may have been copied there from hard disc or flash memory). A memory controller, usually part of a memory management unit that we will explore in Section 4.3, controls external RAM and handles memory accesses on behalf of the CPU.

Within the CPU, an instruction fetch and decode unit (IFDU or simply IFU) retrieves the next instruction to be executed at each instruction cycle. Which instruction to fetch next is identified by an address pointer—which is held in a program counter (PC) in nearly every processor in use today. These items are illustrated in Figure 3.10.

The program counter is normally incremented automatically after an instruction is retrieved, so it naturally "points" to the instruction after the current one in memory, that is, the next one along. In this way, the CPU can execute a program that consists of a sequence of instructions by executing each instruction in turn. Branch instructions override the PC, allowing programs to do more than simply step from one instruction to the next. When a current instruction overrides the PC, it means that the *next* instruction executed will be fetched from somewhere else, not simply the following instruction. Since the PC is a register, overriding it means writing a different binary value into the register. In some processors that binary value is the address of the next instruction to be executed, but many RISC processors like the ARM, normally specify an *offset* to the current address (i.e., how many instructions forward or backward from the current position). This will be described more fully in Section 3.3.2.3.

Once the instruction fetch and decode unit reads in an instruction, it begins to decode that instruction which then flows through steps as shown in the flowchart of Figure 3.11.

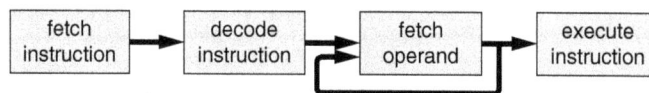

FIGURE 3.11 A flowchart of instruction processing for a typical processor.

3.3.2.1 Instruction Decode

In the ARM, because all instructions can be conditional, the IFU first looks at the condition code bits encoded in the instruction and compares these bitwise with the current condition flags in the processor status register. If the conditions required by the instruction do not match the current condition flags, then the instruction is dumped and the next one retrieved instead.

In the ARM, the simplicity of the instruction set means that the conditional bits of each retrieved word can simply be ANDed with status register bits 28 to 31 (that encode the current condition flags). Box 3.3 explains the quite extensive set of conditional codes available in the ARM.

Looking again at the ARM instruction set, it can be seen that the destination register (for instructions that have a destination) is located in the same place in each instruction word. On decode, the IFU simply takes these 4 bits (used to address the 16 registers) and applies them as a register bank destination address.

3.3.2.2 Fetch Operand

Evidently, the value of the operand being used is not always encoded in the instruction word itself (which would be called an *immediate* operand). The ARM, and many other RISC processors are simplified by being *load-store* architectures where operands in memory cannot be used directly in an operation—they have to be transferred into a register first. The exception is a few data processing instructions that allow immediate values, encoded as part of the instruction, such as MOV (see example in Box 3.4).

The ARM normally prepares operands for an operation either by decoding an immediate value from the instruction word, or by selecting one or more source, and one destination register. The exception is the load (LDR) and store (STR) instructions that explicitly move 32-bit values between memory and a register.

In many other processors, particularly CISC rather than RISC-based, it is possible to execute an instruction that performs some operation on the contents of a memory address and stores the result back to another memory address. Evidently in that kind of processor, the action of moving operands around will require several separate memory accesses (with memory being relatively slow compared to the CPU, and all operands in the same memory device sharing a bus, the fetches and stores have to take turns, and become a significant performance bottleneck). By contrast, RISC processors aim to complete each instruction within a single clock cycle if possible, so the slow access to operands in memory has been disallowed with such architectures for most instructions.

Box 3.3   Condition Codes in the ARM Processor

The ARM, as we have seen in Figure 3.9 reserves 4 bits (bits 31, 30, 29, and 28) for condition codes in every instruction. This means that every machine code instruction can be conditional (although when written in assembly language there may be some instructions which do not take conditionals).

Normally, the condition code is appended to the instruction mnemonic when writing the code. Thus an ADDGT would mean an ADD instruction that only executes when the condition flags in the processor indicate that the result of the last instruction that set the condition flags was greater than zero.

The full set of ARM conditionals is shown in the table below (although strictly the last two are unconditional conditionals!):

Condition nibble	Condition code	Meaning	Conditional on
0000	EQ	equal	$Z = 1$
0001	NE	not equal	$Z = 0$
0010	CS	carry set	$C = 1$
0011	CC	carry clear	$C = 0$
0100	MI	minus	$N = 1$
0101	PL	plus	$N = 0$
0110	VS	overflow set	$V = 1$
0111	VC	overflow clear	$V = 0$
1000	HI	higher	$C = 1, Z = 0$
1001	LS	lower or same	$C = 0, Z = 1$
1010	GE	greater or equal	$N = V$
1011	LT	less than	$N = {\sim}V$
1100	GT	greater than	$N = V, Z = 0$
1101	LE	less than or equal	$(N = {\sim}V)$ or $Z = 1$
1110	AL	always	–
1111	NV	never	–

In practice, these conditionals are very useful when performing calculations, such as the pseudo-code sentence "if the result of the first operation A is negative then do B, else do C." Without ARM-like conditionals for every instruction (i.e., in most CPUs), the pseudo-code example would almost always be implemented with branches, such as "if the result of the first operation A is negative then branch somewhere, do B and branch back, else branch somewhere and do C."

3.3.2.3 Branching

The branch instruction group in the ARM instruction set is, as expected, all conditional—as indeed are branch instructions in nearly all other processors. In the ARM branch instruction (see Figure 3.9), bits 24 to 27 are the unique identifiers that indicate an instruction in the branch group. The L bit distinguishes between a jump or a call (branch-and-link in ARM terminology, where *link* means that the address to return to is placed in the link register LR, which is R14, when the branch occurs). Apart from the 4 bits needed to define the instruction type, 4 bits are needed for condition codes. So there are only 24 bits remaining. These 24 bits are called the *offset*: they indicate where to branch to—which instruction address should be placed in the program counter.

Since the ARM is a 32-bit processor, instruction words are 32 bits wide. However, memory is only byte wide, such that one instruction spans four consecutive memory locations. The ARM designers have specified that instructions cannot start just anywhere, they can only start on 4-byte boundaries: addresses 0, 4, 8, 12, and so on. So the *offset* refers to blocks of 4 bytes.

There are two general methods of indicating addresses to branch to in computer architecture; these are **absolute** and **relative**. Absolute specifies a complete memory address, whereas relative specifies a certain number of locations forward or backward. As computer memory spaces have become larger, specifying absolute addresses has become inefficient due to the large size of individual addresses. Furthermore, it is often unnecessary to specify an absolute address—the principle of locality (Section 4.4.4) states that branch distances (i.e., how many instructions is the branch destination away from the current instruction) will usually be quite small. Small branches only require a few bits to specify as a relative jump, compared to an absolute jump address which would always need to be specified as a complete address (which would be 28 bits in the ARM, since instructions are always on word boundaries, 0x0, 0x4, 0x8, 0xC, and so on).

In the ARM, the jump address is termed an "offset," which means it must be a jump relative to the current program counter location. The branch instruction in Figure 3.9 contains space for a 24-bit offset, so a branch can indicate a jump range of 2^{24} words away in memory which is a 64-MiB range. Of course, the offset has to be signed to allow a jump backward (as in a loop) as well as forward, and so this means a $+/- 32$-MiB jump span.

Is the limited branch range a limitation? Not normally—despite rampant code bloat, even at the time of writing, single programs are not usually 64 MiB long. It is thus likely that the ARMs designers have catered for the vast majority of jump requirements with the instruction. However, a 32-bit memory bus allows up to 4 GiB of memory space to be addressed, which is far larger than the capability of address jumps. So if a 70-MiB jump was required, how could it be accomplished?

In this case, the ARM has a *branch and exchange* instruction. To use this, the destination address is first loaded into a register (which is 32 bits in size), and then this instruction can be issued to jump to the address held in that register. Of course, the question arises as to how the register itself can be loaded with a 32-bit number. Section 3.3.4 will discuss addressing modes, one of which is the immediate constant—a number encoded as part of the instruction. Box 3.4 will also consider how immediate values can be loaded with the MOV instruction.

3.3.2.4   Immediate Constant

Within a 32-bit-sized instruction word, which reserves some bits for conditionals, S bit, some for destination register, and so on; there is not enough space left to specify a 32-bit constant. Therefore, an immediate constant (a value encoded within the instruction word), has to be less than 32 bits.

In the ARM, immediate constants are loaded to a register with the MOV instruction (in the data processing instruction group). An immediate value can be located inside the section labeled "Operand 2" in the ARM instruction set (Figure 3.9). However, not all of the operand is used for holding the constant. In fact, only an 8-bit immediate value is catered for, with the remaining 4 bits used to specify a rotation to the right.

So, although the processor has 32-bit registers, only an 8-bit immediate constant can be loaded in a single operation. However, the very clever rotation mechanism build into the same instruction (with 4-bits reserved for rotation, this can specify 15 positions either left or right) allows a large variety of useful numbers to be loaded, made up from an 8-bit immediate with rotation. Box 3.4 looks in detail at the bitfields present in the ARM processor MOV instruction, to see how these impact the flexibility of one variant of the instruction.

Box 3.4   Understanding the MOV Instruction in the ARM

The MOV is 32 bits long like all ARM instructions. Its structure is shown below:

4-bit cond	0	0	1	opcode	S	Rn	Rd	4-bit rotation	8-bit value

or

4-bit cond	0	0	0	opcode	S	Rn	Rd	immediate/register shift & Rm

The 4-bit condition code is common with all other ARM instructions, the opcode defines the exact instruction in the data processing class, Rn is the first operand register, Rd is the second operand register and, selected through bit 25 = 1, Rm is the third. We will concentrate on the top form of the command, where an 8-bit immediate constant and 4-bit rotation are supplied (the rotation applied *is twice the value specified* in the instruction, and is always a *right* shift with wrap-around—meaning that bits that shift off the end of the word to the right are inserted into the word at the left end). Where the opcode specifies a MOV instruction, the immediate, rotated by the degree specified is loaded into the destination register. Here are some examples:

```
MOV R5, #0xFF    ; Rd = 5, Rn = 0, rotation = 0, value = 0xFF
MOV R2, #0x10C   ; Rd = 2, Rn = 0, rotation = 15, value = 0x43 (loads 0x43>>30)
```

Note: For these MOV instructions, Rn is always set to 0 since it is unused.

Question: How can the processor set a register to 0xF0FFFFFF?
Answer: The programmer would probably write:

```
MOV R0, #0xF0FFFFFF
```

However, the assembler would be likely to complain ("number too big for immediate constant" or similar) since the 32-bit value that is specified cannot fit into an 8-bit register, no matter what degree of shift is required. Some assemblers, and more experienced programmers would know that they can simply convert the instruction to a "move NOT" instead:

```
MVN R0, #0x0F000000   ; Rd = 0, Rn = 0, rotation = 4, value = 0x0F (loads 0xF>>8)
```

As you can see, despite the relatively small immediate value size that can be accommodated within the instruction field, this allied with the instruction flexibility and shift value, can actually encode quite a wide variety of constants.

Many processors work differently. They generally allow at least a 16-bit constant to be loaded immediately encoded in the instruction word. CISC processors often have variable length instructions or use two consecutive instructions: a variable length instruction may be 16 bits long when only an 8-bit constant is to be loaded, or 32 bits long when a 16 or 24-bit constant is loaded. Variable length instructions require the instruction fetch unit to be fairly complex, and thus a more simple method of achieving a similar result is to use two consecutive instructions: the first instruction may mean "load the next instruction value to register R2" so that the IFU simply reads the next value directly into the register rather than trying to decode it. This evidently means that some instructions require two instruction cycles to execute, and imposes a timing penalty, especially in pipelined processors (something we will be exploring in Section 5.2).

In the ARM processor, although the restriction in immediate values exists, in practice many constants can be encoded with an 8-bit value and a shift so that this does not translate to a significant performance bottleneck. The ADSP2181 handles immediate loads in a similar fashion, and has been designed for high-speed single-cycle operation.

3.3.3   Compressed Instruction Sets

Especially in processors with variable-length instructions, *Huffman encoding* is used to improve processor efficiency. In fact, as we shall see later, similar ideas can be used even within a fixed-length processor, but in this case not for efficiency reasons.

Huffman encoding is based on the principle of reducing the size of the most common instructions, and increasing the size of the least common instructions to result in an overall size reduction (i.e., the average instruction size is smaller). Obviously this requires knowledge of the probability of instructions occurring, and then allows the size of the encoded word used to represent those instructions to be inversely proportional to their probability. An example of Huffman coding applied to instruction set design is provided in Box 3.5.

It should be noted that, in the real world, one particular application may have quite different instruction probability statistics compared to the average. For example, programs that perform complicated calculations will have more data processing instructions than those that just react to external events, or "paint" pixels into memory to display on a screen.

Box 3.5   A Huffman Coding Illustration

An example processor has five instructions for which an analysis of the 1000 instruction software program that it runs reveals the following occurrences:

```
CALL 60, ADD 300, SUB 80, AND 60, MOV 500
```

If an equal number of bits were used to represent each instruction in this instruction set, 3 bits would be needed (since that would allow up to seven possibilities). Ignoring any operands, 1000×3 bits $= 3000$ bits are required to represent that program.

The processor designers wish to use Huffman coding to reduce the program size. First they calculate the probability of each instruction (by dividing each occurrence by the total number of instructions):

```
CALL 0.06, ADD 0.3, SUB 0.08, AND 0.06, MOV 0.5
```

Next these are ordered in a list in terms of probability. The lowest two probabilities (CALL and AND) are combined (now denoted **C/A**) and the list reordered:

MOV 0.5	MOV 0.5
ADD 0.3	ADD 0.3
SUB 0.08	**C/A** 0.12
CALL 0.06	SUB 0.08
AND 0.06	

This process is then repeated, until finally there are only two choices left:

MOV 0.5	MOV 0.5	MOV 0.5	MOV 0.5
ADD 0.3	ADD 0.3	ADD 0.3	**C/A/S/A 0.5**
SUB 0.08	**C/A 0.12**	**C/A/S 0.2**	
CALL 0.06	SUB 0.08		
AND 0.06			

Next, traverse the tree from right to left. The bottom two entries in each column are numbered: the upper value is designated binary "1" and the lower is binary "0," and these numbers must be written down when tracing through. Any other column entry can simply be followed left without writing anything more until the original instruction at the left-hand side is reached.

For example, in the right-hand column, a "1" indicates MOV, a "0" indicates any one of CALL/AND/SUB/ADD. Moving left, a "01" now indicates an ADD whereas a "00" is the prefix for any of CALL/AND/SUB. In the next column, "001" indicates either CALL or AND and "000" indicates SUB. Writing all of these out gives the following:

MOV is "1," ADD is "01," SUB is "000," CALL is "0011," AND is "0010." If we look at the number of bits used to represent each instruction, we can see that the most common instruction (MOV) is represented by a single bit whereas the least common (AND) needs 4 bits, so the encoding method seems to have worked in representing the most common instructions with fewer bits. Using the original number of occurrences of each instruction and the number of Huffman bits, we can calculate the new program size:

$$(500 \times 1) + (300 \times 2) + (80 \times 3) + (60 \times 4) + (60 \times 4) = 1820$$

Which is significantly fewer than the 3000 bits we calculated for a fixed 3-bit representation.

As briefly noted previously, many ARM processors contain an alternative 16-bit instruction set called the Thumb or Thumb-2. The Thumb instruction set was designed to improve code density, particularly for embedded applications requiring relatively simple processing. Note however that even though a given memory size can support twice as many Thumb instructions compared to 32-bit ARM instructions, on average more Thumb instructions are required to perform the same function as ARM instructions. For processors which support both ARM and Thumb mode, Thumb instructions are decoded and map internally to ARM instructions (this is probably because there are fewer different Thumb instructions to choose from, or perhaps because ARM mode was developed first). Because of this Thumb mode efficiency is mainly obtained through reducing the bandwidth (bits/second) fetched from external memory, rather than from internal CPU speedups.

The process by which ARM engineers designed the Thumb instruction set is noteworthy since they used a similar idea to Huffman coding. ARM engineers examined a database of example application code, and calculated the number of uses of each instruction for that code. The most common instructions were then made available in Thumb mode, and the binary encoding within the fixed 16-bit word that is used to represent an instruction is length coded based on the number of bits required for the other operands.

Some features of the Thumb instruction set are

- Only one conditional instruction (an offset branch).
- No "S" flag—most Thumb instructions will update condition flags automatically.
- The destination register is usually the same as one of the source registers (in ARM mode the destination and source are almost always specified separately).
- All instructions are 16-bit (but register and internal bus width is still 32-bit).
- Significantly limited addressing scope for immediate and offset addresses.
- Most instructions can only access the lower 8 registers (of 16).

The Thumb instruction set is significantly less regular than the ARM instruction set, although the decoding process (from Thumb instruction fetched from memory to ARM instruction ready to be executed inside the processor) is automatic and very fast. Some example instructions are

16-bit binary instruction bit-pattern			Instruction name	Example
1101	Condition (4 bits)	Offset (8 bits)	Conditional Branch	BLT loop
11100	Offset (11 bits)		Branch	B main
01001	Destination register (4 bits)	Offset (8 bits)	Load memory to register	LDR R3, [PC, #10]
101100001	Immediate (7 bits)		Add to stack	ADD SP, SP, #23

Even from the few instructions shown here, it can be seen that the few most significant bits in the instruction word identify the instruction. These range in size from 3- to 9-bits across the entire instruction set. In the case of the ADD instruction shown, the register it operates on is fixed: it is an add to stack only—the flexibility and regularity of the ARM instruction set, where almost all instructions operate on any registers, is lost—but the most common operations found in software *are* catered for.

It should be noted at this point that the Thumb instruction set, being 16 bits wide, really operates at its best when the interface to external memory is 16 bits, in which case each ARM instruction would require two memory cycles to be retrieved (and thus the processor would run half as fast as it should), whereas the Thumb code could be executed at full speed.

3.3.4   Addressing Modes

Addressing modes describe the various methods of identifying an operand within an instruction. Instructions specify many operations, which may have no operands, one, two, or three. There may, very exceptionally, be instructions with greater than three operands. In most modern processors, common nonzero case examples are

Type	Examples	Operand
Single operand	B address	Address, may be given directly, may be an offset from current position or may be an address in a register or memory location.
Two operands	NOT destination, source	Destination or source may be registers, memory addresses, or memory locations specified by registers. The source may also be a value written directly.
Three operands	ADD destination, source, source	Destination or source may be registers, memory addresses, or memory locations specified by registers. The source may also be a value written directly.

Of course not all possible operand types are suitable for all instructions, and even so may not be available on some processors (e.g., RISC processors, being *load-store*, typically limit the operands of arithmetic instructions to registers, whereas in CISC processors they may be located in memory or elsewhere). A final point to note is the assumption in the two bottom examples above that the first operand written is the destination—which is true for ARM assembly language, but is reversed for some other processors (as we mentioned in Section 3.2.7)—this can be a real cause for confusion when writing assembler code for different processors, and is an enduring occupational hazard for computer architecture lecturers and book authors.

The term "addressing mode" refers to the method of specifying a load or store address, using one of several different techniques. The following table lists the common addressing modes, with ARM-style assembly language examples (although it should be noted that PUSH does not exist in the ARM instruction set, only in the Thumb).

Mode name	Example	Explanation
Immediate	MOV R0, #0x1000	Move hexadecimal value 0x1000 to register R0
Absolute	LDR R0, #0x20	Load whatever is in memory at address 0x20 into R0
Register direct	MVN R0, **R1**	Take content of R1, NOT it and move it to R0
Register indirect	LDR R0, **[R1]**	If R1 contains value 0x123, then retrieve contents of memory location 0x123, and place it in R0
Stack	PUSH R0	In this case, the contents of R0 and push onto the stack (and the assumption is of only one stack)

The following extensions and combinations of the basic idea are also common. ARM-style assembler is again used for ease of illustration (but we should note that the instruction LDR R0, **[R1, R2, #3]** is not supported by the ARM which can handle register indirect, indexed and offset addressing, but not all at the same time).

Name	Example	Explanation (if R1=1 & R2=2)
Register indirect with immediate offset	LDR R0, **[R1, #5]**	The second operand is the content of memory location $1 + 5 = 6$
Register indirect with register indirect index	STR R0, **[R1, R2]**	The second operand is the content of memory location $1 + 2 = 3$
Register indirect with register indirect index and immediate offset	LDR R0, **[R1, R2, #3]**	The second operand is the content of memory location $1 + 2 + 3 = 6$
Register indirect with immediate scaled register indirect index	STR R0, **[R1,R2,LSL #2]**	The second operand is the content of memory location $1 + (2 << 2) = 9$

Various processors including the ARM and the ADSP2181 also offer an automatic way to update registers after they have been used to perform offset addressing. For example, a register indirect access with immediate offset could leave the register used in the access updated after addition of the offset. This is shown in the following examples which assume R1 = 0x22:

```
LDR R0, [R1], #5        Load R0 with content of memory address 0x22
                        and then set R1 = 0x22 + 0x5 = 0x27
LDR R0, [R1, #5] !      Set R1 = 0x22 + 0x5 = 0x27 and then load R0
                        with content of memory address 0x27
```

Note that it is not our intention here to teach the details of the ARM instruction set, but merely to use it as a teaching aid for the underlying addressing techniques.[8]

It is instructive to analyze the limitations that caused CPU designers to provide certain levels of functionality within a processor—and this is rarely more revealing than in the consideration of the instruction set. In this regard, CISC processors tend to be more interesting than RISC processors. Some examples are given below from an imaginary CISC processor, where main memory locations mA, mB, and mC are used for absolute operand storage, and a RISC processor where registers R0, R1, and R2 are used for register direct addressing:

CISC processor: ADD mA, mB, mC ; mA = mB + mC

In this case, once the CPU has read and decoded the instruction, it must read the content of two further memory locations to retrieve the operand values mB and mC, and this probably requires two memory bus cycles. These values must then be transferred by internal bus to the ALU as they are retrieved (and since this is sequential, only one bus is needed). Once the ALU has calculated the result, the value is transferred by bus to a memory interface for writing back to main memory location mA.

The instruction overhead is three external memory cycles in addition to the ALU operation time and the usual time needed to fetch the instruction itself. External memory

---

[8] Those who *do* wish to learn the ARM instruction set are recommended to refer to the book *ARM System Architecture*, by Steve Furber (one of the original inventors of the ARM processor).

cycles are usually far slower than internal ALU operations, and so this is clearly a bottleneck. There is only a need for one internal bus in this processor.

We also note that the instruction word must hold three absolute addresses. With 32-bit memory, this equates to 96 bits, making a very long instruction word. This could be reduced through offset/relative addressing, but would probably still be too big for a 32-bit instruction word.

RISC processor: `ADD R0, R1, R2  ; R0 = R1 + R2`

Now the same operation is performed with registers. All of the operand values are already inside the CPU, which means they can be accessed quickly. Once the instruction has been read and decoded, register `R1` is allowed to drive one internal operand bus and register `R2` allowed to drive the other internal operand bus simultaneously. Both operands are thus conveyed to the ALU in a single very fast internal bus cycle. Once the ALU has calculated the result, an internal results bus will collect the result. `R0` will be listening to this bus and, at the appropriate time, latch the result value from the bus.

The instruction overhead is two fast internal bus cycles in addition to the ALU operation time. In our example description, the CPU must contain three internal buses: two to simultaneously transfer both operands, and one for the result. Other alternative arrangements are equally possible.

The instruction word needs to contain three register values; however, with a bank of 32 registers, only 5 bits are needed to specify each register, and so 15 bits are used in total. This would easily allow the operation to be encoded in a 32-bit instruction.

CISC processor: `ADD mA, mB  ; mA = mA + mB`

Similar to the first example, the CPU must read two external memory locations to retrieve the operand values, requiring two memory bus cycles. It also needs to transfer the result back to memory, and thus execution time is unchanged.

However, the instruction word this time only need contain two absolute addresses instead of three. This would be achievable in a real system, especially if an absolute value is used for the first operand address, and an offset used for the second one.

CISC processor: `ADD mB  ; ACC = mB + ACC`

The CISC processors of the 1980s and earlier commonly utilized accumulators. These were general-purpose registers (the forerunners of the register bank) that were used as an operand for all arithmetic and data processing operations, and to hold the result of those operations. The other operand was almost always an absolute value from memory. In this case, the instruction requires a single value to be loaded from memory prior to the addition, and thus involves a single external memory bus cycle.

The instruction word need only contain a single absolute memory value, which could be achieved by loaded a second instruction word containing the address (thus requiring that two instruction fetches are performed prior to instruction execution).

Stack processor: ADD

This is a special case (that will be explored further in the next section) where a CPU pops the top two stack entries, adds them together and pushes the result back onto the stack. This needs to access a stack which would be quick if it were an internal memory storage block; however, a stack would more normally be located in off-chip memory.

The main benefit with the stack approach is that the instruction does not need to encode any absolute memory addresses. Theoretically, this can make for an extremely small instruction width.

3.3.5 Stack Machines and Reverse Polish Notation

In daily life, most of us employ *infix* notation to represent an operation written on paper (such as $a+b \div c$), where an agreed fixed precedence[9] of operators (that can be overridden using parentheses) determines the order in which the various operations occur. Polish notation (note: not *reverse* Polish notation) was invented by Polish mathematician Jan Łukasiewicz in the 1920s to place the operator before the operands, thus it is a *prefix* notation. By specifying the operand in this way, operator precedence is unimportant, and parentheses are not required.

Reverse Polish notation (RPN) by contrast is a *postfix* notation where the order of the equation completely defines the precedence. This was created during the 1950s and 1960s as an aid to working with a stack-based architecture, and was subsequently introduced and loved (or hated) by two generations of Hewlett-Packard electronic calculator users.

An example of RPN is $bc \div a+$, where the operands b and c are given first, then the command to divide them and hold the result, then operand a is loaded followed by the command to add the previous result to a, and store the new result somewhere. Some further examples are shown below, and in Figure 3.12.

Infix	Postfix
a × b	ab ×
a + b − c	ab + c −
(a + b)÷c	ab + c ÷
(109 − 10)÷9	109, 10 − 9 ÷
(0x1000 × 2) + 0x20	0x1000,2 × 0x20 +

Considering the operations taking place, it becomes evident that using a stack is a very efficient method of performing RPN operations. A stack in this case is a storage device with a single entry/exit point. Numbers can be pushed onto the "top" of the stack, and then popped back off the "top" of it. It is a last-in first-out (LIFO) construct.

An example of a stack operation performing $ab+$ is shown in Figure 3.12, reading from left to right. Some things to note are that only a single push occurs in each step (likely to each take a single cycle in a stack-based processor) although the number of pops required to feed an operation is determined by the number of operands required. For example, an ADD requires two operands, so two POPs are used to load those to the ALU. The result of each operation is PUSHed back onto the top of the stack.

It is also interesting to consider the use of such a stack machine performing complex programming tasks: it seems efficient for simple operations, but sometimes it is possible that the final state of the stack after a sequence of operations may not have the correct

---

[9]Many readers may remember being taught the BODMAS acronym as an aid to remembering precedence during primary school mathematics, standing for **B**rackets, **O**rders (e.g., powers and square roots), **D**ivision, **M**ultiplication, **A**ddition, and **S**ubtraction: see `http://www.malton.n-yorks.sch.uk/MathsWeb/reference/bodmas.html`

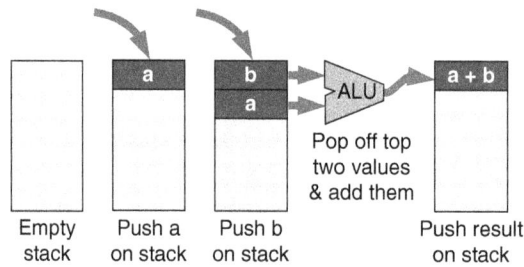

FIGURE 3.12   An illustration of the concept of stack processing. Two operands are pushed in turn on to the stack, and ALU then executes, popping the operands, calculating the sum, and then pushing the result back onto the stack.

results located on the top of the stack. This may be exacerbated by multitasking or interrupt service routines. There must be a way of reordering the stack, such as popping items out and storing into main memory, and them pushing them back in a different order. This could be a very time-consuming process and impacts heavily on the overall performance of a stack machine. This process is also explored in Box 3.6 where reordering is performed to minimize stack usage.

Box 3.6   Recoding RPN Instructions to Minimize Stack Space

Consider the infix expression a + (b x c) which can also be written as (b x c) + a since the order of addition is unimportant to the final result.

For each expression, try to write the equation in postfix notation, and then write out the sequence of stack operations that would be required to execute it. Consider the stack usage for each expression.

It should be clear that writing the equation one way involves a maximum stack depth used of three locations, whereas the alternative way results in a stack use only two locations deep.

It appears that the order of the postfix expression can have a significant impact on the stack resources (and hence hardware resources) needed, although it will not alter the number of steps needed to find a solution.

Not all infix expressions are insensitive to order. Addition and multiplication are, whereas division and subtraction are most definitely not.

3.4   Data Handling

This chapter, up to now, has concentrated on CPU basics—what a computer is and what does it fundamentally consist of. We have mentioned instructions, programs, and so on. As part of this, Section 3.3 considered instruction handling, including some variations on a theme, as well as the important subtopic of addressing modes.

Later, Section 3.5 will present a top-down view of computers. However, in between these two extremes of high-level and low-level there is a more philosophical question regarding the purpose of computers. We can consider a "black box" perspective as an example.[10] Having a black box perspective, we view a computer as a unit that modifies some input data to produce some output data.

Both input and output data could take many forms: commands, knowledge, sensor data, multimedia, and so on. For some systems, input data could consist of a single trigger event. Output data could likewise consist of an actuator switch signal. This is the case in control systems, which often operate with a need for real-time processing of data (real-time issues are considered in depth in Section 6.4). Some systems are data rich—either input or output may consist of dense streams of data, such as digital audio or video. These systems may also need to operate in real time. However, the majority of computer systems are probably general-purpose machines capable of performing both control and data processing tasks with little regard to real-time issues.

The common theme here is clearly data: computers process data, whether that is a single-bit trigger for which timing is critical, or a 1 Tibyte block of multimedia data that completes processing in several minutes. This section is dedicated to this important aspect of computers: what is data, how is it presented, stored, and processed.

3.4.1 Data Formats and Representations

We have discussed number formats in general in Section 2.3, including those of most relevance to computers (unsigned binary, two's complement, and so on). Whatever format is in use, the *width* of the number—the number of bits occupied by one number—can be adjusted by computer architects to either increase the largest magnitude number that can be stored, or to increase the precision. Typically, since computers are byte based, number sizes are in multiples of 8 bits.

Most CPUs have a natural size data format which is determined by the width of the internal buses, for example, byte-wide in the old 6502 processor and 32-bit wide in the ARM. Although the ARM can also handle bytes and 16-bit half-words, it accesses main memory in 32-bit units (assuming a 32-bit bus to memory and operation in ARM rather than Thumb mode), and thus handles 32-bit values no slower than the handing of bytes. Registers, memory locations, most operands, and so on are 32-bit in the ARM.

Programmers typically handle data in memory or in registers through a high-level language such as C or Java (see Section 8.2.1). Although some programming languages tightly define the number format used by data types within the language, that is not really the case for the original version of the C programming language, apart from the definition of a byte, which is always 8 bits in size.

Usually, although it is actually at the discretion of the particular compiler in use, the `int` data type generally matches the natural size of the processor for machines of 16-bit word size and above. Thus, an `int` in a 16-bit machine will normally be a 16-bit number, whereas it will tend to be 32-bit in a 32-bit CPU and 64-bit in a 64-bit CPU.

Programmers beware: if you wish to write portable code, ensure that there are no assumptions made about the exact size of data types like `int`, `short`, and so on. Table 3.1 illustrates the width of several data types for the common gcc compiler targeting

---

[10] For those who have not encountered this term, a "black box" is the name given to something that, when considered as a unit, is defined solely in terms of its inputs and outputs. The idea being that it does not matter what is inside the box, as long as it produces the correct output given the correct input.

C name	8-bit CPU	16-bit CPU	32-bit CPU	64-bit CPU
char	8	8	8	8
byte	8	8	8	8
short	16	16	16	16
int	16	16	32	64
long int	32	32	32	64
long long int	64	64	64	64
float	32	32	32	32
double	64	64	64	64
long double	*compiler specific — may be 128, 96, 80, or 64 bits*			

Note how some of the data types change size between processors, while others remain the same. For a particular implementation, these sizes are usually defined by maximum and minimum representable number specifications in the configuration header file *types.h*. Remember also that the byte order may change between big and little endian processors (Section 2.2).

TABLE 3.1 Comparison of C programming language data type sizes for CPUs ranging from 8 to 64 bits.

different processors.[11] Concerns over the changing nature of some of the original C language data types has led to many developers adopting specific-sized data types, described further in Box 3.7.

Box 3.7 Data Types in Embedded Systems

Although general programs written in languages such as C and C++ will make use of the standard data types shown in Table 3.1, this can cause confusion and even lead to errors when porting code. If a programmer makes an implicit assumption regarding the size of a particular data type, this assumption may no longer be correct when the code is compiled on a different processor.

The situation was actually far worse in the days before widespread adoption of the gcc compiler—compilers sometimes had different compilation modes such as "large memory model" and "small memory model" which could result in the number of bits used to represent variables changing (note that even gcc has command switches which can change this in exceptional circumstances). Cross-compiling for embedded systems, where the target machine may differ from the host compilation machine, makes it doubly important to ensure that any code tested on the host performs similarly on the target.

Perhaps the simplest way to achieve this, and to remain mindful of the limitations of different data types, is to directly specify the size of each type when

---

[11] Note that some compiler implementations will differ, or may not comply to ISO or ANSI C language specifications.

declaring variables. In the C99 programming language (the version of C formalized in 1999) the definitions have been made for us in the `<stdint.h>` header file:

Size	Signed	Unsigned
8	int8_t	uint8_t
16	int16_t	uint16_t
32	int32_t	uint32_t
64	int64_t	uint64_t

The 64-bit definitions (and other odd sizes such as 24-bit definitions) may exist for a particular processor implementation but not for others. Of course, if it exists, it will occupy the sizes given, but otherwise these are optional, so for some machines the compiler will not support anything but the main 8-, 16-, and 32-bit definitions. Writers of code for embedded systems will likely encounter these safer type declarations more often than authors of desktop machine software. The author would encourage those embedded systems developers to use these specific-sized types wherever possible in order to avoid possible porting difficulties later.

Of course, experienced programmers will soon learn that any integer data type in the C programming language (i.e., the top six rows in Table 3.1) can be specified as either signed or unsigned. The default (if neither is specified) is that the data types are signed two's complement.

The `long int` and `long long int` can also be specified as just `long` and `long long`, respectively. On all but the largest machines these will require multiple memory locations for storage.

The char type normally contains a 7-bit useful value, complying with the ASCII standard (American Standard Code for Information Interchange), shown in Table 3.2. Any top-bit-set character (i.e., a char where bit 8 is nonzero) would be interpreted as an extended ASCII character (which are not shown in the figure). Interestingly, characters lower than decimal 32 (space), and including decimal 127 (delete), are nonprintable characters having special values related to their original definitions for teletype terminals. For example, ASCII character 8, \b, is the bell character, which would cause a "beep" sound when printed. A brief Web search can easily reveal the meanings of other special ASCII characters.

ASCII was excellent when computers were effectively confined to English (or American English) speakers, but not particularly useful for other languages. Hence, significant effort has been paid over many years to define different character encodings for other languages. Perhaps the ultimate challenge has been Chinese which has around 13,000 pictograms (individual "letters" called characters): clearly an 8-bit data type is not able to encode written Chinese characters. Many solutions have appeared over the past two decades, most of which use two or more sequential bytes to hold a single character. The current de facto standard encoding is called *unicode*, which has various types but which can use up to four sequential bytes to encode the vast majority of characters, including Chinese, Japanese, Korean, and so on. Although the detail is beyond the scope of this book, the implications are not: early computers were byte sized, and were naturally able to handle byte-sized ASCII characters. These days, it requires a 32-bit machine to handle

Char	Dec	Hex	Char	Dec	Hex	Char	Dec	Hex	Char	Dec	Hex
\0	0	0x00	(spc)	32	0x20	@	64	0x40	'	96	0x60
(soh)	1	0x01	!	33	0x21	A	65	0x41	a	97	0x61
(stx)	2	0x02	"	34	0x22	B	66	0x42	b	98	0x62
(etx)	3	0x03	#	35	0x23	C	67	0x43	c	99	0x63
(eot)	4	0x04	$	36	0x24	D	68	0x44	d	100	0x64
(enq)	5	0x05	%	37	0x25	E	69	0x45	e	101	0x65
(ack)	6	0x06	&	38	0x26	F	70	0x46	f	102	0x66
\a	7	0x07	'	39	0x27	G	71	0x47	g	103	0x67
\b	8	0x08	(	40	0x28	H	72	0x48	h	104	0x68
\t	9	0x09	)	41	0x29	I	73	0x49	i	105	0x69
\n	10	0x0a	*	42	0x2a	J	74	0x4a	j	106	0x6a
(vt)	11	0x0b	+	43	0x2b	K	75	0x4b	k	107	0x6b
\f	12	0x0c	,	44	0x2c	L	76	0x4c	l	108	0x6c
\r	13	0x0d	-	45	0x2d	M	77	0x4d	m	109	0x6d
(so)	14	0x0e	.	46	0x2e	N	78	0x4e	n	110	0x6e
(si)	15	0x0f	/	47	0x2f	O	79	0x4f	o	111	0x6f
(dle)	16	0x10	0	48	0x30	P	80	0x50	p	112	0x70
(dc1)	17	0x11	1	49	0x31	Q	81	0x51	q	113	0x71
(dc2)	18	0x12	2	50	0x32	R	82	0x52	r	114	0x72
(dc3)	19	0x13	3	51	0x33	S	83	0x53	s	115	0x73
(dc4)	20	0x14	4	52	0x34	T	84	0x54	t	116	0x74
(nak)	21	0x15	5	53	0x35	U	85	0x55	u	117	0x75
(syn)	22	0x16	6	54	0x36	V	86	0x56	v	118	0x76
(etb)	23	0x17	7	55	0x37	W	87	0x57	w	119	0x77
(can)	24	0x18	8	56	0x38	X	88	0x58	x	120	0x78
(em)	25	0x19	9	57	0x39	Y	89	0x59	y	121	0x79
(sub)	26	0x1a	:	58	0x3a	Z	90	0x5a	z	122	0x7a
(esc)	27	0x1b	;	59	0x3b	[	91	0x5b	{	123	0x7b
(fs)	28	0x1c	<	60	0x3c	\	92	0x5c			124	0x7c
(gs)	29	0x1d	=	61	0x3d	]	93	0x5d	}	125	0x7d
(rs)	30	0x1e	>	62	0x3e	^	94	0x5e	~	126	0x7e
(us)	31	0x1f	?	63	0x3f	_	95	0x5f	(del)	127	0x7f

TABLE 3.2   The American Standard Code for Information Interchange, 7-bit ASCII table, showing the character (or name/identifier for nonprintable characters), and the representative code in decimal and hexadecimal.

a 4-byte unicode character in a single operation. Similarly, early interfacing methods such as the PC parallel and serial ports (see Chapter 6), were byte based. Memory accesses have often been byte based. The argument has been that a byte is a convenient size for simple counting and for text processing. This argument no longer applies in many cases. Where non-English alphabet systems are concerned, byte-sized processing is nothing more than a historical curiosity.

One final point to note concerning data sizes is the uniformity of the `float` and `double` types. This uniformity is related to the ubiquity of the IEEE754 standard, and the fact that the majority of hardware floating point units comply with the standard (this will be explained a little more in Section 4.6).

3.4.2 Data Flows

Again adopting a black-box view, a computer takes input, processes it, and generates output. Evidently, the requirements of that data are important, in terms of timeliness, quantity, quality, and so on.

Today's computers, and especially many consumer electronic-embedded systems, are heavily user-centric. This means that input, output, or both needs to interact with a human being. Some data also tends to be quite voluminous (video, audio, and so on). Buses, which we will consider more fully in Section 6.1, need to be sized to cope with the required data flows, and systems should also consider human needs. For example, the human sensory organs are often far more sensitive to sudden discontinuities like "blips" and step changes than they are to continuous errors (like white noise). It is usually more annoying for listeners to hear music from a CD player which skips than it is to listen to music in the presence of background noise. Similarly with video: skipped frames can be more annoying than a slightly noisy picture.

Most of the important real-time issues will be explored in Section 6.4. However, at this point we need to stress that computer engineers and programmers should bear in mind the use to which their systems will be put. Embedded computer engineers may have an advantage in that their systems are less flexible, and more generic, and thus better able to satisfy users. Unfortunately, they also suffer the considerable disadvantage that size, cost, and power limitations are more severe, and thus require finer balancing of trade-offs in design.

Technically speaking, data flows through computers use buses. This data may originate from external devices, or some form of data store, be processed in some way by a CPU or coprocessor, and then output similarly either to another external device, or data store.

3.4.3 Data Storage

The memory hierarchy of Figure 3.1 highlighted the difference in memory provision between embedded systems and typical desktop or server systems: with the exception of some systems like NAS (network-attached storage) devices, data in embedded systems are usually stored in flash memory. In desktop systems, it is often stored on a hard disc drive (HDD) or solid state drive (SSD—flash memory masquerading as a HDD) for medium-term storage, or tape/CDROM or DVD (for backup storage).

Data "inside" a computer, is located within RAM, cache memory, registers, and so on. From a programmers' perspective it is either in registers, or in main memory (data in cache memory is invisible to a programmer; it is indistinguishable from data in main memory, except it is accessed more quickly). Data enters memory from external devices or hard discs over buses (Section 6.1) either individually a byte or word at a time, or in bursts, perhaps using a scheme such as direct memory access (DMA—Section 6.1.2). Large amounts of data occupy pages of memory, handled by a memory management unit (MMU—Section 4.3), and small amounts may exist in fixed variable locations, or in a system stack. Since embedded systems often use parallel bus-connected flash memory devices, data in such systems is already directly accessible by the main processor and thus is already considered "inside" the computer.

Data is brought into a CPU from memory for processing, and again may be conveyed as individual items, or as a block. For load-store machines (Section 3.2.3), data to be processed must first be loaded into individual registers, since all processing operations take input only from registers, and output only to registers. Some specialized machines (such as vector processors) can handle blocks of data directly, and some machines have dedicated coprocessing units that can access memory directly, without requiring the CPU to handle loading and storing.

3.4.4   Internal Data

Program written in high-level languages like C need to be compiled before they can be executed (this will be discussed more in Section 8.2.1). It is the compiler that decides how to handle program variables. Some variables, usually those accessed most, will occupy registers during the time that they are being accessed. However, most processors have insufficient registers to store more than a handful of variables, and so most variables will be stored in memory.

Global variables (those that remain accessible throughout a program's execution) almost always have a dedicated memory address during the execution of a program, but other variables are stored in a memory stack. That means that when a program contains a statement such as "index++" and index is a local variable which the compiler decides cannot remain in a register, the compiler dedicates a particular location in stack to the variable index. The pseudo-code operations needed to execute such a statement on a load-store machine would thus be

1. Load the data item at the particular stack offset corresponding to variable index to a register.

2. Increment the value stored in that register.

3. Save that register content to the same stack offset that it was retrieved from.

If there was a subsequent decision to be made on variable index (such as if index > 100 then ...) the compiler knows that index is already occupying a register, so it will reuse that register in the subsequent comparison and decision. Some variables, as we have mentioned, can remain in registers throughout a calculation. It all depends upon how many registers are available, how many variables are in use, and how frequently are these accessed.

Actually the programmer has little control over which variables are to be stored in registers, and which are kept in a stack, although the C programming language keyword `register` asks the compiler to keep a variable in a register if possible. For example, if we wanted to maintain `index` in a register (if possible), we would have declared `index` as

```
register int index=0;
```

Spill code is the name given to the few machine code instructions that a compiler adds to a program to load-store variables between memory and registers. Since memory accesses are far slower than register accesses, spill code not only slightly increases the size of a program, it also adversely affects execution speed. Minimizing spill code has long been a target of compiler researchers and computer architects worldwide.

3.4.5 Data Processing

Adding two 8-bit integers in an 8-bit processor is always going to be a simple proposition, and adding two 8-bit numbers in a 32-bit processor is also relatively simple[12] since both arithmetic operations can be performed with a single instruction.

That single instruction is normally processed very easily in hardware, translating to something like "send the two operands from registers to an ALU and then load the result back into another register."

The situation becomes more interesting when processing larger numbers in a smaller processor, and when performing more complex processing. Let us consider three possibilities in turn: operating on larger numbers then the width of processor, floating point in a fixed-point CPU, and complex numbers.

3.4.5.1 Big Numbers on Small CPUs

Since the C programming language can define 32-bit or even 64-bit data types, this follows that any C compiler for 8-, 16-, or even 32-bit CPUs must be able to support arithmetic and logical operations on numbers larger than the natural size of the processor. The same is true of the vast majority of computer languages—whether compiled or interpreted.

First of all, note that many processors with a certain data bus width actually support higher precision arithmetic. For example, most ARM processors are able to perform a multiplication between two 32-bit numbers. We know that the maximum size of the result of such an operation could be 64 bits. The original ARM multiplier would allow only the lower 32-bit part of that result to be stored to the destination register. However, a "long multiply" instruction on newer ARM processors allows the full 64-bit result to be stored to two 32-bit destination registers. Evidently, the result store part of the operation will take twice as long to complete (but this is a lot less time than trying to determine the upper 32 bits from the operation using other methods).

Let us examine how we can perform a 64-bit multiply on an ARM processor that does not have a "long" multiply instruction (although please note that this may not always be the fastest way to do it):

---

[12]Remember though that sign extension (Section 2.3.8) would need to be performed when placing 8-bit values into 32-bit registers, otherwise negative two's complement numbers may be incorrectly interpreted in the ALU!

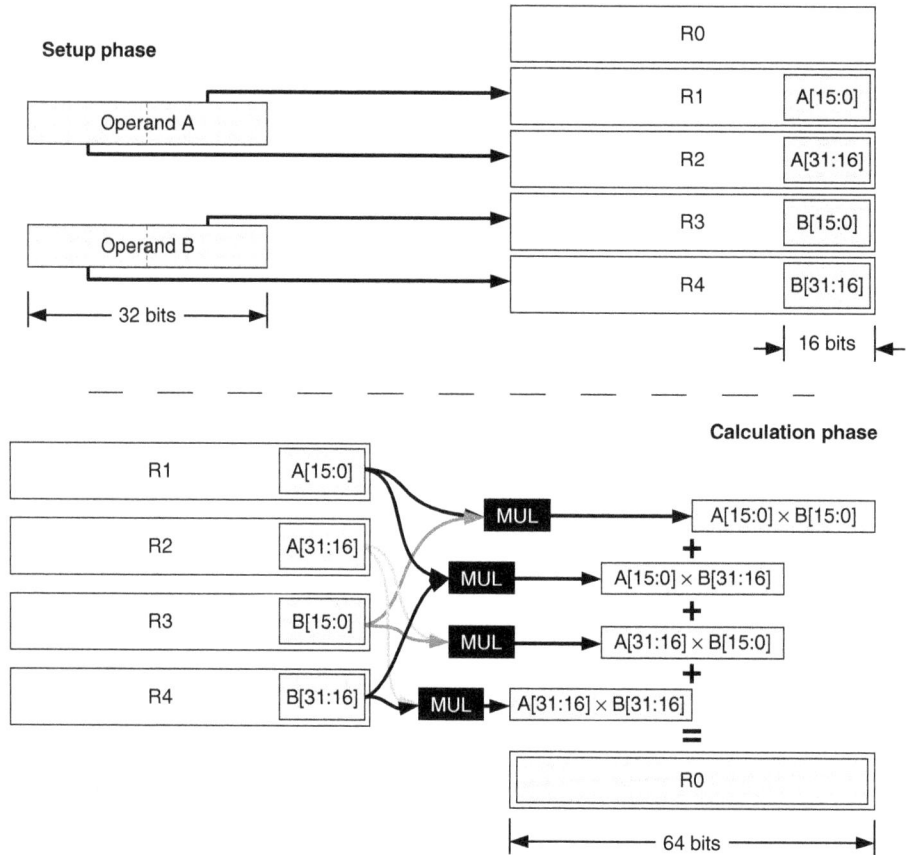

Figure 3.13   A block diagram illustrating the setup and calculation stages of the multistep procedure necessary to perform a 32-bit × 32-bit = 64-bit multiplication using multiplication hardware only capable of returning a 32-bit result (i.e., 16-bit × 16-bit = 32-bit hardware).

1. Load operand 1 lower 16 bits to R1
2. Load operand 1 upper 16 bits to R2
3. Load operand 2 lower 16 bits to R3
4. Load operand 2 upper 16 bits to R4
5. R0 = R1 × R3
6. R0 = R0 + (R2 × R3) << 16
7. R0 = R0 + (R1 × R4) << 16
8. R0 = R0 + (R2 × R4) << 32

This is shown diagrammatically in Figure 3.13, where the loading is shown as a setup stage, and the multiplication and adding is shown as an operation stage. Within this stage, four multiplications, three shifts, and three additions need to be performed to calculate the result.

The clear message here is that lack of the single "long" multiply instruction will entail several additional operations, and possibly registers, to replace it. Of course, there are slightly faster or lower-overhead schemes than the particular one shown, that can work in certain cases; however, for general-purpose multiplication none of these can better the use of a single instruction.

Logical operations on longer data words are quite simple: split the operands, process the logical operations on each part separately, and then reassemble the result. This is because a logical operation on 1 bit in a binary word does not have any impact upon the neighboring bits.

Arithmetic operations require a little more thought than logical operations (but are simpler than multiplication or division). The issue with arithmetic operations is that of overflow: the result of adding two 16-bit numbers may be 17 bits in length. The extra bit (carry) must therefore be taken into consideration when performing the addition of the split numbers. Usually that will involve calculating the lower half of the split first, and then adding this (with carry) to the result of the upper half.

3.4.5.2 *Floating Point on Fixed-Point CPUs*

We had discussed floating point numbers extensively in Section 2.8, and floating point processing in Section 2.9. Most of the time when we discuss "floating point" in computer architecture, we mean IEEE754 standard floating point. This is primarily because most hardware floating point units implement the IEEE754 standard. Meanwhile those processors without floating point capabilities either rely upon compiler support to translate each floating point instruction in a program into much longer and slower fixed-point subroutines, or compile into floating point machine code instructions which are then *trapped* by the processor. This trapping is in effect an interrupt triggered by receipt of an instruction that the processor cannot handle. The interrupt service code then has the responsibility of performing the particular floating point operation using fixed-point code before returning back to normal execution. This is known as floating point emulation (FPE), and is examined further in Section 4.6.1. The first approach can only be used if the programmer knows, at the time of compilation, whether an FPU will be present or not in the machine executing the code, and thus may not be suitable for general software such as on a PC.

When a hardware FPU is not included in a system, the FPE (or compiler) alternative probably will not implement the full IEEE754 standard, since this could be quite slow. Thus the code will end up being potentially less accurate as well as being significantly slower.

Let us refer back, for a moment, to Sections 2.9.1 and 2.9.2, where we had considered the addition/subtraction, and the multiplication of floating point numbers: the addition/subtraction process required a normalization procedure, whereas the multiplication process required a straightforward calculation; however, one containing several subcomputations.

Taking the simple multiplication from Section 2.9.2 as an example:

$$(A \times B^C) \times (D \times B^E) = (A \times D) \times B^{(C+E)}$$

For a machine equipped with FPU, $(A \times B^C)$ and $(D \times B^E)$ would be a pair of 32-bit (for float) or 64-bit (for double) values. These values would be loaded into FPU registers by the CPU, then a single instruction issued by the CPU to trigger the FPU unit to perform the multiplication, and then the answer retrieved from a destination FPU register. By contrast, for a machine without FPU, several fixed-point operations would be required to perform the same calculation:

1. Split off mantissa and exponent A and C, store in R1 and R2, respectively.
2. Split off mantissa and exponent D and E, store in R3 and R4, respectively.
3. Calculate the new mantissa: R1 × R3.
4. Calculate the new exponent: R2 + R4.
5. Normalize exponents.
6. Recombine and store in IEEE754 format.

Clearly a single FPU instruction (in a computer that has an FPU) is preferable to the several fixed-point operations that are needed to replace it in a computer that does not have an FPU.

3.4.5.3 Complex Numbers

Complex numbers, of the form $(a + j.b)$, where $j = \sqrt{-1}$ are frequently used in scientific and engineering systems. Almost all CPUs lack support for complex numbers, and few programming languages cater for them.[13]

Complex number computations on a system with hardware handling only real numbers, just like floating point performed with fixed-precision arithmetic, requires a few steps. Consider the multiplication and addition of two complex numbers:

$$(a + j.b) \times (c + j.d) = (a.c - d.b) + j(a.d + b.c)$$

$$(a + j.b) + (c + j.d) = (a + c) + j(b + d)$$

The complex multiplication requires four real multiplications and two additions. The complex addition is a little simpler, requiring only two real additions. This will require the programmer (or compiler) splitting the operation into steps of several simpler instructions.

A processor with hardware support for complex numbers would possess a single instruction capable of performing these operations. The underlying hardware architecture would actually need to perform all of the splitting, suboperations, and separate multiplies, but this would be handled very quickly within the CPU without requiring separate loads, stores, and data moves.

---

[13]The notable exception is FORTRAN (FORmula TRANslation), the general-purpose compiled language introduced by IBM in the mid-1950s. FORTRAN, updated several times since (the latest being 2003), has natively supported a complex number data type for over 50 years. Among modern languages, there has been some promotion of Java and Python as languages for scientific computation, with a complex number extension. However, both are typically a low slower for complex mathematics processing than FORTRAN is.

3.5   A Top-Down View

3.5.1   Computer Capabilities

Looking at various processors available today, there are a profusion of features, clock speeds, bit widths, instruction sets, and so on. The question arises as to what is needed in a particular computer. Some capabilities are examined below:

3.5.1.1   Functionality

Given that all computable functions can be performed by some sequence of logic operations, the main reason why not all functions are computed in such a way (i.e., as a possibly long sequence of logical operations), is related to efficiency—how long does such a function take to complete, and what hardware resources are required? There is some trade-off in that making a computer simpler can allow faster clock speeds. This argument led to the advent of RISC processors which, being simpler, clock faster—at the expense of having to perform some functions longhand that would be built into a CISC computer as single instructions.

However, it is generally pragmatic to consider how often a particular function is required in software when deciding how to implement it. Put simply, if a function is required very frequently during everyday use, then maybe it is useful to build a dedicated hardware unit to handle it. In this way, an ALU is included in all modern processors, and almost all have hardware multiplier units.

Not only functional operations, but flexibility in the instruction set is an important feature. For example, there may be time-saving instructions available in one design but not another, even when these do not require large amounts of hardware support. Examples are the universality of conditional instructions in the ARM instruction set (Section 3.3.1), and zero-overhead loop instructions in some digital signal processors (shown later in Section 5.6.1).

The internal architecture of a CPU—namely the number of buses, registers, and their organization is also an important consideration for performance. Generally speaking, more buses means more data items can travel around a device simultaneously, yielding better performance. Likewise, more registers can support more software variables that would otherwise need to be stored in slower memory, which again improves performance.

3.5.1.2   Clock Speed

A higher clock speed does not always mean faster operation. For example, it is relatively easy to design a fast ALU, but not at all trivial to design a fast multiply unit. When comparing two processors, clock speed alone is not sufficient to decide which is faster—there are many factors such as functionality, bus bandwidth, memory speeds, and so on, which must be considered: in effect, asking the question "what can be accomplished each clock cycle?" This question is considered in the next section.

3.5.1.3   Bit Widths

Until recently the vast majority of CPU sales were for 4-bit processors, destined to be used in watches, calculators, and so on. These days the profusion of mostly 32-bit processors (generally ARM-based) used in smartphones and network appliances, is tipping the balance toward *wider* processors. Although it might seem a wider processor will result in faster computation, this is only true if the data types being computed make use of the extra width. High-end servers with 64-bit or even 128-bit architectures are available, but if these are being used to handle text (such as 7- or 8-bit ASCII or even 16-bit Unicode), the extra width may well be wasted.

3.5.1.4 *Memory Provision*

The memory connected to a processor is often critical in determining operation speed. Not only the speed of memory access, but the width (together specifying a bandwidth in Mbits or Gbits per second), and technology is important. Other aspects include burst mode access, paging or packetization and single- or double-edged clocking.

On-chip memory also may not always be single-cycle access, but it is likely to be faster than off-chip memory. Given a particular software task that must be run, the amount of memory provided on-chip, and off-chip, must be considered. A cache (Section 4.4) in particular is used to maximize the use of faster memory, and the complexity of hardware memory units tends to influence how well memory use is optimized. In terms of memory, the way software is written and compiled can also result in more efficient use of hardware resources.

3.5.2 Performance Measures, Statistics, and Lies

In order to determine exactly how fast a computer operates, the simplest general-purpose measure is simply how many instructions it can process per second.

MIPS (millions of instructions per second), measures the speed at which instructions, or operations, can be handled. This is a useful low-level measure, but does not relate directly to how "powerful" a computer is: The operations themselves may be very simple such that multiple operations are required to perform a useful task. In other words, a simple computer with a high MIPS rating (such as a RISC processor) may handle real-world tasks slower than a computer with a lower MIPS rating but with instructions that can each perform more work (such as a CISC processor). The *bogomips* rating, which used to be calculated at boot-up on Linux PCs, was a famous attempt to gauge a MIPS score in software—but was unfortunately notoriously inaccurate.

MIPS as a measure is made up from two components, clock frequency f (in Hz), and CPI (cycles per instruction) such that

$$\text{MIPS} = f/\text{CPI}$$

More generally, for a particular program containing P instructions, the completion time is going to be

$$T_{\text{complete}} = (P \times \text{CPI})/f$$

This formula tells us that completion time reduces when CPI is low, f is high or most obviously when P is low (i.e., a shorter program will probably execute faster than a longer one). The trade-off between P and CPI in computer architecture is a revisit of the RISC versus CISC debate, while ever increasing clock frequency is the story of modern CPUs.

The task of minimizing CPI is another aspect of modern computer systems. Up until the 1980s, CPI would be greater than 2, perhaps as much as several hundred in CISC machines. The RISC approach began to shift CPI downward, with the aim of achieving a CPI of unity. ARM family processors typically achieve a CPI of about 1.1, and other RISC processors can do a little better than this.

Later, the advent of superscalar architectures led to CPI values of below unity, through allowing several instructions to execute simultaneously. This, and the inverse of CPI (called *IPC*) will be explored later in Section 5.5.1.

Sometimes floating-point performance is an important attribute, and this is measured in *MFLOPS* (millions of floating point operations per second). In recent times, GFLOPS readings are more commonly quoted, meaning thousands of MFLOPS, and even petaflops, PFLOPS. These values are more indicative of actual performance than MIPS since we are counting useful calculation operations rather than the low-level instructions which comprise them.

Box 3.8   Standardized Performance

In the mid-1980s, the computer industry worldwide saw an unprecedented level of competition between vendors. This was not simply a two-entry race between AMD and Intel, it included thousands of manufacturers selling enormously differing machines—alternative architectures, different memory, tens of CPU types, custom operating systems, 8-bit, 16-bit, and some even more unusual choices.

In the United Kingdom, companies such as Sinclair, Acorn, Oric, Amstrad, Research Machines, Apricot, Dragon, ICL, Ferranti, Tandy, Triumph-Adler, and more battled in the marketplace against IBM, Apple, Compaq, DEC, Atari, Commodore, and others. Claims and counterclaims regarding performance littered the advertisements and sales brochures available at that time. However, with no standard and no baseline, claims were often dubious to say the least.

In response the British Standards Institute (BSI) published a performance standard for computers—testing useful tasks such as integer calculation, floating-point calculation, branching performance, and graphics as well as disc reads and writes. However, at that time the programming language of choice was BASIC (Beginner's All-purpose Symbolic Instruction Set), and hence the standards were written in that language! From today's point of view the graphics and disc tests are also dated: The "graphics" test was actually text being written to the screen or VDU (visual display unit) in the parlance of the time. This was important for many users interested in nothing more than word processing. Also disc reads and writes were to floppy discs—a great advance on the tape drives used for most home computers at the time—hard discs (usually known as Winchester drives in those days) were simply too expensive and not even supported on most machines available at the time. Far more common was saving programs to cassette tape.

Today, new hardware and software are often tested with a battery of performance measuring software that is far removed from the BSI standard, but follows the same rationale. While there are some measures of pure number-crunching ability, other measures such as "screen refresh rate for playing my favorite game" and "time taken to sort 1 million rows of random numbers in a spreadsheet" are also common. After all, more users are interested in playing games that are graphically intensive than in how quickly they can calculate π to 100 decimal places.

Benchmarks are so important that several companies exist to provide such services (Box 3.8 explores the background and necessity of having such benchmarks). *BDTi* is one example which published comparative speeds for several digital signal processors (DSPs). Their measures are skewed toward outright calculating performance, something which is the mainstay of the DSP market.

Otherwise, *SPECint* and *SPECfp* benchmarks compute integer and floating-point performance directly. These are obtainable in source code format from the Standard Performance Evaluation Corporation (SPEC), for a fee, and can be compiled on an architecture to assess its performance. Each measure is calculated from a set of algorithms that have to be run, and results combined. Generally, a year is provided to indicate test version. Thus SPECint92 is the 1992 version of the integer standard.

The SPEC measures themselves incorporate two earlier measures known as *Dhrystone* and *Whetstone*, both originating in the 1970s and measuring integer and floating-point performance, respectively. Many other performance metrics exist, and may be used to assessing performance for various tasks (such as graphics rendering, real-time performance, byte handling, and so on).

Unfortunately, it is a well-known fact that, given any single performance measure, computer architects can tweak an architecture to yield a high score at the expense of other unmeasured operations. Furthermore, none of these measures really reflect the overall completion time of anything but the simplest tasks running in isolation. Many issues intervene in the real world to confuse results, such as interrupted tasks, operating system calls, varying memory speeds, disc speeds, multitasking, and cache performance.

In computing, a cache (covered in detail in Section 4.4) is a small block of very fast memory provided in a system having main memory that is much slower (cache is usually on-chip, main memory is usually on a separate chip to the CPU, or in a distant part of the IC connected by a bus). Any program running directly from the cache will execute quicker than one running from slow main memory. Why this is relevant is that in the past, at least one processor vendor has deliberately designed a cache just big enough to hold an entire performance measure algorithm (i.e., the entire SPECint or Dhrystone program) so that it runs much faster than it does on a competitors' machine.

In such an example, if the main memory were made to run 10 times slower, the performance score would not change significantly since the program that does the measuring runs from cache and not main memory. Obviously, this gives a performance measure that is not realistic for bigger programs. In fact, such a machine would yield a faster performance score than a competitor with a smaller cache but significantly faster main memory—one which would in reality probably perform real-world tasks much quicker.

Given significant performance-altering factors such as those we have mentioned, it is clear that the world of benchmarking is fraught with difficulty. A system designer is thus urged to be careful. In practice, this may mean understanding device operation in detail, building in large safety margins, or prototyping final code on a CPU before committing firmly to use that CPU. Although it is rare in industrial projects for software to be working before the hardware is complete (usually the software is the last part to get working), if there is some way software can be ready first—or at least prototype code—the approach of prototyping it on actual CPUs is strongly recommended.

3.5.3   Assessing Performance

Section 6.4.4 will discuss completion times and execution performance for real-time and multitasking systems, but here we consider estimation of performance. In order to underscore the need for accurate performance estimation, we will consider a real example from industry:

An embedded design group working on a portable device needed hardware to run an algorithm requiring 12 MIPS of processing power. A 32-bit CPU rated at providing 40 MIPS when clocked at 40 MHz was chosen to execute this. In an attempt to reduce design risks, the designers obtained a development board, loaded a Dhrystone measure on to the board and checked actual performance themselves before committing to that processor as the design choice.

During the design process, they realized that on-chip memory was insufficient to hold all of their software, so they added external DRAM memory. Due to the small size of the CPU package, and the low number of pins, the external memory bus was limited to being 16 bits wide. External memory accesses were therefore 16 bits wide instead of 32 bits.

Having completed their hardware design and built the system, they loaded up the code and found it would not execute in the time required. Where had they gone wrong?

First, the Dhrystone measure fitted into fast on-chip memory and so could run at full speed, whereas their wanted algorithm was too large to fit in on-chip memory and therefore had to be stored in DRAM instead. Not only were DRAM accesses themselves slower than internal memory accesses, but DRAM needs a "time out" occasionally to refresh itself. During that time-out, all memory accesses by the CPU are stalled.

Finally, the 16-bit interface meant that two memory reads were now required to fetch each 32-bit instruction—two 16-bit accesses were also required to read in every 32-bit data word. This meant that, when executing a program from DRAM, the CPU needed to spend half of its time idle: every odd cycle it would fetch the first half of the instruction, in the even cycles it would fetch the second half of the instruction, and only then begin to process it.

The 16-bit interface effectively dropped the 40 MIPS down to 20 MIPS, and the lower speed of DRAM access plus refresh time reduced the 20 MIPS performance further to around 9 MIPS.

The solutions were unpleasant: either switch to using very fast external memory (SRAM) which was perhaps 20 times as expensive, or upgrade to another CPU with either (i) faster speed or (ii) a wider external memory interface, or both. Designers chose neither—they chose to add a *second* CPU alongside the first to handle some of the processing tasks.

This example underscores the necessity of matching performance requirements to hardware. In general, there are two approaches to this: the first one is through a *clear understanding* of the architecture and the second is through *careful evaluation* of the architecture. In both cases, the architecture referred to is not only the CPU, it also includes other important peripheral elements such as main memory and bus bandwidth.

Gaining a *clear understanding* of software requirements means having fixed software that needs to be run on a system, analyzing that software to identify its contents (particularly any bottlenecks) and then matching the results of that analysis to available hardware. At the simplest level this might mean avoiding an integer-only CPU when most calculations need to be done in floating point.

This approach is commonly taken for DSP systems and will include a close look at memory transfers, placement of variable blocks into different memory areas that can be accessed simultaneously (see Section 4.1.4), also input and output bottlenecks, and mathematical operations which are typically the major strength of such processors. Slow setup, user interface, and control code are generally ignored in such calculations—except in the sizing of overall program memory requirements.

At this point, it is useful to note that most, if not all, software developments end up overrunning initial program memory use estimates. Clever coding can often bring down data memory use, and can reduce processing requirements, but can seldom save significant amounts of program memory. Unlike desktop computer designers, embedded designers do not have the luxury of providing for RAM expansion: this must be fixed at design time. In such cases, it is wise to significantly overestimate memory needs up-front.

The second approach mentioned of matching required performance to hardware, is through *careful evaluation*. This does not require detailed architectural understanding, but does require detailed levels of testing. Ideally, the final run-time software should be executed on candidate hardware to evaluate how much CPU time it requires. A list of other tasks to be performed should also be made, and checked to see whether those can fit into whatever spare processing time remains. Software profiling tools (such as GNU *gprof*) will identify any bottlenecks in the run-time code, and make clear which software routines require large amounts of CPU time.

It is important to run any test a number of times (but do not average the results for systems where timing is critical—use the worst case result). It is also important to increase program size sufficiently to force it out of cache or on-chip memory, if appropriate, and to enable whatever interrupts and ancillary tasks might be needed in the final system.

If, as is sometimes the case, the target software is already running on another machine, it is possible to compare its execution on that machine to execution on another—but only after considering all important architectural factors as discussed in these last two chapters. In such instances, compiling and comparing a suite of standard benchmarks on both machines will help, assuming that the benchmarks chosen are ones of relevance to the target software.

The world is full of examples where designers have estimated processor performance and/or memory requirements incorrectly (including one example designed for an Asian industrial manufacturer in 1999 by the author: a portable MP3 player that could only replay 7 seconds of MP3 audio at a time, due to memory bus bandwidth restrictions. Fortunately for the author, a faster speed grade processor became available and was used instead).

You have been warned—beware the pitfalls of performance estimation, evaluation, and measurement. Above all, remember to read the small print below manufacturers' performance claims.

3.6 Summary

In this chapter, the basics of the microprocessor have been covered, starting with the functionality of a CPU, the ability to control this with a program, the need to transfer this program (and store it somewhere).

A control unit needs to keep a processor on track, managing operations and exceptions, and being directed in turn by the computer program through a sequence of instructions. Control units can be centralized, or distributed with timing from a state machine, a microcode engine, or using self-timed logic.

Each instruction in a program is part of an allowable instruction set that (depending on your point of view) describes the operations capable of being performed by that processor, or which specifies the microprocessor behavior. Such behavior includes data transfer through internal buses to various functional units. Having laid the foundation for CPU design here and in the previous chapter, in Chapter 4 we will delve into the internal arrangements and functional units of most mainstream CPUs—and attempt to relate that to the programmer's experience.

3.7 Problems

3.1 If the assembler instruction LSL means "logical shift left," LSR means "logical shift right," ASL means "arithmetic shift left," and ASR means "arithmetic shift right" then what are the results of performing the following operations on signed 16-bit numbers:

1. 0x00CA ASR 1
2. 0x0101 LSR 12
3. 0xFF0F LSL 2
4. 0xFF0F LSR 2
5. 0xFF0F ASR 3
6. 0xFF0F ASL 3

3.2 An analysis of representative code for a fictitious RISC processor, which has only eight instructions, finds the following occurrences of those instructions:

Instruction	Number of occurrences
ADD	30
AND	22
LDR	68
MOV	100
NOT	15
ORR	10
STR	60
SUB	6

1. If each instruction (excluding operands) is 6 bits long, how many bits does the program occupy?
2. Use the information in the table to design a Huffman coding for the processor.

Calculate the number of bits needed to store the program using the Huffman coding instruction set.

3.3 Show the sequence of stack PUSHes and POPs during the execution of the following reverse Polish notation (RPN) operations, and translate each into infix notation:

  (i) ab+

  (ii) ab + c×

  (iii) ab × cdsin+−

Consider the maximum depth of stack required to perform these operations.

3.4 A ROT (rotate) instruction is similar to a shift, except it wraps around—when shifting right, each bit that drops off the LSB end of the word is moved around to become the new MSB. When shifting left, each MSB that drops off is moved around to become the new LSB.

The ROT argument is positive for left shifts and negative for right shifts.

So imagining a process that has a ROT instruction but no shift, how can we do arithmetic and logical shifting?

3.5 Translate the following infix operations to reverse Polish notation (RPN):

  a) (A and B) or C

  b) (A and B) or (C and D)

  c) ((A or B) and C) + D

  d) C + {pow(A, B) × D}

  e) See if you can perform the following translation in three different ways:
     {C + pow(A, B)} × D

3.6 Calculate the maximum stack usage (depth) for each of the three answers to part (e) above.

3.7 Translate the following RPN notations to infix:

  a) AB + C + D×

  b) ABCDE + × × −

  c) DC not and BA++

3.8 Given the following segment of ARM assembler:

```
        ADDS R0, R1, R3
        BGE step2
        ADD R2, R1, R6
        BLT step3
step2   ADD R2, R3, R6
step3   NOP
```

Rewrite the code to use conditional ADDS to remove the need for any branch instructions.

3.9 In ARM assembly language, determine the least number of instructions in each case to perform the following immediate loads (hint: make use of the MOV and MVN instruction):

   a) Load a value `0x12340001` to register R0
   b) Load a value `0x00000700` to register R1
   c) Load a value `0xFFFF0FF0` to register R2

3.10 Identify the sequence of operations, in a RISC processor, that is required to add the contents of two memory addresses $m1$ and $m2$ and store the result to a third address $m3$.

3.11 Scientists discover a new type of silicon memory cell. Semiconductor engineers design this into a new memory chip. Identify six factors that computer architects would look at when deciding whether to adopt this new technology for mass storage in an embedded video player.

3.12 Consider the following fictitious instructions and decide whether they are from a RISC or CISC processor:

   a) MPX: Multiply the content of two memory locations, then add the result to an accumulator (a special on-chip register with fast access).
   b) BCDD: Perform a binary-coded decimal division on two registers, format the result in scientific notation and store as ASCII to a memory block ready for display to the screen.
   c) SUB: Subtract one operand from another and return the result as a third operand. The operands and result are register contents only.
   d) LDIV Rc, Ra, Rb: Perform a 100-cycle long division of Ra/Rb and place the result in register Rc.

3.13 Write a pseudocode program in microcode to perform any two of the instructions from the previous problem. Assume an internal RISC-style architecture and registers.

3.14 What is a load-store architecture, and why would computer designers adopt such an idea?

3.15 In a simple computer pipeline, what process normally follows the instruction fetch stage?

3.16 For a fictitious 32-bit processor, the hexadecimal machine code instruction, for the assembler command to store a word `0x1234` in memory location `0x9876` looks like this:

```
0x0F00 1234 088D 9876
```

By examining the machine code instruction, determine whether this processor is likely to be capable of absolute addressing, and justify your answer.

3.17 Another fictitious processor, this time an 8-bit CPU, has eight registers. In this processor, is it possible to have instructions that specify two operand registers and a separate result register?

3.18   Assuming ARM-style assembly language (but not necessarily an ARM processor), identify the type of addressing represented in the following instructions:

a) MOV R8, #0x128

b) AND

c) STR R12, [R1]

d) AND R4, R5, R4

e) LDR R6, [R3, R0, LSL #2]

f) LDR R2, [R1, R0, #8]

g) STR R6, [R3, R0]

3.19   Which processor is likely to be faster at processing 32-bit floating point data: a 900 MHz 32-bit floating point CPU, or a 2-GHz 16-bit integer-only CPU?

3.20   When writing code in the C programming language on different processors, is a byte always represented as 8-bits? How about the short and int—what size are these, and are they always the same?

CHAPTER 4

Processor Internals

C hapter 2 covered much of the low-level numerical calculations performed by computer, but also the definitions of computer functional units and classifications of some connectivities. Chapter 3 began to form this information into cohesive units with different functions that were able to execute sequences of instructions as specified by a programmer, since we know that computers, and indeed CPUs, can be divided logically into a number of functional units performing different tasks.

This chapter will extend beyond the basic high-level discussion of what goes into a CPU, and begin to focus on the largest, most prominent, and most important of the internal units that are commonly found in modern processors. We will look in more detail at what tasks those units perform and how they are able to do so. This discussion mainly covers the arithmetic logic unit (ALU), floating-point unit (FPU), memory management unit (MMU), and memory cache unit. However, before embarking upon that discussion, we first consider the issue of how the units are wired up through buses.

It is time to assess the actual architecture—specifically the interconnected bus structure—of units within a CPU.

4.1 Internal Bus Architecture

4.1.1 A Programmer's Perspective

From a programmers' perspective, the internal bus architecture of a processor can be seen in two main, but related, ways. First is in the degree of flexibility of register use. This is evident in the set of possible registers that can be used as operands in a particular instruction: in the ARM for instance, where a register operand is allowed, *any* register from its register bank can be named:

```
ADD R0, R1, R2  ; R0 = R1 + R2
```

Any register could be used—we could even use the same register:

```
ADD R0, R0, R0  ; R0 = R0 + R0
```

Many processors do not have this flexibility, or are less *regular*. Second, there is the issue of how much work can be performed in a single instruction cycle, and this is normally implicit in the instruction set itself. Again looking at the ARM, there are at most two register input operands, and a single register result operand associated with any arithmetic or logic instruction:

```
ADD R0, R1, R2  ; R0 = R1 + R2
```

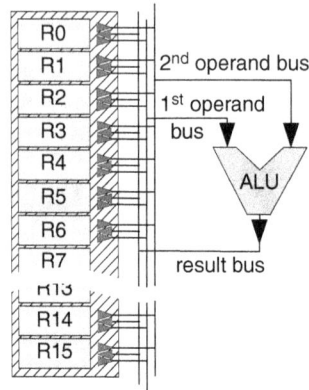

FIGURE 4.1 A schematic diagram of an ALU and a bank of registers interconnected with a three-bus arrangement.

Thinking about the means of transporting data from a register to the ALU and back again: if this all happens in a single cycle, it implies that both inputs and the output have their own buses (since only one operand can travel along one bus at any time). One bus will convey the content of R1, one will convey the content of R2, and another bus conveys the result from the ALU back to register R0.

Taking the two observations together implies that all registers connect to all buses, and there are at least three main internal buses.

The arrangement concerning registers and ALU that we can deduce from a brief examination of the instruction set is shown in Figure 4.1. This is actually a simplified schematic of the ARM processor internal interconnection arrangement. The arrows indicate controllable tristate buffers, acting as gates controlling read and write access between the registers and the buses. Control logic (described in Section 3.2.4) is not shown.

4.1.2 Split Interconnection Arrangements

The ARM is justly famed for its regularity and simplicity. Some other processors are less friendly to low-level programmers: where the ARM has a bank of 16 identical registers with identical connectivity,[1] it is more usual to assign special meanings to sets of registers. One common arrangement is to dedicate several *address registers* to holding and handling addresses, whereas the remainder are *data registers*. It is easy where there is such a split to imagine an internal address bus that only connects to those registers dedicated to handling addresses. In the ARM, where every register can hold an address (since it uses indirect addressing, explained in Section 3.3.4), every register must also have connectivity to the internal address bus.

In some processors, such as the ADSP21xx, there is no bank of registers—there are instead specific registers associated with the input and output of each processing element. This means that when using a particular instruction, the low-level programmer has to remember (or look up) which registers are allowed. Sometimes an instruction

---

[1] In fact, registers R14 and R15 are the link register and program counter, respectively. These require internal connections that other registers will lack which are not really evident through examining the instruction set. Registers also vary in their shadowing arrangements.

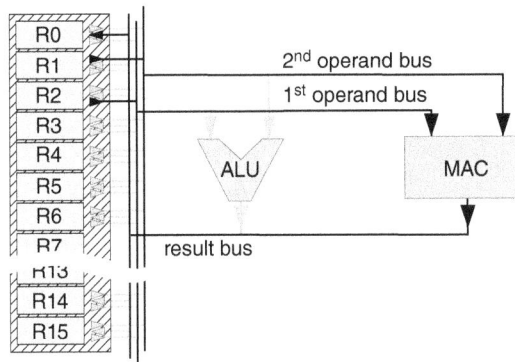

FIGURE 4.2 A schematic diagram of an ALU, a MAC, and a bank of registers interconnected with a three-bus arrangement. The ability to convey two operands simultaneously to a single functional unit is highlighted.

has to be wasted to switch a value from one register to another to perform a particular function—although clever instruction set design means that these inefficiencies are quite rare. These days, such architectures are uncommon among general-purpose processors, but still found in some digital signal processors (DSPs) such as the ADSP21xx.[2]

So why would designers go to such trouble, and complicate the instruction set? The answer requires us to take a snapshot of the internals of a processor as it performs some function. In this case, we will look at the ARM as it performs two instructions:

```
MUL R0, R1, R2   ; R0=R1 + R2
ADD R4, R5, R6   ; R4=R5 + R6
```

using hardware which is shown diagrammatically in Figure 4.2.

The snapshot of time represented in Figure 4.2 shows data being output from R1 and R2 simultaneously on the two operand buses (dark color), flowing into the multiply-accumulate unit (MAC), and the result flowing over the results bus back into register R0.

The thing to notice during this snapshot is that registers R3 onward, and the ALU are all sitting idle. When CPU designers see resources sitting idle, they tend to wonder if it is possible to utilize them—in this instance, to see if there is a way of using the ALU and the MAC simultaneously. One answer is to partition the design as shown in Figure 4.3.

In the arrangement shown, both the MAC and the ALU have their own buses—both input and result, and by extension, their own set of preferred registers. Thus as long as the programmer remembers to use R0 to R3 when dealing with the MAC, and R4 to R7 when dealing with the ALU, both of the example instructions:

```
MUL R0, R1, R2   ; R0=R1 + R2
ADD R4, R5, R6   ; R4=R5 + R6
```

can be performed simultaneously in a single cycle.

---

[2] The "xx" means that there are various serial numbers in the ADSP21 family which share these characteristics, such as the ADSP2181, ADSP2191, and so on.

FIGURE 4.3   A schematic diagram of an ALU, a MAC, and a bank of registers interconnected with a three-bus arrangement. This is similar in resource use to Figure 4.2 although in this case bus partitioning has been used to allow the two functional units to transfer their operands simultaneously.

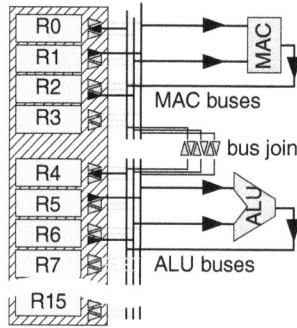

This process is probably the underlying thinking below the design of the ADSP21xx hardware, squeezed by designers for every last performance gain.

4.1.3   ADSP21xx Bus Arrangement

In the ADSP21xx, every processing element is limited to receiving its input from only a few registers, and outputting a result to another small set. This means there are many internal buses, but means that many operations can be performed very quickly in parallel.

A simplified diagram of some of the many internal buses within the ADSP21xx is shown in Figure 4.4. In this figure, PMA is program memory address, DMA is data memory address, both being address buses that index into the two blocks of memory

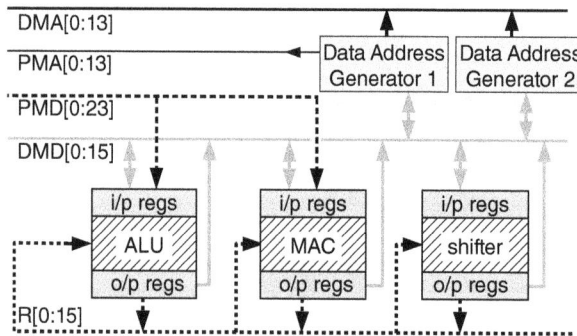

FIGURE 4.4   A simplified diagram of the ADSP internal bus arrangement.

(program and data) which also indicate that this is basically a Harvard architecture processor (see Section 2.1.2), but actually goes a step beyond this in its partitioning of address spaces. PMD and DMD are program and data memory data buses, respectively. Note the bus sizes: not only does this ADSP have a complex internal bus interconnection arrangement, but the bus widths, and widths of the interconnects differ.

The diagram shows that the ALU and the MAC, but not the shifter, can receive input operands from the 24-bit PMD bus, but all can receive input, and output, from the 16-bit DMD bus.

4.1.4 Simultaneous Data and Program Memory Access

A topic that is very important in areas such as signal processing is the consideration of how fast external data can be brought into a computer, processed, and then output. Signal processors typically operate on streams of such data—whether such data is high-fidelity audio or wideband wireless signals or some other kind of information.

Signal processing operations tend to be some form of digital filter. This can be considered as a time series of samples, $x[0]$, $x[1]$, $x[2]$, and so on being the input values at time instant 0 (which we can think of as "now"), one sample previously, and two samples previously, respectively. $y[0]$, $y[1]$, $y[2]$ are the output values at those corresponding times. If this were audio data, then x and y would be audio samples, probably 16 bits, and if they were sampled at 48 kHz, the time instants would each be $1/48,000 = 21\,\mu s$ apart.

Without delving too deeply into digital signal processing (DSP), we can say there are two general filter equations: the finite impulse response (FIR) filter and the infinite impulse response (IIR) filter. FIR outputs are obtained by multiplying each of the previous n samples by some predetermined values and then adding them up. Mathematically this is written

$$y[0] = \sum_{i=0}^{n-1} a[i] \times x[i]$$

So the current output $y[0]$ depends on n previous input values multiplied by the filter coefficients $a[\,]$ then all are summed together. The number of previous values defines the *order* of the filter. A 10th-order filter would be defined by setting $n = 10$ and predetermining ten $a[\,]$ values. An *adaptive FIR filter* would be one in which the $a[\,]$ values are changed from time to time.

The IIR filter, by contrast, makes the output value dependent upon all previous outputs as well as previous inputs:

$$y[0] = \sum_{i=0}^{n-1} a[i] \times x[i] + \sum_{i=1}^{n-1} b[i] \times y[i]$$

This includes the use of a further set of filter coefficients, $b[\,]$. IIR filters can also be adaptive and are generally able to perform the same work as FIR filters but with a lower *order* (which means a smaller value of n). This strong filtering action comes at a price, and that is mainly observed by IIR filters becoming unstable.

The art of designing high-performance DSP processors is to make these equations able to operate as quickly as possible, with the goal of being able to calculate a value $y[0]$ in

FIGURE 4.5 A block diagram of Harvard architecture internal memory access in a DSP augmented by the ability to add external shared memory.

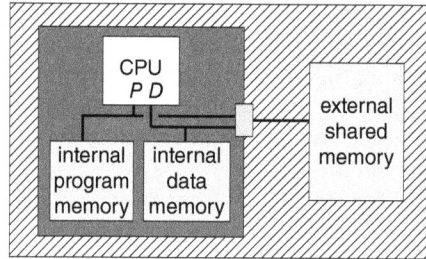

as few clock cycles as possible. Looking back at the equation for the FIR filter, we can see that most of the work is done by the following low-level operation:

```
ACC:= ACC + (a[i] × x[i])
```

The act of multiplying two values and adding to something already there is called multiply-accumulate. This requires an accumulator, usually abbreviated to "ACC."

Now we need to relate that calculation to the hardware of a DSP. There are many subtleties that could be discussed here, and using this operation, but in this case the most important aspect is the memory access arrangements.

Consider the block diagram in Figure 4.5 showing a DSP containing a CPU and two memory blocks, plus a block of external shared memory. The device seems to have an internal Harvard architecture (separate program and data memory and buses), but connects externally to a block of shared memory: a type of arrangement that is very common, with the internal memory being static RAM (SRAM), and sometimes having SDRAM (synchronous dynamic RAM) externally for the main reason that it is far less expensive than SRAM (refer to Section 7.6 for details on memory technologies and their features).

On-chip memory uses short internal buses and is generally extremely fast, sometimes able to access contents in a single cycle. Occasionally, a block of two-cycle memory is also provided: this is twice as slow as single-cycle memory since it requires two clock cycles between requesting data and it being made available.

Ignoring the memory speed for now, and referring back to the multiply accumulate calculation, we need to feed the multiplier with two values: one being a predetermined coefficient, $a[]$, and the other being an input data value $x[]$. Given a shared bus, these two values cannot be obtained/transferred simultaneously. However, given the internal split buses in the diagram, they can both be fetched together, and begin to be multiplied in a single cycle—if they were stored in separate on-chip memory blocks. Overall this will probably be a multi-cycle operation: one cycle to load and decode the instruction, the cycle following that to load the operands, and then one or more cycles to operate on those. However, given fast single-cycle on-chip memory, it is possible for the operand fetch to occur as part of an internal instruction cycle.

Usually anything that traverses an off-chip bus is slow compared to data following on-chip paths, and this is one major driving factor behind the use of cache memory (explored later in Section 4.4). Where the external memory device is SDRAM, there will almost always be an on-chip cache to alleviate the issue that, however fast SDRAM is, there is always a two- or three-cycle latency between requesting a single memory value and it being provided.

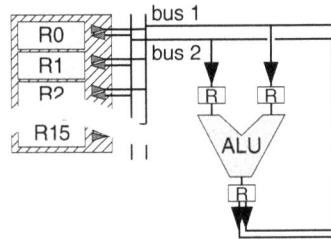

FIGURE 4.6   A dual-bus connection between an ALU and a register bank.

4.1.5   Dual-Bus Architectures

Taking a step backward, a large hardware saving is made by minimizing the number of buses: buses are bundles of parallel wires that must be routed through an IC, which cost in terms of buffers, registers, and interconnects. They are expensive in silicon area and consume prime "real estate" on chip. It is entirely possible to reduce area (and thus cost) by moving to a two-bus architecture and beyond that to a single-bus architecture (Section 4.1.6).

This is one case where our investigation does not parallel computer architecture evolution. The reason is that a three-bus architecture is actually more sensible and easier to explain than using a single bus. Tricks are required when buses are fewer—tricks that have been used in silicon before the 1980s but which nevertheless complicate the simple view of a bus as a path between the source and destination of operands and results. All examples in this and the next section are fictitious: they present something like the ARM architecture, but with different bus arrangements. Original reduced bus designs, such as the venerable 6502 processor did not have the luxury of a register bank, let alone a multiplier. Therein lies the problem: Silicon area was too limited to allow a nice architecture, or sometimes even a time-efficient architecture. In many cases it was simply sufficient that the design could be manufactured and could work. With space for only 3 general registers, the 6502 designers were never going to be able to shoehorn in another parallel bus—they would have added some more registers instead.

Figure 4.6 presents a register bank connected to an ALU using a two-bus arrangement. There are three registers, or latches shown clustered around the ALU (actually making this very similar to the 6502—ignoring the larger register bank of course).

In order for this, and the following examples to make sense, it is necessary to remember something about the ALU. That is the propagation delay timings. When we present stable electrical signals at the two input arms of the ALU, we need to wait for a certain length of time before the answer appearing at the bottom of the ALU is valid. Some control logic (not shown) would be present to instruct the ALU as to exactly what arithmetic or logic operation it should be performing, and this is assumed constant here. But the length of time we have to wait depends on the exact operation being performed—and the maximum (worst case) time is the one that determines how fast we can clock the circuitry based around this ALU. In a modern system this delay may be something like one or two nanoseconds.

That delay is accounted for, but the problem here is that there is effectively no minimum delay: What this means is that as soon as one of the input signals is removed, or changes, the result can start to become corrupted. The consequence of this is that the input operands must remain in place driving the ALU as the result is collected and stored. Only then can the input operands change.

Hence, the registers on the ALU input arms. Without at least one register there is no way for a two-bus architecture to drive an ALU with input operands and simultaneously collect the result. With one or two registers present there are several alternatives that may save on hardware slightly, but the more general is the following sequence of events performing:

```
ADD R0, R1, R2  ; R0 = R1 + R2
```

Each numbered step is at a monotonically increasing time instant.

1. Set up system, clear buses, and set ALU functionality switch to "ADD."

2. Allow register R1 to drive bus 1 (by turning on register output buffer).

3. Allow register R2 to drive bus 2 (by turning on register output buffer).

4. Latch bus 1 value into first ALU operand register;
   latch bus 2 value into second ALU operand register.

5. Turn off R1 register output buffer (bus 1 becomes free);
   turn off R2 register output buffer (bus 2 becomes free);
   wait for worst-case propagation delay through ALU.

6. Latch ALU result into ALU output buffer.

7. Allow ALU output buffer to drive one bus.

8. Latch content of that bus into register R0.

9. Turn off ALU output buffer (both buses become free and system is ready to perform the next operation).

It can be seen that the very simple ADD command actually comprises a number of steps that must be performed in hardware. These steps add up to something like eight time periods, ignoring ALU propagation delay. In a three bus design (Section 4.1.1) such an add would require only three time periods.

As an aside, the complexity of these steps even for a simple ADD instruction goes some way toward explaining the importance of a control unit inside a CPU to manage this process (Section 3.2.4). Imagine the control complexity needed for a large multi-cycle CISC instruction.

FIGURE 4.7   A single-bus connection between an ALU and a register bank.

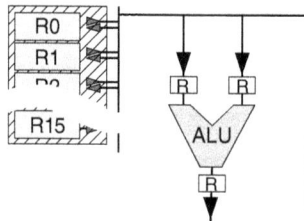

4.1.6   Single-Bus Architectures

The case of a single-bus architecture can be extrapolated from the section above. Again using a fictitious ARM-style processor as an example, the architecture may look similar to that shown in Figure 4.7.

Note the architectural simplicity of the design, which belies the operational complexity of the multistep operation of such a system. Again we consider adding R0 = R1 + R2 with each numbered step being at a monotonically increasing time instant.

1. Set up system and set ALU functionality switch to "ADD."

2. Allow register R1 to drive bus (by turning on the register output buffer).

3. Latch bus value into the first ALU operand register.

4. Turn off register output buffer for R1;
   allow register R2 to drive bus (by turning on the register output buffer).

5. Latch bus value into the second ALU operand register.

6. Turn off register output buffer for R1;
   wait for worst-case propagation delay through ALU.

7. Latch ALU result into ALU output buffer.

8. Allow ALU output buffer to drive the bus.

9. Latch content of the bus into register R0.

10. Turn off ALU output buffer (bus becomes free and system is ready to perform the next operation).

Comparing the sequence above to that for a two-bus system in Section 4.1.5, the two extra steps, and resulting reduction in efficiency are noticeable. One common improvement made historically to single bus architectures was the addition of a very short and inexpensive result feedback bus as shown Figure 4.8.

Again there are several alternative arrangements to perform this functionality, but all allow the result of an ALU calculation to be fed back to the input of one arm of the ALU. This would be useful when performing accumulation, or when following one arithmetic or logical operation after another. In this case the register on the left-hand arm of the ALU became known as the *accumulator*. Older low-level programmers came to know and love the accumulator: it would be the basis for almost every operation, the most used register

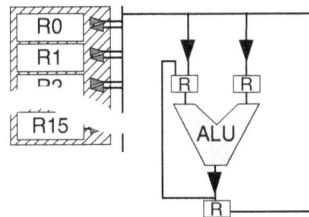

FIGURE 4.8   A single-bus connection between an ALU and a register bank as in Figure 4.7 but augmented with a single feedback link from ALU output to one of the ALU input latches.

in the entire system—the programmer's friend. Talking to many of this dying breed, I have come to learn that many, like me, mourned the death of the accumulator, killed by RISC and CISC advancements alike. A quote that sums it all up from New Zealand engineering management guru Adrian Busch is "if there's no accumulator, it ain't a real CPU."

4.2 Arithmetic Logic Unit

4.2.1 ALU Functionality

Clearly, an arithmetic logic unit (ALU) is the part of a computer capable of performing arithmetic and logical operations. But what exactly are these? An example of ALU operations defined from the instruction sets of two common processors may give some indication:

- ADSP2181: Add, subtract, increment, decrement, AND, OR, EOR, pass/clear, negate, NOT, absolute, set bit, test bit, toggle bit. There are limits on which registers can be used as input, and only two registers are available for output.

- ARM7: Add, subtract, increment, decrement, AND, OR, EOR, pass/clear, NOT. Any register can be used as input, and any register as output.

In general, the ALU performs bitwise logical operations, tests, and addition or subtraction. There may be other functions performed by the ALU that are derivatives of these, and using multiple ALU operations enables a great deal of other functions to be performed.

A basic ALU, performing addition or subtraction, can be constructed from a number of single-bit slices operating in a chain, similar (in the add/subtract case) to the carry-propagate adder of Section 2.4.2, and drawn in Figure 4.9. In this case, where control or function-select logic is not shown, eight separate single-bit ALUs operate bitwise with carry, on two input bytes to generate a result byte. The operation being performed is

```
R = ALU_op(A, B)
```

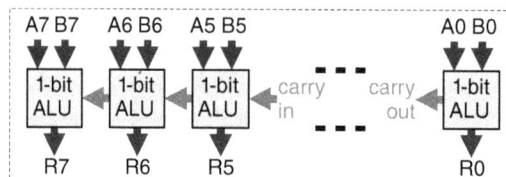

FIGURE 4.9 A block diagram of the parallel bitwise functional chain of parallel 1-bit units that comprise a word-width ALU.

Some 4-bit examples of ALU operations are given below (Appendix B presents the bitwise truth-tables for each type of gate, and their widely used logic symbols). Note that EOR, exclusive-OR, is also commonly written as XOR:

1001	AND	1110	=	1000	Bitwise AND
0011	AND	1010	=	0010	Bitwise AND
1100	OR	0001	=	1101	Bitwise OR
0001	OR	1001	=	1001	Bitwise OR
0001	ADD	0001	=	0010	Addition
0100	ADD	1000	=	1100	Addition
0111	ADD	0001	=	1000	Addition
	NOT	1001	=	0110	Negation
0101	SUB	0001	=	0100	Subtraction
0110	EOR	1100	=	1010	Exclusive-OR

From the background work in Chapter 2, we know that addition and subtraction are not entirely parallel bitwise operations. By that we mean that the nth bit result of an addition depends not only on the nth bits of each input operand, but also on all previous bits, $n, n-1, n-2 \ldots 0$. In fact, arithmetic operations between two values in general are not accomplished in a bit-parallel manner, but logical operations between two values are.

Knowing what types of functions an ALU performs in typical devices, and having explored some examples, it may now be instructive to perform a low-level design of an ALU to explore how it operates.

4.2.2 ALU Design

The block symbol traditionally used for an ALU is shown in Figure 4.10 with n-bit input operands A and B and n-bit result output indicated.

Function select is normally a bit-parallel control interface that identifies to the ALU the operation being performed. Status information includes whether the answer is positive,

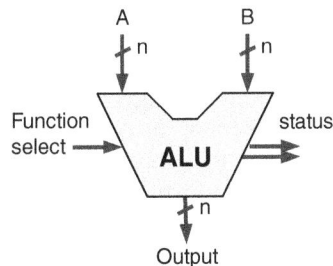

FIGURE 4.10 The block symbol normally used to represent an ALU showing n-bit operand inputs A and B, function select logic, and finally both n-bit result output and status flag output.

negative, equal to zero, includes a carry or is an overflow. In some processors, these values are abbreviated to N, Z, O,[3] and C.

Before		Operation	Afterward	
R1	R2		R0	Flags
5	5	SUB R0, R1, R2	0	Z
8	10	SUB R0, R1, R2	−2	N
Assume that the registers are 8-bit for the next two. An 8-bit register can hold 0 to 255 unsigned or −128 to 127 in two's complement signed binary.				
255	1	ADD R0, R1, R2	0	C
127	1	ADD R0, R1, R2	128 (unsigned), −128 (signed)	O, N
−1	1	ADD R0, R1, R2	0	Z, C

Remember that, for 8-bit numbers, `01111111 + 0000001` will always equal `10000000` in binary. The question is how you interpret this. The input numbers are 127 and 1, but the output is −128 if interpreted in two's complement, or +128 if interpreted as an unsigned number. Without any further information, only the programmer will know which meaning was intended.

The overflow (O) flag is intended as a help when using two's complement numbers. To the ALU there is no difference between these and unsigned numbers; however, the ALU will inform the programmer using the O status flag whenever a calculation has resulted in a potential two's complement overflow. If the programmer is dealing with unsigned numbers, it is safe to ignore this; however, when the numbers are two's complement, this has to be taken as an indication that the answer cannot be represented in this number of bits: it is too large in magnitude.

For the ALU that we will design here, we will ignore the status apart from a simple carry indication, and will perform AND, OR, and ADD only. We will consider that it is a bit-parallel ALU, and design just a single bit in the chain (since all the bits should be equal).

The resulting design, drawn in logic would look similar to the schematic representation in Figure 4.11. Box 4.1 builds upon this design to calculate the propagation delay that such a device would exhibit.

Box 4.1   Exploring ALU Propagation Delays

Let us say for the sake of argument that each logic gate has a propagation delay of 4 ns: that is the amount of time measured from when a new value is input to the gate to when the new output result stabilizes (if it does change).

Examine the ALU diagram in Figure 4.11 (ignoring the function select signals) to look for worst-case longest paths. Both inputs A and B go through two blocks of gates. The block on the top left has only two rows, but the full adder at the bottom right has the inputs flowing through four gates before reaching the output on the right. They have to go through three gates to reach C_{out}.

[3] "V" is often used to represent the overflow flag instead of "O" which might be confused with a zero.

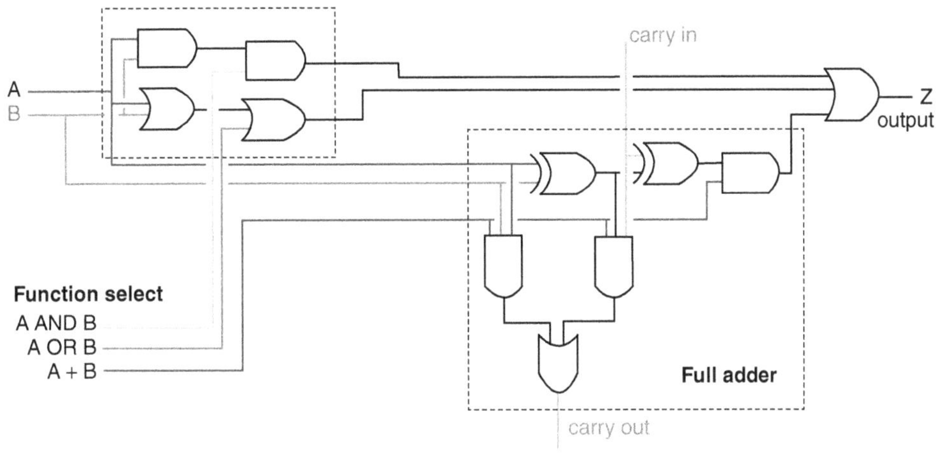

FIGURE 4.11 A schematic representation of the logic devices and connectivities within a single-bit slice of a typical ALU.

On the other hand, the carry in has to flow through two gates before it reaches the carry out, and three gates until it reaches the output, Z. This is summed up as

A/B to Z: $4 \times 4\,ns = 16\,ns$
A/B to C_{out}: $3 \times 4\,ns = 12\,ns$
C_{in} to Z: $3 \times 4\,ns = 12\,ns$
C_{in} to C_{out}: $2 \times 4\,ns = 8\,ns$

Let us use these figures to find a worst case propagation delay (and hence maximum operating speed) for a 4-bit ALU:

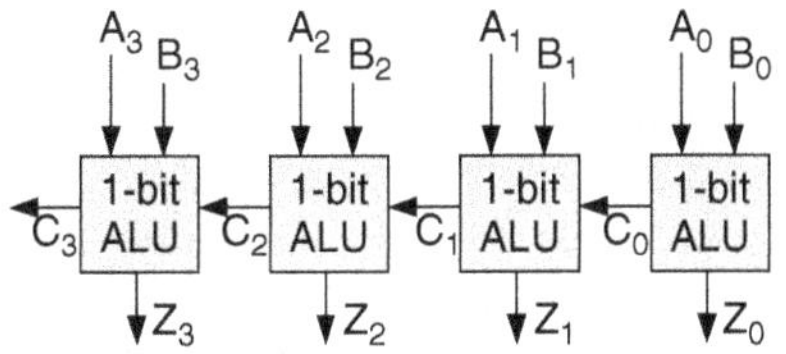

This is adding $A + B = Z$, and since it is an add, we need to account for the carry propagate.

We can now trace the worst-case propagation path which is the input at the right-hand side of the ALU, through each carry in turn, to the most significant ALU. Since the delay from any input to the Z output is more than the delay to the carry out, the worst case is thus the sum of

Bit 0: A/B in C_{out} 12 ns
Bit 1: C_{in} to C_{out} 8 ns
Bit 2: C_{in} to C_{out} 8 ns
Bit 3: C_{in} to Z 12 ns
Total: 40 ns

If this was being clocked at maximum rate, the clock period could not exceed 40 ns to ensure that a correct and final output was generated for each input. Of course, sometimes the correct output would appear much sooner than that, but there is no easy way to determine, in advance whether the output will appear quickly or slowly. It is therefore necessary to always wait for the known worst-case delay of $1/40$ ns $= 25$ MHz.

This is not a fast clock rate for a modern processor. It may therefore be necessary to either use faster gates, allow the adder to take two clock cycles to complete, or employ some tricks to speed up the adder. One such trick is the carry predictor, or look-ahead unit that was introduced in Section 2.4.3. This is quick, but can occupy a significant amount of logic when the number of bits that the adder operates on is large.

4.3   Memory Management Unit

A memory management unit (MMU) allows the physical memory available to a computer to be organized in a different logical arrangement as far as the CPU is concerned. The hardware resides between CPU and main memory, on the memory-access bus. The logical memory arrangement is also known as *virtual memory*. This was invented at Manchester University in 1962, and is sometimes called *paging* memory.

4.3.1   The Need for Virtual Memory

Virtual memory provides the CPU with a very large space of memory that user programs can access. In reality, the physical memory is much smaller, and the current *page* of memory being used by the CPU must be loaded into the physical memory on demand. Many modern operating systems, such as Linux, rely on virtual memory.

Virtual memory allows a program (or sequence of programs) that are larger than available RAM to be executed on a computer. Of course, this could be accomplished with clever programming and access to a large memory space such as hard disc, but an MMU allows programs to be written as if memory is continuous and large, and the MMU takes care of where, exactly, the program is located, and takes care of exactly how much physical RAM there really is in a machine.

The original rationale for virtual memory was the great disparity in cost between fast expensive RAM and slow inexpensive hard disc. Using virtual memory allows a computer with lower cost smaller RAM to behave as if it were a higher cost machine with more memory, the only difference being that some memory accesses are slower.

With an active MMU, the average memory access speed will reduce as compared to pure RAM, and that is because hard disc is far slower. This is seen as an acceptable penalty to pay in order to have a large memory space.

Note that the secondary storage is not necessarily hard disc—it could be any storage media that is more spacious and slower than the main RAM—including slower flash memory.

4.3.2   MMU Operation

In modern MMU systems, unused pages are usually stored on hard disc, which is far larger than the physical memory, but much slower.

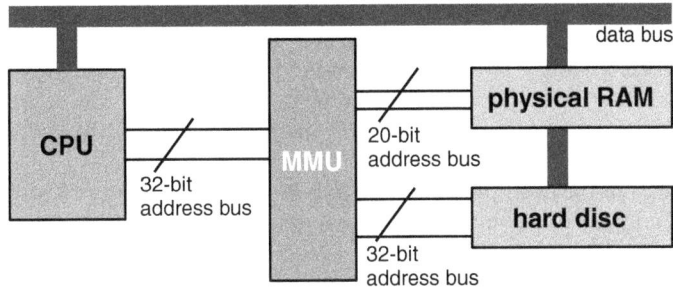

FIGURE 4.12 A memory management unit (MMU) is shown connected between a CPU and both physical memory and a hard disc. While the data bus connects these elements directly, the MMU adjusts the address bus signals "visible" to the various components.

An example of simple MMU connectivity is shown in Figure 4.12. In this figure, as far as the CPU is concerned, the system has a 32-bit address space (and can therefore address something like 2^{32} memory locations, or 4 Gibytes of memory); however, the memory in our example is only 20 bits wide (2^{20} memory locations or 1 Mibyte). The MMU hides this from the CPU.

Memory is split into *pages*. If we assume that a page is 256 Kibytes in length (a typical value), then main memory can hold 4 pages, but the CPU can access up to 16,384 pages.

The MMU loads new pages into RAM and stores unused pages to hard disc (which is big enough to hold all of the logical memory). If the CPU requests a page that is not loaded, then the MMU first retires an unused page from RAM (stores it back to hard disc), and then loads in the requested page from hard disc.

To know which page to retire, the MMU needs to track which pages are being used, and ideally chooses an unused page for retirement. This is a similar idea to what happens in memory caching (described later in Section 4.4). Two look-up tables are used to keep track of what is currently in RAM and what is currently on hard disc, and these are known as physical RAM contents table and disc memory contents table, respectively.

Within the MMU, if the CPU requests lookup of a memory location that resides on a page that is already in RAM, this is known as a hit. If the page containing that memory location is not already in RAM, this is a page fault or miss. This operation can be seen in Figure 4.13 (also refer to the worked example in Box 4.2).

Box 4.2 MMU Worked Example

page no.	loaded?	@RAM address
16383	✗	n/a
16382	✗	n/a
1	✔	0×0100
0	✔	0×0000

The physical RAM contents table in a simple CPU probably looks similar to the figure shown at the bottom of the previous page. In this case, there is a line in the table corresponding to every logical page address in the computer. A parameter indicates which of these pages is currently loaded into RAM and, if so, at what RAM address.

Notice in the example table that page 0 is at RAM address 0 and page 1 is at RAM address 0x0100. Now we know that the pages can be placed anywhere within RAM, but in this case we can see that the page size may be 0x0100 locations (256). This corresponds to 8 bits of the address bus, and would allow the 8-bit line number to be anything between 0 and 255.

We can also see that there are 16,384 pages: we would need 14 bits to represent this many pages. This gives us an indication of the memory size on the CPU: $14 + 8 = 22$ bits. Eight bits of the address represent the line number and the remaining 14 bits the page number. With 22 bits there will be $2^{22} = 4$ Mibytes of memory (assuming each location is a byte). We can confirm that since $16,384 \times 256 = 4,194,304$ as expected.

Note: This also tells us that in such a computer the conversion from CPU logical address to line and page number is simple: the bottom 8 bits are the page number, and the top 14 bits are the page number.

The sequence of events needed when a CPU is requesting a read from memory location X is shown below:

1. CPU placed address X on the address bus, then asserts a read signal.

2. MMU signals CPU to wait while it retrieves the contents of address X.

3. MMU splits address X into page number, and line number within that page.

4. MMU interrogates physical RAM contents table.

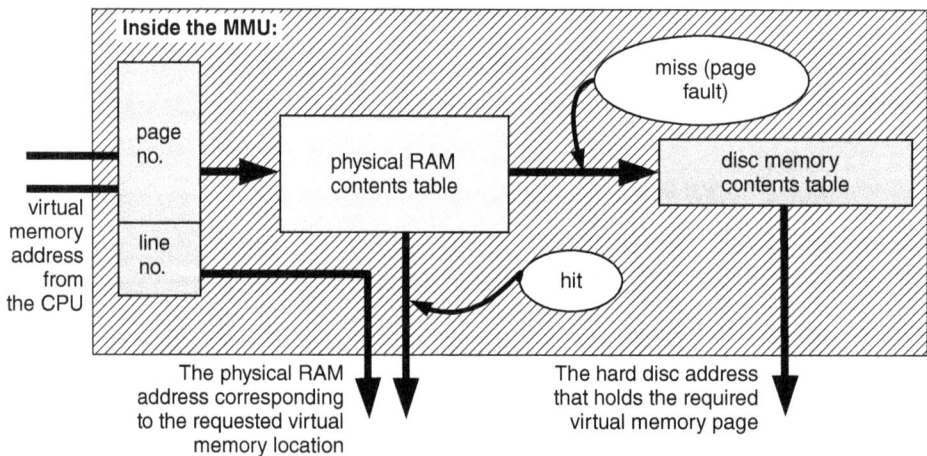

FIGURE 4.13 A block diagram of the internal units and connectivity between them for a simple MMU, showing the consequence of a miss and a hit, respectively.

- If the required page is loaded (a hit), this block outputs the physical RAM address of that block. The physical RAM address, combined with the line number within the block, forms the address in physical RAM to be retrieved.

- If the required page was not loaded (page fault), then the page number is passed to the disc memory contents table. This looks up the hard disc address of that page, and then loads the entire page into RAM. Since the page is now in RAM, the contents of address X are not retrieved in the same way as for a page hit.

- Note that since physical RAM is not infinite in size, there must be a process to retire pages back into hard disc, and indeed a process tracks the usage of pages to know which page is to be retired.

5. The MMU outputs the contents of memory location X on the data bus and signals to the CPU that the data is ready.

The CPU clearly must wait for a longer time to retrieve a value from memory when a page fault occurs. Hard disc may be hundreds of times slower than RAM, and the lookup process itself may be relatively slow despite manufacturers' best efforts to create a fast system. This wait is sometimes called a *stall time*.

It should be noted that some critical programs (e.g. those responding to interrupts) cannot want to wait for the retrieval time involved in page faults. In that case, the variables or programs that are speed-critical are placed into a special page that is locked into physical RAM, and in fact page attributes allow advanced MMUs to handle pages in several ways. Most modern operating systems locate interrupt service routines and low-level scheduling code in such locked pages.

The method of storing pages of memory on slow hard disc for use later, and loading them into RAM as required, seems a logical method of allowing users to experience a larger memory than they actually have available. However, the difficulties lie in actually implementing such as system: What methods are used to indicate which page gets retired when a new one needs loading, and how big should the pages be? The next two sections consider these problems.

4.3.3   Retirement Algorithms

If a new page is loaded from hard disc to physical RAM, unless RAM happens to be empty, space has to be made by saving one of the pages that is already loaded, back to hard disc (and then updating the physical RAM contents table).

Different algorithms can be used to decide which page is *retired* back to the hard disc:

- LRU or least recently used, where the LRU page is retired.
- FIFO or first-in first-out, where the oldest loaded page is retired.

Both algorithms have their advantages and disadvantages. Users of Microsoft Windows operating systems on older machines may be familiar with disc thrashing—the process whereby the hard disc seems to be continually operating. This is said to be due to the choice of a particularly bad algorithm for retiring pages. Consider a program loop that is so large its code is spread across multiple pages. In this case, just moving from the bottom of the loop back to the top of the loop may result in a page fault if, in the meantime, the page holding the top of the loop had been retired.

Worst case is a large program with variables scattered across many pages. If a short piece of code writes single values to each of those variables, then the pages containing them will have to be in RAM, maybe having to be loaded in specially, just for a single write. In this case, the compiler and operating system have failed to optimize the program by clustering memory locations.

The problem of retirement is similar to that faced by the memory cache, which will be discussed in Section 4.4.

4.3.4 Internal Fragmentation and Segmentation

Inefficiency results if an entire page needs to be reserved for a single memory location within that page. Or worse if a program is slightly larger than one page so that just a few lines of code are stored on an otherwise empty page such that the program takes up two memory pages but is actually more like one page long.

In both cases, the precious fast RAM of the computer will be made to contain unused spaces. Furthermore, the long and slow process of retiring pages and loading new ones into RAM will be performed each time, for mostly meaningless data. This is termed internal fragmentation.

One response to internal fragmentation has been to reduce the size of pages. However, that makes the LUTs in the MMU large, and eventually causes the look-up process itself to become a bottleneck to MMU operation.

A more recent response has been to introduce memory segments—variable length pages, but also pages that are able to grow and (in some cases) shrink on demand during program execution. A C language program may use one segment for local function variables and one for global variables. Another segment could contain the program stack. Although the C programmer need not be concerned with low-level details, the underlying operations would be to access variables by segment number and location within that segment (line). This is called a two-dimensional memory.

One advantage of such segmentation is that segments can be protected from each other. Program memory segments may be executable whereas data memory segments are not, such that erroneous attempts to branch into data memory would result in an error (rather than the total machine crash common of older operating systems and computers). Similarly, a rogue program storing variables to an incorrect location would not be allowed to overwrite the memory of another application.

4.3.5 External Fragmentation

Segmented memory spaces are more complicated because they need routines to keep track of both the size and the contents of each segment in addition to the various location contents tables. However, they are more efficient than the original paged systems because they do not suffer from internal fragmentation in the way mentioned in Section 4.3.4.

Unfortunately, they suffer from *external* fragmentation instead as shown in Figure 4.14. Working from left to right, (1) an original program is loaded, occupying four segments in memory; (2) the operating system wants to access some new memory in segment 5 so it retires a segment (in this case it chooses segment 3); (3) segment 5 is then loaded.

In (4), segment 1 is retired to hard disc and in (5) the operating system wishes to access segment 3 and thus has to reload it.

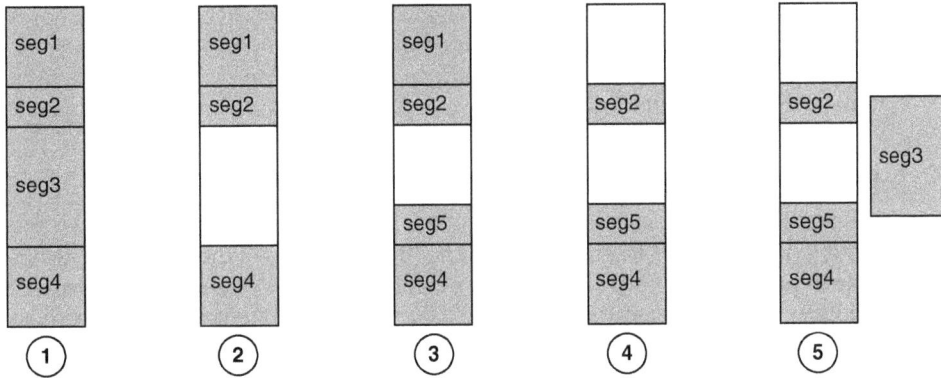

FIGURE 4.14   An illustration of external fragmentation: five steps in memory segment loading and unloading result in a memory map having sufficient free space but insufficient contiguous free space to reload segment seg3.

At this point, there is clearly sufficient empty space in RAM for segment 3, but it is not continuous empty space. There are two responses possible. One is to split segment 3 into two parts and load wherever it can be fitted in, and the second is to tidy up memory and then load segment 3. The first response would work in this instance, but could quite quickly become very complex, and would in time actually contribute to the problem because there will be more and more smaller and smaller segment parts. For that reason, the second response is usually used. The tidying process is called compaction, and it is performed before loading segment 3, as illustrated in Figure 4.15.

Since compaction obviously takes some time, it should be performed only when necessary.

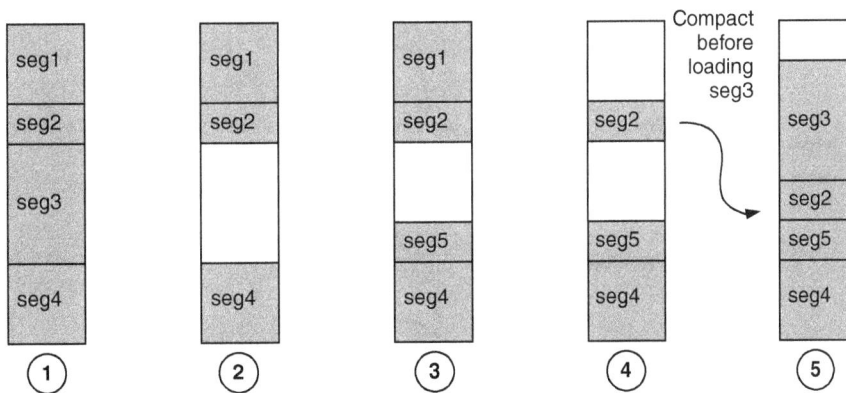

FIGURE 4.15   The same series of memory operations are performed as in Figure 4.14 but in this case a compaction operation before reloading seg3 allows reorganization of memory contents sufficient to allow the segment to be loaded.

There are a profusion of segment management algorithms available as this has been an active research field for many years. Common among them is the need to track used and unused portions of memory, and the ability in some way to perform compaction. Some of the simpler algorithms default to always performing compaction if a gap appears.

4.3.6 Advanced MMUs

The MMU hardware shown in Section 4.3.2 works well for fixed page sizes, but what about with segmented memory? Remember that the speed of the physical RAM contents table in particular is very important to overall memory access speed—all requested locations must be searched for in this block. For segmented memory it is not sufficient any longer to simply divide the address bus into two and consider the bottom few bits to be line and the top few bits to be page, because now the pages have different sizes. This means that the contents table becomes a complex contents-addressable LUT.

Such LUTs have lookup time proportional to size, and so the bigger the table gets, the slower it is. The problem is that, in order to reduce external fragmentation, the system needs to cope with some fairly small segment/page sizes. Consider the example of the UltraSPARC II. This supports up to 2200 Gibytes of RAM, but has a minimum page size of 8 Kibytes. This means worst case that there could be 200,000 pages in the system. A LUT capable of storing information on each of these pages would be very slow: It would mean that all memory accesses, in physical RAM or not, would be considerably slowed down by the lookup process.

The solution is to introduce a small, fast LUT for commonly used pages and store the less commonly used pages in a slower LUT (or RAM). This is effectively caching the contents table and is termed a translation look-aside buffer (TLB). It has other names such as translation buffer (TB), directory look-aside table (DLT), and address translation cache (ATC). It is shown in Figure 4.16.

In general, the TLB technique is relatively expensive in terms of the additional hardware and memory required compared to the performance improvement achievable, and thus tends to be reserved for processors used in speed-critical applications.

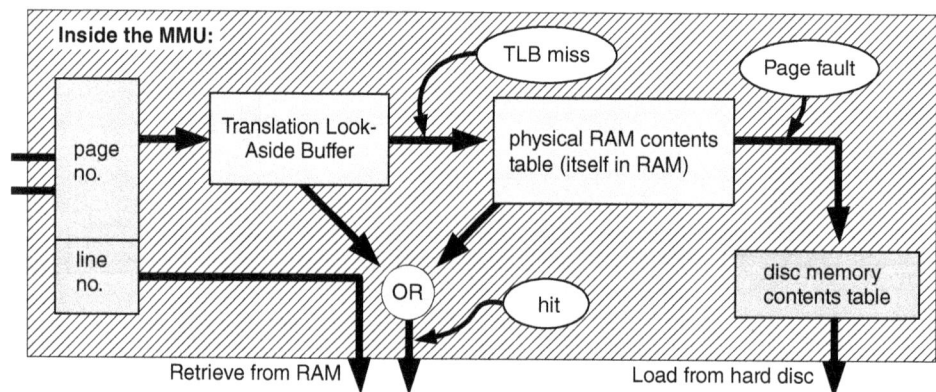

FIGURE 4.16   A block diagram of MMU operation using a TLB. Contrast this to the non-TLB case in Figure 4.13.

4.3.7 Memory Protection

There are some remaining benefits that an MMU can provide the system designer beyond the ability to swap pages into and out of physical memory and store them on hard disc. Actually the price of RAM has dropped year by year to the point where very few software applications require more RAM than can be fitted inexpensively into a desktop computer (despite the best efforts of code-bloating software writers). For embedded processors too, the MMU is often still present even when there is no physical offline storage space, such as hard disc. The question is why do system designers persist in building MMUs when the original purpose of these has largely disappeared?

The main reason is memory protection. Since the MMU sits between main memory and the processor, it is capable of scanning and modifying addresses very quickly without processor intervention. The MMU is capable of raising a signal to the processor alerting it of any problems (such as "you have asked for an address that doesn't exist"). In the ARM, for example, this would be through an interrupt signal called *data abort* for a data fetch, or *prefetch abort* if it happened when requesting an instruction. Special low-level handlers would be written as part of an operating system to (attempt to) deal with these if and when they occur.

Looking at the issue from a software perspective, the system programmer can set up the MMU to restrict access to various portions of memory, or flag other portions of memory as allocated or not allocated. Compiled code usually has a number of program and data areas—program areas are not usually writable, but data areas are. When applied to an MMU, a given program area that is currently being executed from will have a certain set of other memory areas it can read from and write to.

In most modern operating systems, user code does not have indiscriminate access to write to every memory location—it can only write to its own allocated memory areas. This prevents, for example, a mistake in user code from corrupting the operating system and crashing the computer.

Non-OS code cannot write to system control registers and cannot overwrite the data areas allocated to other programs. This is vital to system security and reliability.

One of the most important traps is to protect the memory area at address zero. Several very common coding mistakes (see Box 4.3) result in reads from, or writes to, address zero. In Linux, a compiled C language program that attempts to do this will exit with a *segmentation fault* error.

Box 4.3   Trapping Software Errors in the C Programming Language

For readers familiar with the C programming language, it is worth knowing that compilers will usually initialize newly defined variables to zero. This helpfully allows easy trapping of several errors occurring at zero:

```
int *p;
int x;
x=*p; //since p is set to NULL (0), a read from here will
    trigger a data abort
```

Defining a block of memory with library function `malloc()` will fail if, for example, there is insufficient memory space left to claim. On failure, malloc will return NULL, and NULL has the numerical value 0.

```
void *ptr=malloc(16384);
//we forgot to check the return address to see if malloc
    failed
*ptr=20; //if ptr holds NULL (0), this will trigger a data
    abort
```

Similarly, there is the issue of calling a function which has a runtime allocation

```
boot_now()
{
    void (*theKernel)(int zero, int arch);
    ...

    ...
    printf("Launching␣kernel\n");
    theKernel(0, 9);
}
```

In this code (taken from an embedded system bootloader, see Section 9.5), the function `theKernel()` is defined in the first line, and should point to a memory address where the OS kernel has been loaded; however, the programmer has forgotten to add this in. By default it will thus be set to zero. Launching the kernel will jump the code execution to address zero, resulting in a prefetch abort.

As an aside, note that the values 0 and 9 passed to the function (for an ARM) are simply going to be stored into registers `R0` and `R1` before the branch occurs. If the kernel does reside at the address specified and is embedded Linux, it would execute—decompressing itself, and then set up the system based on these values that it finds in `R0` and `R1`.

4.4   Cache

Cache memory is close to the CPU and has very fast access speed, but is usually expensive. If cost were not an issue, computer designers would employ only fast memory in their systems. As it is, this would be uneconomical for all but the most expensive supercomputers.

Cache fits into the memory hierarchy shown in Section 3.2.2, where memory near the top of the hierarchy is fastest, smallest and most expensive, and memory toward the bottom is slowest, largest (in storage terms), and cheapest.

Cache attempts to increase average access speed for memory accesses in contrast to the MMU which allows a larger memory space to be accessed, but in so doing, it actually reduces average access speed. Unlike the MMU, a cache does not require any operating system intervention, but like the MMU, it is transparent to the applications programmer.

There need not only be a single cache—there can in fact be different levels of cache operating at different speeds. The highest-level caches (close to the CPU) are usually implemented as fast on-chip memory. These tend to be small (8k for some ARMs and the 80486), and the size tends to increase as the caches approach main RAM. A good illustration of the concept of a cache in a real (but now outdated) system is in the Pentium Pro processor, described in Box 4.4.

Box 4.4   Cache Example: The Intel Pentium Pro

Intel's Pentium Pro was innovative in its day, packaged with a 256-Kibyte cache in the same chip package as the CPU but on separate silicon. Unfortunately this approach, shown diagrammatically below, was found to be unreliable and ultimately led to the discontinuation of the Pentium Pro as a product line.

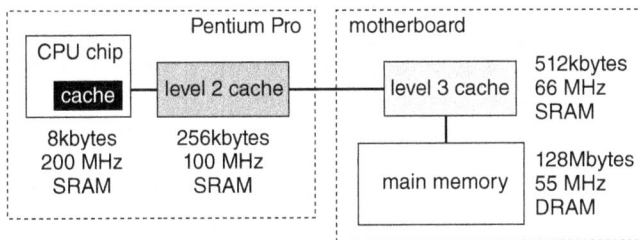

```
           Pentium Pro          motherboard
  ┌─────────────────────────┐  ┌──────────────────────────┐
  │ ┌──────────┐            │  │                 512kbytes │
  │ │ CPU chip │            │  │ ┌─────────────┐ 66 MHz    │
  │ │ ┌──────┐ │ ┌────────┐ │  │ │level 3 cache│ SRAM      │
  │ │ │cache │─┼─│level 2 │─┼──┼─│             │           │
  │ │ └──────┘ │ │ cache  │ │  │ └──────┬──────┘           │
  │ └──────────┘ └────────┘ │  │        │        128Mbytes │
  │  8kbytes      256kbytes │  │ ┌──────┴──────┐ 55 MHz    │
  │  200 MHz      100 MHz   │  │ │ main memory │ DRAM      │
  │  SRAM         SRAM      │  │ └─────────────┘           │
  └─────────────────────────┘  └──────────────────────────┘
```

In the diagram it can be seen that the relatively fast CPU has a small amount of level 1 cache (8 Kibytes) built in. Level 2 cache is in the same package, roughly half the speed but 32 times as large. Level 3 cache is fast SRAM located on the motherboard, slower still and larger than the L2 cache. Finally, main memory capacity is huge by comparison, but significantly slower. It is implemented in DRAM (dynamic RAM), a low-cost high-density technology that is typically much slower than SRAM.

Note: Today, cache systems will still look quite similar but there may be extra zeros on each of the RAM sizes, and perhaps even another level of cache. Main memory would have transitioned through SDRAM (synchronous DRAM) to RDRAM (Rambus), or DDR (double data rate) RAM or beyond (see Section 7.6).

Split caches can be used separately for data and instructions, necessary for caching in Harvard architecture processors (those that have separate memory for data and program storage, see Section 2.1.2), but often also advantageous for von Neumann architecture processors too. For example, the innovative DEC StrongARM processors (long since replaced by Intel XScale ARM-based architecture) were ARM based and therefore had an internal von Neumann architecture; however, they used a Harvard architecture cache. This allowed the two cache parts to be optimized for different behavior: program memory accesses tend to be sequential in nature whereas data memory accesses tend to jump among clusters of locations, and different caching schemes and architectures suit each behavior differently.

Similarly to virtual memory, a cache miss is when the required data is not in the cache and has to be fetched from slower memory. As before, some data has to be retired first, and possibly some compaction takes place.

The hit ratio is the proportion of requested locations that can be found in the cache, and is therefore the primary measure of cache performance. This can be maximized by good cache organization and an efficient caching algorithm, based on the use of the cache.

There are a number different forms of cache organization that significantly affect the cost and performance of the cache. Three of the more common ones, the *direct cache, set-associative cache*, and *full associative cache* are outlined in the following sections.

Note that in modern CPUs, caches actually read blocks of memory, perhaps 32 or 64 bytes at a time, rather than single-memory locations. For simplicity, most of the examples given in this section will consider that a cache entry holds only a single memory location. In the more realistic case of memory blocks, the tag address in cache is the start address of the block, and the cache controller knows that m consecutive memory locations must be cached in one cache line. The advantage of caching blocks in this way is that modern memory such as SDRAM or RDRAM is much more efficient when asked to load or save blocks of sequential memory locations than it is when handling multiple single addresses.

4.4.1   Direct Cache

In this scheme, each cache location can hold one line of data from memory. Each memory address corresponds to a fixed cache location, and as the cache is much smaller than the memory, each cache location corresponds in turn to many memory locations.

Therefore, when the direct cache is requested to return a particular memory address content, it only needs to check in one cache location for the correct tag. The cache location is taken from the lowest n bits of the memory address (assuming the cache and memory widths are equal) such as the 32-bit example below:

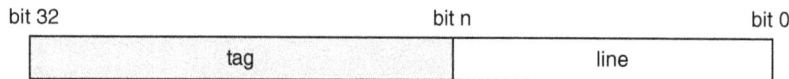

bit 32	bit n	bit 0
tag	line	

The split between tag and line is conceptually similar to the page and line split in the MMU (Section 4.3). The number of locations in the direct cache is equal to the number of lines. Every page (tag) has the same number of lines, so if a value from one page (tag) is cached, it is placed in cache at the location specified by the line.

Each cache location actually contains a number of fields: a dirty/clean flag indicates if the cache value has been updated (but not yet stored in main memory). A valid bit indicates if the location is occupied. A tag entry indicates which of the possible memory pages is actually being cached in that line. Finally, the cache stores the data word that it is caching from RAM.

The direct cache algorithm is as follows:

CPU reads from memory: Split the required address into TAG and LINE. Check the cache at the LINE location and see if the TAG entry matches the requested one. If it does, read the value from the cache. If the TAGs do not match then look at the dirty flag. If this is set, first store the current cache entry on that line back to main memory. Then read the main memory value at the required address into that cache line. Clear the dirty flag, set the valid flag, and update the TAG entry.

CPU writes to memory: There is a choice depending on exactly how the cache is set up to operate:

- *Write through* writes the value into the cache line (first storing any dirty entry that was already there), and also writes the value into main memory.

- *Write back* does not store into main memory (this will only happen next time another memory location needs to use the same line), just stores to cache.

- *Write deferred* allows the write into the cache, and sometime later (presumably when there is time available and the CPU is not waiting, the cache line is written back to main memory).

Whenever the cache value is written to main memory, the dirty flag is cleared to indicate that the main memory value and cache value are the same, called cache-memory coherence.

With the *write through* scheme, if the memory location being written to is not already in the cache, it is possible to directly store the data to memory, bypassing the cache. This is called *write-through with no write-allocate* (WTNA). Where the value is *always* stored to cache irrespective of being written to memory as well, it is termed "write-through with write-allocate."

The main advantage of the direct cache is its lookup speed. For every memory address in main RAM, there is only a single location in cache that needs to be interrogated to decide whether that address is being cached. Unfortunately, this very advantage is also a problem—every cache line corresponds to many real memory locations. Box 4.5 presents an example of the direct cache access.

Box 4.5 Direct Cache Example

The diagram below represents a direct cache currently in use within a simple microcomputer system.

	valid	dirty	tag	data
line 1023	✔	☹	0000	0000 2001
line 1022	✔	☺	0001	FFFF FFF1
line 2	✔	☺	0100	0000 0051
line 1	✗	☺	XXXX	XXXX XXXX
line 0	✔	☹	0000	1A23 2351

The cache has 1024 lines (corresponding to 10 bits of the address bus), and each line stores two flags, a tag entry, and the actual cached data. The smiley characters indicate dirty (sad) and clean (happy) entries, respectively.

On system start-up, all entries are clean but invalid, like line 1. This probably means that line 1 has not been used in this cache since the system last reset.

Line 0 on the other hand is valid, so it must be caching real data. It is dirty, so the data must have changed recently and the new data not yet written back to main RAM. With a tag of 0, line 0 must hold the cached value for CPU address 0, and the latest content for that location is the 32-bit value 0x1A23 2351.

Since there are 1023 lines in cache, line 0 could have been caching addresses 0x400 (1024), 0x800, 0xC00 instead.

Line 2 is also valid but clean, meaning that the data it holds is the same as the data in RAM that it is caching. The location it is caching is line 2 from page (tag) 0x100. Since the line indicates the bottom 10 bits of the address bus, the actual address being cached in that line is (0x100 << 10) + 2 = 0x40002, and the data there (also that in main RAM currently) is 0x51.

Finally, line 1023 is valid but dirty, meaning that the data it holds has been changed since the last write to RAM. With a tag of 0, this is caching address location (0x0 << 10) + 1023 = 0x003FF.

4.4.2 Set-Associative Cache

The problem with the direct cache was that address locations 0, 1024, 2048, 3072..., etc. all compete for one cache line. If we run software that happens to use addresses 0, 1024, and 2048 then only one of these can be cached at any one time.

To improve on this, an n-*way set-associative cache* allows *n* entries to each line. In some ways it looks like *n* banks of direct cache operating in parallel.

In a two-way set-associative cache, there are two possible locations that can cache any main memory address (this type of cache is illustrated through an example in Box 4.6). When reading from such a cache, the process can still be quick—equivalent to interrogating two LUTs (and the interrogation can be performed in parallel). This technique is commonly used, for example the original StrongARM processor from Digital Equipment Corporation contained a 32-way set-associative cache.

Box 4.6 Set-Associative Cache Example

The diagram below represents a two-way set-associative cache currently in use within a simple microcomputer system.

line	valid	dirty	tag	data	valid	dirty	tag	data
1023	✔	☹	0000	0000 2001	✔	☺	0015	0110 2409
1022	✔	☺	0001	FFFF FFF1	✔	☺	0002	0000 0003
2	✔	☺	0100	0000 0051	✗	☺	XXXX	XXXX XXXX
1	✗	☺	XXXX	XXXX XXXX	✔	☹	0006	FFF1 3060
0	✔	☹	0000	1A23 2351	✔	☺	0004	4A93 B35F

This cache bears a strong resemblance to the direct cache of Box 4.5, but with two entries for each line (being two way set-associative). The cache has 1024 lines (corresponding to 10 bits of the address bus). The smiley characters as before indicate dirty (sad) and clean (happy) entries, respectively.

On system start-up, all entries are clean but invalid, like line 1 left-hand side and line 2 right-hand side. This probably means that those entries have not been used since the system was last reset.

The difference between direct and set-associative caches can be illustrated with reference to line 0. On the left it holds the same as in the direct cache example of Section 4.4.1; however, in this case, the same line is simultaneously caching a memory location from page (tag) 4. This entry is dirty-valid, indicating the value has changed in cache and not been written back to main RAM. The cached data is the 32-bit value 0x4A93 B35F and this is the latest available content for address $(0x004 << 10) + 0 = 0x1000$.

As with all caches, values may need to be retired before a new location is cached. The question is which way of the n-ways is chosen for retirement? This can be seen to be similar to the choice given in the MMU case, and again there are a choice of algorithms for retirement, covered in Section 4.4.4.

4.4.3   Full-Associative Caches

If we run software that happens to use addresses 0, 1024, and 2048 but does not use addresses 1, 1025, 2049, then direct or set-associative caches line 0 will always be busy, with cached locations being swapped in and out. Cache line 1 will by contrast always be empty.

A full-associative cache improves on this because it allows any memory location to be mapped into any cache location. In this case, the cache TAG holds the full address of its content (rather than just the page).

The problem is that when this cache is asked to retrieve a location, every cache entry TAG must be checked—in other words, every line in the cache needs to be examined. In the direct case, only one TAG needed to be checked. In the n-way set-associative cache, only n TAGs had to be checked.

So although the chances of getting a good hit/miss ratio are better with a full-associative cache, the operation of the cache itself is slower due to the increased checking required. This is a similar problem to that faced by the physical RAM contents table in an MMU.

4.4.4   Locality Principles

The storage patterns of variables being loaded and unloaded are heavily dependent on the use to which the cache is put, but in general, in a computer with a few general-purpose programs running, there are two well-defined characteristics: those for data memory and program memory. These lead on to a well-known term in computer architecture, which is the *principle of locality*. There are actually two locality principles, the first being *spatial locality*, which refers to items clustered by address. The second, *temporal locality*, refers to items clustered in time.

These can be visualized by looking at a computer memory map and coloring data variables used within the past few thousand clock cycles. If a computer is frozen during operation, there will probably be a few very well-defined clusters of highlighted memory addresses and large areas of currently unused memory. Freezing again after a few seconds would show different clustered areas of active memory. The operation of a good cache would attempt to place as much as possible of the highlighted clusters into fast cache memory and thus speed up average program execution time.

If the visualization method were applied to program memory instead, there would be some sequential blocks of highlighted memory flowing like ribbons through memory.

The principle of *spatial locality* states that at any one time, active items are probably located near to each other in terms of their address. For program memory this is due to the sequential nature of program instructions, and for data memory due to the way a compiler will cluster defined variables into the same memory segment.

The principle of *temporal locality* states that an item that has recently been accessed is more likely to be accessed again than any other locations. For program memory, this can be explained through looping constructs, whereas for data memory this may be the repetitive use of some variables throughout a program.

Both principles of locality are illustrated in Figure 4.17, where three memory pages are shown, as snapshots of memory usage, at several instants progressing through time. The density of memory usage is shown by the shading of the rectangular blocks within the pages, and the memory addresses are indicated by the position within the rectangular page. It can be seen that temporal locality results in a gradual move between different memory clusters as time progresses. Spatial locality means that memory accesses tend to cluster together. Note that variables (or stack items) stored across several pages may be active at any one time, because different types of item could reside in different pages (in particular, data and program items would be unlikely to share a memory page).

The implication of locality is that in general it is possible to predict roughly which memory locations are likely to be accessed in future. The function of a good cache is to use this information to cache those locations and therefore to increase average access speed.

4.4.5   Cache Replacement Algorithms

A replacement algorithm keeps track of locations within an operating cache. It operates when a new location is requested but the appropriate parts of cache are full, meaning that some location already in cache must be replaced by the new location. If the appropriate location in cache is "dirty" (in other words it had been written to but had not been saved back to RAM since then), then the data must be saved to RAM prior to being overwritten. By contrast, clean cache entries can be replaced straight away since they will by definition hold the same value as the cached location in RAM. Of course which is an appropriate location is a function of the cache organization: a full-associative cache will not restrict location, but direct or set-associative caches limit which line (or lines) a memory address can be cached in.

The issue remains, however, that if a line that has just been retired back to RAM is requested a short-time later, it will have to be loaded back in again, and possibly require the retirement of more data in the process, and this is a time-consuming process.

A good cache is one that minimizes the number of loads and unloads required or, put another way, maximizes the hit ratio. One way to do this is to ensure that the correct data

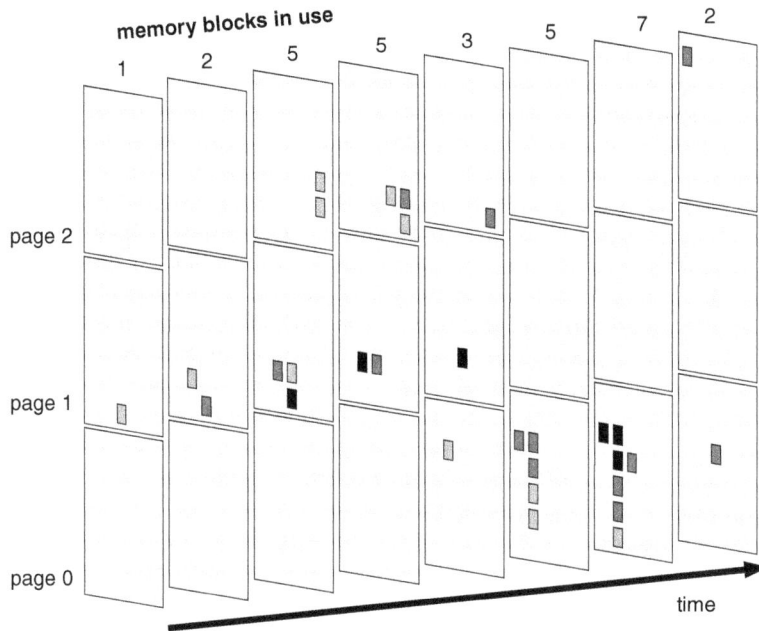

FIGURE 4.17   An illustration of both principles of locality, showing how memory use (indicated by the darkness of blocks of memory on several memory pages) changes over time. Temporal locality is illustrated by the way clusters of memory that are active at one time tend to also be in use during the next time instant, but differ from those in use at a much later time. Spatial locality is illustrated in the way clusters of active memory reside in particular areas within each page, rather than being scattered evenly. The number of blocks within the pages that are used at any one time is shown at the top of the figure.

(defined as the least useful data) is retired, and this is the job of the cache replacement algorithm. There are a few common algorithms worthy of mention:

- LRU (least recently used) scales in complexity with the size of the cache, since it needs to maintain a list of which order each entry was used in. The next item to be retired will come from the bottom of the list. LRU tends to perform reasonably well in most situations.

- FIFO (first-in first-out) replaces the location that has been longest in the cache. It is very easy to implement in hardware, since each loaded line identifier simply goes into an FIFO, and when an item needs to be needs retired, the ID at the output of the FIFO is the next one chosen. It is less effective than LRU in cases where some memory location is repeatedly used for a long time while other locations are used only for a short time.

- LFU (least frequently used) replaces the LFU location. It is more difficult to implement since each cache entry needs to have some form of counter, and circuitry to compare all the counters. However, LFU performs very well in most situations.

- Random is very easy to implement in hardware: just pick a (pseudo-) random location. Surprisingly this technique actually performs reasonably well.

- Round robin (or cyclic) will take turns retiring cache lines. It is common in *n*-way set-associative caches, where each of the *n* ways is retired in turn, its chief advantage is ease of implementation, but performance is poor for smaller caches.

Remember that caches must be *FAST*, and since these algorithms will need to keep track when any line is accessed, and be called when a replacement is needed, they need to be implemented in such a way that they do not limit the performance of the cache: A perfect replacement algorithm is no use if it slows the cache down to the same speed as main RAM. These algorithms will need to be implemented in fast hardware rather than software, and the implementation complexity is therefore an issue.

Boxes 4.7 and 4.8 present worked examples of how cache replacement algorithms operate for some simple sequences of reads and writes.

Box 4.7   Cache Replacement Algorithm Worked Example 1

Q. A computer system has cache and main memory states as shown in the diagram to the right. At reset, the cache is entirely empty but the main memory had locations filled with the values *aa*, *bb*, *cc*, and up to *ii* as shown. Each cache line can cache one memory address.

If the LRU replacement algorithm is used with a write-back system, and the cache is full associative (and filled from the bottom up), trace the actions required, and draw the final state of the cache after the following sequence of operations:

(1) Read from address 0.

(2) Read from address 1.

(3) Read from address 0.

(4) Read from address 2.

(5) Read from address 3.

(6) Read from address 4.

(7) Write 99 to address 5.

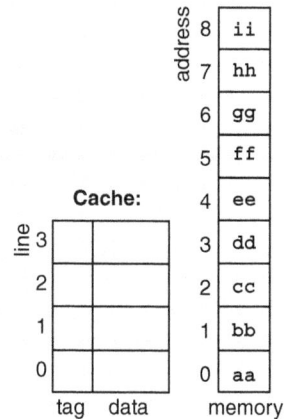

A. We will work step by step through the operations and draw the state of the cache in full after steps 5 through 7 to illustrate the actions, in the following diagram:

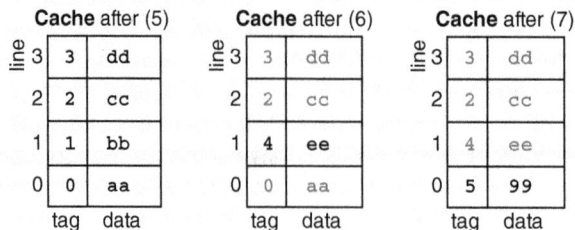

First, (1) is a miss because the cache is empty. So the value *aa* is retrieved from memory and placed in cache line 0 with tag 0 (since a full-associative cache tag is the full memory address). (2) is also a miss, and this would result in *bb* being placed in cache line 1. (3) is a hit—address 0 is already present in line 0, so no further action is required. (4) is a miss, and would result in *cc* being written to cache line 2. (5) is similarly a miss, and this would cause cache line 3 to be filled.

At this point the cache is full, so any new entry will require a retirement. Since we are using LRU, we need to take account of the last time each entry is accessed. (6) is a miss, so the value in memory location 6 must be loaded into cache. Looking back, the least recently used line was line 1 in step (2) and not line 0 in step (1) because we accessed line 0 after loading line 1, in step (3). Step (6) therefore stores the memory address 4 data, *ee*, to line 1.

Finally, step (7) involves a write from CPU to memory. Since we have a write-back system, this value must be placed in the cache too. Applying the LRU algorithm again, we see that line 0 is this time the least recently used location, and this is therefore replaced with the new data (it is not retired because we did not write to it since it was loaded).

Box 4.8   Cache Replacement Algorithm Worked Example 2

Q. A computer system has cache and main memory as shown in the diagram on the right. At reset, the cache is empty but main memory has locations filled with the values *aa, bb, cc* up to *ii* as shown. Each cache line can hold two memory addresses (in other words it is two-way set associative). If the FIFO replacement algorithm is used with a write-back system, trace the actions required, and draw the final state of the cache after the following sequence of operations:

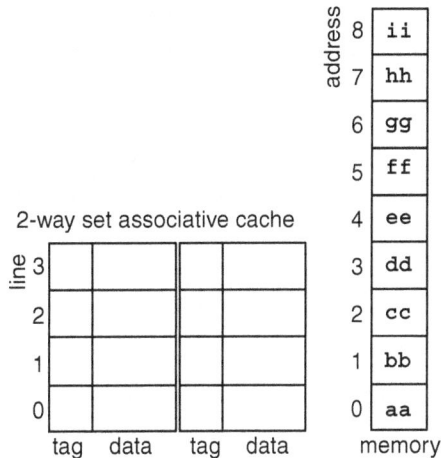

(1) Read from address 0.

(2) Read from address 1.

(3) Read from address 0.

(4) Read from address 2.

(5) Read from address 3.

(6) Read from address 4.

(7) Write 99 to address 5.

(8) Write 88 to address 8.

A. First, it is important to determine the tag range. Since the cache has four lines, memory address range {0-3} resides in tag area 0, {4-7} in tag area 1, {8-11} in tag area 2, and so on. Memory addresses 0, 4, and 8 map to line 0. Then addresses 1, 5, and 9 map to line 1, and so on.

Working step by step through the operations now, (1) will result in a miss, and cause *aa* to be loaded in to cache line 0. For the sake of readability we will fill the left-hand "way" first. (2) is also a miss, and will fill line 1. (3) is a hit and cause the value in cache line 0 left-hand side to be read out. (4) and (5) are also cache misses and will fill lines 2 and 3, respectively, with data *cc* and *dd*. At this point every line in the left-hand side of the cache has been filled. So step (6), a read miss from address 4 will cause data *ee* to be placed in cache. Address 4 maps to cache line 0, and line 0 left-hand side is full, so it will be written in the right-hand side. Note that address 4 is in tag area 1.

(7) is a write to address 5, which maps to cache line 1 with a tag of 1. We have not accessed address 5 so this is a miss, and will cause the written data *99* to be placed in the spare part of cache line 1, namely the right-hand side. The state of the cache at this point is shown in the diagram below (left).

Cache after (7)

line				
3	0	dd		
2	0	cc		
1	0	bb	1	99
0	0	aa	1	ee
	tag	data	tag	data

Cache after (8)

line				
3	0	dd		
2	0	cc		
1	0	bb	1	99
0	2	88	1	ee
	tag	data	tag	data

The final step (8) is to write *88* to address 8. Address 8 maps to cache line 0, and is in tag area 2. This must be placed in cache since a write-back scheme is in use. However, cache line 0 is full. One entry therefore needs to be retired. Applying the FIFO scheme, the "first in" must be removed. For the case of line 0, the first of the two choices to be loaded was the left-hand side, so this is replaced by the *88* (shown in the right hand diagram above).

4.4.6 Cache Performance

The time taken for a hit equates to the time taken to test for a hit (to access the cache LUT) plus the time required to retrieve the value from the cache and return to the requesting CPU. It is assumed that updating the runtime part of a replacement algorithm does not add to this timing. Since the cache is, by definition, fast, then time taken to test for a hit should be minimized.

The time taken for a miss is a little more involved. This first requires time to test for a hit (access the cache LUT), then to run the replacement algorithm, then to check for a dirty flag on the chosen line. If set, the time required to retire this value back to main RAM must be factored in. Finally, the time taken to load the requested value from main RAM to cache plus the time taken to retrieve this from cache.

Let us say that cache location M_1 has access time T_1 for a cache hit, but for a cache miss we need to transfer word M_2 from main memory into cache M_1, with transfer time T_2.

Assume that hit rate, H is the number of cache hits divided by the number of requests. Then *overall access time* T_S is given by

$$T_S = H \times T_1 + (1 - H)(T_1 + T_2) = T_1 + (1 - H)T_2$$

As T_1 is much smaller than T_2 (of course, a hit is *much* faster than a miss), a large hit ratio is required to move the total access time nearer to T_1 (in other words to try to achieve $H \approx 1$).

If C_1 is the cost per bit in the cache memory of size S_1 and C_2 is the cost per bit in main memory of size S_2 then the average cost per bit is given by

$$C_S = (C_1 S_1 + C_2 S_2)/(S_1 + S_2) = C_1 S1/(S_1 + S_2) + C_2 S_2/(S_1 + S_2)$$

Considering that $C_1 \gg C_2$, the cache has to be small, otherwise it is prohibitively expensive. Cache design is all about the three-way trade-off between cost, speed, and size (size because low-level cache normally has to fit on the same silicon die as a CPU, sharing valuable space).

Access efficiency is defined as $T_1/T_S = 1/\{1 + (1 - H)(T_2/T_1)\}$ which can be considered to be the ratio between the theoretical maximum speed-up if the hit ratio is 1.0 divided by the actual average access speed derived previously. Some typical values of access efficiency for several values of T_1/T_S with respect to hit ratio are given in Box 4.9.

Box 4.9   Access Efficiency Example

Look at some typical values of access efficiency for values of T_1/T_S against hit ratio:

	T_2/T_1		
	5	10	20
0.6	0.33	0.20	0.11
H **0.8**	0.50	0.33	0.20
0.9	0.67	0.50	0.33

These are typical figures for some real CPUs: A 75-MHz ARM7 with 16-MHz memory will have T_2/T_1 approximating to 5 and (with a good cache over fairly benign or predictable program executions) may achieve a 0.75 hit ratio. Other systems with much faster cache would extend this. For the case of multilevel caches, the analysis can be repeated to account for T_3 and T_4, etc. Of course, if the programs being executed all managed to fit within cache, the hit ratio would reach 1.0.

Note that having a huge cache is not unknown. This is effectively the approach taken in some DSPs: A large provision of very fast single cycle internal RAM allows CPU operation at full speed without waiting for memory accesses. A popular example is the Analog Devices ADSP2181 with 80 Kibytes of fast on-chip memory. In this case, users are willing to pay the cost of a large block of RAM tied closely together with the CPU, for the benefit of the performance it allows (all operations—including memory accesses—completing within a single cycle).

Note that there are various techniques for improving cache performance, such as predictive read-ahead and adaptive replacement algorithms. A good full-associative cache may provide a hit ratio of up to 0.9, although this might be in a specialized system, and achievable only with a small program size.

4.4.7 Cache Coherency

Cache coherency is ensuring that all copies of a memory location in caches hold the same value. We took account of this by simply specifying clean/dirty and valid/invalid flags in the examples shown so far. However, coherency is particularly difficult and important in shared memory multiprocessor systems.

Imagine the case of a shared variable used by two CPUs, A and B. If it is read by both CPUs, it would then end up cached by both. Now if one of those CPUs, say A, changed the variable (by writing to it), the variable as stored in CPU A's cache will be updated. In a write-through system, the new value of that variable is also immediately written back into memory, so memory will then be up-to-date. However, CPU B still has the old value of the variable in cache. If CPU B reads that variable, it will be a cache hit and will use the old value from its cache, rather than the correct latest value from RAM. The fact that CPU B is now reading an incorrect variable is termed a "coherency" issue: The cached item inside CPU B is not *coherent* with the other stored values of that variable.

An example of parallel computer system is shown in Figure 4.18, which could be extended with many more processors. Since bus bandwidth is shared between CPUs, it would quickly become a performance bottleneck, and so the individual cache sizes are made large in order to minimize accesses to shared RAM (and hence bus usage). However, this only exacerbates any coherency problems.

There are a number of techniques in use in modern computer systems to alleviate this problem. A common solution begins with what we term "snooping." Snooping is the process where a cache "listens" to accesses on the shared bus by other caches. This can provide two pieces of useful information: First, when another cache reads a location that is also cached locally, and second, when another cache writes back to memory to a location that is cached locally.

With the information gleaned through snooping, an intelligent cache controller can take some form of action to prevent coherency issues. For example, invalidating a corresponding local cache entry when another cache writes to that location in shared RAM. In fact, there are a number of methods of handling the issue, although something called the *MESI protocol* is one of the most popular.

The MESI protocol, named after its states (modified, exclusive, shared, and invalid), is based around the state machine shown in Figure 4.19. An (S) or (E) after the read misses in the figure indicates that when the value was fetched from main memory, another cache

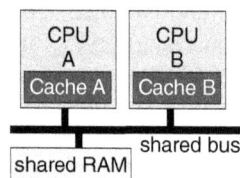

FIGURE 4.18 The connectivity of two CPUs each with individual cache, to a shared bus architecture with shared main RAM.

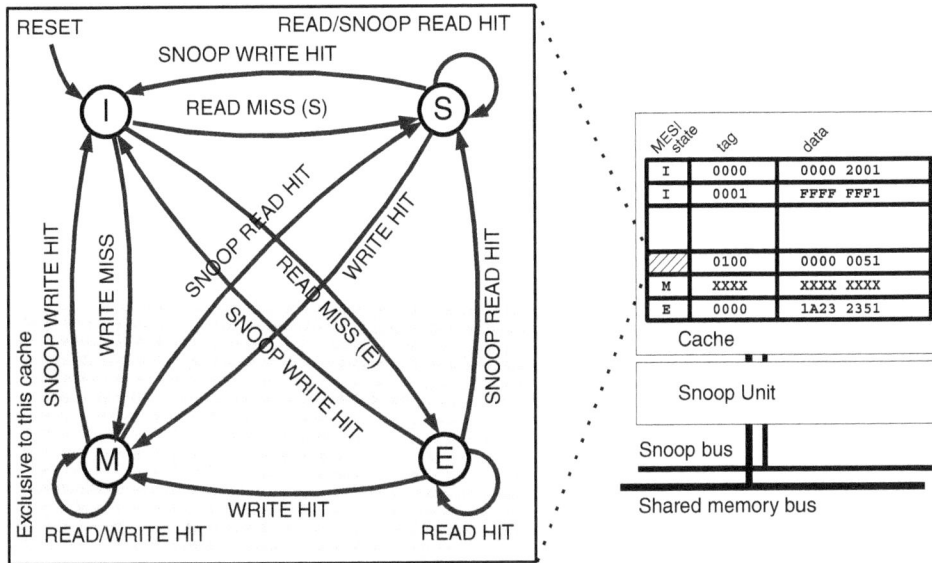

FIGURE 4.19 The MESI protocol state transition diagram (left) and a section through cache memory showing where the MESI state identifiers would be located for a particular cache line.

snoop unit indicated that it was also holding a copy (and hence an S for shared) or no other unit indicated it was using it (hence an E for exclusive). It can be seen, therefore, that snoop units have the responsibility to not only listen to other accesses, but also to inform other processor cache units when they themselves are caching a value that the other cache requests.

Each cache line can have one of four states associated with it (instead of the two states of valid/invalid and dirty/clean):

- I is invalid, indicating that the data in that line is not correct or is not caching anything.

- S is shared, meaning that *another* CPU is probably also caching this value. A cache can determine this by snooping the shared memory bus. The value in cache is the same as the value in main memory.

- M is modified in that the value has been updated. This means that any other caches holding this value will actually be holding old data.

- E is exclusive, an indication that no other cache is currently holding this value, but that it is the same value as would be found in main memory.

If this scheme is used in a shared-memory multiprocessor system, then each CPU has its own cache, and each cache uses the MESI protocol for control. The usual line number and tags are still required for each line in cache, but the valid/clean flags are replaced with two flag bits that specify the state (invalid, modified, exclusive, or shared).

On reset, all cache lines are set to invalid. This means that any data in the cache lines is not correct.

Readers may wish to refer to Box 4.10 for a worked example of the MESI protocol operating in a dual-processor shared-memory system.

Box 4.10   MESI Protocol Worked Example

To illustrate the operation of the MESI protocol in a twin-CPU shared-memory system, the state of the system will be followed through a representative sequence of events. The CPUs are named A and B and their caches begin from reset (so all entries start in the I state).

CPU A reads from location X in shared memory. Since cache is all invalid, this will be a read miss, and cause the value to be retrieved from main memory. Cache B will snoop the bus, see the transfer, look internally, and see that it is not caching location X and therefore keep quiet. Looking at the state diagram, and applying this to cache A, a read miss from state I with no snoop information will lead to state E.

Now imagine that CPU B also reads location X after that. There is nothing inside cache B and hence it is a read miss. Cache B reads the value from shared RAM, but cache A snoops the bus. Cache A looks internally and sees that it is also caching location X. Cache A will then indicate on the snoop bus to cache B that it is holding location X. Cache B will continue to read the value, but since it is a shared read, the state diagram indicates we must follow the read (S) from state I to state S. Similarly, inside cache A there was a snoop read hit, and so the state of the cache line holding location X moves from E to S. At this time, both caches hold location X and both are in the shared state.

Next, imagine that CPU A writes to location X. Given a write through scheme (where any write is committed directly to main memory), cache A realizes that this is a write hit, which from state S moves the line state to E. Cache B snoop unit is monitoring the bus and determines a snoop write hit. Since it was also in state S, this will take it to state I, which means invalid. This is correct since the value that it was caching is no longer the latest value—the latest value is in the *other* cache and now back in main memory.

4.5   Coprocessors

There are certain classes of computational task that are better performed with hardware that is not arranged as a standard CPU. A common example is the processing of floating point numbers, usually faster when handled using a dedicated FPU than with a CPU (early PCs did not provide for floating point calculations in hardware: Some readers may remember the sockets provided on Intel 80386-powered PC motherboards for the Intel 80387 floating point coprocessor, and alternatives). In fact, since the earliest computers, there have been occasions where special-purpose hardware has been used to perform certain functions separately from the CPU, leaving the CPU for general-purpose computing.

Probably the most prominent example of this technique outside the handling of floating point numbers is Intel's MMX extension to the Pentium range of processors, later extended and renamed as *streaming SIMD extensions* (SSE). However, there are others—many modern embedded processors contain dedicated coprocessing units for functions such as encryption, audio or video processing, and even dedicated input-output handling.

We will examine MMX and SSE later in Section 4.7, but for now, we will consider the most prominent example—the floating point unit: something which every modern desktop computer contains built into their CPU, but which is less often found within processors designed for embedded systems.

4.6   Floating Point Unit

Floating point, as covered in Chapter 2, is the conveyance of numerical information using a mantissa and exponent, for a particular base system. As was explained, IEEE754 standard floating point is by far the most common representation, widely adopted within the computing industry.

Because of this standardization, devices which implement the standard do not change as often as the other parts of a computer system within which they are used. As an example, the Intel 80486[4] and Pentium processors contained an on-chip FPU that was basically unchanged from the original version that appeared in the mid-1980s as the 80387. This was a separate coprocessor chip for the 80386. In those days a desktop PC could be bought with or without an on-board FPU, and most PCs without FPU could be upgraded by purchasing the chip and inserting it into an empty socket on the motherboard, as mentioned previously.

There was a reason (and still is) for not supplying floating point capabilities, and that is due to the nature of FPUs: large in silicon area and power hungry. Especially for embedded and battery-powered systems, it is often preferred to use a processor with no floating point capabilities, and to write all algorithms in fixed-point arithmetic, or to use a higher-level language and employ a software floating-point emulator.

In use, the CPU loads operands into special registers which are shared between the main CPU and the FPU (whether this is a separate chip or on the same silicon), and the FPU activated through issuing a special instruction. The FPU will then read the shared registers and begin processing the required instruction. Sometime later, the FPU returns the result to the special register area and can inform the CPU through an interrupt that the process has finished. Many modern processors include the FPU inside an execution pipeline so that the extra interrupt is not required (pipelines will be covered further in Section 5.2).

The FPU can generally not access data in memory or on shared buses directly, it can only operate on what is loaded to those special purpose shared registers by the main CPU, as a slave processor. These registers are long enough to hold multiple IEEE754 double precision numbers, although internally to the FPU, the extended intermediate formats are used (see Section 2.9.3).

In more recent 586-class processors and above, these registers are shared with an MMX unit or its descendent the SSE family (Section 4.7). This means that the main CPU loads the values to the registers and then activates either the MMX or the FPU. So in many

---

[4]Some 486-class processors had no floating point capabilities, particularly those made for low-power applications.

586-class processors, *MMX and floating point could not be used together*, programmers had to choose one mode or another at any particular time.

As an aside, the limitation of allowing FPU or MMX led to the development of the AMD 3DNow! Extension containing 21 new instructions effectively allowing AMD processors to interleave floating point and MMX instructions in the same piece of code. This then prompted Intel to develop the streaming SSEs which we will discuss further as another example of a coprocessor in Section 4.7. For an alternative approach, consider the development of the ARM FPU in Box 4.11.

Box 4.11   An Alternative Approach—FPU on the Early ARM Processors

An alternative approach was taken by ARM engineers in the early years, as described in the book *ARM System Architecture* by Steve Furber:

Engineers first surveyed a large amount of common software to find out what type of floating point operations were used most commonly. Employing the RISC design methodology, they implemented these most common instructions in silicon to design the FPA10, a floating point coprocessor for the classic ARM3 device. The FPA10 has a four-stage pipeline that allows it to process operands every cycle, and to have up to four calculations simultaneously performed.

Less common instructions are not implemented in silicon—these are computed either purely in fixed-point software on the ARM, or partly in software and partly inside the FPA10 instructions.

4.6.1   Floating Point Emulation

As we have seen, the FPU is a device capable of operating on floating point numbers. Usually it provides the standard arithmetic, logic, and comparison functions, along with multiplication. Often division and other more specialized operations (such as rounding) are also supported. Most FPUs comply with the IEEE754 standard, which defines their operations, accuracy, and so on.

Programmers writing in high-level languages (i.e., the majority) will access an FPU whenever they use floating point data types in their programs. For example, in the C programming language these types are almost always those we had identified in Section 3.4.1, namely:

float—a 32-bit single precision floating point number comprising sign bit, 8-bit exponent and 23-bit mantissa
double—a 64-bit double precision floating point number comprising sign bit, 11-bit exponent and 52-bit mantissa

There is one further floating point data type in C, that is meant to be higher precision than the double precision type, and that is the long double. However, long double appears to be less standard (as was mentioned briefly in Section 3.4.1), in that it ranges from being the same as a double, through the IEEE754 extended intermediate format (see Section 2.9.3) and up to a true quad-precision number.

However, although "floating point" usually means IEEE754 compliance, it does not necessarily have to. As noted in Section 3.4.5.2, this holds only when the underlying hardware available is IEEE754 compatible. In some embedded systems, where power and size are at a premium, designers made a pragmatic choice to provide floating point

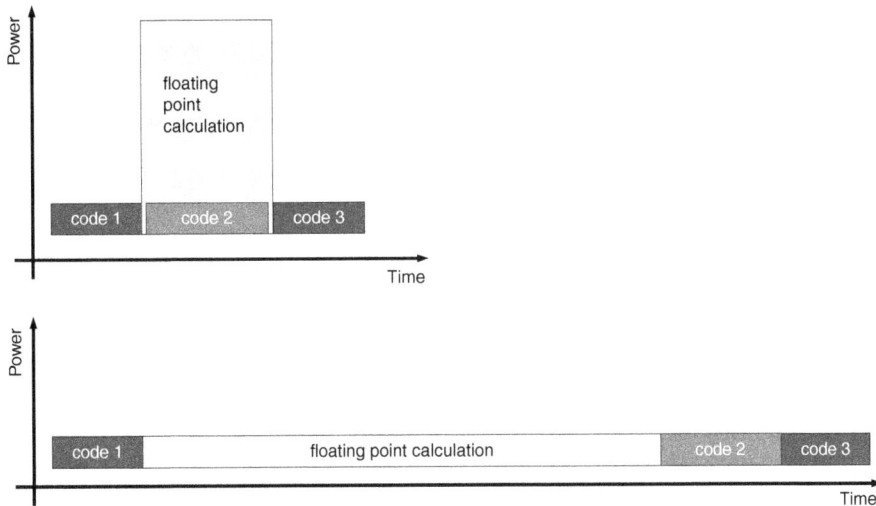

FIGURE 4.20   A diagram illustrating the trade-off between (i) executing floating point calculations in a dedicated hardware FPU while fixed point code continues executing in the main CPU (top diagram), and (ii) executing floating point calculations using FPE code which takes longer, but is less power hungry (bottom diagram).

with slightly less accuracy than IEEE754. From the point of view of the programmer, the data types of `float` and `double` still exist; however, the accuracy of the calculations using these may differ.

Where hardware support for floating point is not available, in other words in the absence of an FPU, instructions specifying floating point operations will be picked up by the CPU, causing an interrupt (or trap—see Section 3.4.5) and handled by specialized code. The code that replaces an FPU is called a floating point emulator (FPE).

Quite often, FPE code is sub-IEEE754 in precision. The time taken to calculate IEEE754 operations using multiple fixed point instructions is so time consuming that it is a trade-off between speed and accuracy. Designers usually prefer to increase speed as much as possible.

Another aspect of this trade-off is illustrated in Figure 4.20, where a processor having a hardware FPU is shown (top) compared to a fixed point processor (bottom). The same code is executed on both. In the relatively unlikely event that all other factors are equal (i.e., the only difference between the two is in the presence of an FPU coprocessor in the first case), the FPU-enabled processor can pass the floating point operations over to the FPU, which consumes a significant amount of power while it operates, while the main CPU performs other, unrelated functions. Once the floating point calculations are completed, the result is passed back to the CPU and operation continues.

In the case of the fixed point processor, the floating point calculations must be emulated by FPE code running on the main CPU. Since there is no coprocessor in this case, there is no possibility for the floating point code to be executed in parallel with other code. Obviously the program will then execute more slowly, even if the FPE code is as quick as the FPU. However, usually an FPE function is several times, maybe 10 or more times, slower than execution in the FPU.

In terms of energy consumed—which is the important measure in portable electronics where battery life is concerned—energy is shown by the shaded areas in the figure: power multiplied by time. Although the FPU consumes significantly more power than the fixed-point CPU, it does so for a shorter period of time,[5] and thus may well be more energy efficient than floating point emulation. Of course, as we had noted previously, in such a situation the system engineers may well decide to employ lower accuracy floating point routines to speed up the calculations. Even more preferable would be for programmers to refrain from using floating point operations in the first place, and this is often a target of embedded systems developers. Programmers could consider using `long` integers, or choose to program in fractional (Q-format) notation (see Section 2.3.7).

4.7   Streaming SIMD Extensions and Multimedia Extensions

Multimedia extensions (MMX) was the name given by Intel to a hardware multimedia coprocessor for the Pentium processor. The MMX unit was actually an SIMD (single instruction, multiple data) machine as defined in Section 2.1.1. In use, a set of numbers are loaded into the MMX registers, and then a single MMX instruction can be issued to operate on the data in every register, in parallel. An example of this type of processing would be for eight integers to be shifted right by two places simultaneously, or for four of the registers to be added to the other four, with the result overwriting the contents of the first four. There are many variations on this theme, but the important aspect is that each of the separate operations will occur simultaneously, triggered by a single instruction.

After Intel released the MMX, competitors Cyrix and AMD soon offered similar accelerators for their devices, whereas others such as ARM and SUN created custom-designed equivalents for their RISC CPUs. These hardware devices were offered on-chip rather than as an external coprocessor, and derived from the observation that processing of multimedia data often involves the repeated application of relatively simple arithmetic actions across a large amount of data.

4.7.1   Multimedia Extensions

An example of the type of processing that MMX technology was designed to accommodate would be the color adjustment of an area on a display screen. If each pixel of displayed data on the screen is a byte or word then adjustment of color may simply be an addition of a fixed value to each of these words, or may be a logical masking operation. Whatever the exact operation is, it must be repeated uniformly across a large number of pixels, perhaps 1280×1024 pixels or more. If this was performed on a standard CPU, there would be $1280 \times 1024 = 1.3$ million repeated additions.

With the addition of an MMX unit, a CPU can load blocks of data into the MMX unit, then perform an arithmetic operation to all data items within that block simultaneously. Meanwhile the CPU itself is free to perform other actions. It is easy to see that if the MMX unit has 16 entries, the time required to process all pixels can be reduced by a factor of around $1/16$.

---

[5]This assumes that when the FPU is not calculating it remains turned off and thus does not consume power—unfortunately this assumption is not always true in practice.

4.7.2 MMX Implementation

The argument for an MMX extension was convincing, especially in light of the growth in multimedia processing requirements in PCs during the years of MMX development. However, pertinent questions were to find out how to best implement this type of processing, and exactly what type of processing to support.

In the case of the Intel Pentium, the implementation problem was primarily that Intel required any new Pentium to be backward-compatible with early 8088 and older 16-bit software as used in DOS, and even some surprisingly modern versions of Microsoft Windows. There was thus very little possibility of expanding the capabilities of x86 CPUs through changing its instructions—this would have meant that new software could not operate on older machines, something that customers would *not* be happy with (this type of compatibility change needs to be made more gradually, needing time to sink in with customers). In addition, the number of registers could not just suddenly grow from one Pentium version to the next because this would invalidate the process of context save and restore used in older software.

However, Intel engineers found two clever ways to accomplish their aims. The first was to give the Pentium an extra instruction which would place it in MMX mode (and they released simple code that would allow programmers to first check for MMX capabilities and then run one version of code for machines with MMX or another one for machines without MMX). In MMX mode, extra 57 new instructions were then made available for MMX processing. Older software would not use this mode and hence not experience the extra instructions. The second innovation by Intel engineers was to reuse the registers of the FPU for holding the MMX data. In normal mode, these were used by the FPU, but in MMX mode they could now be used for MMX processing.

Unfortunately, programmers did not adopt MMX en masse. There were criticisms relating to the fact (mentioned in Section 4.6) that selecting MMX mode completely removed floating point capability. Ultimately, this led to the AMD 3DNow! inspired SSE. However, before we jump into a discussion on SSE in Section 4.7.4, let us examine how these systems work in general, starting with the venerable MMX.

The logical structure of the MMX unit, showing its eight registers, is shown in Figure 4.21 (although it should be noted that this figure is highly stylized—the actual MMX is rather more complex than the one drawn here). Note the bus looping from the

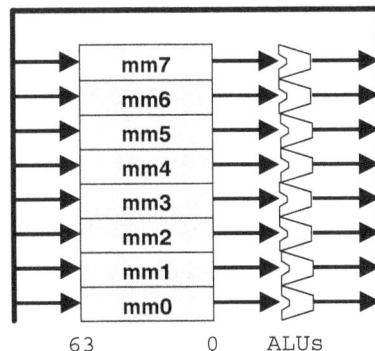

FIGURE 4.21 The MMX registers, parallel functional units (looking like little ALUs) and bus interconnections shown for an MMX-enabled Intel CPU.

mm7
mm6
mm5
mm4
mm3
mm2
mm1
mm0

63   0   ALUs

output of the eight ALU blocks back into the registers. This is a simplified representation of the internal structure of the MMX unit, but serves to illustrate the parallel nature of the paths from each of the registers. Each line is a separate bus.

In MMX mode, there are eight registers of width 64 bits. (Why 64 bits? Well, remember that 64 bits are needed to represent the double precision floating point values that are normally held in these registers in FPU mode.) Instructions operate in parallel and are all from register to register, except the load and store instructions.

Although each register is 64 bits in size, it can hold either 8 bytes or four 16-bit words or two 32-bit double words or a single 64-bit quad word. This is under the control of the programmer and leads to significant flexibility in creating MMX code.

Arithmetic, logical, comparison, and conversion operations are supported, and these can be applied to whatever data size is known to exist within the transfer registers. Of course, it is the programmer's responsibility to load the correct sized data and choose the correct operations to apply to this data.

4.7.3   Use of MMX

To use MMX capabilities on a suitably equipped Pentium processor, it is first necessary to check whether the CPU can enter MMX mode (and there is a simple backward-compatible mechanism to do this). If it can, then MMX mode processing can continue, otherwise, code must be provided to perform the same function using the CPU capabilities alone. This will obviously be far slower, but is needed for backward compatibility in every portable program.

However, the speed gains for specialized programs using this technology are very significant: real-life testing of MMX capabilities for image processing has shown that MMX optimized code could be at least 14 times faster than non-MMX code in test software under Linux.

4.7.4   Streaming SIMD Extensions

MMX was actually an Intel-specific name for single-instruction multiple data (SIMD) extensions to the x86 instruction set, originally launched in 1997. AMD introduced their hardware extensions under the term 3DNow! but had added floating point capabilities to the integer-only hardware from Intel. Not to be outdone, the battle hotted up with streaming SIMD extensions (SSE) of various types from Intel, and enhanced 3DNow! from AMD.

SSE provides 70 new instructions for the SIMD processing of data and makes available eight new 128-bit registers.[6] These can contain the usual integer values, but now of course allow the use of floating point:

- Four 32-bit integers
- Eight 16-bit short integers
- Sixteen bytes or characters

---

[6] In 64-bit mode this doubles to sixteen 64-bit registers.

- Two 64-bit double precision floating point numbers
- Four 32-bit single precision floating point numbers

SSE has actually evolved considerably from its initial incarnation through SSE2, SSE3, SSE4, and even a proposal for SSE5. Each iteration has brought new capabilities, new instructions, and new capabilities for programmers to learn. Interestingly, from SSE4 onward, support for using the old MMX registers has been discontinued.

SSE4 introduced some fast-string handling operations, also many floating point operations, such as parallel multiplies, dot products, rounding, and so on. There is also now some degree of compatibility between Intel and AMD versions (perhaps more than there is to previous generations of x86 processors), but the ongoing evolution of these capabilities allied with some aggressive marketing tactics make direct comparisons of the capabilities of the two leading x86-style processors quite difficult.

4.7.5   Using SSE and MMX

With so many versions, and differing compatibilities among different CPU ranges, let alone between manufacturers, software tools have tended to lag behind capabilities. Many compilers do not support these coprocessors by default, or at best provide sparse support across the range of possible hardware inclusions (preferring to restrict support to only the most common options). Although the situation has improved significantly in recent years, especially with the availability of compilers from Intel themselves which presumably support these extensions, the programming tools do not yet tend to take full advantage of this specialist hardware.

Also the need to write several versions of code specialized for various different processors has meant that use of these SSEs has tended to be confined to instances of specialized software, rather than general releases of commercial operating systems and applications. However, they exist and are available, especially in desktop or server machines, for absolute maximum processing performance.

4.8   Coprocessing in Embedded Systems

Few embedded systems utilize x86 style processors these days, notwithstanding low-power variants such as the Atom; by far the largest proportion are ARM based or use similar lower-power RISC CPUs. Even among the x86 processors, few have full SSE capabilities (since these coprocessors have a reputation for being power hungry). However, those that do may have an advantage over their use in desktop and server systems. The reason being that many embedded systems run controlled or dedicated software, compared with the situation in desktops that can run literally any software. While desktop systems need software to be backward compatible (and as such require code for SSE, code for MMX, code for SSE4, and bare x86 code in case of no extensions at all), in an embedded system, the programmer often knows in advance exactly what hardware is available and can develop his or her software appropriately.

The converse is also true—knowing what software is to be run can provide the opportunity to modify or create custom hardware. As an illustration of this process, in Section 4.6 we had met the FPA10, the main ARM floating point coprocessor, which was designed based on an analysis of the most common software requirements.

There are many other coprocessors in use within embedded systems, apart from the FPUs and MMX/SSEs already mentioned. Consider the following ARM-specific coprocessors:

- *Jazelle:* The name seems to be the "J" from the Java language, added to a Gazelle, bringing to mind a swift and agile execution of Java code. This is precisely the aim: the ARM engineers who designed Jazelle have created a hardware unit able to directly process many Java instructions (bytecodes) without interpretation, leading to speed and efficiency improvements. A branch to Java (BXJ) instruction enters Jazelle processing, allowing the CPU to natively execute most of the common bytecodes (and trap the rest for execution in optimized software routines).

- *NEON-advanced SIMD:* Similar to Intel's SSE, this is a 64/128 bit SSE with a very complete instruction set able to process packed integer and floating point operations in parallel. This is probably what SSE would have been if it had been designed from the bottom up, cleanly, for a modern processor (instead of evolving from the MMX addition to a 30-year-old semi-backward compatible slice of silicon history).

- *VFP:* A vector-coprocessor for ARM processors enhanced with floating point capabilities (VFP stands for "vector floating point"). This is used for matrix and vector computation—repetitive sequential operations on arrays of data.

Remember back in Section 3.2.6, we had discussed the different rationale behind RISC and CISC processors? CISC processors were presented as the bloated lumbering endpoint in an evolutionary process which packed more and more functionality into individual CPU instructions. RISC, by contrast, was lean and swift.

RISC instructions tend to be very simple, but quick. The argument being that even though more instructions are required to do anything useful, those instructions can execute faster and thus overall performance increases compared to a CISC approach. However, the use of a coprocessor can allow an RISC processor: small, lean, and fast, to hand off specific computational tasks to a separate processing unit. Thus some of the application-specific instructions available to a CISC processor could be handled by an RISC coprocessing unit.

A further refinement, bringing to mind the dual-mode method that Intel used for the early MMX, is to have a coprocessor that is reconfigurable. This allows the silicon resources used by the coprocesssor to be adjusted to suit the computation required at any particular time. Evidently there will be a cost to the reconfiguration—it will take both time and energy—however, the benefits of having fast accelerated processing for some complex computation, could easily outweigh this.

For embedded systems designers, probably the prime example of this would be within a field programmable gate array (FPGA). A "soft core" processor, residing in an FPGA, is one written in a high-level hardware description language such as Verilog. One of the prime features of FPGAs is their reconfigurability. Many of the free, and commercial, soft cores that are available already implement a coprocessor interface, and several researchers have experimented with attaching reconfigurable processing units to these. It is likely that the importance of these approaches to embedded systems will continue to be explored, and consequentially grow in adoption. We will consider these kinds of devices further in Section 7.14.

4.9 Summary

This chapter investigated the internal elements commonly found within today's general-purpose microprocessors. All of these include the means to transfer data through internal buses to and from various functional units such as ALU, FPU, or other coprocessors and accelerator units that may be connected.

An MMU and cache may be present within the system, and can be thought of as residing on the address and data buses between the processor core and the outside memory system. A cache acts to speed up average memory access time by predicting future memory recall patterns and storing some past memory accesses that match predicted future accesses. Meanwhile an MMU has two important roles, the first of which is to allow the use of virtual memory which expands the allowable address range and storage space of the processor, and the second of which allows memory page and segments to be defined and used—an important benefit of which is the memory protection between running processes (something that prevents a rogue process from overwriting the private memory of other processes, or a kernel, and thus prevents—of at least reduces the chances of—crashing). The cost of using virtual memory is in a performance hit: it tends to reduce the average memory access time.

While the contents of this chapter are commonly found implemented in modern CPUs and are considered standard functional units and capabilities in general-purpose processors, Chapter 5 will turn our attention toward improving performance—common speedups and acceleration techniques. In the headlong rush by CPU manufacturers to have either faster and faster or lower and lower power devices (but rarely with both characteristics simultaneously), some interesting methods have arisen and been adopted as we shall see.

4.10 Problems

4.1   Referring to the ALU design in Section 4.2.2, if each logic gate has a 10-ns propagation delay between any input and any output, what would be the maximum operating frequency of your ALU?

4.2   Referring to the single bit ALU in question 4.1.

   a) Show how four of them can be combined to make a 4-bit ALU for unsigned numbers.

   b) Show how the design could be modified to perform the computation using twos' complement signed numbers.

4.3   The following pseudo-code segment is executed on an RISC processor:

```
loop i=0,1
read X from memory address 0
read Y from memory address i
Z=X+Y
write Z to memory address i+1
```

The processor takes one cycle to complete all internal operations (including cache accesses). Saving data from cache to RAM takes four cycles. Loading data from RAM to cache takes four cycles (+1 cycle to continue from cache to CPU).

Assuming that the system has a **direct** cache which is initially empty, how many cycles are required for this code if the cache uses the following policies?

a) Write back

b) Write through with no write allocate (WTNWA)

c) Write through with write allocate (WTWA)

4.4   You have a small von-Neumann computer with a data cache that can be switched between two-way set-associative and direct. It can hold a single data word in each of its 512 cache lines, and all data transfers are word sized. The following algorithm is to be run on the processor.

```
define data area A from address 0 to 1023
define data area B from address 1024 to 2047
set R0=512, R1=address 0, R2=address 1024
{
lp  [R1]=R0+R0      ;save to address stored in R1
    [R2]=[R1-1]+[R1]
    R1=R1+1
    R2=R2+1
    R0=R0-1
    if R0>0 then goto lp
}
```

a) Which cache organization would be best if the system operates with a write-back protocol?

b) Name three cache-entry replacement algorithms, and comment on their hardware complexity.

c) The algorithm given is run just after a reset that clears the cache, and it iterates twice. If the system uses a direct cache with write-through (and write-allocate), taking 10 ns for CPU-cache transfer and 50 ns for cache-RAM transfer, answer the following:
   (i) What is the hit rate?
   (ii) What is overall access time for two iterations?

4.5   Rewrite the algorithm of the previous question to improve hit rate (hint: adjust the data area definitions rather than the loop code itself).

4.6   An advanced music/photo device uses virtual memory to allow the CPU to access 1 Gibyte of logical memory space, although the system only has 1 MiB of RAM. The OS programs the MMU to allow a fixed page size of 4 Kibytes. The byte-wide RAM has a 20 ns access time, while the hard disc is limited by its IDE interface to 2.2 Mibytes per second data transfer. The RISC CPU has 32-bit instructions.

a) How many pages can reside in RAM at one time?

b) How many wires must the MMU-to-RAM address bus contain?

c) How much time is required to read each instruction from RAM?

4.7 Using the information from question 4.6, calculate how much time is necessary to load a page from disc to RAM (or from RAM to disc), and use this to determine two possible timings for the CPU to retrieve and instruction from a retired memory page.

4.8 The MMU-RAM address bus in the previous question is not wide enough to accommodate more memory. Name three (hardware or software) methods of overcoming the address-bus size limitation and connecting more memory on that physical interface.

4.9 A dual-processor machine has a block of shared memory and a snoop bus. Write-back caches in each of the processor modules implement the MESI protocol, starting with all cache lines in the invalid (I) state.

Trace the cache states through the following sequence (X, Y, and Z are not equal):

1. CPU1 reads from RAM address X.

2. CPU1 writes to address X.

3. CPU2 reads from address Y.

4. CPU1 reads from address Y.

5. CPU1 writes to address Y.

6. CPU2 reads from address X.

7. CPU2 reads from address Z.

8. CPU1 writes to address Z.

4.10 Consider the block diagram of an ALU and three registers connected in a three-bus CPU as shown below. Assume that this diagram is complete except for a memory interface to each bus, and that memory transfers are much slower than register data movements. X, Y and Z are three memory locations.

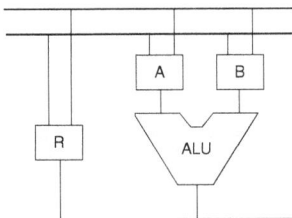

a) Draw arrows on the diagram to indicate allowable data directions for each bus connection.

b) How efficient is the operation $X + Y$?

c) How efficient is the operation $X + X$?

d) How efficient is the operation $(X + Y) + Z$?

e) Suggest an alternative connection arrangement to improve efficiency.

4.11   Identify four basic logic operations that every ALU is likely to be capable of performing (excluding shifts) and the two main arithmetic operations.

4.12   Identify the three different types or directions of bitwise shift that are usually implemented within even simple CPUs, excluding rotate instructions (and can you explain why the question asks for just three types rather than four).

4.13   Following the propagation delay example of Box 4.1, determine the propagation delay of an 8-bit ADD, and an 8-bit AND operation. In each case assume the function select signals are correct and unchanging (so they do not affect the timings at all). What would be the maximum clock speed of this device if the ALU was expected to operate in a single cycle?

4.14   If cache memory can improve processor performance, can you think of any reasons why manufacturers would not simply sell ICs with massive on-chip cache blocks?

4.15   Calculate the overall access time for a computer system containing a direct cache having a 10 ns access time for hits and a 120 ns access time for misses, when the hit ratio is 0.3.

4.16   Assuming the computer designers of the machine in question 4.15 wish to increase performance. They have only three things that they can change in the system (and since each change costs money they only want to do one of these things, and much choose the best one). Determine which of the following would most improve overall access time in that system:

a) Fit faster main memory, with a 100 ns access time.

b) Fit faster cache memory, with a 8 ns access time.

c) Increase the hit ratio to 0.4 by squeezing in a much bigger cache with a better arrangement and cleverer replacement algorithm.

4.17   Assume a small 16-bit embedded system primarily executes integer code, but sometimes needs to quickly process a block of floating point data. This can either be processed using a dedicated FPU, could be executed in an FPE, or the code converted so that it uses very large integers instead. Discuss the main factors influencing whether the choice of processing solution for this device should contain an FPU or not.

4.18   Chapter 3 introduced the concept of relative addressing. Briefly discuss how this is related to the principles of spatial and temporal locality as explained in Section 4.4.4.

4.19   In the context of cache memory, what is the meaning of "write-through with write-allocate" (WTWA) and how does this differ from "write-through with no write-allocate" (WTNWA)? Which would be more appropriate in a system outputting vast amounts of transient graphical data to a memory-mapped display?

4.20   In an embedded system that has a full development and debugging software suite, an experienced programmer has set a memory *watchpoint*[7] on address 0x0000 in RAM while trying to debug code which occasionally crashes. However, your code, data, and variables are located elsewhere in memory: You certainly did not fix any variables or code to be at address 0x0000. Can you think of a reason why he/she should be interested in this address which is not supposed to be used anyway?

---

[7] A *watchpoint* is a location in memory that the debug software will constantly monitor, and will halt program execution whenever the contents of this address change.

CHAPTER 5

Enhancing CPU Performance

R eaders of books like this one seldom read through everything in previous chapters before delving into a section or chapter that they are interested in. We authors understand this phenomenon, and indeed many of us encourage our readers to "surf" sections from different textbooks to gain a variety of perspectives on the subject matter. So if you are one of those people—perhaps your eye was drawn to the topic of making computers go faster—then welcome to the chapter that describes the "go faster" preoccupation of computer architects and programmers alike since the dawn of the computing age. This chapter can be read stand-alone but it does draw together a number of elements (or maybe loose ends) from previous chapters, such as the generational view of computers, emerging forms of parallelism, the RISC versus CISC debate, the layered view of computer organization, and Flynn's classification.

For those who did read sequentially through the various previous chapters, congratulations on having reached this point. I hope that a picture is emerging in your minds of an evolutionary process in computer design, with a history of incremental advances in performance. In designing new CPUs, blocks of functionality tend to be aggregated together to form a prototype CPU and then evaluated (see Section 3.5.2). The performance limiting blocks are identified and then adjusted or speeded up, with small improvements being common. Truly revolutionary change, by contrast, is rare.

In most cases, the motivation behind designing a new CPU has been performance, and is ultimately driven by the sales department. However, in embedded systems power consumption (which is related to battery lifetime) is a further significant driver, and is sometimes a reason to *not* adopt a particular performance technique. Other motivations include reliability, correctness, programmability (ease of programming), and security, and these have led to some interesting and innovative CPU designs.

Everybody wants a faster computer. It has been said that there are no speed limits on the information superhighway, and in most cases users feel that more speed means less wasted time. The same could be said of transport—who would say no to a Ferrari or Aston Martin? However, the author is skeptical of this idea—seeing his own students wasting more time with faster computers than their peers did with sluggish machines a generation ago! For embedded systems, especially those requiring real-time processing, there is no doubt that greater speeds lead to greater functionality. For the desktop, however, the suspicion is that much of the speed, memory, and storage increases are swallowed by the bloatware of software developers, particularly in regard to the operating system. Still, the mythical "performance" target has always been a major driver in the computer industry, and one which has yielded some extremely interesting (wild and wonderful) solutions. In this chapter, we will consider many of the more common methods of improving CPU performance and a few clever but less common ideas.

5.1   Speedups

For much of computer history, the primary method of increasing performance was adjusting clock speeds to make them faster and faster. There are two factors which limit that approach—namely logic propagation delay (how long it takes logic gates to finish processing) and power dissipation. Making gates faster usually require making silicon feature sizes smaller (i.e., thinner and smaller strips of semiconducting silicon within an IC). Smaller feature sizes cost more to design, are more difficult to manufacture, and more tricky to operate once they have been manufactured (e.g., power supply stability requirements increase). Packing more gates together increases the amount of waste heat generated in an IC, while increasing clock speed causes very significant increase in the amount of heat generated per gate. All of those factors means it becomes progressively more difficult to improve on existing CPU designs, which are often already operating close to the edge of current achievable technology.

Other designers looked elsewhere, and ideas such as RISC processing began to emerge and take hold as alternative ways to improve performance. Some companies concentrated on increasing the word size, from 4 bits and 8 bits through 16 bits to 32 bits. More recent designs have been 64 bits, 128 bits, and even 1024 bits (covered further in Chapter 12).

Not only did clock speeds increase, but an emphasis was placed on achieving more in each clock cycle, and this led to parallelism and pipelining (and occasionally a combination of both).

The now defunct SUN Microsystems took a different approach with their Java processors which revisit CISC processor design rationale, but this time from a software perspective (and neatly integrate ideas of stack-based and RISC processors into the bargain). They designed the PicoJava and similar processors bottom-up to accommodate the Java language, rather than a language translated to run on the processor, which was the approach adopted by almost everyone else. Ultimately, this software-first approach seems to have achieved only moderate commercial success, or perhaps it is another idea whose time is yet to come.

The intention of this chapter is to cover a number of design ideas and approaches that have been explored and adopted into the mainstream, and who trace their ancestry (or rationale) more to profit motive than to academic ideals—in the fight to get faster and cheaper parts out to the customer as quickly as possible. We begin with the biggest and most common speedup, pipelining.

5.2   Pipelining

Sometimes attributed more to modern industrial manufacturing techniques than to computer evolution, pipelining improves processing *throughput* rather than the time taken to complete *individual* instructions (in fact, this may even increase yet result in better performance). It allows the different stages of instruction processing to overlap—and thus process multiple slower instructions concurrently, giving an overall throughput increase.

fetch instruction	→	decode instruction	→	fetch operand	→	execute instruction

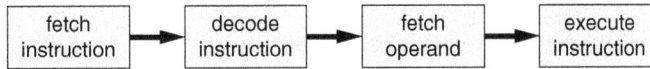

FIGURE 5.1 A flowchart of four stages of instruction processing in a simple CPU.

The *throughput* is the average number of operations performed per second: the cycles per instruction benchmark of Section 3.5.2. This measure is much more important than how long each individual instruction takes. To appreciate this fact, let's consider a typical CPU instruction processing flowchart as shown in Figure 5.1.

In this example, every instruction is handled in four stages which we will assume are all of a single clock cycle duration. An instruction must traverse all four stages to complete, thus requiring four clock cycles.

A non-pipelined machine would grab and process one instruction, then wait for that instruction to complete before starting on the next one. We use something called a *reservation table* to visualize this:

fetch instruction	$Inst_1$				$Inst_2$				$Inst_3$
decode instruction		$Inst_1$				$Inst_2$			
fetch operand			$Inst_1$				$Inst_2$		
execute instruction				$Inst_1$				$Inst_2$	
Clock cycles:	1	2	3	4	5	6	7	8	9

The different functional units for handling an instruction are listed down the left side of the table, and the clock cycles are shown along the bottom. Inside the table, we indicate what's happening in that cycle. The table shown covers nine successive clock cycles.

$Inst_1$ is fetched in the first cycle, then decoded, then its operand fetched and finally the function encoded in that instruction is executed. $Inst_2$ then begins its journey.

But think of this reservation table a different way: if we consider the rows as being resources and the columns as time slots, it is clear that each resource spends a lot of time slots doing nothing. It would be far more efficient if we allowed instructions to overlap, so that resources spend more of the time doing something. Let us try it out:

fetch instruction	$Inst_1$	$Inst_2$	$Inst_3$	$Inst_4$	$Inst_5$	$Inst_6$	$Inst_7$	$Inst_8$	$Inst_9$
decode instruction		$Inst_1$	$Inst_2$	$Inst_3$	$Inst_4$	$Inst_5$	$Inst_6$	$Inst_7$	$Inst_8$
fetch operand			$Inst_1$	$Inst_2$	$Inst_3$	$Inst_4$	$Inst_5$	$Inst_6$	$Inst_7$
execute instruction				$Inst_1$	$Inst_2$	$Inst_3$	$Inst_4$	$Inst_5$	$Inst_6$
Clock cycles:	1	2	3	4	5	6	7	8	9

The most obvious effect is that instead of getting to the start of $Inst_3$ in the nine clock cycles, the overlapping now covers nine instructions: It processes three times faster. It does this without having to increase clock rate or change processing order, simply by allowing the possibility of overlapping instructions.

This overlap is called *pipelining*. It is a technique used to speed up almost all modern processors. Control of functional units becomes more complex, but the speed gain tends to outweigh this disadvantage. To determine the actual amount of speedup, refer to the analysis in Box 5.1.

Later some more of the difficulties introduced by pipelining will be examined, but first we take a look at different types of pipeline.

Box 5.1   Pipeline Speedup

There are two useful measures of a pipeline: degree of *speedup* and *efficiency*. Let us consider a program that consists of s sequential instructions, each instruction needing n clock cycles to complete.

In a non-pipelined processor, the program execution time is simply $s \times n$ cycles.

Now let us pipeline this processor into n stages, each of a single clock cycle. How long will the program take to execute?

Well the first instruction takes the usual n cycles from start to finish, but then each subsequent instruction completes a single cycle later, so the total time is $n + (s - 1)$ cycles.

Speedup S_n is the ratio of non-pipelined to pipelined operation:

$$S_n = \frac{sn}{n + s - 1}$$

Looking at this, it seems that as $s \to \infty$ then $S_n \to n$ meaning that the bigger the program is, the more efficient it is (because no matter how fast the pipeline is, it starts empty, and ends with a single instruction—the final one—inside). In other words, the starting and ending conditions are less efficient.

So a measure of efficiency, on the other hand, must take account of these start and end conditions. Efficiency is the total number of instructions divided by the pipelined operating time:

$$E_n = \frac{s}{n + s - 1}$$

But does not this look similar to the speedup equation? Yes! $E_n = S_n/n$ and this is also the same as throughput, which is the number of instructions completed per unit time.

5.2.1   Multifunction Pipelines

Pipelines do not have to be simple strings of functions (*unifunction*), they can allow different instructions to be handled differently, as in *multifunction* pipelines. In fact this is common, but increases the complexity of control. Consider an example shown in Figure 5.2.

In the top pipeline of Figure 5.2, the first instruction needs to fetch something from memory in order to complete, and thus it needs to use the "fetch operand" unit. At the bottom the same pipeline is shown at a later time, executing a different instruction. This one does not require an operand fetch (since the immediate value 3 is encoded as part of the instruction and therefore already inside the CPU). So the "fetch operand" pipeline unit is unnecessary in this case. However, this does not mean that the pipeline skipped

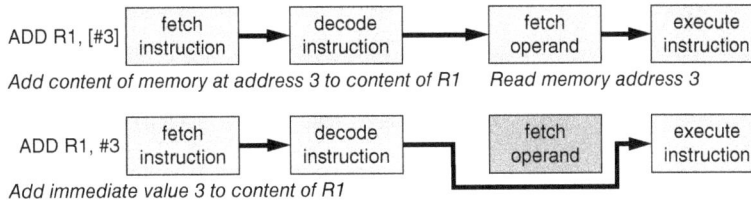

FIGURE 5.2   A flowchart of four stages of instruction processing in a simple CPU for two assembler instructions. The top instruction utilizes every stage of the pipeline whereas the bottom instruction does not need to fetch an operand from memory, and thus skips the third stage. This illustrates the concept of a multifunctional pipeline where different instructions are handled differently, based upon their needs.

a stage and the second instruction was executed more quickly…consider the reservation table below where these two instructions are executed sequentially:

fetch	ADD R1, [#3]	ADD R1, #3	$Inst_3$	$Inst_4$	$Inst_5$	$Inst_6$
decode		ADD R1, [#3]	ADD R1, #3	$Inst_3$	$Inst_4$	$Inst_5$
fetch			ADD R1, [#3]	**NOP**	$Inst_3$	$Inst_4$
execute				ADD R1, [#3]	ADD R1, #3	$Inst_3$
cycles:	1	2	3	4	5	6

Clock cycle 4, for the second instruction is marked as a NOP (no operation). It would not be possible for the CPU to immediately skip from "decode instruction" to "execute instruction" because, in cycle 4, the hardware that performs the "execute instruction" is still handling the previous instruction (ADD R1, [#3]).

This illustrates an interesting point: This pipeline needs to cater for all instruction types, but is limited by the slowest instruction. In a non-pipelined processor, simple instructions could be executed very quickly, and difficult ones more slowly. But a pipelined processor generally takes about the same length of time to process anything, unless some very advanced techniques are used.

Designers need to be careful with pipelines. The very argument for having a pipeline was so that processing elements are kept busy for more of the time; however, we now see NOPs creeping into the reservation table. NOPs indicate an unused, or wasted resource for a cycle. Judicious examination of instruction requirements and the frequency of occurrence of instructions is needed to ensure that the design of the pipeline minimizes these wasted slots.

5.2.2   Dynamic Pipelines

By definition also a *multifunction* pipeline, a *dynamic pipeline* does not simply bypass an unused function, but allows alternative paths to be taken through the pipeline, depending on the instruction being processed and the current state of the processor.

This is illustrated in the example of Figure 5.3, where four unnamed pipelined units (T_1 to T_4) process three instructions which traverse the pipeline through different paths. Not shown is the complex switching control required for this, and delay elements that must be used to slow down instructions which skip units (such as instruction 3 bypassing pipeline unit T_2).

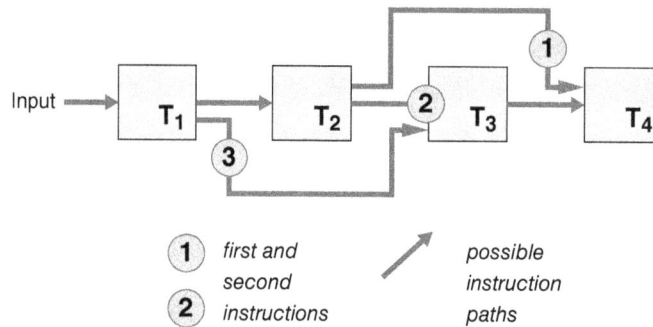

FIGURE 5.3 A dynamic pipeline allows different instructions to follow various paths through the pipeline depending upon their execution needs.

The delay elements would also be dynamic: They would only need to be activated to ensure that instructions arrive in order at the pipeline units. As an example, instruction 3 is about to "catch up" instruction 2, and would therefore need to be delayed by one clock cycle to stop both colliding at T_3. Whereas instruction 1 has skipped pipeline unit T_3, but is not about to "catch up" any other instruction, and therefore does not need to be delayed.

The interested reader will note that some processors are intelligent enough to decide for themselves which instructions need to be processed in order by pipeline units, and those that can be executed out of order will not be delayed unduly.

5.2.3 Changing Mode in a Pipeline

Everything that has been written up to now assumes that each instruction that flows through the pipeline is independent, and that an instruction can enter the pipeline even before the previous instructions have completed.

Evidently, these assumptions are not always true. We will consider three cases which impact the operation of a pipeline in this and the next two sections.

First, there is the changing of mode that can occur in some processors, triggered by receiving a mode change instruction, and meaning that all subsequent instructions are treated differently. Some examples of this are

1. In the ARM CPU where a totally new instruction set can be enabled (the 16-bit Thumb instruction set rather than the 32-bit native ARM instruction set).

2. In some processors (including the ARM), which switch between big and little endian operation. The first few instructions may be stored as little endian, then comes the mode switch, and then the rest are stored as big endian.

3. In some DSPs such as TMS320 series fixed point processors, which change mathematical mode perhaps to turn on or off sign extension, affecting all subsequent instructions.

Although these instructions do occur, they are relatively infrequent. The first two, for example, are likely to be issued at the start of a program only, and the third one would be once per block of mathematical processing.

Due to the sparse nature of these instructions, most processors will simply *flush the pipeline* once they are required to execute one. That means that all subsequent instructions already being handled in the pipeline will be discarded, and the pipeline must begin again as if it were empty. In logic terms, this is a very easy solution, although drastic. If affects pipeline efficiency, but is rare enough in most programs that it is irrelevant to performance.

Consider the example reservation table below, being hit by some kind of mode change instruction which we will call *ChM*. It is clear that, although instructions 3, 4, and 5 were already being handled by the pipeline, these are discarded, the CPU is switched to its new mode in cycle 6, and then these instructions have to be fetched again.

fetch instruction	$Inst_1$	ChM	$Inst_3$	$Inst_4$	$Inst_5$	X	$Inst_3$	$Inst_4$	$Inst_5$
decode instruction		$Inst_1$	ChM	$Inst_3$	$Inst_4$	X		$Inst_3$	$Inst_4$
fetch operand			$Inst_1$	ChM	$Inst_3$	X			$Inst_3$
execute instruction				$Inst_1$	ChM	X			
Clock cycles:	1	2	3	4	5	6	7	8	9

This type of reservation table could be the result of a sequence of instructions such as the one that follows:

```
Inst1:    ADD   R0, R0, R1
Inst2:    MODE big_endian
Inst3:    SUB   R4, R1, R0
Inst4:    NOP
Inst5:    NOP
Inst6:    NOP
```

In this example, instructions 3 to 5 are encoded as big endian. This is not evident from the listing, but would become apparent when viewing the program in memory, or examining a hexadecimal dump of the executable code.

Once the mode change is made the pipeline would have to be flushed and the following instructions reloaded.

In newer processors, this would be performed automatically by the CPU, but in older pipelined processors, this may not be automatic, and would have to be done by the compiler (or even by a programmer hand-crafting the assembler code). In the example, it is fairly easy to perform in software by changing the order of the program:

```
Inst1:    ADD R0, R0, R1
Inst2:    MODE big_endian
Inst4:    NOP
Inst5:    NOP
Inst6:    NOP
Inst3:    ASUM R4, R1, R0
```

Otherwise, a sequence of NOP instructions would need to be inserted after the mode change instruction. As an aside, ideally the NOP instruction would be encoded the same if read in big or little endian. For example, instruction words 0x0000 and 0xFFFF would

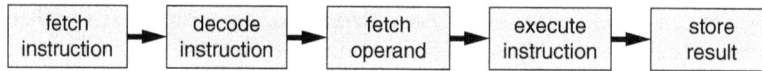

FIGURE 5.4   The sequence of processing a five-stage pipeline.

always be 0x0000 and 0xFFFF, respectively, no matter in which order the bytes were arranged, making it irrelevant what encoding is used for those instructions between when the mode change instruction is read, and when the mode change actually occurs.

5.2.4   Data Dependency Hazard

In the same way that a mode change can cause problems by changing the state of the processor part way through a program, the same is true of the continuous changes to internal registers and memory locations when a program is running. Under some circumstances this can complicate matters.

Consider for example the following code sequence:

```
ADD R0, R2, R1   ; R0=R2 + R1
AND R1, R0, #2   ; R1=R0 AND 2
```

In this example the second instruction relies upon the result of the first instruction to have been written to R0 before it can be read. But in a pipeline this may not always be true. Examine the artificial pipeline construction of Figure 5.4.

The major difference here to what we encountered previously is the addition of a final pipeline stage that stores the result of whatever calculation has occurred. This is added firstly to help illustrate the data dependency issue, and secondly because many processors really do contain such a final stage.

A reservation table of the two lines of program code given above is reproduced below. Note that under the table there is an indication of the contents of register R0 during each time slot:

fetch instruction	ADD R0	AND R1				
decode instruction		ADD R0	AND R1			
fetch operand			ADD R0	AND R1		
execute instruction				ADD R0	AND R1	
store result					ADD R0	AND R1
Clock cycles:	1	2	3	4	5	6
R0	x	x	x	x	R2+R1	R2+R1

What is important to understand is that the second instruction, the AND, makes use of the content of register R0 as its operand (R1=R0 AND 2) and this *operand fetch* is the third stage in the pipeline (in bold). In the example shown, the operand fetch for the second instruction occurs in cycle 4, but that is before the first instruction has written its result to register R0 (which occurs only by cycle 5).

As matters stand, the second instruction will therefore perform its operation with an incorrect value from R0.

This is called a RAW (read-after-write) hazard since register R0 is supposed to be read *after* it is written, but instead was read *before* it was written by the previous instruction.

If you look carefully at the example, there is another hazard there. In this case, a WAR (write-after-read) anti-dependency on register R1. The first instruction reads R1, the second instruction writes to R1, the hazard being to ensure that the first instruction has finished its read before the second instruction performs its write. With the example pipeline shown, this hazard could not occur, but in certain advanced dynamic pipelines with out-of-order execution, it is something to be aware of.

There is also such a thing as a WAW (write-after-write) hazard: an example of this is shown in Box 5.2.

Box 5.2   WAW Hazard

Believe me, this type of hazard is easier to explain than it is to pronounce. A WAW hazard occurs when two nearby instructions write to the same location, and a third instruction must read from that location. It must perform the read neither too early nor too late.

Here is an example:

```
ADD R0, R2, R1    ; R0=R2 + R1
AND R1, R0, #2    ; R1=R0 AND 2
SUB R0, R3, #1    ; R0=R3 − 1
```

There is a WAW hazard on R0. Without drawing a reservation table, it should be evident that the second instruction operand fetch must occur after the result store in the first instruction, and before the result store in the second instruction.

Note that after this code segment, R0 contains the final value, so the write to R0 by the first instruction is simply a temporary store. It could be changed to any other register, or eliminated through data forwarding (see later in Section 5.2.10).

WAW hazards sometimes occur in memory systems, where the write back to RAM is slower than the read. Usually it is the responsibility of the cache hardware to ensure that the hazard does not turn into a real problem.

5.2.5   Conditional Hazards

Given that some instructions can execute conditionally, there is the question of when the conditions are checked to determine if execution should occur. Here is an example code segment:

```
ADDS R0, R2, R1     ; R0=R2 + R1 & set condition flags
ANDEQ R1, R0, #2    ; R1=R0 AND 2 if zero flag set
```

Remember that in the ARM processor, an "S" at the end of an instruction tells the processor that the result of that instruction should update the condition flags (namely zero flag, negative flag, carry flag, and overflow flag, all stored in the CPSR register on an ARM processor—Box 5.3 describes the types of conditional flags possible). The second instruction is conditional—the "EQ" indicates that this instruction should only happen if the result of the previous condition-setting instruction was zero (in this example, if and only if register R0 is zero).

Box 5.3 Conditional Flags

Although some processors have slightly different combinations and names, the following set of condition flags are most commonly found in commercial CPUs:

N or negative flag: The result of the last condition setting operation was negative.
Z or zero flag: The last condition setting operation resulted in a zero.
C or carry flag: The last condition setting operation generated a carry-out.
V or overflow flag: The last condition setting operation caused a sign change.

Some examples of these flags and how they can change are shown in the following code segment. Note the use of the "S" flag at the end of an instruction that determines whether or not it will cause the condition flags to be updated.

Instruction	Meaning	N	Z	C	V
MOV R0, #0	set R0=0	0	0	0	0
MOV R1, #2	set R1=2	0	0	0	0
SUBS R2, R1, R1	R2=R1−R1 (result is zero)	0	1	0	0
SUBS R3, R0, R1	R3=R0−R1 (result is negative, 0xFFFFFFFD)	1	0	0	0
SUB R2, R1, R1	R2=R1−R1 (result is zero but "S" flag not set)	1	0	0	0
ADDS R4, R1, R1	R4=R1+R1 (result is positive, 0x4)	0	0	0	0
ADDS R5, R4, R3	R5=R4+R3 (0x4 + 0xFFFFFFFD)	0	0	1	0
MOV R8, #0x7FFFFFFF	the largest positive 32-bit signed number	0	0	0	0
ADDS R9, R8, R1	R9=R8+R1 (result is 0x80000001)	0	0	0	1

Note that zero is usually regarded as a positive number, rather than negative, and that the carry and overflow flag interpretation is used differently depending upon whether we interpret the operands as signed or unsigned numbers. If dealing with signed numbers, the overflow flag is important, whereas only the carry need be considered for unsigned numbers. Please refer to Section 2.4 for more information.

Next we can start to populate a reservation table from the example code:

fetch instruction	ADDS R0	ANDEQ R1	$Inst_3$	$Inst_4$		
decode instruction		ADDS R0	ANDEQ R1	$Inst_3$		
fetch operand			ADDS R0	ANDEQ R1		
execute instruction				ADDS R0		
store result						
Clock cycles:	1	2	3	4	5	6
NZCV	0000	0000	0000	0000		

By the end of cycle 4, the first instruction has been executed and the condition flags updated. Note that the second instruction has already entered the pipeline, even though it is not clear at the present time whether it should be executed or not—the choice was to either allow it to enter the pipeline, or *stall* the pipeline, waiting until the first instruction completes. Many processors would use *speculative execution* in this way to load and process the second instruction anyway. Once the conditional flags are known, a decision is made whether to terminate the second instruction or keep it.

We can now complete the reservation table on the basis that the result of the first instruction was not zero, and the second instruction therefore was not executed (or rather it was executed, but the result was ignored):

fetch inst.	ADDS R0	ANDEQ R1	$Inst_3$	$Inst_4$	$Inst_5$	$Inst_6$
decode inst.		ADDS R0	ANDEQ R1	$Inst_3$	$Inst_4$	$Inst_5$
fetch op.			ADDS R0	ANDEQ R1	$Inst_3$	$Inst_4$
execute inst.				ADDS R0	X	$Inst_3$
store result					ADDS R0	X
Clock cycles:	1	2	3	4	5	6
NZCV	0000	0000	0000	0000	0000	0000

Since the zero flag was not set by cycle 5, the second instruction was effectively executed as if it is a NOP. This results in an entire wasted diagonal in the reservation table. By contrast, if the pipeline had waited for the first instruction to complete before fetching the next instruction, this would have occurred in cycle 5, and there would have been three wasted diagonals instead.

At this point, the reader should probably be thinking in terms of "What extra pipeline functionality is needed to support this type of speculative execution?" We will leave further discussion of that until Section 5.7, apart from a short illustration in Box 5.5.

5.2.6 Conditional Branches

The ARM has an instruction set where all (or almost all) instructions are capable of conditional operation. However, most processors support conditional execution with branch instructions only, and use these to alter program flow. Here is an example of a conditional branch:

```
loop:   MOV R1, #5          ; R1=5
        AND R4, R3, R1      ; R4=R3 AND R1
        SUBS R2, R0, R1     ; R2=R0 − R1
        BGT loop            ; if result positive, branch
        NOT R3, R4
```

The important lines are the BGT (branch if condition flags greater than 0) and the line before this which sets the condition flags. Evidently, there is no way of knowing whether the branch should be *taken* or not until the SUBS instruction has finished and the condition flags updates.

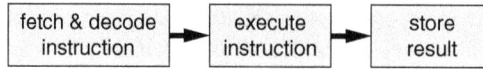

FIGURE 5.5   A flowchart of a very simple three-stage pipeline, where instruction fetch and decode are performed in a single step, and no stage specified for operand fetch.

Let us run this program through just a small and simplified three-stage pipeline, as shown in Figure 5.5.

We then use this pipeline to "execute" the sequence of operations (up to the branch) in a reservation table:

fetch & decode instruction	MOV	AND	SUBS	BGT					
execute instruction		MOV	AND	SUBS	BGT				
store result			MOV	AND	SUBS	BGT			
Clock cycles:	1	2	3	4	5	6	7	8	9

During cycle 5, the result of the SUBS is now known, the condition flags updated, and the branch instruction is being executed, thus the next instruction can be fetched in cycle 6, but this gives a wasted diagonal in the pipeline, which was ready to accept the next instruction in cycle 5.

fetch & decode instruction	MOV	AND	SUBS	BGT	**X**	NOT			
execute instruction		MOV	AND	SUBS	BGT	**X**	NOT		
store result			MOV	AND	SUBS	BGT	**X**	NOT	
Clock cycles:	1	2	3	4	5	6	7	8	9

To reduce this waste, many processors, as mentioned in Section 5.2.5, are capable of speculative execution. That means they will start by fetching the next (NOT) instruction anyway. If the branch is to be taken, this is deleted from the pipeline, and if not, execution continues as normal. Here is a reservation table for *speculative execution*, but where the speculation is incorrect:

fetch & decode instruction	MOV	AND	SUBS	BGT	NOT	MOV			
execute instruction		MOV	AND	SUBS	BGT	NOT	MOV		
store result			MOV	AND	SUBS	BGT	**X**	MOV	
Clock cycles:	1	2	3	4	5	6	7	8	9

Speculative execution, of course, does not always speculate correctly: when it is correct, the pipeline operates at full efficiency, but where it is wrong, there is a loss of efficiency, but no worse than without speculative execution. There are many weird and wonderful techniques to improve the correctness of speculative execution hardware (see Box 5.4).

Box 5.4   Branch Prediction

Given that some CPUs can speculatively execute a branch, it is possible for it to speculate either way—branch taken or branch not taken. A correct speculation results in no loss of efficiency (but an incorrect one results in wasted cycles).

For some CPUs, they always speculate fixed one way, such as "not taken." Then a compiler can improve performance if it organizes code such that "not taken" is more common than "taken."

More intelligent CPUs keep track of past branches. If most of them were taken then they assume "taken" for subsequent speculations, otherwise "not taken." This is called a *global predictor*. More advanced hardware keeps track of individual branches—or more commonly tracking them by the lowest 5 or 6 address bits, so there is a "cache" of 32 or 64 branch trackers with perhaps several branches aliased to each tracker. This is a *local predictor*.

The most complex hardware combines a *global predictor* with several *local predictors* and in such cases impressive prediction rates can be observed. As expected this is a fertile area of performance-led research, but by far the best results are obtained when both the compiler and the hardware work together.

We will explore these topics further in Section 5.7, but note a simple example of speculative hardware in Box 5.5.

5.2.7   Compile-Time Pipeline Remedies

One more point remains before we look to branch remedies, and that is the amount of efficiency reduction caused by pipeline stalls. This obviously depends on the pipeline construction and length, but consider how the two are related.

Three-stage pipelines are rare in modern processors. These days seven, eight, or more stages are commonplace, and wildly complex customized pipelines even more so. The single wasted diagonal in our three-stage example can become a troublesome seven-stage pipeline stall, dragging down processor performance and efficiency. Maybe this explains the amount of time and effort that has been spent on improving pipelines in recent years.

Compile-time tricks to improve pipeline performance range from the trivial to the highly complex. Illustrating one of the more trivial but useful methods, consider the code example from Section 5.2.6:

```
loop:   MOV R1, #5          ; R1=5
        AND R4, R3, R1      ; R4=R3 AND R1
        SUBS R2, R0, R1     ; R2=R0 − R1
        BGT loop            ; if result positive, branch
        NOT R3, R4
```

The problems with this code were that there was no way to know whether the branch should be taken or not before the following instruction is due to be fetched. So it must either wait to be fetched, or fetched speculatively.

But in this case, we could reorder the code to separate the condition-setting instruction (SUBS) and the conditional instruction (BGT) a little further as follows:

```
loop:   MOV R1, #5        ; R1=5
        SUBS R2, R0, R1   ; R2=R0 − R1
        AND R4, R3, R1    ; R4=R3 AND R1
        BGT loop          ; if result positive, branch
        NOT R3, R4
```

In this instance, the reordering does not change the outcome (because the AND does not depend on anything that the SUBS changes, and likewise the SUBS does not depend on anything that the AND changes). The result will be the same—but look at the reservation table:

fetch & decode instruction	MOV	SUBS	AND	BGT	NOT			
execute instruction		MOV	SUBS	AND	BGT	NOT		
store result			MOV	SUBS	AND	BGT	NOT	
Clock cycles:	1	2	3	4	5	6	7	8	9

Whether we take the branch or not, the condition flags are updated by the SUBS at the end of cycle 3, and the branch needs to be decided before cycle 5. There is thus sufficient time between the condition flags changing and the branch so that there does not need to be a delay waiting for the conditions to change—and execution can continue at full efficiency.

The changing of code to suit a pipeline can also be performed for the other hazards—data and mode changes. When reordering is not possible (perhaps because of two sequential branches, or many dependencies), then the compiler is able to either insert a NOP, or simply assume that the pipeline is sufficiently intelligent that it will stall for a short time automatically. This is a reasonable assumption with modern processors, although some of the early pipelined machines relied upon compilers inserting NOPs in this way for correct execution.

Box 5.5   Speculative Execution

Over the years, many forms of speculative execution have been developed. Most notably being the split-pipeline from IBM which, at every conditional branch followed *both* branch paths simultaneously using two identical pipeline paths. One of these pipeline paths would be flushed once the conditions for the original branch were resolved. This machine could thus guarantee absolutely no loss in efficiency caused by conditional branching—but at a substantial hardware cost.

Moving down the ability range, is the *probabilistic branching* model, in which the processor would keep track of how often a branch would be *taken*, as described in Box 5.4, entitled branch prediction, and explored more deeply in Section 5.7.

Despite some very advanced hardware on specialized machines, many speculative branch systems simply fix their speculation to "always take a branch" or "never take a branch." Compilers then have to take note of this and arrange the code appropriately. To do this they need to order the code to attempt to maximize the proportion of the time that the guess made by the processor is correct. Again, much research has been conducted in this active and important area.

5.2.8   Relative Branching

Examining some of the reservation tables that have been discussed or given in examples above, it is clear that the various pipeline stages are performed by different functional units, and so the reservation table can indicate which of those functional units are busy at any particular time.

The execution stage includes the ALU (alongside whatever other single-cycle numerical engines are fitted—that is, not an FPU which is usually multi-cycle). At first glance, it may seem that the ALU has no use during a branch instruction.

However, if a branch target address requires calculating, then perhaps the ALU can be used to perform that calculation? Indeed this is the case for a *relative branch*. That is, a branch forward or backward by a set number of locations (see Box 5.6, and refer also to Chapter 3). These branches are relative to the program counter (PC), and require a certain address offset to be added to the PC, and then the PC to be set to this new value.

Box 5.6   Relative Branching

In the ARM processor, instructions are 32 bits in size (as are both the address and data buses, except in the earliest ARM processors that used a 26-bit address bus). Given that each location on the 32-bit address bus can be specified as an address in any instruction such as a branch, then it should be clear that 32 bits are needed to represent any address in full.

It is thus impossible to store a 32-bit address within a branch instruction if some of the other instruction bits are used for other information (such as identifying the instruction words as being a branch, and identifying any conditions on the branch). Thus *absolute* addressing is not used in the normal ARM branch instruction. Instead, *relative* addressing is used.

The value stored inside the branch instruction word is therefore a signed offset that needs to be added to the current program counter (PC) to determine the location of the *branch target* address.

In fact, the ARM encodes branch offsets as 24-bit signed numbers. By remembering that addresses are on a byte-by-byte basis, but instructions are 4 bytes in size, and specifying that all instructions are aligned to a 4 byte boundary address (such as 0, 4, 8, 12, 1004, and so on) then the lowest 2 bits of any *branch target* address will always be zero, so these two bits need not be stored in the instruction.

In other words, that 24-bit number counts instructions backward or forward from the PC rather than individual bytes. This is a $+/-$ 32 Mibyte range: a huge overkill at the time of the ARMs original design when desktop computer memory rarely exceeded 512 Kibytes, but less impressive in today's world of code-bloat.

In fact, the branch becomes an addition just like the add instruction:

```
ADD PC, PC, #24
```

would move 24 address location forward. And similarly:

```
ADD PC, PC, #-18
```

would move 18 locations backward. Looking again at the previous reservation table, it should be clear that, when a relative branch occurs, whether it is conditional or not, the processor cannot fetch the next instruction until the branch has completed the "execute" stage of the pipeline, and thus the address to fetch it from is determined. Here is an example:

```
ADD R2, R0, R1    ; R2=R0 + R1
B +24                    ; branch 24 locations forward
NOT R3, R4            ; R3= NOT R4
 . .                         . .
 . .                         . .
(24 locations beyond the Branch)
SUB R1, R0, R1    ; R2=R0 − R1
```

The simple three-stage reservation table for this unconditional relative branch is as follows:

fetch & decode instruction	ADD	B	X	SUB					
execute instruction		ADD	B	X	SUB				
store result			ADD	B	X	SUB			
Clock cycles:	1	2	3	4	5	6	7	8	9

This throws open again the whole nature of pipeline efficiency. Even when a branch is not conditional but is relative, it seems that the pipeline must stall. There are two solutions: one is to include a dedicated ALU solely for relative branch calculations and the other is discussed in the next section.

5.2.9   Instruction Set Pipeline Remedies

Since the compiler can reorder code (as described in Section 5.2.7) to separate a condition setting instruction and a branch, it is possible to enforce this in the instruction set. Hence the *delayed branch*, as used in original MIPS processors and some older Texas Instruments DSP processors.

The delayed branch operation does exactly what it says: It delays the branch by a number of cycles—exactly enough cycles to completely solve any problems caused by relative branching, or delays due to condition-setting instructions near to a conditional branch. In the author's opinion, it does this at the expense of the unfortunate assembly language programmer. Having written code for both the processors mentioned, he learned that it is sometimes tempting to negate the improvements generated through the delayed branch mechanism by dropping a couple of NOPs after the instruction, for sanity and safety's sake. As we shall see, failure to observe the delay causes bizarre code problems that have tripped up the best of programmers.

Here is an example of the delayed branch in use:

```
loop:   MOV R1, #5          ; R1=5
        SUBS R2, R0, R1     ; R2=R0 − R1
        BGTD loop           ; conditional branch,delayed
        AND R4, R3, R1      ; R4=R3 AND R1
        NOT R3, R4          ; R2=NOT R4
        NOP
```

As in some of the previous examples, this is a conditional branch. It is also a relative branch, such that the assembler will encode the "BGTD loop" as "BGTD-2" since the loop label is two instructions before the branch instruction, so at run-time, the machine would actually be doing PC=PC-2 if the branch is to be taken.

Since the branch is delayed, it is necessary to know by how much, and this information would be found with details of the instruction set. We shall assume that the branch is delayed by two instruction. What this means is that the branch would not occur at the program line containing the BGTD instruction, it would actually occur two lines later—namely between the NOT and the NOP. Let us examine a reservation table in Table 5.1.

In this 12-cycle table, the loop has been run through twice. During the first iteration (in bold) the branch was taken, during the second iteration the branch was not taken. The first time the branch instruction is encountered, in cycle 3, it loads into the pipeline, and being conditional, waits for the previous condition flag-setting instruction (SUBS) to complete. Although the branch is to be taken, the next two instructions (AND and NOT) are loaded anyway, and the branch not taken until cycle 6, where the PC returns to the MOV instruction at the *loop:* label.

The second iteration sees absolutely identical operation, except that the NOP instruction follows the NOT rather than the MOV to indicate that the branch was *not* taken in this instance.

Concerning the relative nature of the branch, the first iteration of BGTD claims execution during slot 4 (and hence access to the ALU to perform the branch target calculation), easily in time to provide a complete branch target for the PC to load the next instruction from for slot 6.

There are no spaces in the reservation table, indicating full efficiency, whether branches are taken or not, conditional, unconditional, relative, or absolute.

From an assembly-language programmer's perspective, it is important to remember that the AND and the NOT will always be executed irrespective of whether the branch is taken. Confusing? Yes, that is why the following is all too common:

```
BD somewhere
NOP
NOP
```

This might help the low-level programmer who forgets the branch is delayed, but with such code the efficiency gains possible through using a delayed branch instruction will of course be lost. A compiler, on the other hand, would take care of the delayed branches automatically.

5.2.10 Run-Time Pipeline Remedies

Consider again the list of hazards discussed in Section 5.2.4, namely write-after-write, read-after-write, and write-after-read. These can be handled through compile-time

Fetch & decode	MOV	SUBS	BGTD	AND	NOT	MOV	SUBS	BGTD	AND	NOT	NOP	
Execute instruction		MOV	SUBS	BGTD	AND	NOT	MOV	SUBS	BGTD	AND	NOT	NOP
Store result			MOV	SUBS	BGTD	AND	NOT	MOV	SUBS	BGTD	AND	NOT
Cycles	1	2	3	4	5	6	7	8	9	10	11	12

TABLE 5.1 A reservation table capturing 12 clock cycles of the delayed branch example code listed in Section 5.2.9.

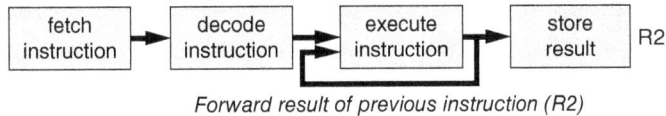

Forward result of previous instruction (R2)

FIGURE 5.6 A four-stage pipeline with data forwarding to send the result of one instruction directly into the execution unit for the following instruction, without storing it to a register (R2) and then reloading it immediately.

measures although, most pipelined processors will automatically handle such hazards without compiler intervention. These processors use run-time methods to resolve hazards that are likely to be troublesome.

If $O(i)$ is the set of output locations (including registers, memory addresses, and condition flags) affected by instruction i, and $I(j)$ is the set of input locations affecting instruction j, then a hazard between instructions i and j will exist if

$$O(i) \cap I(j) \neq \emptyset \text{ for RAW hazard}$$

$$I(i) \cap O(j) \neq \emptyset \text{ for WAR hazard}$$

$$O(i) \cap O(j) \neq \emptyset \text{ for WAW hazard}$$

In general, such hazards can be resolved through forwarding: fetch-fetch, store-store, and store-fetch. Look at a RAW hazard example:

```
ADD R2, R0, R3     ; R2=R0 + R3
AND R1, R2, #2     ; R1=R2 AND 2
```

The hazard is on R2, which must be written by the first instruction before it is read by the second instruction (something that, given a long pipeline, may not necessarily always be true). However, we can imagine a separate path in hardware that takes the output of the first instruction and feeds it directly into the input of the second instruction, only writing the result to R2 afterward as illustrated by the separate feedback path from the output of the execution unit (EX) to one of its inputs in the block diagram of Figure 5.6.

This effectively bypasses the store result (SR) stage and mathematically would be equivalent to performing the following transformation.

```
R2 = R0+R3; R1=R2 & 2  →  R1=(R0+R3) & 2; R2=R0+R3
```

Forwarding is also used to improve speed of execution through, for example, reducing the number of reads and writes to slow off-chip memory by making greater use of on-chip registers.

For example, the following code:

```
LDR R0, [#0x1000]    ; load R0 from mem. address 0x1000
ADD R2, R0, R3       ; R2=R0 + R3
LDR R1, [#0x1000]    ; load R1 from mem. address 0x1000
ADD R3, R2, R1       ; R3=R2 + R1
```

Can easily be replaced by

```
LDR R1, [#0x1000]    ; load R1 from mem. address 0x1000
ADD R2, R1, R3       ; R2=R1 + R3
ADD R3, R2, R1       ; R3=R2 + R1
```

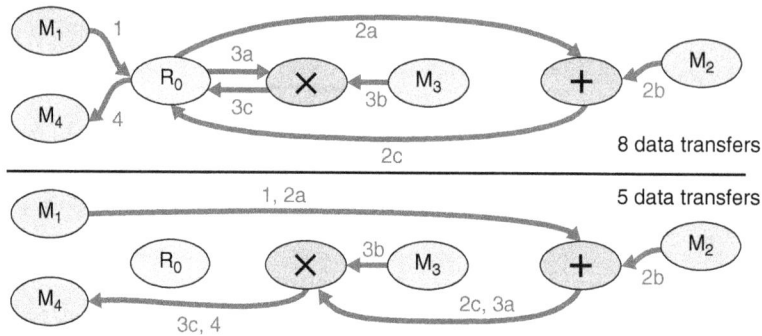

FIGURE 5.7   An example of a simple arithmetic calculation performed without data forwarding (top), and with the use of forwarding to reduce memory save/load operations (bottom).

This example of fetch-fetch forwarding improves execution speed by 25 percent at no cost except either a compile-time optimization stage or a run-time code optimizer. Store-store forwarding would perform a similar task for writes to memory.

Note here, that sometimes multiple reads or writes are communicating with a memory-mapped peripheral such as a UART,[1] where it is entirely reasonable to have multiple writes to the same address (such as serial byte out register), something that would be wasted if it were memory! In the C programming language, such memory pointers would be marked with the keyword *volatile* to prevent a compiler from optimizing them out (the reason for this is described later in Section 7.8.2). For run-time code organization, an intelligent processor (or well set-up memory areas) would detect such addresses as being outside regular memory pages, and thus not optimize in this way.

A final data forwarding example is provided by the following artificial code segment:

```
Instruction 1   LDR R0, [m1]        ; load R0 from address m1
Instruction 2   ADD R0, R0, [m2]    ; R0=R0 + content of address m2
Instruction 3   MUL R0, R0, [m3]    ; R0=R0 × content of address m3
Instruction 4   STR R0, [m4]        ; store R4 to address m4
```

This is represented in the upper half of Figure 5.7 showing eight data transfers involved in the operation, and again in the lower part of the figure as an optimized code section involving only five data transfers in total. In both cases, the instruction to which the transfer is associated is identified. The numerical result of the operations would be the same for both, and the original source code is identical, but speed of execution and resource usage will differ markedly. At run time, the forwarding rules can be determined and applied to accelerate the execution of code by minimizing time consuming and resource-hogging data transfers.

The downside with run-time remedies is that they cost: extra hardware has to be present for them to work, and this increases power consumption, size, and therefore the price of each and every processor that is made. However, for pure processing performance, or when backward compatibility is needed, ruling out compile-time speed ups, run-time methods alone are suitable.

[1] UART: universal asynchronous receiver/transmitter, usually called simply a *serial port*.

5.3   Complex and Reduced Instruction Set Computers

Section 3.2.6 introduced the debate between RISC and CISC architectures, and presented RISC processors as the culmination of an evolutionary process that began with a simple control unit, moved through microcode and then applied the microcode (simple instructions) approach to the entire CPU, resulting in a RISC approach.

So this led to a RISC processor being defined as a CPU with fewer and simpler instructions than normal. Typically 100 instructions were regarded as the upper limit for a RISC processor. However, over the years since their introduction, several more distinctive features of these devices have come to the forefront, as listed below. Be aware though that there are no hard and fast rules here—much is down to the marketing department of the design company.

- *Single-cycle execution*: All instructions are supposed to complete in a single cycle. Not only does this improve processor design difficulties and promote regularity in the instruction set, but it also has the side benefit of reducing interrupt response times (discussed later in Section 6.5). In practice, many RISC processors adhere to this loosely, for example in the ARM, the load/store multiple registers instruction (LDM/STM) can take multiple cycles to complete.

- *No interpretation of instructions*: There should be no need for an on-chip interpreter, since instructions should relate directly to the actual physical hardware available on the processor.

- *Regularity of instruction set*: A glance at the instruction set of a common CISC processor will reveal little commonality between instructions. Bit-fields in the instruction word may mean totally different things from one instruction to another. Some instructions can access one register, others cannot. This is troublesome to the assembly language programmer, but also acts to increase the size of the on-chip instruction decode unit. RISC processors by contrast should have a very regular instruction set that is easy to decode.

- *Regularity of registers and buses*: One way to help achieve the regularity in the instruction set is to maintain a (preferably large) bank of independent registers, all of which are identical in scope and operation. In some CISC processors it is necessary to visualize the internal bus structure to work out how to transfer a value from one functional unit to another using the minimum number of instructions. In a RISC processor by contrast this should be simple: If one register can "see" the value, then all registers can "see" it equally as well.

- *Load-store architecture*: Since memory is far slower than registers, it is far more difficult in a fast clock cycle to load a memory location, process that location, and then store back to memory. In fact, the best way to prevent the external memory accesses from forming a bottleneck is to ensure that, when an external load or store occurs, nothing else happens to slow that instruction down. Thus there is precisely one instruction to load from memory and one instruction to store to memory. All data processing instructions should operate on registers or immediate values only.

As mentioned, there are few rules: There is no global certification authority to decide what is RISC and what is CISC, and many modern designs pragmatically borrow from both camps.

5.4   Superscalar Architectures

The evolution of pipelining in performance-led processors naturally resulted in ever-increasing degrees of pipeline complexity, despite the simplifications promised by the RISC approach. Multifunction dynamic pipelines became more involved, with more customized handling of particular instructions and thus increased control requirements.

Coupling ever-increasing pipeline complexity with the consequent growth in opportunities for hazards, the hazard detection and resolution steps within runtime hardware became more important. These led to significant hardware resources required for the management of pipelines.

5.4.1   Simple Superscalar

One pragmatic alternative to greater pipeline complexity then emerged—an arrangement with a very simple linear pipeline, but augmented with multiple functional units in the execution stage. In this scheme, instructions are issued sequentially, but may follow different paths in the process of execution.

Often the execution stage is the most time-consuming part of the pipeline, and of course in a pipeline the slowest stage is the bottleneck. For this reason, in a superscalar pipelined system, the instruction fetch unit issues instructions into the pipeline at a faster peak rate than any one individual execution element can process them. Multiple copies of execution elements then accept instructions in turn. Such a system is shown in the five-stage pipeline of Figure 5.8.

This approach was pioneered in DSPs which had more than one multiply-accumulate unit (MAC), but only became formalized as a superscalar approach when applied in general-purpose CPUs.

In the diagram shown in Figure 5.8, the floating point unit (FPU) has been inserted into the superscalar pipeline. FPU devices are notoriously slow: placing an FPU in a linear pipeline (with constant instruction clock rate) would result in a very slow processor, but in a superscalar machine, an instruction issued to the FPU would continue executing in parallel with other instructions being handled by the ALUs, multipliers, and so on. Some recent superscalar machines have eight ALUs and 16 MACs, or several ALUs and four FPUs.

A reservation table is given in Table 5.2 for an example superscalar pipeline. In this example, there is a single fetch and decode unit issuing one instruction per clock cycle.

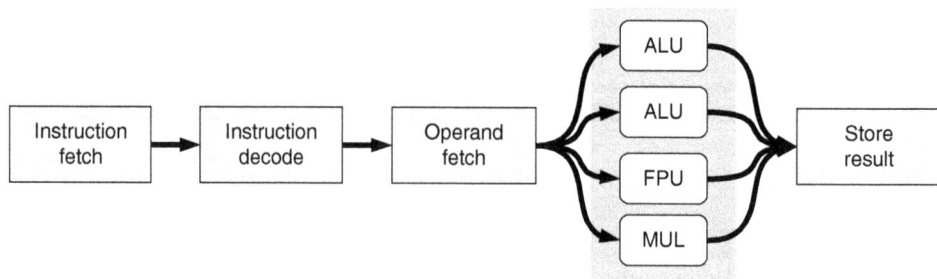

FIGURE 5.8   A five-stage superscalar pipeline showing a fairly conventional pipeline augmented with multiple functional units in the execution stage.

Fetch & decode	ADD	SUB	AND₁	FADD	NOT	MUL₁	MUL₂	–	–	NOR	AND₂	NOT
ALU		ADD		AND₁		NOT				NOR		
ALU			SUB								AND₂	
FPU					FADD							
MUL							MUL₁			MUL₂		
Store result				ADD	SUB	AND₁		FADD	NOT	MUL₁	MUL₂	
Clock cycles	0	1	2	3	4	5	6	7	8	9	10	11	12

CPU resources represented in this pipeline are as follows:

1 Fetch & decode unit, issuing up to one instruction per clock cycle.
2 ALU units, each taking two cycles to complete each instruction.
1 FPU unit, taking three cycles to complete the FADD instruction.
1 MUL unit, taking four cycles to complete each instruction.
1 Store result unit, taking a single cycle to store the result of each instruction.

TABLE 5.2 A reservation table capturing 12 clock cycles of operation for an example superscalar pipeline. Note that the second MUL instruction in cycle 6 stalls the fetch unit until cycle 10 when the MUL unit is free.

Instructions are issued to four functional units (two ALUs, one FPU, and one MUL unit). A single store stage then completes the pipeline. Examining Table 5.2, it should be noticeable that the instruction fetch unit issues instructions faster than any of the individual pipelined execution units can operate—and also that the stored results could be out of order compared to the input sequence. Not all machines are able to cope with out of order execution, usually requiring complex runtime hazard-avoidance hardware. In fact, we will examine one machine that excels at handling out-of-order execution, the Tomasulo method, in Section 5.9, and another method called *scoreboarding* is briefly described in Box 5.7.

Box 5.7   Scoreboarding

A central "scoreboard" is used to keep track of the dependencies of all issued instructions, and to allow any instruction which has no dependencies at that time to be issued, irrespective of its order in the original program. Let us consider how this works.

On instruction issue, the system determines the source and destination operand registers specified by that instruction. It then stalls until two conditions are met: (i) any other instruction writing to the same register has completed and (ii) the required functional unit is available. These conditions counteract WAW hazards and structural hazards, respectively.

Once an instruction is issued to a functional unit, operand(s) are then fetched from the instruction source register(s). However, the fetch process stalls until the completion of any current instructions that will write to the source register(s). This solves RAW hazards.

Having collected all operands, the instruction is then executed (and of course the scoreboard continues to keep track of that instruction until it completes).

Finally, the instruction completes and is ready to write its result to the destination register. However, at this point, the write will stall if there are any earlier instructions that have been issued but have not yet fetched their operands, and which the current instruction would overwrite. In other words, if an earlier instruction is still stuck somewhere waiting to execute, and this needs to read from register Rx, but the current instruction is about to write to Rx, then the current instruction will be delayed until the earlier instruction becomes unstuck and completes its reading of Rx. This mechanism avoids WAR hazards.

Although the example program in Table 5.2 is rather short, it can be seen that the instruction output rate is less than the instruction input rate: eventually this system will have to pause the issuance of instructions to wait for pipeline elements to become free. The system thus requires the ability to maintain a higher peak instruction handling rate than the average rate which it achieves when executing real-world code. It is quite possible that when benchmarking such a processor, the manufacturer might choose an instruction sequence that happens to run at peak rate rather than a realistic average rate (we have briefly met such benchmarking issues in Section 3.5.2).

Unfortunately this simple view is not the end of the story—and in fact is not the most common view of a superscalar system. For that we need to consider the issuing of multiple instructions in a single cycle, discussed below in Section 5.4.2.

5.4.2   Multiple-Issue Superscalar

In Section 5.4.1, we had considered adding multiple functional units to a scalar pipeline. This does not quite create a full superscalar machine but rather something that is more competent than a scalar machine.

An advance on the simple superscalar machine is the ability to issue multiple instructions per cycle. That is, instead of issuing one instruction per cycle to multiple functional units, we issue multiple instructions per cycle to multiple functional units.

The block diagram of Figure 5.8 may not change significantly; however, the realization of multiple instructions issued per cycle leads to a different reservation table. This can either appear similar to those drawn previously, albeit with two "spaces" for fetch and decodes each cycle (and for store), or in an entirely different form we will encounter a little later.

The following execution table shows two fetches per cycle. The fetch units are feeding three execute units, which in turn are supported by two result store units. Interestingly, there are two apparent gaps in the pipeline operation, in cycles 3 and 5, respectively. During both of those cycles, the second fetch and decode unit cannot fetch a new instruction. The reason is that in each case, the previously fetched instruction has not yet been issued since the required functional unit (execute unit 1 in both instances) is occupied. Instead, it stalls that fetch unit until the needed functional unit is available.

Fetch & decode instruction	I_1	I_3	I_5	I_6	I_8	I_9	I_{11}
Fetch & decode instruction	I_2	I_4	–	I_7	–	I_{10}	I_{12}
Execute unit 1		I_1	I_3	I_4	I_6	I_7	I_{10}
Execute unit 2		I_2				I_8	
Execute unit 3				I_5			I_9
Store result			I_1	I_3	I_4	I_6	I_7
Store result			I_2		I_5		I_8
Clock cycles:	1	2	3	4	5	6	7

Throughout this text we have drawn many reservation tables similar to the one above to illustrate pipeline operation. However, there are other ways of drawing reservation tables. An example is shown in Figure 5.9. This shows instructions being issued sequentially from the top down, and time along the horizontal axis. In this case, there are no blockages in the pipeline and so instructions are both issued in order and retired in order; however, this may not necessarily be the case in reality.

5.4.3   Superscalar Performance

Superscalar architectures are characterized by the speed at which they issue instructions compared to how quickly they process them. In theory a superscalar machine does not need to be pipelined, but in reality all, or almost all of them are pipelined.

Everything depends, of course, upon what constitutes the measure of performance (see again Section 3.5.2). We had already noted that superscalar machines need to be capable of high-speed instruction issue, even though the average issue rate in practice may be significantly below this peak—depending primarily upon the occupancy of the execution units. Taking these constraints into consideration, issue rate can be improved

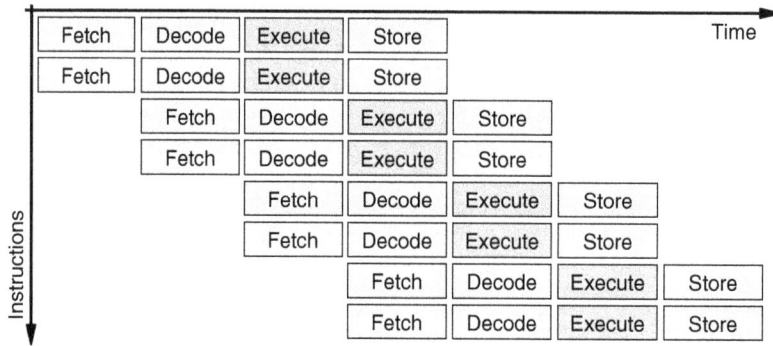

FIGURE 5.9   An alternative reservation table format showing instructions being executed sequentially from top down, moving forward in time from left to right. A vertical line drawn through the table identifies the operation at a specific time—as it also does in the reservation tables drawn so far.

through compiler settings to interleave instructions for different functional units and depends strongly upon the nature of the task being computed.

At best, and clearly seen in Figure 5.9, a superscalar machine is actually handling instructions in parallel. It is, therefore, a form of parallel computer (something that we will explore more fully in Section 5.8).

5.5   Instructions per Cycle

Instructions per cycle (IPC) is a very important measure of how quickly a processor can execute a program, at least in theory. It is not a measure of the amount of work that can be achieved per cycle—because that depends upon how powerful the instructions are and also upon how intelligent the coder/compiler is. Thus, it is not particularly relevant for comparing the execution speed of code on different machines.

However, IPC is a useful indicator, when averaged carefully over representative code, of the raw processing ability of an architecture. In fact, the ratio of average to peak CPI could be seen as an "honest measure"—an average close to the peak value indicates an architecture that is very well optimized for the code being executed.

5.5.1   IPC of Difference Architectures

Different types of processors aim for different IPC scores and naturally fall into certain operating regimes as a result of their structures:

- *CISC processors* exhibit an IPC far below 1.0. This is because instructions tend to be relatively time-consuming and there was historically little attempt to simplify instructions in such machines.

- *RISC processors*, by contrast, exhibit an IPC which approaches 1.0, although for various reasons they may not quite achieve such a score. Reasons for falling short include the use of occasional lengthy instructions (such as multiply, divide, and so on), and the need to wait for loads/stores to/from slower external memory—RISC machines are almost always load-store machines (Section 3.2.3). This latter effect is particularly prevalent in some DSP devices when using external memory. Pipelining can help to push the IPC of a RISC processor even closer to 1.0.

- *Superscalar processors* as we have seen, aim to issue multiple instructions in parallel. Where there are n issue units (or up to n instructions issued each cycle), then IPC approaches n. However, as we have seen in Section 5.4.2, sequences of instructions that require the same functional unit (hardware dependencies) or having unresolved data dependencies will often cause pipeline stalls. Clearly the more often a pipeline stalls, the lower will be the achieved IPC.

- *VLIW[2]/EPIC[3] processors*, both of which are discussed later in Section 5.10, aim for an IPC which is significantly greater than 1.0, and are useful in niche areas, typically related to media or signal processing.

- *Parallel machines* may include two or more processor cores insider a computer, each of which has a lower IPC, but when operating in parallel exhibits a higher throughput. In effect, the IPC of the entire computer would approach the IPC per core multiplied by the number of cores. We will discuss this a little more below.

Improving IPC has been a major focus for many processor designers in recent years, and has been the prime tool that computer architects have used for increasing performance.

The case of parallel machines is particularly interesting, and currently relevant due to the push of major processor manufacturers such as Intel, at the time of writing, toward ever higher levels of parallelism. We will consider parallel processing approaches more fully later in Section 5.8, but suffice it to say that manufacturers appear to have reached a point with increasing clock speed, and in terms of architectural complexities, of decreasing returns: further efforts in either direction do not translate to a commensurate increase in performance. In other words it has become increasingly difficult to push the performance envelope using approaches that change silicon gate structures, or employ clever architectures within CPUs.

As will be revealed as we progress further through descriptions of the remaining items in the above list, the approaches chosen by modern computer architects are increasingly offloading responsibility for increased performance onto compilers and software. Let us recap a little to illustrate this observation: CISC processors performed many functions in hardware. By contrast the RISC approach simplified (and speeded up) hardware by providing simpler instructions. RISC meant that more software instructions were often required, but these could be processed faster. So RISC programs are typically longer than CISC programs, and the compiler has to work just a little harder to create them. Superscalar systems then included some limited parallelism within the pipeline, but issues of handling data dependency became important to prevent pipeline stalls, and thus to achieve good performance, compilers had to take dependencies into account and have an intimate knowledge of the capabilities of the superscalar pipeline. VLIW and EPIC,

---

[2]VLIW: very long instruction word.
[3]EPIC: explicitly parallel instruction computing.

as we will discover, are far more complex to program than anything we have discussed up to now, relying totally upon compiler-level scheduling.

So also with parallel machines. Although the processors themselves may be simple, their interactions can become complex. Furthermore it is debatable whether the current generation of software engineers is really able to think and program "in parallel." Beyond the programmer, there is little debate that the most popular programming languages are not at all optimal when producing code for parallel processors. It seems that two things are needed before parallel machines can be fully exploited: (i) a new generation of programmers who are naturally able to write parallel code and (ii) a new generation of programming languages and tools to support them.

Just one further note on parallel processing. Although "going parallel" has been the pragmatic response of processor manufacturers to continual demands for increased performance, achieving the promised speedups is largely left to programmers. For individual programs, this speedup is elusive. However, for server and desktop machines in particular, running advanced multitasking operating systems such as Linux, it is very common for several *threads* (tasks) of execution to be running simultaneously. Parallel machines can apportion different threads to different processors, and although the individual threads do not execute any faster in terms of CPU time, they will complete quicker because they no longer get time-sliced and pre-empted as frequently by other tasks. In embedded systems, where typically fewer tasks are running, or perhaps only one major task is active at a time, there is less advantage in moving to a parallel processing solution. In these systems, a significant speedup would only be evident if the critical tasks themselves were "parallelized."[4] This brings the argument back to good parallel-aware tools and languages being written by parallel-aware programmers.

5.5.2 Measuring IPC

As we have already noted in Section 5.4.3 and elsewhere (including through our discussions in Section 3.5.2), performance measures in computing are notoriously unreliable at predicting real performance. An engineer wanting to execute a known algorithm can simply try the algorithm on several architectures to determine which is fastest. However, any prediction of the performance of nonspecific code depends upon so many factors that it may be more useful to follow ballpark figures such as dividing a quoted IPC average by the instruction clock frequency and then multiplying this figure by the number of instructions that need to be executed.

As the size and generality of the program increases, the more accurate this type of determination will be. Bear in mind though that embedded systems are more normally characterized by a small fixed collection of computational tasks—by contrast the code running on desktop and server machines can seldom be predicted at design time.

The question arises as to whether quoted IPC figures are accurate. In any architecture, there are certain tricks that could be used to enhance quoted IPC figures. It is instructive to consider some of these:

- Quote peak IPC figures rather than average IPC, meaning absolutely best case figures are given.

---

[4]Parallelized: made to run in parallel.

- Quote average IPC, but averaged only over selected test code, not over representative code.

- IPC figures given are for execution from internal memory only.

- IPC has been calculated using external memory, but with the operating clock set very slow (so that the speed of the memory does not affect the figure—the cycle itself is slower).

- Slow instructions have not been used (or are rarely used) in the code chosen to evaluate IPC and interrupts have been disabled.

- Known slow sequences of instructions have been removed.

Every architecture has its advantages and disadvantages, and through the descriptions in this book the reader can begin to appreciate some of these. However, choosing a processor for a particular computational task is often an art rather than a science: it may require intuition. In the view of the author, ignore the sales and marketing information relating to performance. Performance is seldom the most critical criteria and will be outweighed by the ease of programming, expandability, available support and development tools, and product lifetime, in addition to other more technical characteristics.

5.6   Hardware Acceleration

Much of the silicon area in a modern CPU is dedicated to accelerating basic processing operations. Acceleration methods include using a fast cache, adding extra buses to the architecture, pipelining, and incorporating dedicated numerical processing units.

Originally, processors contained only a basic ALU for number processing (and it can be shown that *all* processing operations can be performed with an ALU alone—at least if execution speed is not important). Later, multiply-accumulate units were added to speed up multiplication operations, which were previously performed using repeated additions.

Floating point hardware, now deemed mandatory in desktop computers, was originally an extra-cost option requiring insertion of a separate chip. Alongside floating point, desktop processors now routinely contain SIMD hardware (see Section 2.1.1), and are beginning to incorporate various accelerators for wireless networking capabilities.

Other processing accelerators include those for graphics manipulation, cryptography, communications, and data compression. It seems that the profusion of these units will continue to increase, as will their application-specific nature—especially in dedicated embedded system-on-chip processors.

On the other hand, there are structural improvements to increase processing speed that are not data-processing related. Several have been considered previously, such as pipelining (Section 5.2), caches (Section 4.4), multiple bus architectures (Section 4.1), and customized instructions (Section 3.3). In this section, several further generic architectural support methods are considered.

5.6.1   Zero-Overhead Loops

Many algorithms consist of loops, such as *for()*, *while()*, or *do()*. Generally, the loops require some sort of overhead. Take as an example a loop that iterates a given number of times. This is a simple construct in most programming languages.

```
i=20;
while (i-- > 0)
{
        <do something>
}
```

This requires a sequence of steps to its operation:

1. Set $i = 20$.
2. Compare i and zero.
3. Branch to instruction after loop if equal.
4. $i = i - 1$.
5. Perform the body of the loop.
6. Branch back to the start of the loop (step 2 on this list).

The loop condition could be checked either before or after the body of the loop is executed depending on the type of loop, but what is clear is that when the item inside the loop is very simple, there is a large overhead. Consider the following example from DSP code implementing a digital filter:

```
for(i=20;i>0;i--)
  y=y+x[i]*coeff[i];
```

The body calculation in the loop, although it appears complicated, can be executed in a single instruction in a modern DSP processor. However, if the six-step loop sequence above is applied, this code will take 1 instruction to set up, then 20 iterations of steps 2 to 6, that is, up to 101 instructions in total.

Since many DSP loops are tight and small like the one illustrated, DSP designers recognized the inefficiency of needing so many extra instructions to support looping and developed the zero-overhead loop (ZOL) concept.

Here is an assembler-style example from the Texas Instruments TMS320C50:

```
        set BRCR to #20
        RPTB loop -- 1
        ... <body of loop>
loop    ... <now outside loop>
```

In this case, there is a single instruction required to preload the BRCR loop counter, and then a single instruction to launch the looping. The DSP will examine the address of the program counter, and when it reaches a value of (loop – 1) will automatically reset it to the start address of the loop. It will do this 20 times in total. For the 20-iteration loop, this now requires only 22 instructions to complete rather than the 101 needed if the ZOL support was missing.

Analog Devices have a similar concept in their ADSP2181:

```
        set  CNTR to #20
        DO loop UNTIL LE
        ... <body of loop>
loop    ... <now outside loop>
```

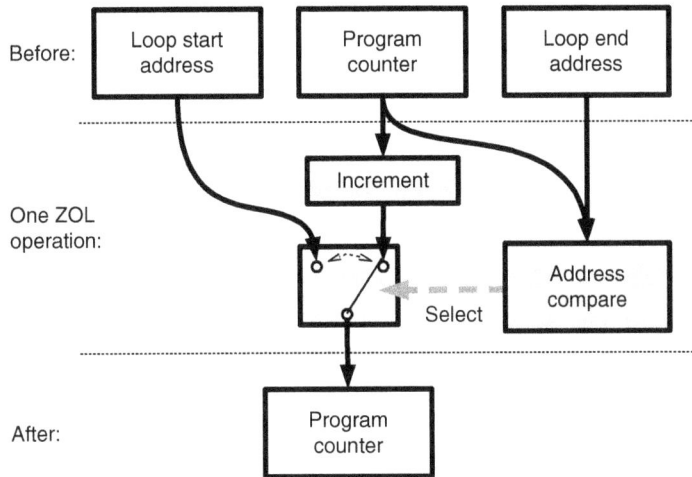

FIGURE 5.10 A block diagram showing the hardware needed to implement a zero-overhead loop within a processor, and how that interfaces with the various registers including program counter and loop setup registers.

It can be seen that the operation principle is the same—but provides the possibility of different loop end conditions (LE meaning "less than or equal to"—there are 15 other possible conditions). Section 5.6.2 will expand on the addressing capabilities of the ADSP2181 beyond this.

The hardware required for ZOL support in this way is relatively simple, as can be seen from the block diagram in Figure 5.10.

Namely, somewhere to store the program address at the start of the loop, somewhere to store the address of the end of the loop, a way to determine when the PC reaches the end of the loop (an address comparator), and a way to branch back to the start of the loop by reloading the PC from the loop start address. In addition, there needs to be a means of holding the loop counter, decrement this, and determine when the end condition is reached (e.g., the loop counter had reached zero).

One complication that can occur is when the contents of the loop is not a simple instruction, but perhaps calls other functions which themselves contain loops. Thus there may be a need to *nest* the loops. In the ADSP the ZOL registers are actually incorporated inside a stack to allow automatic nesting of loops with the minor proviso that loop end addresses cannot coincide. By contrast the TMS lacks supporting hardware, and so such nesting has to be accompanied through manually saving and restoring the loop registers.

The second complication is that, although the two ZOL examples were written in assembly language, most code these days is written in C. The C compiler therefore has to recognize opportunities for using the ZOL hardware. Simple C constructs such as the *while* and *for* loops above, and the following, are easy to map to ZOL:

```
k=20;
do{
    <something>
} while (k-- >0)
```

Note that these examples all have loop counters that count downward. In the TMS there is no way for a loop counter to increment, only decrement, so code such as

```
for(i=0;i<20;i++)
{
    <do something>
}
```

would need to be converted to a downward counting loop (i.e., the counter is decremented from 20 to 0) in the final assembly language code, always assuming that the particular compiler in use is sufficiently intelligent.

Still the onus is on the software programmer to ensure that C code is structured so that it can take advantage of ZOL hardware. In such hardware it is best to avoid any loop increment or decrement other than by one, and to avoid using the loop index for performing arithmetic within the loop.

Given that simple loops can have zero overhead, the old embedded code guideline of merging separate loops together wherever possible is not always true. In fact, it may be detrimental if (perhaps due to a shortage of temporary registers) it forces loop variables to be stored to external memory during the loop.

In the ADSP2181 it is possible to have an infinite hardware loop, but an exit condition can be evaluated manually as part of the loop body. This can actually be highly advantageous for C programs since it generalizes to all possible loop constructs.

This type of loop acceleration hardware is called a PC trap. More complex hardware exists to perform similar tasks as described in the next section.

5.6.2   Address Handling Hardware

While the ARM processor has only a single bank of general-purpose registers, many processors differentiate between registers depending upon whether they are for storing data or addresses. In fact, several of these processors force this distinction through having different widths of data and address bus.

The Motorola 68000 series CPUs, although having uniform 32-bit registers, makes a distinction between the eight data registers D0-D7 and seven address registers A0-A6. Although any value can be contained in these, many addressing modes only apply when the address is stored in the correct set of registers, and similarly many processing instructions cannot store a result directly to an address register. Programmers can take advantage of dedicated hardware attached to the address registers to perform increments and decrements of address values either before or after access, and perform indexing, but if more complex address calculations are required, they are likely to have to move an address from the A to a D register, perform the arithmetic, and then move the result back to an A register.

The ADSP21xx series of DSPs extends this approach further through the use of data address generators (DAGs). There are two of these in the ADSP2181, each containing four I, L, and M registers:

I0	L0	M0
I1	L1	M1
I2	L2	M2
I3	L3	M3

DAG1

I4	L4	M4
I5	L5	M5
I6	L6	M6
I7	L7	M7

DAG2

Each index (I) register contains an actual address used to access memory, the L registers holds a memory region length to correspond with those addresses, and the M registers hold modification values.

In assembly language, a read from memory is accomplished through syntax such as

```
AX0 = DM(I3, M1);
```

This means that a value is read from data memory address pointed to by I3, that value is stored in register AX0, and then register I3 is modified with the content of register M1. If this new value of I3 exceeds the length register L3 + initial I3, then the value in I3 will be stored modulo the initial I3 value (the initial I3 value means the start address of the buffer). If the length register L3 is set to zero, then there is no change to the content of I3. Some examples will clarify this arrangement (see examples box), but first note that nowhere in the instruction was L3 mentioned. That is because the I and L registers operate in pairs, whereas by contrast the M registers are independent: within each DAG, any M register can be used to modify any I register, but the M registers in one DAG cannot modify I registers in the other DAG. Box 5.8 presents three examples of the ADSP21xx ZOL hardware in action.

Box 5.8 ZOL Worked Examples

Example 1: Let us use a made-up hybrid of ADSP assembler to access memory, and ARM-style assembler for everything else. The exact addresses used in this (and the other) examples are for illustrative purposes only, usually they would be allocated by linker since there are certain constraints not discussed here (but covered in the ADSP21xx manual).

```
       MOV I0, #0x1000    ; Set I0=0x1000
       MOV L0, #0x2       ; Set L0=2
       MOV M0, #0         ; Set M0=0
       MOV M1, #1         ; Set M1=1
loop:  AX0 = DM(I0, M0)   ; Load AX0
       ADD AX0, AX0, #8   ; AX0=AX0+8
       DM(I0, M1) = AX0   ; Store AX0
       B loop
```

Next we will construct a table showing the values of I0 as the loop is executed:

	After this instruction	I0 is this:
1	MOV M1, #1	0x1000
2	AX0 = DM(I0, M0)	0x1000
3	ADD AX0, AX0, #8	0x1000
4	DM(I0, M1) = AX0	0x1001
5	B loop	0x1001
6	AX0 = DM(I0, M0)	0x1001
7	ADD AX0, AX0, #8	0x1001
8	DM(I0, M1) = AX0	0x1000
9	B loop	0x1000

Note how the value in I0 is first modified by M0 in the second row, but as M0 contains zero, it is unchanged. In the fourth row, I0 is modified by M1. Since

M1 = 1 this is an increment. Again, the modification by M0 in row 6 is followed by another increment in the eighth row. Here, however, I0 hits `0x1002`, and since L0 = 2, this ends the circular buffer, and the address therefore wraps back to `0x1000`.

Example 2: L1 has been loaded with 0, I1 contains `0x1000` and M0 is `0x10`.

Successive readings from I0 using `AX0=DM(I1,M0)` will see address register I0 holding the following successive values: `0x1000`, `0x1010`, `0x1020`, `0x1030`, `0x1040`, `0x1050`, and so on. Since L1 holds zero, there will be no wrap around.

Example 3: In this case L4 has been loaded with 50 and I0 = 0, M4 = 2 and M5 = 10. This corresponds to a circular buffer of size 50 locations, starting at address 0. The following loop is executed:

```
loop:   AX0 = DM(I4, M5)
        AY0 = DM(I4, M4)
        B loop
```

As this loop progresses, I4 will take the following values:

0, 10, 12, 22, 24, 34, 36, 46, **48, 8,** 10, 20, 22, . . . , and so on. The numbers of interest are highlighted, that show that from 48, the index I3 would normally increment by 10 to become 58, but since L4 holds 50, this has exceeded buffer length, and the register must therefore wrap around to the start, hence being 8.

Undoubtedly the ADSP has very capable and advanced address handling, but consider the addressing modes of Section 3.3.4 which are based on those available in the ARM processor. In fact the ADSP does not really have any capability beyond those addressing modes.

Thus the DAG, and it's extra hardware are useful in maintaining circular buffers, and performing synchronized addressing changes (for example backward and forward in predefined steps), but beyond those efficiency gains do not fundamentally improve processor performance. The cost of this efficiency gain is the silicon area needed for hardware such as that shown in Figure 5.11 for one of the DAG units in the ADSP2181 DSP.

It can be seen from the figure that, since at most one of the registers in each DAG is accessed per instruction cycle, that each of L, I, and M are accessed through shared buses. The DMD bus is data-memory-data which conveys data operands, and links to data memory (see Section 4.1.3 for more detail on the unusual internal bus architecture of the ADSP device). In addition to its other abilities, DAG1 (not shown) is able to bit-reverse its address outputs: a significant performance improvement for performing fast Fourier transform (FFT) calculations and several other signal processing techniques.

With separate data and program memory on-chip in the ADSP21xx serviced by independent buses, and with the dual DAGs, the device is capable of accessing two DAG-addressed operands in memory indirectly, with dual post modification and wraparound. Once accessed, these two operands can be processed and stored, all in a single instruction cycle. The ARM, by contrast, is functionally capable of performing the same operations, but cannot hope to achieve as much in a single cycle. Having said that, the ARM does not ever need to perform two separate addressing operations in a single instruction (since being load-store has at most only one address operand)—see Box 5.9 for more detail.

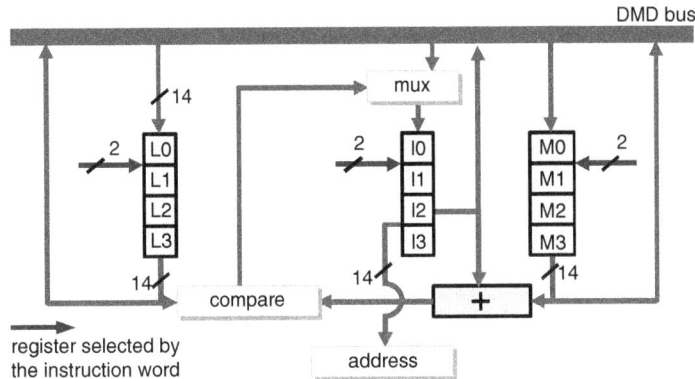

FIGURE 5.11 A block diagram of the second data address generator (DAG2) hardware within the ADSP2181 digital signal processor, showing how internal length registers L0-L3, index registers I0-I3 and modifier registers M0-M3 are wired up to a dedicated address adder, and to the internal DMD (data memory data) bus.

Box 5.9 Address Generation in the ARM

Being a RISC design, the ARM minimizes special handling hardware for addresses, but through simplicity tries to streamline instructions so that nevertheless they process very quickly.

As discussed in Section 3.2.3 the ARM has a load-store architecture with one data load and one data store instruction (actually there is also a *swap* instruction intended for multiprocessor systems). The address to load or store from can be indexed with pre- or postoffset (increment or decrement), and can be direct or indirect.

The ARM utilizes the main ALU and shifter for address calculations since these are free for a pipeline slot during load or store (see Section 5.2.8). This also provides an advantage over the DAG of the ADSP in that the main ALU and shifter are more flexible than the dedicated ALUs in the DAG.

Here is an example of that flexibility:

```
LDR   R0,   [R1, R2, LSL#2]
```

This loads the value at memory location (R1 + (R2×4)) into register R0. LSL means "logical shift left," an addressing calculation that is not accessible to the DAGs of the ADSP21xx, despite their evident capabilities.

Note finally that there are no alternate or shadow DAG registers (described below in Section 5.6.3) in the ADSP21xx. This means that use of DAGs is dependent on such factors as program context and interrupt servicing: it is likely that direct hand-coding of assembly language is needed to exploit these address handling accelerators to the full.

5.6.3 Shadow Registers

CPU registers are part of the *context* of the processor that is viewable by a running program. Other aspects of context include status flags and viewable memory.

When a program thread is interrupted by an external interrupt signal (this topic is explained in Section 6.5.1), an interrupt service routine (ISR) is generally run that responds to that interrupt appropriately. Once the ISR completes, control returns to the original program. "Control" in this sentence basically referring to where the program counter is pointing. The program may be happily stepping line by line through some assembler code, then an interrupt causes it to fly off to an ISR which it then steps through and completes, before returning back to the original program, continuing as if nothing had happened.

Thinking about this process, it is evident that when interrupts are enabled, an ISR could be triggered in between any two instructions of any program! It is therefore vitally important that the ISR, when it returns, tidies things up so that the context is exactly the same as it was when the ISR was called.

A few years ago, programmers would have to perform what is called a context save at the start of the ISR, and then a context restore prior to exit. The save would be to push each register in turn onto a stack, while the restore would be to pop these back off again in the reverse order. This might mean an overhead of 20 or 30 lines of code that would need to be run inside the ISR even before it could do anything useful.

To remove this overhead, the concept of a shadow register set was developed. This being a second set of registers, identical in every way and operation to the main set, but which can be utilized as required (and thus altered) inside an ISR without changing the content of the original registers visible to the main program. On the TMS320C50 for instance, once an interrupt occurs, the processor jumps to the relevant ISR and automatically switches to shadow registers. When the ISR finishes, a special return instruction causes a jump back to wherever the PC was before the interrupt, and switches back to the original registers.

With such shadow registers there is no need to perform a manual context save and restore at the beginning and end of an ISR. Any piece of code can be interrupted without any overhead. However, if there is only a single set of shadow registers, interrupts cannot be nested. That means that one interrupt cannot interrupt another.

5.7   Branch Prediction

In Section 5.2.6 we had investigated the phenomenon whereby pipeline performance will often reduce as a result of branching. We had seen that branching per se can be problematic, and is exacerbated by issues such as conditional branching hazards and relative branching. We also briefly met the idea of performing branch prediction (Box 5.4) and allowing speculative execution (Box 5.5) as methods of reducing this *branch penalty*.

In this section, we will first summarize the reasons for branch-induced performance loss and then discuss methods of branch prediction allied with the capability of speculative execution to mitigate against such losses. As we progress, consider the sheer ingenuity involved in some of the methods we present (and the hardware costs involved), and let this bear testament to how much branch-induced performance loss is a thorn in the flesh to computer architects.

In an ideal world, we could train programmers to avoid branch instructions, but until that happens, specialized hardware presented in this section will continue to be needed and continue to be the focus of CPU performance research.

5.7.1 The Need for Branch Prediction

First, let us recap on some of the issues related to branching. Consider the following code executing on a four-stage fetch—decode—execute—store pipeline:

```
i1       ADD R1, R2, R3
i2       B loop1
i3       ADD R0, R2, R3
i4       AND R4, R2, B3
loop1:   STR R1, locationA
```

Without constructing a reservation table, let us follow the first few cycles of operation:

- *i1* is fetched.

- *i2* is fetched while *i1* is decoded.

- *i3* is fetched while *i2* is decoded and *i1* is executed. At the end of this cycle the CPU "knows" that *i2* is a branch.

At this point, instruction *i3* has already been fetched and is in the pipeline. However, correct operation would require the instruction at label `loop1` to be the next one executed because *i2* is a branch. *i3* therefore has to be deleted from the pipeline, and the correct instruction fetched. This will cause a "bubble" in the pipeline, consequentially reducing efficiency.

We had also met the issue of relative branching in Section 5.2.8: the very common arrangement where the branch target address (i.e., the address of the next instruction to fetch) is stored within a branch instruction as a relative offset to the current program counter (PC) address. The CPU thus has to perform an ALU operation to add this offset to the PC to obtain the address from which it can fetch the next instruction.

In our example above, if the address to branch to (in this case the address of the instruction at label `loop1`) has to be calculated, this will require another cycle *after* the branch instruction has been decoded. Most likely, processors using this technique will then immediately clear the pipeline and perform the branch. As a sequence of operations, that would look like the following:

- *i1* is fetched.

- *i2* is fetched while *i1* is decoded.

- *i3* is fetched, *i2* is decoded, and *i1* is executed.

- *i4* is fetched, *i3* is decoded, *i2* is executed (which means that the branch target address is calculated using the ALU) and the result of *i1* is stored.

- The result of *i2*, the branch target address, is stored—but to the PC rather than to another register, and the remainder of the pipeline is reset (thus discarding *i3* and *i4*).

- The instruction at the calculated branch address is then fetched.[5]

---

[5] It should be noted here that many processors would have fetched this instruction in the previous cycle by directly outputting the calculated address from the ALU onto the address bus—a form of data forwarding—while simultaneously loading it into the PC.

Up to now, we have not even mentioned the conditional branch hazard situations, where the pipeline needs to wait for the resolution of a previous condition-setting instruction before deciding if a branch should be taken or not.

We had mentioned the role of speculation in alleviating the problems associated with branching. To recap, *speculative execution* means always beginning to execute one path while waiting for the outcome of the conditional, and sometimes also for the address calculations to complete. Before the path being speculatively executed is allowed to complete, the processor fully determines whether that speculation was correct (in which case the speculatively executed instructions can complete) or incorrect (in which case these instructions, and their results, are trashed).

Some processors speculate deterministically—for example, they always speculate that a branch is taken, or perhaps always that it is not taken. Of course, in the absence of any other effects, such a technique cannot really hope to be correct more than 50 percent of the time. Wherever possible, it also makes sense for a compiler producing code for such a CPU to arrange the generated code so that the speculative path is more commonly taken.

In effect, speculation is guessing: betting that a particular path is taken. A correct guess pays off because usually in this case the processor will have experienced no pipeline stall. An incorrect guess will probably cause a pipeline stall while the remains of the speculative execution are cleared from the pipeline.

A refinement of speculation is *branch prediction*, which means making a more intelligent guess based on information such as

- Past behavior

- Code region/address

- Hints put in the code by the compiler (e.g., a take-don't take bit—TDTB[6])

Dynamic Branch Prediction usually relies on some measure of past behavior to predict a future branch. This was summarized previously in Box 5.4. When the CPU sees a branch it (i) uses a predictor to very quickly make a decision of which path to speculate on. Later when the actual branch outcome is known it (ii) updates the predictor, to hopefully refine the prediction decisions in future to continually improve accuracy.

We will investigate seven different prediction methods in turn, discussing their operation and performance:

- Single T-bit

- Two-bit predictor

- The counter and shift registers as predictors

- Local branch predictor

- Global branch predictor

- The gselect predictor

- The gshare predictor

---

[6] A take/do not take bit (TDTB) is inserted in the program code by a smart compiler to tell the speculation unit what it believes to be the most likely branch outcome at this position. Remember that the compiler has more knowledge available to it than the branch unit—the compiler can "see" into the future, knows the full extent of loops, functions, and programs, and knows what the next instructions will be down each of the alternative paths.

Following these subsections, hybrid schemes will be considered (Section 5.7.9) and then the refinement of using a branch target buffer (Section 5.7.10).

5.7.2 Single T-Bit Predictor

In the very simple single T-bit prediction scheme, a flag "T" is set to 1 whenever a branch is confirmed as taken, and 0 when not. This is updated whenever the CPU has just completed every branch instruction, that is, after all conditionals and other factors have been resolved. The T-bit global predictor has very low hardware overheads—just one bit being used to predict the behavior of the entire CPU.

Whenever a new branch instruction is encountered, the pipeline speculates by following the state of the T bit. In other words, if the last branch was taken ($T = 1$), the predictor is indicating that the next one will also be taken. If the last branch was not taken ($T = 0$), the prediction is that the next branch will not be taken either. This is not a particularly intelligent scheme, but can work surprisingly well—especially with compiler support. Primarily, it is good method where many simple loops exist within the code being executed.

For example, consider the following ARM-style assembler code with initial conditions R1=1 and R2=4:

```
i1   loop:   SUBS R2, R2, R1   ; R2=R2-R1
i2           BGT loop          ; branch if result >0
```

Now we will "run" this code through a CPU that has a global T-bit predictor, in order to ascertain how well the predictor copes with the simple loop case:

trace:	i1	i2	i1	i2	i1	i2	i1	i2
R1	1	1	1	1	1	1	1	1
R2	3	3	2	2	1	1	0	0
T-bit	–	1	1	1	1	1	1	0
branch	–	T	–	T	–	T	–	NT
correct	–	–	–	Y	–	Y	–	N

Starting in the leftmost column of the trace table,[7] after instruction *i1* has completed the first time, the register contents will be as shown since R2 has been decremented from 4 to 3 by the subtraction. In the next cycle, *i2*, the branch instruction, will be taken since the result of the SUBS was greater than zero. On this first loop, the predictor is assumed uninitialized and therefore cannot predict anything accurately.

As the trace progresses, the loop repeats two more times and then exits (by virtue of not taking the branch back to the beginning of the loop during the final cycle). By the second loop the predictor has learned that the previous branch was taken, and therefore correctly predicts that the branch will be taken. Likewise, the prediction during the third loop is correct. Upon reaching the branch instruction for the final time, however, the prediction is incorrect.

---

[7]This trace table cannot take the place of a full reservation table because it does not represent what is happening within the pipeline at a particular time—neither does it indicate how long it takes to execute each instruction—but simply as an indicator of the state of the system after each instruction has completed in order.

In general it can be seen that the first branch in such a loop might not be correctly predicted, depending upon the state of the T-bit predictor prior to executing this code. The final branch will be incorrectly predicted, but within the body of the loop—no matter how many times it repeats—the prediction will be correct. This holds true for any size simple loop: no matter what code is placed in between *i1* and *i2*, as long as it contains no other branches, the prediction will be as we have described.

Unfortunately, however, loops are rarely as simple as this. There will often be other branches within the loop code. Let us illustrate this, again with another simple example:

```
i1   loop:   SUBS R2, R2, R1   ; R2=R2−R1
i2           BLT error          ; branch if result < 0
i3           BGT loop           ; branch if result > 0
```

We will again "run" this code through a CPU that has a global T-bit predictor, in order to ascertain how well the predictor copes with the simple loop case. In this case, we will assume an initial condition of R2=3 in order to reduce the number of columns a little:

trace:	*i1*	*i2*	*i3*	*i1*	*i2*	*i3*	*i1*	*i2*	*i3*
R1	1	1	1	1	1	1	1	1	1
R2	2	2	2	1	1	1	0	0	0
T-bit	–	0	0	1	0	0	1	1	0
branch	–	NT	T	–	NT	T	–	NT	NT
correct	–	–	N	–	N	N	–	N	N

In this case, performance is not so good: The predictor fails to correctly predict *any* of the branches. Unfortunately, such a result is all too common with the simple T-bit global predictor. As we can see in subsequent sections, this can be improved by either predicting with a little greater complexity, or by applying separate predictor to each of the two branch instructions. First though, let us investigate doubling the size of the predictor.

5.7.3   Two-Bit Predictor

The 2-bit predictor is conceptually similar to the T-bit, but uses the result of the last *two* branches to predict the next branch, instead of just the last one branch. This method uses a state controller similar to that shown in Figure 5.12. This is also referred to as a bimodal predictor.

Since there are now four states (i.e., two bits to describe the state), one would expect this predictor to be more accurate than the single T-bit. While this is so in general, it is in the nested loop case—the instance where the single T-bit did not fare very well—that the 2-bit predictor can provide better performance.

To illustrate this, we will use the same code as used in the previous section:

```
i1   loop:   SUBS R2, R2, R1   ; R2=R2−R1
i2           BLT error          ; branch if result < 0
i3           BGT loop           ; branch if result > 0
```

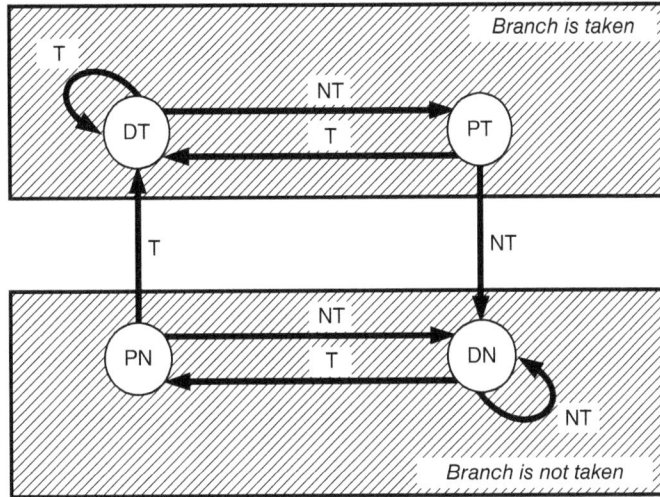

FIGURE 5.12 The state diagram of a 2-bit predictor, showing states DT (definitely take), PT (probably take), PN (probably not take), and DN (definitely not take). When in the first two states the prediction is that the branch will be taken, in the second two states the prediction is that the branch will not be taken. Following the resolution of the branch conditions, the predictor state is updated based upon the actual branch outcome: T (taken) or NT (not taken).

This time we will "run" the code through a CPU that has a global 2-bit predictor. We will again assume an initial condition of R2=3, R2=1 and that the predictor is initialized to state "DT":

trace:	i1	i2	i3	i1	i2	i3	i1	i2	i3
R1	1	1	1	1	1	1	1	1	1
R2	2	2	2	1	1	1	0	0	0
predictor	DT	**PT**	DT	DT	PT	DT	DT	PT	DN
branch	–	NT	**T**	–	NT	T	–	NT	NT
correct	–	N	**Y**	–	N	Y	–	N	N

The trace table is similar to those in Section 5.7.2, and care should be taken to read the table accurately. Remember that each column shows the state of the processor after the instruction indicated has completed, and that there is no timing information implied in the table, simply the sequence of operations. As an example of reading the table, find the first time instruction *i2* has executed. In this column, we see that it has left R1 and R2 unchanged, but since it is a branch that is NOT taken, it will have shifted the predictor state from "DT" to "PT" (shown in bold). When instruction *i3* has completed, since it is a branch that IS taken (shown in bold), it will have shifted the predictor state back to "DT." When *i3* began, the predictor state was still "PT," thus the prediction was that the branch would be taken, and in fact this was a correct prediction. This correct prediction is shown in bold as the result on the bottom line. Thus remember to compare the branch outcome in any particular column with the prediction shown in the previous column when determining prediction accuracy.

While this predictor has clearly not particularly excelled in its prediction of all branches, it has correctly predicted one of the branches in the loop every cycle apart from the termination cycle. This is halfway between the result shown by the T-bit predictor and a perfect result.

Let us now explain this rationale a little more closely: it seems a single-bit predictor has some problems, which can be partially solved by using a 2-bit predictor. If 2-bit predictor has problems, can these be solved through applying more bits? Well, the answer is potentially "yes" because in general spending more resources on the problem results in better performance. However, the quest is to use as small a hardware resource as possible while improving performance as much as possible.

At this point we need to recognize that it is quite difficult to ever predict the outcome of *i3* based upon the previous outcome of some other branch instruction *i2*. Much better to predict the future outcome of *i2* based upon the past history of *i2* and to predict the future outcome of *i3* based upon the past history of *i3*. In other words, to somehow separate the prediction of the different instructions. In fact, this is what we will encounter starting with the bimodal predictor in Section 5.7.5. However, first we will look at using even more bits for our predictors.

5.7.4 The Counter and Shift Registers as Predictors

A simple saturating counter can be incremented each time a branch is taken and decremented each time a branch is not taken. The counter saturates rather than wraps-around, so that a long sequence of branches which are taken will lead to the counter hitting maximum and staying there.

For such a counter, the branch prediction is simply the state of the most significant bit (MSB). That is effectively giving the majority, since the MSB becomes "1" once the counter is half of its maximum value or above, and is "0" when below half its maximum.

The counter is fairly simple hardware, but it can take a long time to "learn" when switching from a normally taken to normally not taken loop and does not work well on a branch within a nested loop.

A similar-sized item of hardware is the shift register. An *n*-bit shift register holds the results of the past *n* branches. Whenever a branch instruction is resolved by the processor, the result is fed into the shift register with the contents shuffling along to accommodate it and the oldest stored value being discarded. For example, with a "1" representing a branch that was taken and a "0" representing a branch that was not taken, a shift register storing the result of the past eight branch instructions with a sequence NT, NT, NT, T, T, NT, T, NT would contain 00011010. If another branch was then taken, the shift register would be updated to 00110101 by shifting every bit along to the left, discarding the leftmost "0" and appending the new "1" to the least significant bit position.

We do not investigate either of these techniques in isolation because they are more normally used when combined together in a prediction mechanism which employs some locality. Four of these are now discussed in turn.

5.7.5 Local Branch Predictor

A simple observation in low-level code is that some branches are almost always taken, and some are almost never taken. It seems that the global T-bit and global 2-bit predictors

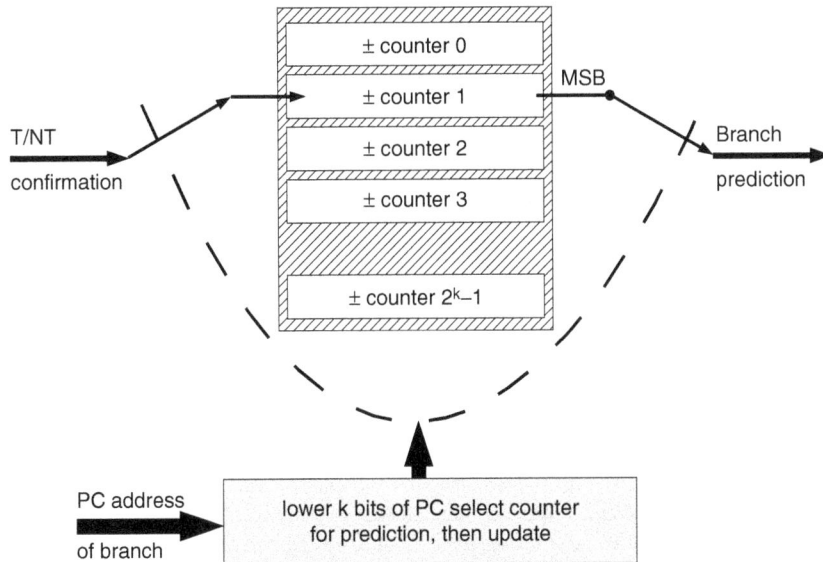

FIGURE 5.13 A block diagram of a local branch predictor, showing a bank of saturating branch history counters which are indexed based upon the lower bits of the address of any particular branch (thus the different counters can each map to different groups of branch instructions). The most significant bit of each counters indicates the branch prediction as explained in Section 5.7.4.

treat all branches within a CPU in the same way. A more sensible scheme would be to predict different branches locally, rather than globally. This also relates back in some way to the principle of locality of Section 4.4.4: for example, it is reasonable to assume that the branching behavior in library code would be different to that in user code and thus both should be predicted differently. Even within user code, different regions of a program would also naturally exhibit different branch patterns.

As mentioned previously it would seem possible to have a T-bit predictor (or 2-bit predictor) for each individual branch. However, there are potentially thousands or even millions of branch instructions in some code. The hardware needed to enable this would be quite significant.

So then perhaps there is some compromise between a single global predictor, and one predictor for each branch. This gives rise to the concept of a bank of predictors. In some ways, this mirrors the hardware arrangement of cache memory (Section 4.4), and also suffers from a similar problem: look-up time. Using such a system, whenever a branch instruction is encountered, the predictor for that branch would need to be "looked up" and the prediction determined. With more and more predictors to be searched, the look-up time becomes longer and longer, eventually maybe even exceeding the cycle time of the pipeline. Thus the emphasis of computer architects is actually on having fewer predictors—but making their operation more intelligent.

An arrangement of saturating branch history counters (Section 5.7.4) is shown in Figure 5.13. Instead of having one counter predictor for all branches, there are 2^{k-1} separate counter predictors, each predicting branch instructions at different addresses.

Since the lower k bits of the address bus[8] are used to select which counter is used for prediction (and of course which counter will later be updated by the outcome of a particular branch instruction once it is resolved), a branch located at address 0 will be predicted by counter 0, a branch located at address 1 will be predicted by counter 1, and so on. If there are only eight counters, then counter 0 would also predict branches at address 8, 16, 32, 64, ..., and so on.

Note that the bank of predictors could instead be a bank of T-bit or bimodal predictors rather than saturating counter predictors. The important thing is that the principle of locality has been brought into play: prediction is based, at least in part, upon address location. We can illustrate the operation of this system using the code we had previously tested for the global T-bit and 2-bit predictors:

```
i1   loop:   SUBS R2, R2, R1   ; R2=R2−R1
i2            BLT error         ; branch if result < 0
i3            BGT loop          ; branch if result > 0
```

This time we will "run" the code through a CPU that has a local branch predictor as shown in Figure 5.13. We will again assume an initial condition of R2=3, and that the predictor counters are 4-bits in size and are initialized to 0111 prior to execution. Instruction $i1$ is located at address zero:

trace:	i1	i2	i3	i1	i2	i3	i1	i2	i3
R1	1	1	1	1	1	1	1	1	1
R2	2	2	2	1	1	1	0	0	0
c0	**0111**	0111	0111	**0111**	0111	0111	**0111**	0111	0111
c1	0111	**0110**	0110	0110	**0101**	0101	0101	**0100**	0100
c2	0111	0111	**1000**	1000	1000	**1001**	1001	1001	**1010**
branch	–	NT	T	–	NT	T	–	NT	NT
correct	–	Y	N	–	Y	Y	–	Y	N

The table this time shows three predictor counters (c0, c1, and c2) which are mapped to the addresses of instructions $i1$–$i3$ since the code begins at address zero. In this case, predictor counter c0 never changes because there is no branch instruction at address 0 to update it. The other two counters are updated as a result of the completion of the branch instructions which map to them. The predictor which is selected at each address is shown in bold font.

In each case of a branch instruction the prediction is made by examining the MSB of the corresponding prediction counter from the column before the current instruction (since as always, the columns contain the machine state *after* the respective instruction completes, but the prediction is sought *before* the instruction begins).

The performance of the predictor is rather different from that encountered previously. The first branch instruction is correctly predicted during each loop. The second branch instruction is incorrectly predicted during the first and last loops, but within the loop body—no matter how many times it repeats or how many non-branch instructions it contains—is always correctly predicted. This should be seen as a significant improvement over the case in Section 5.7.3.

[8] Some processors, such as the ARM, count addresses in bytes but have instructions which are larger. In this case, since instructions are actually at addresses 0, 4, 8, 16, address bus bits A0 and A1 will always be set to zero for any instruction in the ARM. These bits are thus ignored and the address bits used by this and subsequent local predictors begin at A2.

Unfortunately, the story does not end here because while this predictor is quite capable, it suffers from aliasing effects as illustrated by example in Box 5.10.

Box 5.10   Aliasing in Local Prediction

Let us execute the following assembly language code in a processor that has a four-entry local predictor array containing 3-bit saturating counters:

```
0x0000   loop0   DADD R1, R2, R3
0x1001           BGT loop1
0x1002           B loop2
  . . .    . . .    . . .
0x1020   loop1   DSUB R3, R3, R5
0x1021           B loop0
```

We will assume that on entry R2=0, R3=2, R5=1, each of the counters c0, c1, c2, c3 is initialized to 011 and that the code exits with the branch to loop2.

address	outcome	branch	predictor	correct
0x0000	R1 ← 2			
0x1001		T	c1 ← 100	N
0x1020	R3 ← 1			
0x1021		T	c1 ← 101	Y
0x0000	R1 ← 1			
0x1001		T	c1 ← 110	Y
0x1020	R3 ← 0			
0x1021		T	c1 ← 111	Y
0x0000	R1 ← 0			
0x1001		NT	c1 ← 110	N
0x1002		T	c2 ← 100	Y

In this table, the address of the instruction just executed is shown in the leftmost column. Next is the outcome of the instruction (i.e., whether any registers were changed). The third column indicates, for branch instructions, whether they were taken or not taken. Each branch outcome will involve the update of a predictor counter in the next column, while the final column tallies the success of the predictor.

Overall, the prediction was fairly successful; however, the most important point to note is that only two predictor counters are used. Counter c1 has actually aliased to represent two branch instructions—at addresses 0x0001 and 0x0021, respectively. Thus we have hardware capable of local prediction, but we are essentially not utilizing it effectively. In order to more effectively "spread" the available counters among the branches, we need to introduce some other mechanisms. Two of these are described in Sections 5.7.7 and 5.7.8.

5.7.6   Global Branch Predictor

The basic global branch predictor is an attempt to improve upon the basic local branch predictor in one particular way. This is namely in the ability to introduce *context* into the branch prediction. We have already seen how the principle of locality has been brought

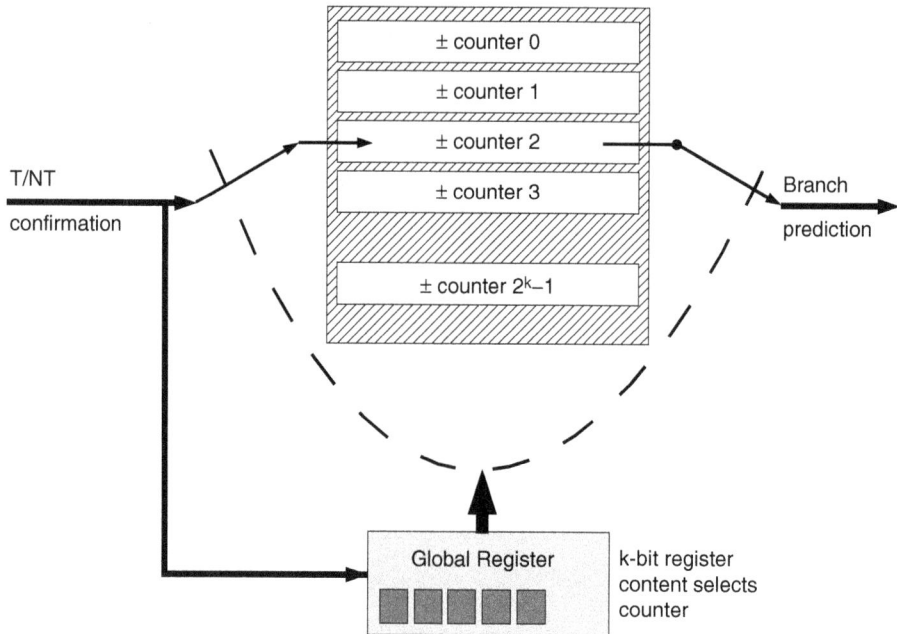

FIGURE 5.14 Block diagram of a global predictor, showing a bank of counters which are indexed based upon the content of a shift register, which stores the outcome of the previous k branches. As usual, each counter increments when a branch related to it is taken, and decrements when the branch is not taken. Counters saturate rather than wrap-around, so that their most significant bit indicates the branch prediction. The global shift register updates after each and every branch instruction outcome is resolved.

into branch prediction, but the aliasing issue in local branch prediction has branches located in totally different types of software aliased to the same predictor.

In the global branch predictor, a global shift register is used, instead of the least significant address bits, to index into an array of counter predictors (both of these elements were briefly described separately in Section 5.7.4). The overall structure is as shown in Figure 5.14, and appears very similar to the local predictor, with the exception of the counter select mechanism as we have discussed.

Since the counter selected to predict a particular branch is chosen based upon the outcome of the past k branch instructions, this scheme is in some ways predicting a branch based upon *how it was reached* rather than where it is located in memory. In other words, it is more like a trace-based selector.

In some circumstances this prediction select mechanism is obviously sensible: for example, a simple library routine can be called many times from different areas of code. How it behaves (in term of its branching behavior) when called could naturally depend upon what it is asked to do, which in turn depends upon how it was called (and from where). The observation from examining many execution traces of common software is that some quite complex sequences of branches may be executed repetitively. Using this predictor, where the sequence of branches select the predictor, it is considered more likely that individual counters would map more closely to individual branches.

We can examine the operation of the global predictor with another simple example:

```
i1   loop1   ADD R1, R1, R2
i2           BEZ lpend
i3           SUB R8, R8, R1
i4           B loop1
i5   lpend   NOP
```

We will assume that on entry R1=3, R2=-1, R8=10, and that there is a 4-bit global register (GR, initialized to 0000), hence 15 counter predictors, each 3 bits and initialized to 011.

address	outcome	branch	GR	predictor	correct
i1	R1 ← 2		0000		
i2		NT	0000	c0 ← 010	Y
i3	R8 ← 8		0000		
i4		T	0001	c1 ← 100	N
i1	R1 ← 1		0001		
i2		NT	0010	c2 ← 010	Y
i3	R8 ← 7		0010		
i4		T	0101	c5 ← 100	N
i1	R1 ← 0		0101		
i2		T	1011	c11 ← 100	N
i5			1011		

The construction of this table is similar to those in previous sections, and the GR value is shown in full—there is only one GR and it is updated after every branch instruction. Although this code loops around three times, the interesting fact is that none of the branches aliases to the same counter predictor. Even the subsequent invocation of the same branch instruction has no history in this example.

In general, it shows that the aliasing problem has largely been avoided and that the branch instructions have been "mixed up" among the counter predictors, but unfortunately the past history has been lost: We could have used that history to predict the branches at *i2* and especially *i4* very well.

It has to be said that in much larger examples than this tiny piece of code, the predictor performs quite well: Figures of over 90 percent accuracy for large global predictors running loop-based benchmark code are not unheard of. However, the basic objection stated above remains: much of the locality information has been lost. We therefore now consider two predictors in turn that combine both the global register trace-based behavioral selection with the address-based local selection.

5.7.7 The Gselect Predictor

The gselect predictor, shown in Figure 5.15, updates the global predictor by also considering the address of the branch to be predicted. In fact, the *k*-bit index which chooses the particular counter predictor (or T-bit or bimodal predictor) to consult for a particular branch is made up from an *n*-bit global register concatenated with the lowest *m* bits of the program counter.

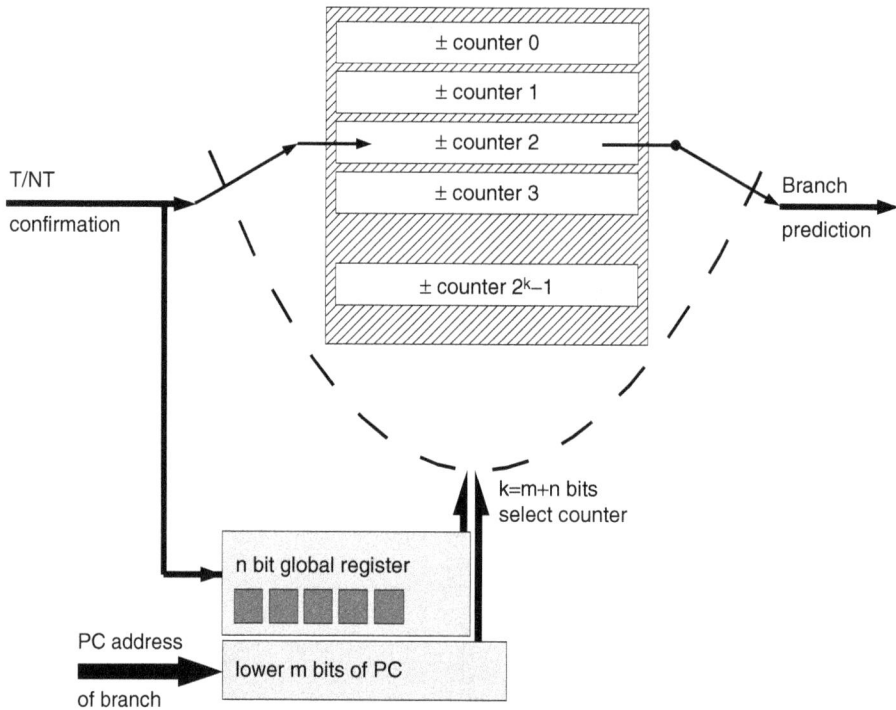

FIGURE 5.15 Block diagram of a gselect predictor, showing a bank of counters that are indexed based upon both the content of a shift register that stores the outcome of the previous n branches and also the lower m bits of the address bus. As usual, each counter increments when a branch related to it is taken, and decrements when the branch is not taken. Counters saturate rather than wrap-around, so that their most significant bit indicates the branch prediction. The global shift register updates after each and every branch instruction outcome is resolved.

For example, where $k = 10$ is made from a 4-bit global register, G, and 6 bits from the address bus, A, the 10-bit index would then be

G_3	G_2	G_1	G_0	A_5	A_4	A_3	A_2	A_1	A_0

Gselect is reportedly well suited for reasonably small banks of individual predictors— which probably indicates its suitability for a resource-constrained embedded system. Where the bank becomes larger, perhaps $k > 8$ the similar gshare scheme, discussed in the next section, may perform better.[9]

---

[9] Remember when discussing performance that it is highly dependent upon many factors, not least of which is the particular code that is to be executed. While we can predict performance in general, there is no substitute for actually testing out the schemes with real code.

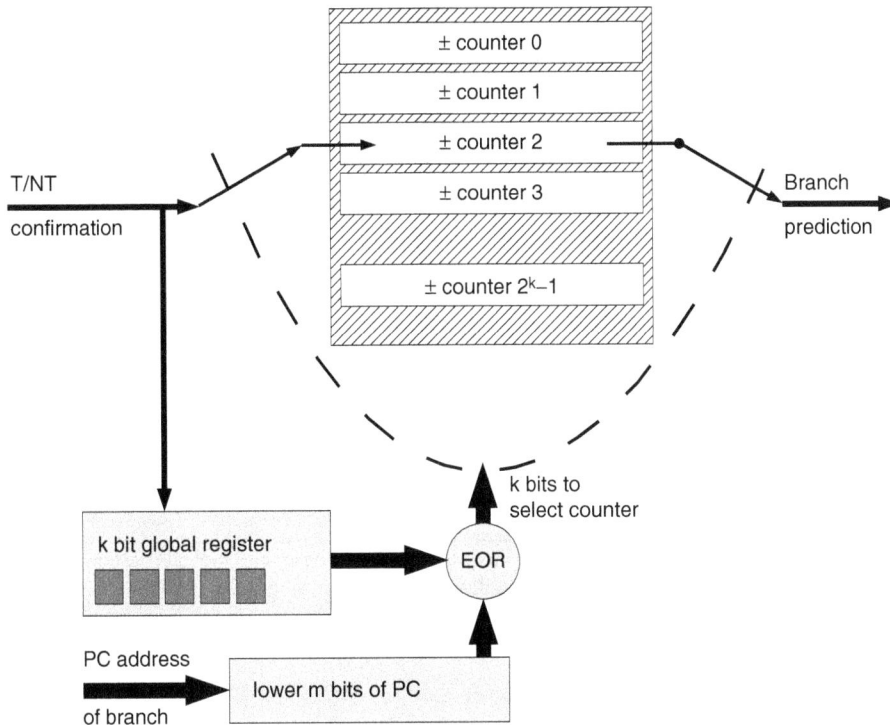

FIGURE 5.16 Block diagram of a gshare predictor, showing a bank of counters that are indexed based upon both the content of a shift register that stores the outcome of the previous k branches and also the lower k bits of the address bus (in this case, the counter is selected as the exclusive OR of these two k-bit values). As usual, each counter increments when a branch related to it is taken, and decrements when the branch is not taken. Counters saturate rather than wrap-around, so that their most significant bit indicates the branch prediction. The global shift register updates after each and every branch instruction outcome is resolved.

5.7.8 The Gshare Predictor

The gshare predictor is simply a refinement of the gselect predictor of Section 5.7.7. Compare the gselect block diagram in Figure 5.15 to that of the gshare in Figure 5.16: The only difference is that the gshare uses the exclusive-OR of a k-bit global register and the k lowest bits of the program counter to index into the array of individual predictors.

Gshare, like gselect and the global branch predictor can exceed 90 percent accuracy if correctly set up and tuned. However, the beauty of both gshare and gselect is that relatively small bank sizes can perform well. Small bank sizes (i.e., fewer individual prediction counters) means that the lookup process can be very quick. Gshare can outperform gselect in most situations apart from very small bank sizes as it does a better job of distributing branch instructions among the individual prediction counters: In other words gshare is more likely to see an even distribution of branches to the counters whereas gselect may see just a few counters aliasing to many branch instructions.

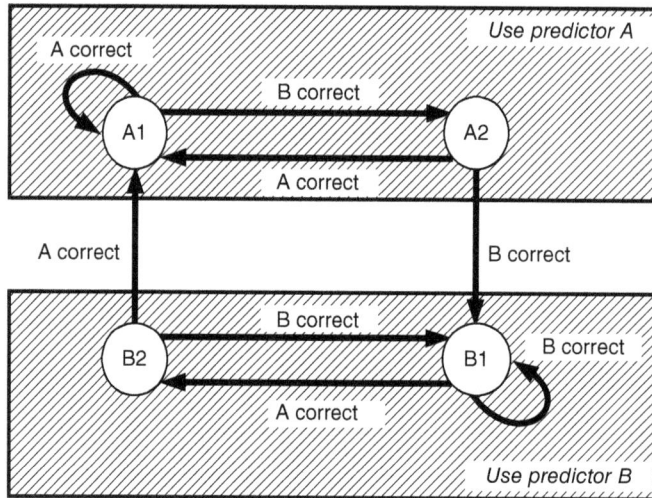

FIGURE 5.17   Two different predictors, having unique characteristics that suit certain types of code in particular can be combined. One way of doing that is by using a 2-bit state machine—very similar to the 2-bit predictor—to select the best prediction method to use. In this state machine, if both predictors are correct in any state, we can assume that no transition takes place.

5.7.9   Hybrid Predictors

If we pause to think about it, there is a strong likelihood that branch characteristics will probably be different for different programs. Up to now we have presented many schemes and discussed some of their particular advantages and disadvantages. The emphasis has been on choosing a branch prediction scheme that seems to work well. However, independent testing of all of these schemes in the academic literature shows that certain types of code are more likely to work better with certain predictors. Thus perhaps it is useful to combine predictors.

This is precisely the approach of the hybrid predictors. These allow multiple branch predictors to be created, along with logic to select the best one. A scheme for selecting between two predictors A and B is shown in Figure 5.17 (and looks rather like the bimodal predictor of Section 5.7.5). In this scheme, the A/B selector is used to keep track of the prediction accuracy of predictors A and B. Whichever predictor is most accurate will quite quickly be chosen as the overall predictor in the system.

We would expect that different programs, or even different regions within programs, would gravitate toward different predictors, and that is precisely what happens in practice.

One famous example of a hybrid predictor is found in the Alpha 21264 processor. A block diagram of this is shown in Figure 5.18. In the block diagram, an A/B predictor is shown, which selects either a global predictor or a two-level local predictor.

The global predictor uses a 12-bit history of previous branches to select one of 4096 2-bit predictors. This predictor is accurate to branch behavior—in other words it is sensitive to *along what path* a particular branch instruction was reached (refer to Section 5.7.6).

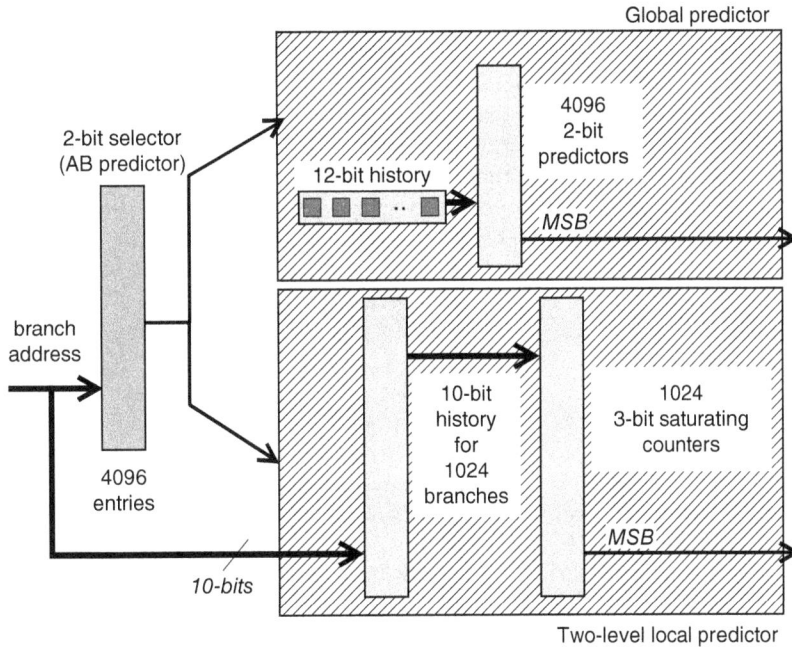

FIGURE 5.18 The hybrid predictor used in the Alpha 21264 processor, shown here as a block diagram, uses a state machine similar to the AB predictor of Figure 5.17 (see leftmost block) to choose between either a global predictor or a two-level local predictor, yielding excellent prediction performance.

The local predictor uses the lowest 10 bits of the address bus to select one of 1024 10-bit shift registers. This shift register is a local version of the global register . . . it keeps track of the history of branches occurring at the current 10-bit address. Do not be confused that both the address and the shift register size are 10 bits, they could have been any size.

This local shift register value is then used to choose one of 1024 3-bit saturating counters: individual prediction counters. The prediction value is the MSB of those counters.

The predictor in the Alpha 21264 uses both a multilevel structure (for local prediction) and a dynamic selection between two very different predictors. It would seem to ally almost all of the prediction elements discussed up to now.

However, we need to ask how well this performs. Given that a limited amount of hardware within a CPU can be "spent" on branch prediction, it is appropriate to wonder whether this amount of hardware would be better spent on one type of predictor, or on another type—or even on improving some other aspect of the pipeline.

In this case, that question was answered for us back in 1993, the year that the Digital Equipment Corporation (DEC) Alpha 21264 CPU branch prediction unit was being designed. Tests indicated that this hybrid approach outperformed both an equivalent-sized global predictor and an equivalent-sized local predictor. In fact, the branch prediction accuracy of this processor is an amazing 98 percent on real-world code—a figure that is hard to beat even in the most modern CPUs.

5.7.10   Branch Target Buffer

As we have seen in the previous sections, branch predictors can quite accurately know whether a particular branch will be taken or not. Returning to the reasons for wanting to predict a branch, remember that this is to improve the chance that code executed speculatively is the correct code, and will not need to be flushed from the pipeline.

One of the main reasons that we need to speculatively execute code is that when a branch is to be taken, a target address which is stored as a relative offset within the branch instruction requires an ALU to add this offset to the PC before the target address can be determined. This process needs an ALU, and in a machine without a dedicated address ALU, the only time the shared ALU is available for the address calculation is during the pipeline slot when the branch instruction is in the "execute" stage. We had seen this way back in Section 5.2.8.

However, even if we correctly predict whether a branch is taken or not, we still need to perform this address calculation. In other words, we might be able to predict very quickly, but then we have to wait for the calculation to take place (or at least perform both in parallel—in which case we need to wait for the slowest of the operations).

So computer architects came up with an ingenious idea: Why not store the target address in the predictor? Instead of simply predicting take/do not take, why not predict the entire target address? After all, there is only one place a branch instruction can branch to, and if we are storing a history of branch behavior, we could easily store the branch target address at the same time.

This is what a branch target buffer (BTB) does.

Using a BTB means we do not have to wait for the branch target address computation in the ALU if we predict right and have executed the current branch at least once before. The decision flowchart within the BTB is given in Figure 5.19. If we need to execute a branch prediction, we first consult the BTB. If we get a BTB hit (i.e., this branch instruction has an entry in the BTB, meaning we have "seen" it previously), then we simply load the BTB target address into the PC and begin executing from that address.

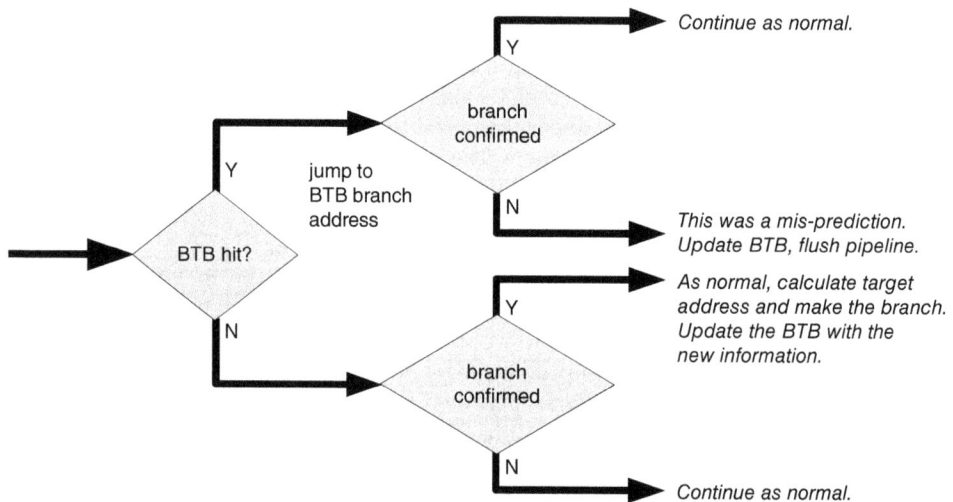

FIGURE 5.19   A simple flowchart illustrating the operation of a branch target buffer (BTB).

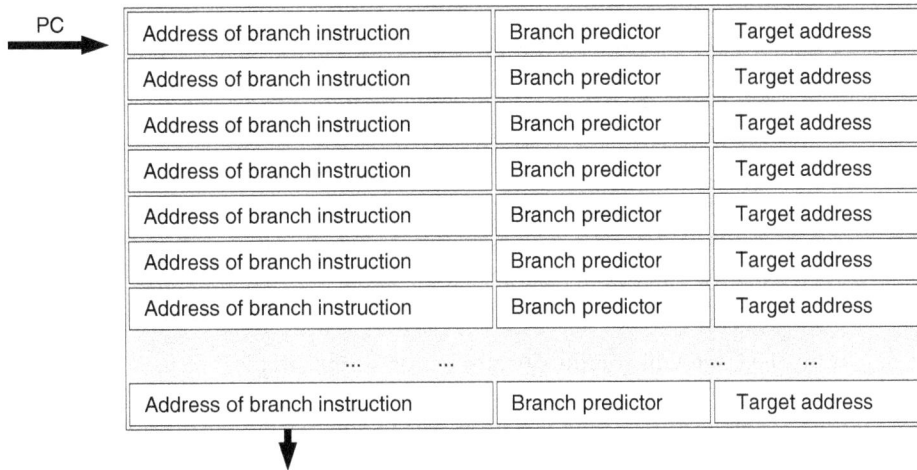

PC →	Address of branch instruction	Branch predictor	Target address
	Address of branch instruction	Branch predictor	Target address
	Address of branch instruction	Branch predictor	Target address
	Address of branch instruction	Branch predictor	Target address
	Address of branch instruction	Branch predictor	Target address
	Address of branch instruction	Branch predictor	Target address
	Address of branch instruction	Branch predictor	Target address
	... ...		... ...
	Address of branch instruction	Branch predictor	Target address

FIGURE 5.20 The branch target buffer (BTB) is organized in a similar way to cache memory, and in fact performs a similar function of reducing the average access time of instructions contained within it.

As soon as the branch is resolved (immediately for an unconditional branch, or after the condition-setting instruction has completed for a conditional branch), we know whether to continue with the speculation, or to flush the pipeline, update the BTB, and fetch the correct instruction.

If we did not have a BTB hit then we speculate "not taken." Once the branch has been resolved, if it should have been taken, we update the BTB with the branch target address, flush the pipeline if we have speculated, and then jump to the correct address to continue execution.

In actual fact, the contents of the BTB, shown in Figure 5.20, appear very similar to a cache memory (Section 4.4) with a tag made up from the branch instruction address, an entry to store the branch prediction (using any of the prediction algorithms that we have presented so far), and the target address. Like a cache, the BTB can be full associative, set associative, or employ more exotic associativity schemes.

However, this is not the end of the story regarding the BTB. There is one further innovation of note: Consider what happens when the CPU branches to the target address . . . it then loads the instruction found there into the pipeline. Around the time that it finishes decoding and executing that instruction, the previous branch will have been resolved so this instruction is either kept or flushed.

But we can speed this process up a little further by storing the actual target *instruction* in the BTB rather than the address of the target instruction. The pipeline then speculates on a BTB hit by loading that stored *instruction* directly into the pipeline. It does not need to fetch the instruction first.

5.7.11 Basic Blocks

There is one further refinement to the BTB technique of Section 5.7.10, which is worthy of note, and that is to deal in code blocks rather than individual instructions. In fact, moving beyond the abilities of the single-instruction BTB actually requires us to work on blocks of code. There are three types of code block in common use within the computer architecture and software architecture fields:

- *Basic blocks* are sequences of instructions that are to be executed sequentially with no branches in or out, that is, one entry point, one exit point.

- *Superblocks* are a trace (execution sequence) of basic blocks with only one entry point but possibly several exit points.

- *Hyperblocks* are clusters of basic blocks similar to superblocks in that they have only one entry point, but possibly more exit points. Hyperblocks differ in that they can contain several trace paths (i.e., more than a single control path).

In this text, we will confine our discussion to the simplest of these, basic blocks, as applied within block-based BTB schemes. Imagine a BTB, or even a memory cache, that store, and can feed, blocks of code into the pipeline. For a pipeline able to reorder or execute out of order, this allows for maximum flexibility and yields an excellent performance improvement.

Basic blocks are easily formed as the string of instructions between branches and branch targets, and a program trace can identify which path is traversed through a connected graph of basic blocks. An example path through a set of basic blocks is shown in Figure 5.21.

At first we saw how to predict branches as taken/not taken. Next we predicted branch target address. Then we predicted the branch target instruction. Now we can predict basic block sequences.

FIGURE 5.21   A set of interconnected basic blocks (lines of code in between branches) are traversed during execution of a program.

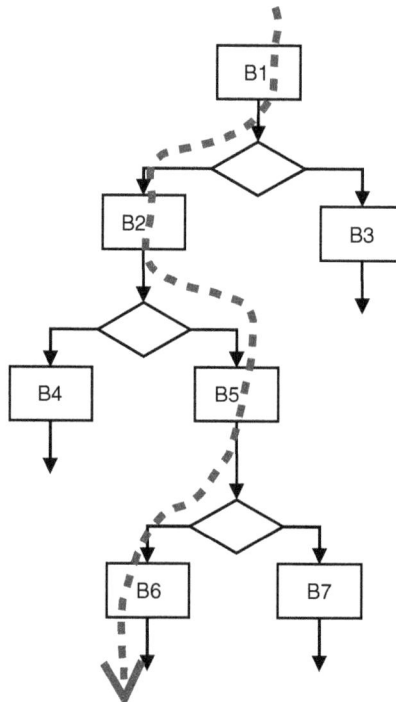

Recurring and frequent sequences of basic blocks are identified and hopefully cached so we can very quickly issue the instructions. For example, with reference to Figure 5.21, a block-BTB could directly issue the instructions contained with B1, B2, B5, B6 into a pipeline with no branching necessary—assuming we had correctly predicted the trace path through the blocks.

Of course, we still need to check the branches were correct and flush the pipeline if we got a prediction wrong. In real code, there may be several basic blocks (BB) involved, each potentially containing several tens of instructions (the average BB size is approximately seven instructions, but of course varies widely based upon the computation being performed, the processor and the compiler).

The trace cache is updated over time and whenever the CPU hits the root BB (B1), a branch prediction algorithm predicts the ongoing path. If this matches the second entry in the trace cache (B2) then this is a hit and the CPU then starts to follow the basic block contents from the trace prediction (which themselves can be cached).

This system was in fact used in the Pentium 4, but with the additional feature that instead of caching the BB instruction contents, it caches the already decoded instructions themselves, that is, not only can we bypass the pipeline "fetch" stage, we can also bypass the "decode" stage.

5.7.12   Branch Prediction Summary

An enormous amount of effort has been spent on trying to keep a hungry pipeline continually "fed" with instructions to process.

All of this effort on branch prediction, speculation, and so on is about making sure instructions are issued at as fast a rate as possible. Pipelining and instruction level parallelism are means of making sure that instructions get executed in as fast and efficient manner as possible.

We should note that no single performance refinement method alone is supreme—for maximum performance a pragmatic selection of these techniques working in well-tuned harmony is the ideal.

One more caveat is that few hardware speedups can make up for a poor compiler,[10] and conversely effort spent in creating a good compiler can provide more benefit than some of the speedup techniques alone.

5.8   Parallel and Massively Parallel Machines

Section 2.1.1 introduced Flynn's classification of processors into four groups characterized by the handling of instructions and data, namely

- SISD—single instruction, single data
- SIMD—single instruction, multiple data
- MISD—multiple instruction, single data
- MIMD—multiple instruction, multiple data

---

[10] The author recommends and uses GCC, the GNU Compiler Collection, himself.

By and large, up to this point we have considered only SISD machines—the single microprocessors typically found in embedded systems and traditional desktop hardware. We also introduced some elements of SIMD found in MMX and SSE units (Section 4.5) and in some ARM-specific coprocessors (Section 4.8). We will skip MISD, which is most often used in fault-tolerant systems such as those that perform calculations on data multiple times, and compare the results from each calculation—something that Section 7.10 will discuss more thoroughly. So the next form to consider after SIMD is MIMD.

At the time of writing, current trends in the processor industry are to extend machines beyond SISD, through SIMD and on to MIMD. MIMD machines are thus becoming more prevalent. We already discussed some common coprocessors in Section 4.5, where a main CPU is augmented by an external functional unit capable of performing various specialized functions. Here we take matters one step further and consider the case of identical processors working together, in parallel, in an MIMD arrangement.

Actually, there are several levels of parallelism that can be considered in computers, since the term "parallel machines" is very loosely defined. Let us briefly run through the scale of these levels:

- *Bit-level parallelism* relates to the size of word that a computer processes. An 8-bit computer processes 8 bits in parallel, but four times as much data can potentially be handled in a 32-bit machine through multiplying the word size four times.

- *Instruction-level parallelism*—as we have seen in many cases, different instructions can be overlapped and processed simultaneously, providing there are no data dependencies between them. Pipelining is a simple example, but superscalar machines, coprocessors and Tomasulo's algorithm (Section 5.9) are others.

- *Vector parallelism* relates to SIMD machines that process not just single words of data, but entire vectors at one time. SSE and MMX are examples of this type of parallelism.

- *Task parallelism* means that entire tasks, or program subroutines and functions, can be executed simultaneously by different hardware. We will discuss this more throughout this section.

- *Machine parallelism* describes the huge *server farms* used by companies such as Google and Amazon. These are buildings containing hundreds or even thousands of separate computers, each operating toward a certain computational goal, in parallel. We will consider this type of system in Section 12.2.

Each of these levels of parallelism is illustrated diagrammatically in Figure 5.22, showing the encapsulation of bitwise manipulation by instructions into higher and higher levels of parallel activity.

In a discussion of parallel processing it is also useful to distinguish the characteristics of what needs to be processed in terms of "coupling." *Loosely coupled* parallel processing means that different parallel threads of execution have few dependencies, and can largely be executed independently. These are very easy to operate in parallel—independent processor cores can handle each task separately. An example might be two different Google search requests, from two independent users, running on two machines in a Google server farm. On the other hand, *tightly coupled* tasks are very interdependent. They may need to share data, communicate frequently and have situations where one

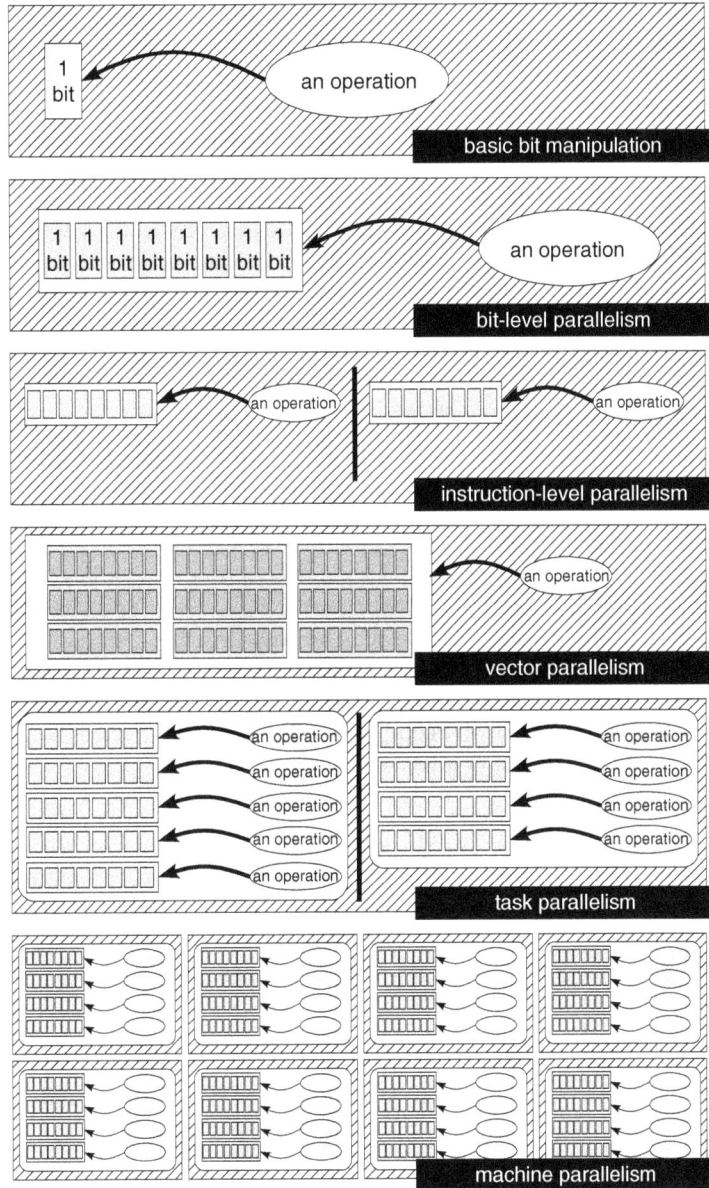

FIGURE 5.22   Starting with basic bitwise manipulation, higher and higher levels of parallelism are achieved by encapsulating and repeating the basic operations in parallel.

task is dependent upon input from the other task. It would be better to run these tasks on the same machine so that communications between the tasks does not become a bottleneck to performance. Naturally, machine architectures can then be either loosely or tightly (closely) coupled to match these tasks.

In terms of computer architecture, the more relevant forms of parallelism are those toward the top of the list given previously. We have already touched upon most of the categories shown, and will consider large-scale machine parallelism again later, but for now let us turn our attention to the middle ground—task parallelism. This is higher level than superscalar and vector approaches, but lower level than machine parallelism: It is of growing importance to the architecture of desktop-sized computers and likely to be similarly influential in the field of embedded computer architecture, in the years to come.

There are two major motivations for parallelism which we will discuss in turn. The first is due to the gradual evolution of SISD machines with additional functional units into true MIMD parallel machines. Second is the deliberate adoption of parallelism for reasons of improving raw performance. We will explain both motivations in turn in the following subsections.

5.8.1   Evolution of SISD to MIMD

SISD machines are easy to write programs for—from a programmers' perspective there is usually only one thing happening at any time, and programs execute sequentially, following whatever branches are to be taken. In the early days of stored-program computers this was precisely what the designers required of a computer: load todays program and execute it. Tomorrow they would load a different program and execute that. Switching from one task to another would involve replacing a stack of punched cards.

However, in the decades during which computers have started to find widespread acceptance, software has progressed from predominantly calculation-based (accounting, simulation, mathematical formulae), through control (monitoring sensors and adjusting actuators in machinery) toward complex multitasking, often involving multisensory and immersive human-computer interfacing.

Where once computers were expected to fulfill a single task at any one time, today's machines (desktop *and* embedded) are almost always required to handle multiple concurrent tasks. Such different tasks often imply a range of timing and operating requirements. Section 6.4 will introduce and discuss tasks from a real-time software perspective, but here it is necessary to appreciate that software often requires different code to be executed at different times. Each piece of code can be encapsulated into a separate task, and thus different tasks, performing different functions, can run on the same computer at different times.

Often these tasks will have strong individual characteristics, and there may even be conflicting requirements between tasks.

In general, when faced with two (or more) conflicting requirements, system architects often respond by partitioning a system—with separate sections of hardware and software dedicated to fulfilling different requirements. The partitioning is almost always done in software: Two tasks handle different processing aspects, but share the same CPU, but may also be in hardware where there are two processors and each handle one task.

A simple example to illustrate conflicting requirements would be a desktop machine running a mouse-controlled windowing desktop display, simultaneously with an MP3 (compressed music file) playback system. In this scenario, the MP3 playback requires some mathematical processing, and handling of streams of audio. The important requirement here is that individual samples of audio are output on time—any delay in a sample being output will create a "click" or maybe even more annoying sounds. A system designer, realizing this, may grant the MP3 playback task a high CPU priority so it runs frequently and is seldom waiting for other tasks to finish. Unfortunately the

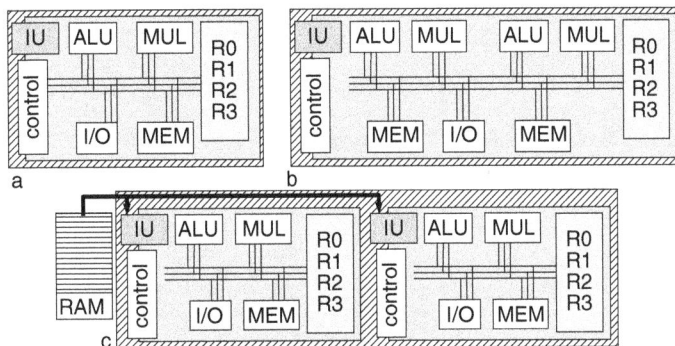

FIGURE 5.23 Block diagrams of (*a*) a basic SISD processor comprising four functional units (ALU, multiplier, I/O block, and memory unit), a control unit, and an instruction fetch/decode unit (IU) augmented by a bank of registers. In diagram (*b*), there are extra functional units shown, moving toward an SIMD machine. In diagram (*c*), a full shared memory MIMD machine is shown containing two complete processors on a single chip.

user, controlling the MP3 playback with a mouse, might then find that the mouse pointer movements are not smooth. The solution may well be to increase the mouse pointer priority higher than the MP3 priority, or better still, to employ a system capable of adjusting priority dynamically.

There is, however, a third option: The use of an MIMD machine which allows a single hardware device to contain two (or more) separate streams/tasks of instruction and data, and execute these simultaneously. There is no longer any need to time-share on a single processor, but two processors inside the same device, with shared memory and peripherals, can effectively partition tasks.

The hardware choices can be illustrated by examining Figure 5.23*a* which shows a basic SISD processor, a shared-memory MIMD machine, and an intermediate form capable of SIMD processing. This basic SISD machine has an ALU, multiplier, I/O block, memory unit, control unit, and an instruction fetch/decode unit (IU). A bank of four registers hangs off an internal three-bus arrangement. Given two software tasks, each would have to time-slice, running on the same hardware. In Figure 5.23*b*, extra functional units have been added to transition the design toward an SIMD machine where calculations could potentially be performed in parallel—and which might allow two tasks to be merged together in software. However, this machine has a clear bottleneck in the internal bus arrangement, given that these have not been upgraded from the SISD system. In Figure 5.23*c* where a shared-memory MIMD machine is illustrated, each individual CPU has an independent internal bus system, allowing for true parallelism. This comprises essentially two complete processors on a single chip; however, the bottleneck in this case would probably be accesses to shared external memory.

As software fragments further into separate threads, and designers reach limits of ever-increasing clock speed, data width, and so on, the next logical performance improvement is toward increasing parallelism—SISD to SIMD and thence to MIMD.

In the world of embedded computing, one prominent example of the expanding lineup of dual-core solutions is the ARM946 dual core platform (DCP). This is based upon dual ARM9 processor cores integrated onto a single chip with shared memory

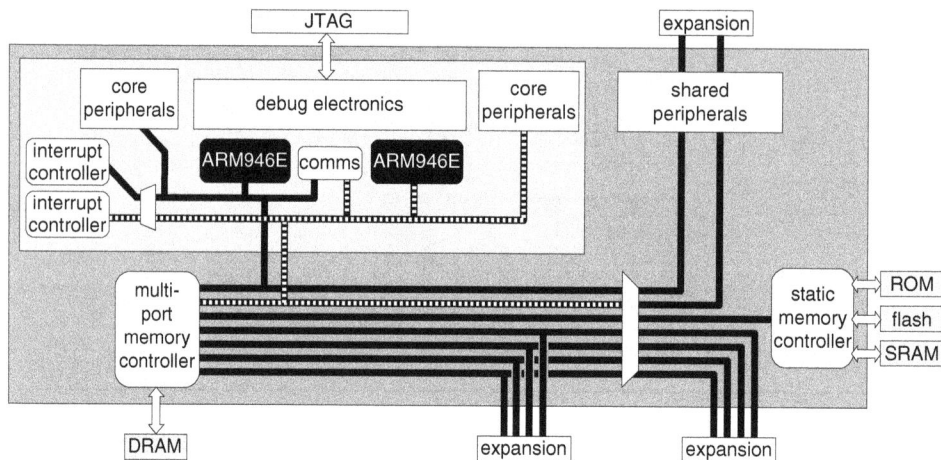

FIGURE 5.24 Block diagram of the ARM946 dual-core platform.

interfaces and an on-chip communications interface. Figure 5.24 shows a block diagram of the device architecture.

This device is advertised as being a loosely coupled, pre-integrated dual-core architecture supported by simultaneous debug and program trace in hardware. A large amount of software and firmware is available that is compatible with the system, and operating system's support is readily available for the ARM9. Such support in software would typically include the ability to execute different software threads (tasks) in parallel on the two processing cores, arbitrating between these through the hardware communications port (labeled "comms" in the figure).

Although this device is being discussed in a section devoted to parallel processing, it is better characterized as a dual-core device rather than a parallel machine. Two processor cores are much easier to synchronize than multiple independent units, and in this case most of the core peripheral devices are simply replicated twice.

Since this system is undoubtedly targeted at embedded products, one possible partitioning of a system would be user-interface code running on one processor, being triggered as and when necessary by user intervention, and media processing (with critical timing requirements) running on the second processor. Or perhaps MP3 decoding on one processor, and wireless Ethernet processing on the other one, for a wireless LAN-based audio device.

Whatever the application, dual-core devices such as these are currently becoming more popular, and look set to create, and occupy a significant niche in the worldwide processor market. Most likely greater numbers of cores will be clustered together in future, and this is due to the continual and perceived ongoing need for increased performance.

5.8.2 Parallelism for Raw Performance

We have already mentioned the pressure on computer designers to increase performance. The well-known Moore's law has passed into the public consciousness so well that consumers expect ever-increasing power from their computers, and consequently from their computer-powered devices.

Perhaps more concerning is that *software* writers have also learned to expect that computer power (and memory size) will continue to grow year on year. It is traditional for computer architects to direct some blame toward programmers—and has been ever since the profession of programmer split from the profession of computer designer during the early years of computer development. Most computer designers (the author included) believe that they can do a far better job of programming their machines than the software engineers who actually do so.

Whether such beliefs are tenable or not, the increasing size of software (often known as "bloat" by computer architects) and decreasing speed have consumed much of the performance gains made by architectural improvements, clock rate increase, clever pipelining techniques, and so on. A typical desktop machine of 2009 has a speed[11] that is at least 50 times faster than the computer that the author was using a decade ago. Unfortunately, the current machine does not feel 50 times faster—webpages still load slowly, saving and loading files is still annoyingly slow, and booting the operating system still takes around 10 seconds. Clearly, there are other factors at work beside CPU improvements, including the limiting speed of connected devices such as the Internet, hard discs, and so on. Software-wise there is nothing major that the current computer can do that the old one could not, and yet the operating system has bloated out from being tens of mebibytes in size to over one gibibyte.

This is not to apportion blame on software developers, it is simply to state the fact that software has increased in size and complexity over the years: running much of today's software on a decade old computer is unthinkable, and in many cases would be impossible.

From a position where software grew in step with increases in computer speed and processing capacities, we now have the situation where it is the software itself that is the driving factor for increased computer speed.

Whatever the reasons and driving factors, manufacturers do feel significant pressure to continue increasing performance, and this has driven many responses such as increasing clock speed, increasing IPC, and so on (see Section 5.5.1). Unfortunately, it is becoming increasingly difficult for manufacturers to improve performance using these means—it takes more and more effort and complexity to see performance increase by smaller and smaller amounts. Manufacturers have therefore turned to parallelism to increase performance. It is much easier to design a relatively simple processor and then repeat this 16 times on a single integrated circuit (IC) than it is to design a single processor using all of the resources on that IC that is 16 times faster. It is also easier to use two existing processors in parallel than it is to build a single new processor that executes twice as fast as an existing one.

In theory, having more processors or execution units running in parallel will speed up computation—but only if that computation is inherently able to be executed in parallel pieces. Given m parallel tasks, each requiring T_m seconds to execute, a single CPU will execute these in $m \times T_m$ seconds.

Where there are more tasks than execution units, n, so that $m > n$ then these tasks will be executed in T_m seconds. Thus the speedup achieved is $\{m \times T_m\}/\{T_m\} = m$ times a single execution unit which is called *perfect speedup*. Of course, this equation does not

---

[11]Speed in this case is measured by the execution rate of a simple code loop—namely the infamous Linux bogomips rating of Section 3.5.2.

account for message passing overheads, or operating system support needed for parallel processing, and assumes no data dependencies between tasks.

In general, for a program comprising a fraction f of parallel tasks, and taking T_p seconds to execute sequentially, sequential tasks require a time of $f \times T_p$ and parallel tasks a time of $(1 - f) \times T_p$. Assuming no overhead, parallel execution using m execution units would thus mean the total time is reduced to $(1 - f) \times T_p/m + f \times T_p$, as speedup equals original execution time divided by the parallel execution time:

$$\text{speedup} = n/\{1 + (m - 1) \times f\}$$

where f is 0 (i.e., there is no sequential component) the result indicates perfect speedup as before. The relationship shown, and speedup calculation, is known as Amdahl's law and indicates the potential gains achievable through parallel processing.

5.8.3 More on Parallel Processing

Symmetrical multi-processing (SMP) systems are those that have two or more identical processing elements connected to a block of shared memory. There are many variations on this theme, including shared cache, individual cache (which may well use the MESI cache coherency protocol—Section 4.4.7), and so on. The alternative is asymmetrical multiprocessing, a term which is not really in such common use, but could refer to something as simple as a coprocessor. One of the more common SMP systems, up to quad core at the time of writing, is Intel's Core architecture. The Core 2 duo dual-core is shown in Figure 5.25, where its symmetrical nature should be immediately apparent, as is the central role of shared memory (specifically L2 cache) in this architecture.

Multicore machines combine two or more processing elements (usually entire CPUs) onto a single integrated circuit (IC). Actually some dual- or quad-core ICs advertised as multicore actually contain two separate silicon dies within a single IC package (this makes it a multichip module, or MCM). As the number of cores increases, at some point the device can be referred to as a *many-core* machine. It is relatively easy for designers to build both multicore and many-core machines using soft cores within an FPGA.

Homogeneous architectures are those in which all cores within a machine are identical. In many ways, this is easier to design and program for; however, sometimes *heterogeneous* architectures are more promising—these are machines comprising one or more different cores. This allows cores to be included which can specialize in different types of processing.

One of the neatest examples of a heterogeneous multicore machine is the Cell processor from IBM, Sony, and Toshiba. This processor, which powers several supercomputers, and (arguably of more world impact) the millions of Sony PlayStation III consoles worldwide (but not PlayStation IV), is a remarkable example of combining the power of several unremarkable processors into a remarkable multicore processor.

The Cell (actually more properly known as the Cell Broadband Engine Architecture), is shown diagrammatically in Figures 5.26 and 5.27.

The Cell processor consists of eight identical, and fairly simple, SIMD architecture processors called synergistic processing elements (SPE), augmented with one IBM Power-architecture power processing element (PPE), which is very similar to an off-the-shelf IBM PowerPC RISC processor. The idea being that the eight SPE are basic number-crunchers controlled by the PPE, which will probably host an operating system.

In itself the Cell processor is not at all appropriate for many embedded systems due to its size, power consumption, and thermal dissipation, although it does represent an

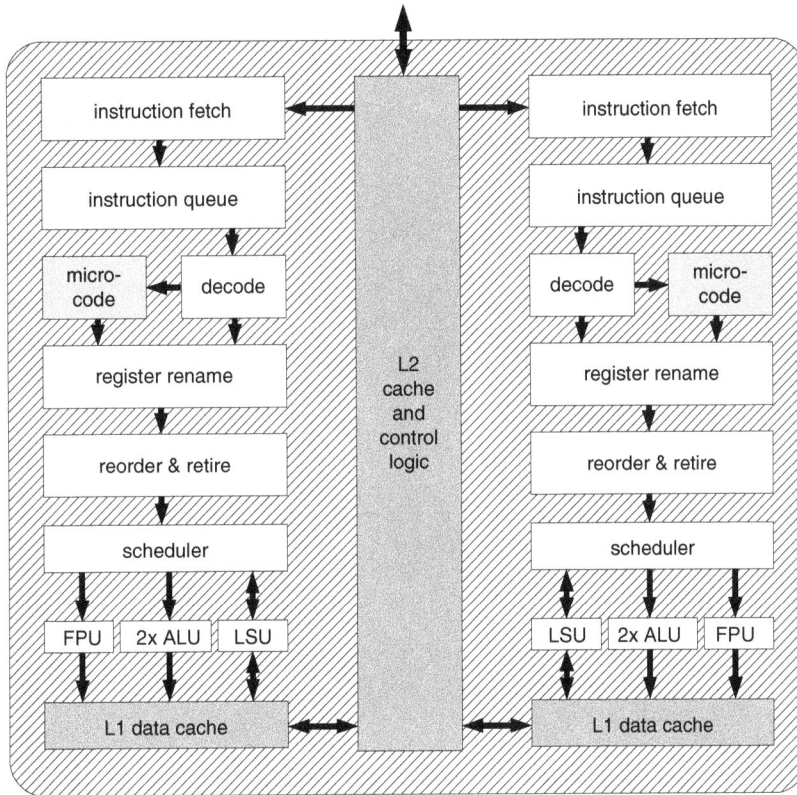

FIGURE 5.25 A block diagram of the internals of the Intel Core architecture, showing a symmetrical two-core device with two identical processing units (including the full superscalar pipelines, instruction handling hardware, and so on), sharing a level 2 cache which connects outward to the system bus.

interesting approach to computer architecture that is likely to make future inroads into the embedded world. Apart from physical and electrical factors, it has also become apparent that the development tools available for the creation of applications on the heterogeneous Cell processor have hindered its adoption. It has been reported that much of the code running on the SPEs has to be hand-coded, the partitioning process between SPEs and PPE, and indeed between the individual SPEs also requires the application of human ingenuity. Until these activities can be better automated, or assisted by development tools, the Cell will probably remain an attractive, but niche, product.

Cluster computers, most notably Linux Beowulf, comprise entire computers, each with individual rather than shared memory (and often individual hard disc storage too). This will be discussed along with the similar topics of *grid* and *cloud* computing in Section 12.2. At the time of writing, it is interesting to note that several of the fastest supercomputers in the world[12] (all of which are clusters), are built from IBM Cell processors.

---

[12] The latest list of the world's fastest machines, updated every 6 months, can be viewed at www.top500.org.

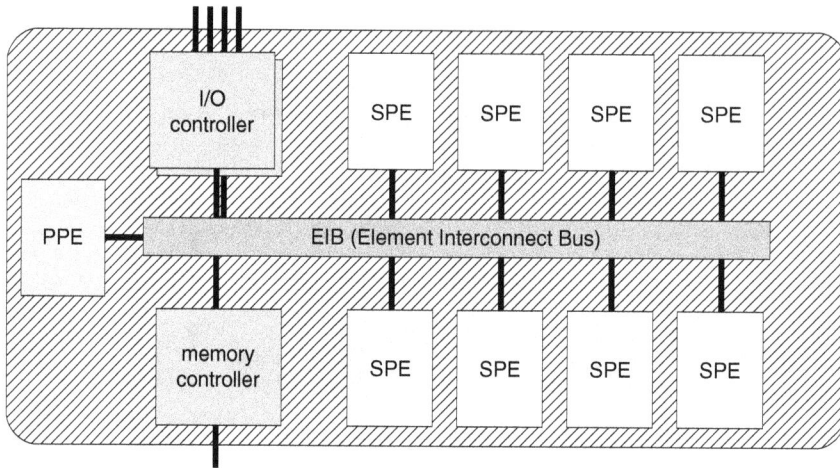

FIGURE 5.26   Block diagram of the Cell Broadband Engine Architecture, showing eight synergistic processing elements (SPE) hanging off an element interconnect bus (EIB), along with the obligatory memory and I/O interfaces, plus a single IBM Power-architecture power processing element (PPE).

Pervasive computing is the idea that computers are everywhere around us, and this is extended by the term *ambient intelligence,* which describes those computers acting in concert in some way (which required connectivity and control) to provide services for us. More recently the term *everywhere* computing has come to describe the kind of instant connectivity and 24/7 availability that we demand from our technology. In fact most of us *are* currently surrounded by computers—many of them very small embedded devices—but

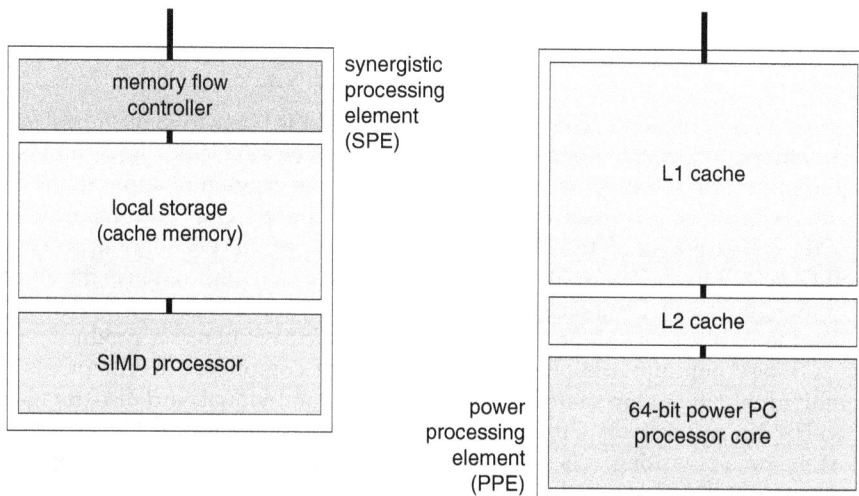

FIGURE 5.27   A look inside the PPE and one of the SPEs from the Cell Broadband Engine Architecture processor, showing the important position of local memory/cache in the design.

at present these do not tend to cooperate. While they might link by Bluetooth or similar technologies, it is not instant connectivity and is far from providing a seamless service. However, the computer and networking that underpins the ideas of ambient intelligence are very much within the realm of modern devices, the control aspects—the operating system, management software, and service layers do not exist at present.

Cloud computing is a little more down to earth, being an interconnected arrangement of distributed machines that cooperate to share processing and storage. The cloud can be dynamically changing, with some machines becoming unavailable, some becoming available, others disappearing or reappearing, and yet the overall computing service itself continues. In some cases *virtualization technology* is used by the different connected machines to make them appear uniform within the cloud, since a virtualized computer instance can run on different hardware devices with ease. Then the cluster of virtualized machines, along with a (probably distributed) control interface, forms the cloud. The analogy also matches the way in which the Internet is depicted as a cloud in many network diagrams. When this type of arrangement is more formalized it can be referred to as *grid computing*, which is a form of organized cluster where the networked machines may be distributed and could actually be PCs sitting on other people's desks. The analogy of a grid comes from the idea of a power grid, which interconnects producers and consumers so as to share resources and balance loads more evenly.

Cloud computing has become big business. Companies operating the largest clouds will sell computing time priced in terms of CPU seconds, or similar. Their sales come from the fact that the entire cost of these services should be lower than it would cost their clients to run their own computers, while the availability (e.g., uptime) should be better. Some of the server farms which provide cloud services are huge, occupying areas larger than a football field. Not many of them look as beautiful as the Barcelona Supercomputer, a cluster machine pictured in Section 1.4.

The *embedded cloud* comes from the idea of pervasive computing being everywhere, and always connected. Given that we are surrounded by such an always-on computing resource, why not use this? Instead of designing power hungry CPUs inside a portable embedded device (e.g., mobile phone) to do the kind of complex processing that people demand more and more of, we can have a low power and relatively simple mobile CPU connected wirelessly to more powerful devices that do the "grunt" work.

This is the idea currently used for most cases of speech recognition in mobile devices. Over the past few decades, the performance of automatic speech recognition (ASR) has become better and better, exceeding the recognition ability in recent years, of human beings (i.e., ASR systems can now understand typical speech better than most humans can in a similar situation). This has been achieved using big data approaches, which use massive data sets to train complex decoding engines. In some cases many decoding engines can run in parallel to improve performance. All of this performance came at great computation cost. In loose terms, we can say that the more CPU processing power and more data we can throw at the ASR problem, the better the resulting performance can be. Unfortunately, our mobile devices are tiny, in processing power terms, and are also limited by the amount of electrical power they can use. If we did all ASR processing on our mobile devices, performance would be poor (it might understand only 70 percent of what we say). That is why Google and others prefer to use the mobile device to only record a clip of speech rather than understand it. This clip is encoded on the device, transmitted to a remote server farm, which uses powerful ASR algorithms to recognize the content, and then send back the reply to the mobile device.

This story is repeated for video, games, music storage, document conversion, optical character recognition, and numerous other so-called mobile services. In each case the computing happens in powerful remote computers, with the result being conveyed wirelessly to the mobile device.

This is an attractive vision of offloading complex processing remotely, and helps enormously to overcome the power limitations of portable or embedded systems (battery technology has not progressed at the same rate as processor power has increased). However, the wireless technology necessary to achieve such a vision in a reliable and cost-effective way, still requires some improvement at the time of writing. Furthermore, it is interesting to note that all of the most advanced wireless links are themselves very complex and computationally intensive. In some cases, the computing needed for the wireless link itself is more computationally intensive than the calculation being offloaded!

Mobile multiprocessing is well and truly established today. Current mobile devices contain a plethora of different sized CPUs: Apart from the main (probably multi-core) processing engine, modern smartphones contain a separate CPU to handle power management, another handles global positioning satellite (GPS) tracking, another does audio management, a multitude of GPU cores handle the display, another CPU handles touchscreen and haptic interface and yet another will ensure security to the device (and more besides). Just focusing on the main CPU itself, these are predominantly multi-core devices that used to be very much homogeneous (i.e. a cluster of identical cores, as with the ARM946 in Figure 5.24). However, in recent years there have been power consumption benefits in moving to partially homogeneous devices. One of the most memorable examples—both in terms of naming as well as in impact—is ARMs big.LITTLE computing architecture.

The growing number of devices that employ the big.LITTLE architecture combine one of more fast processor cores ("big") with one or more lower speed cores ("LITTLE"). The beauty of the arrangement is that both are ARM cores, and both employ the same basic ARM architecture (Section 3.2.7), memory access methodology, bus accessing, and so on. Rather than being just two-core devices, the architecture is designed for many-core devices with two sets of cores, one set being LITTLE and one set being big. So, for example, an eight-core device with four faster and four slower cores.

Big cores, such as the Cortex-A72, are powerful, accelerated and fast number crunchers. They will be operational when the device is turned on and running at full speed. Since they are fast and powerful, they are also relatively power hungry, so battery life would suffer if these were continually operating. Hence, a set of LITTLE cores, such as the Cortex-A53 will be operational when the device is on a power saving mode, or when it is performing only background tasks. Depending upon how the big.LITTLE architecture is arranged with respect to shared memory (and how this relates to the process switcher in the operating system), the technology can support several different operating modes. All big.LITTLE devices can operate in all-big or all-LITTLE modes, where one set of cores is turned on and the other set is turned off. Some devices partitioned their processor cores into big and LITTLE pairs, with each pair sharing their own memory and being able to separately switch between big and LITTLE operation. In this case, for an eight core device there might be one big core running simultaneously with three LITTLE cores that are handling background processing, that is, a 1:3 arrangement (and the device would support the possible configurations 0:4, 1:3, 2:2, 3:1, and 4:0). The newest and most flexible big.LITTLE devices are able to support all cores being individually turned on or off. For an eight core device with four big and four LITTLE cores, this means that every combination from "all turned off" to "all turned on" is possible.[13]

5.9 Tomasulo's Algorithm

Before we leave CPU enhancements, let us wind the clock back a little more than 40 years to an innovation found in the IBM System/360. Although we have constantly stressed the evolutionary development of computer technology throughout this book, we have acknowledged the occasional revolutionary idea. Tomasulo's algorithm is one of those, and one which may have relevance to embedded systems (as we will discuss in Section 5.9.3).

Robert Tomasulo was faced with performance limitations of instruction dependencies stalling programs running in the floating point coprocessor unit that he was designing for the IBM System/360. He thus designed an ingenious method of allowing limited out-of-order execution (nonsequential execution) to help "unblock" many pipeline stalls. This method has been known since then as "the Tomasulo algorithm," although it is perhaps better described as a method rather than an algorithm.

5.9.1 The Rationale behind Tomasulo's Algorithm

Before we discuss exactly how it works, let us just examine the need for something like the Tomasulo algorithm. The problem goes back to our discussion of data dependencies in Section 5.2.4, where we saw that any instruction that uses the output from a previous instruction as its input, needs to wait for that previous instruction to be completed before it can itself be processed. Put more simply, an instruction cannot be executed until its input operands are available.

We had seen how one of the compile-time remedies (Section 5.2.7) to the problem is to reorder instructions, so that neighboring instructions, as far as possible, have no dependencies. Another method is to allow out-of-order execution, so that the CPU, rather than simply waiting for a dependency to clear, takes a future, unrelated instruction and executes that instead. This allows the CPU to remain busy by executing some future instructions without unmet dependencies (if any are available). For this, of course, the CPU needs to fetch ahead of the current instruction—and this is a very strong motivator behind having good branch prediction/speculative execution because otherwise the processor cannot fetch beyond a conditional branch, and reordering would be limited to small segments of code between branches.

Tomasulo got around these problems by allowing instructions to be "issued" from the instruction queue with unresolved operands, in this case called *virtual operands*, instead of waiting for them to be resolved. These instructions proceeded to reservation stations

---

[13] Moving beyond the basic big.LITTLE architecture, ARM is promoting DynamIQ big.LITTLE at the time of writing. This combines dynamic voltage and frequency scaling of individual cores with the most flexible big.LITTLE arrangement. In addition to allowing some cores to be turned on while others are turned off, DynamIQ big.LITTLE supports individual cores operating at fractions of their maximum speed. For example, an eight-core system could have two big cores running at full speed, one at 25 percent speed and one turned off, while all four LITTLE cores are running at different speeds—and this could change fluidly from moment to moment as processing requirements evolve, ensuring an optimum balance between power efficiency and user experience.

(depending upon the functional unit they were destined for), where they waited until the virtual operands are resolved before being handled by their functional unit. This meant that the instruction queue was not blocked by each and every data hazard, although some persistent hazards could still block the issuing of instructions.

It is interesting to compare this approach to advances in the healthcare industry. Twenty years ago, patients arriving at a hospital would wait in a large room for a doctor to become available, sometimes for several hours. A doctor would then see them, and often specify additional investigations, such as blood tests. While these were being undertaken the patients would remain in the waiting room until the test results returned and they could finally proceed to a specialist.

Today, the procedure is normally for all patients coming into hospital (the instruction queue) to be seen quite quickly by a triage nurse who then decides where to send the patients to. Patients are directed to smaller specialist clinics, with their own waiting rooms (reservation stations). They may have blood or urine tests performed, waiting until these test results are available and the specialist doctor is free before entering the consultation room (functional unit).

5.9.2 An Example Tomasulo System

We will now examine a Tomasulo method processor. A Tomasulo-style arrangement for a dynamically scheduled machine with common data bus is considered with reference to Figure 5.28. Separate reservation stations (RS) handle the various functional units: four are shown. These are fed from an instruction queue (IQ), and all are connected to various buses and banks of registers (one register bank is dedicated for integer values, and one for floating point values).

First let us examine how this system works. Initially, IQ contains a sequence, or string of instructions which would normally be issued to functional units in the sequence in which they are listed. Each instruction consists of an opcode plus one or more operands. The IQ issues instructions, in sequence, to empty slots in the appropriate RS. For example, an *ADD.D* instruction (double precision addition) would be issued by the IQ to the RS feeding the FP ALU unit. If the appropriate destination RS is full, then the instruction queue stalls for that cycle and does not issue anything.

Evidently the size, or depth, of each RS is a design parameter in such systems. The ideal situation is to maintain several slots free in the RS, such that the IQ can issue instructions into one of those slots.

When the IQ issues an instruction (i.e., the opcode plus the operands, if any), it checks for dependencies among the operands. It checks if the instruction being issued requires any operand from prior instructions that have not yet executed, in other words, for unresolved data dependencies. If a dependency does exist, the instruction is issued to the RS, but with a "virtual" operand in place of the missing one. If no dependency exists, then the instruction is issued with real (resolved) operands.

Each RS works independently of the others, and can issue an instruction every cycle to its functional unit if (i) the instruction operands are all resolved and (ii) the functional unit is not currently busy.

Generally each functional unit takes a different length of time to process its instructions, so the RSs will empty at different rates. If an RS holds more than one instruction with fully resolved operands, so that more than one instruction can be issued, the oldest one should normally be issued first. The common data bus (CDB) writes results back to

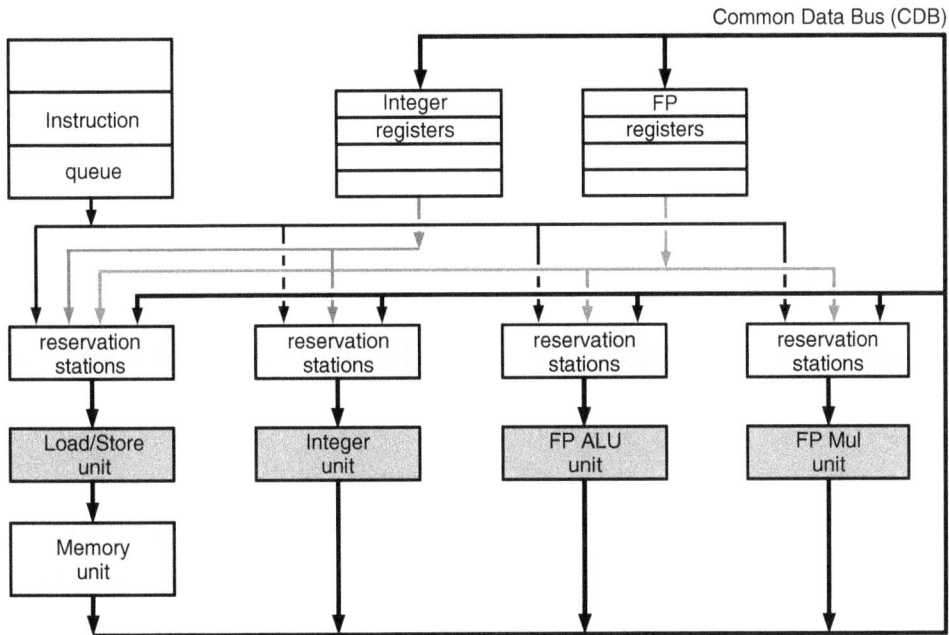

FIGURE 5.28   A block diagram of a general CPU structure implementing the Tomasulo algorithm, showing an instruction queue at the top left feeding instructions to four reservation stations each dedicated to a functional unit via several dedicated buses. Two register banks and a common data bus (CDB) feed operands into the functional units. The outputs of all functional units are also connected to the CDB, as are the register banks.

registers (i.e., it is a load-store machine). But the CDB can carry only one result per cycle, so if two instructions complete in the same cycle, the oldest one needs to be written back first on the CDB.

Every RS continually "listens" to the CDB. Any RS holding an instruction with a virtual operand will be looking for the register writeback that will resolve that operand and make it real. When it sees what it is waiting for, it grabs the value from CDB to resolve its virtual operand. Of course, this does mean that the CDB has to carry more than just the result value and destination register, it has to carry something to inform the instructions waiting in the RSs whether this particular result is the one that resolves their dependency (because an instruction waiting for a value to be written to register R3 may "see" several CDB writebacks to register R3—only the writeback that is immediately before that instruction in the original program is the one that conveys the correct value).

A unique tag is provided to each operand issued from the instruction queue, and this tag is carried through the RS, through the functional unit, and is conveyed on the CDB along with the result writeback from that instruction. Subsequent instructions dependent upon the output of this first instruction are issued with their virtual operands as we have seen: but these virtual operands contain two items of information—the register name plus the tag value. The dependent instruction that "listens" to the CDB is actually "listening" for a writeback to the correct register that has the correct tag value.

Let us illustrate this entire process with an example. We will define a Tomasulo machine, as shown in Figure 5.28, with the following timing specifications:

Load/store unit:	5 cycles to complete
Floating point adder:	2 cycles to complete
Floating point multiplier:	2 cycles to complete
Integer unit:	1 cycle to complete
Reservation station depth:	1 instruction
Instructions issued per cycle:	1
Number of registers:	32 gpr + 32 fp

The following embedded code is going to be executed on this machine:

```
i1   LOAD.D fp2,(gpr7, 20)
i2   LOAD.D fp3, (gpr8, 23)
i3   MUL.D fp4, fp3, fp2
i4   ADD.D fp5, fp4, fp3        ; meaning fp5=fp4+fp3
i5   SAVE.D fp4, (gpr9, 23)     ; meaning save fp4 in address (gpr9+23)
i6   ADD gpr5, gpr2, gpr2
i7   SUB gpr6, gpr1, gpr3
```

A full reservation table showing the program operation is provided in Table 5.3, where the instruction flow can be seen from the queue, through the reservation stations, and into functional units when virtual operands are resolved. Results are written back to registers using the CDB.

Note that the program sequence of instructions *i1* to *i7* is not at all reflected in the out-of-order completion sequence: *i1, i6, i7, i2, i3, i4, i5* shown on the CDB. Interestingly, if we had manually reordered the code segment to minimize execution time on a simple pipelined processor, we may well have resulted in the same execution sequence—instructions *i6* and *i7*, having no data dependencies with other instructions, would have been pulled forward to separate those instructions that do have dependencies.

One final point to note here is that the main cause of latency in this execution is the load store unit (LSU). Of course, the specification indicated that loads and stores each required five cycles (something that is not at all excessive for a modern processor, although the use of on-chip cache memory could speed up some of them). Given the specification, it is to be expected that the LSU is a bottleneck.

A possible way of overcoming the bottleneck may be to consider adding a second LSU (either having its own reservation station or working off the existing LSU RS). Of course, no matter how many LSUs there are, reordering of load-store operations is the major way of resolving such bottlenecks in a Tomasulo machine. However, readers should be aware that dependencies exist in memory access also, and the Tomasulo algorithm does not resolve these. Consider the small code example above: Although the three addresses mentioned appear different, they may not be in practice. The three addresses are as follows:

```
i1   read from (gpr7, 20)
i2   read from (gpr8, 23)
i5   write to (gpr9, 23)
```

Instruction *i1* reads from address (gpr7 + 20). If gpr7 happens to hold the value 1003 then the address read from would naturally be 1023. Similarly, if gpr8 happened to hold the value 1000 then the address read from in *i2* would also be 1023, causing a

	1	2	3	4	5	6	7	8	9	10	11	12	13	14	15	16	17	18	19	20	21
IQ	i1	i2	i3	i4	i5	i5	i5	i6	i7												
RS:lsu		i1	i2	i2	i2	i2	i2	i5	i5	i5	i5	i5	i5	i5	i5	i5					
LSU			i1	i1	i1	i1	i1	i2	i2	i2	i2	i2					i5	i5	i5	i5	i5
RS:alu									i6	i7											
ALU										i6	i7										
RS:falu					i4	i4	i4	i4	i4	i4	i4	i4	i4	i4	i4	i4					
FALU																	i4	i4			
RS:fmul				i3	i3	i3	i3	i3	i3	i3	i3	i3	i3								
FMUL														i3	i3	i3					
CDB								i1			i6	i7	i2			i3			i4		

TABLE 5.3 A reservation table showing the Tomasulo machine operation beginning with a program stored in an instruction queue (IQ), issuing into several reservation stations (RS) for a load-store unit (LSU), arithmetic logic unit (ALU), floating-point ALU (FALU), and floating-point-multiply unit (FMUL). Completed instructions are written back to the register banks using the common data bus (CDB). Instructions waiting for virtual operands to be resolved, and during multi-cycle processing in functional units, are shown in grey.

Common Data Bus (CDB)

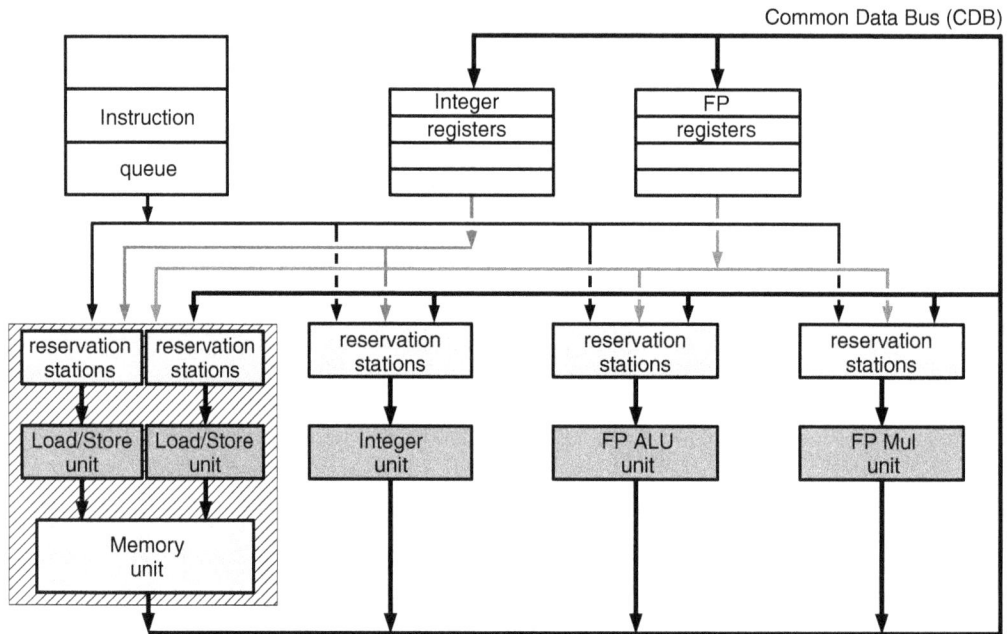

FIGURE 5.29   A modification of the basic Tomasulo machine of Figure 5.28 to incorporate dual memory load-store units and reservation stations.

read-after-read hazard: not a particularly worrisome possibility, but still one that could be optimized if detected early.

Perhaps of more concern is the fact that if gpr8 happens to equal gpr9 then *i2* and *i5* form a WAR hazard (as described in Section 5.2.4). In the current code segment, with only a single LSU, it is not possible for *i5* to be reordered to before *i2*, and therefore no major problem will be caused by this occurrence. However, that is more due to the good fortune that *i5* has a register dependency than anything else.

Let us examine this claim by changing the code and the machine to highlight the problem. In this case, *i5* becomes SAVE.D fp1, (gpr9, 23), so that it has no register dependencies with the rest of the code. We will also add a second LSU and RS, shown in Figure 5.29, and run the following program:

```
i1   LOAD.D fp2,(gpr7, 20)
i2   LOAD.D fp3, (gpr8, 23)
i3   MUL.D fp4, fp3, fp2
i4   ADD.D fp5, fp4, fp3      ; meaning fp5=fp4+fp3
s5   SAVE.D fp1, (gpr9, 23)   ; meaning save fp1 in address (gpr9+23)
i6   ADD gpr5, gpr2, gpr2
i7   SUB gpr6, gpr1, gpr3
```

In this case, the reservation table of the modified machine, running the new code, is given in Table 5.4. It is quite interesting to note that the addition of another LSU has significantly improved program execution—the entire program now completes in 15 cycles instead of 21, and is far more compact.

	1	2	3	4	5	6	7	8	9	10	11	12	13	14	15	16
IQ	i1	i2	i3	i4	s5	i6	i7									
RS:lsu1		i1				s5	s5*									
LSU1			i1	i1	i1	i1	i1	s5	s5	s5	s5	s5	s5			
RS:lsu2		i2														
LSU2				i2	i2	i2	i2	i2								
RS:alu							i6	i7	i7*							
ALU								i6	i6*	i7						
RS:falu					i4	i4*	i4*	i4*	i4*	i4*	i4*	i4*				
FALU													i4	i4		
RS:fmul			i3	i3	i3	i3	i3	i3								
FMUL										i3	i3					
CDB								i1	i2	i6	i7	i3	i2	s5	i4	

TABLE 5.4 A reservation table for the Tomasulo machine as in Table 5.3, but with two LSUs and reservation stations, and a slightly modified program being executed. Instructions that are waiting for a "space" to be made available, either in the CDB, in the RS or in a functional unit, are marked with an asterisk, as in "i6*" waiting at the output of the ALU during cycle 9 because the result from instruction i2 is occupying the CDB during that slot.

The speedup is a good thing, but let us consider the memory accesses in more detail now. Note that s5 enters the first LSU during cycle 8 and begins to write to memory address (gpr9 + 23). Simultaneously, i2 is still reading from address (gpr8 + 23). Clearly, if gpr8=gpr9, the read and the write will be to the same location. As instruction i2 is reading from that location, instruction i5 will be writing to it. Quite likely the value read by instruction i2 would be corrupted, or would be incorrect.

Effectively, this type of problem is occurring because there is no mechanism here for handling hazards on memory addresses. It is unlikely that a simple solution exists to such problems, except in tracking, and resolving, memory access addresses early, or in enforcing in-order execution of unresolved memory reads and writes.

5.9.3   Tomasulo in Embedded Systems

As mentioned, Tomasulo designed his method for a large mainframe computer, the IBM System/360, specifically model 91. We actually included a photograph of this monster in Chapter 1 (refer to Figure 1.5). Why then is such a method included in a computer systems book emphasizing embedded systems?

The first reason is that out-of-order execution methods are typically not trivial to implement, and for those designing CPUs for use in embedded systems, out-of-order execution may not be something that they are likely to consider possible. However, Tomasulo's method trades resources (extra registers in the reservation stations) for improved performance. It also does not rely upon more sophisticated techniques such as branch prediction, superscalar pipelining, and so on. Basically it puts out-of-order execution within reach of fairly simple CPU designs.

Second, the Tomasulo algorithm makes distributed decisions concerning instruction execution. There is no real bottleneck in the instruction issue unit, and this is not really limited by clock speed (in fact Tomasulo's method is easy to extend with multiple functional units, requiring only quite minor adjustments to structure). The distributed nature of the system suits an FPGA. The one main bottleneck in the Tomasulo algorithm is the CDB which must stretch to every reservation station, and to every register in every register bank. However, this type of global bus is readily implemented inside an FPGA, and to some extent is more convenient than having many "shorter" parallel buses.

Finally, we had noted in our examples in Section 5.9.2 how additional functional units (in this case a second memory load-store unit) could be added to optimize performance, although we noted the particular memory address dependency issues for the case of additional LSUs. Within embedded systems it is more likely that addresses of variables and arrays can be fixed at compile time, and not necessarily be specified relative to a base register, something that would solve the problems associated with having additional LSUs. More importantly, it is often possible to know in advance what software will be running on an embedded system, and with this knowledge determine in advance what types of functional units are necessary (and indeed how many there should be).

5.10   Very Long Instruction Word Architectures

Very long instruction word (VLIW) architectures promise to speed up CPU execution by exploiting as much instruction-level parallelism (ILP) as possible. The key point is that they achieve this by shifting much of the responsibility for efficient instruction sequencing and hazard (dependency) checking from the processor to the compiler.

5.10.1   What Is VLIW?

Conventional RISC architectures, shown in Figure 5.30 (top) attempt to fetch and execute a single instruction per cycle, and we already know from Section 3.2.6 that the rationale behind the RISC approach is to handle many simple instructions per second. Instructions are the same size, constrained in a load-store architecture, and are generally executed in a single cycle—this simplicity, similarity, and conformity means that the architecture remains regular, low complexity, and efficient. While this works well, it leads to internal functional units being idle most of the time, because each instruction typically only activates a single functional unit, and therefore only one functional unit is active per cycle. This is not good from an efficiency perspective—the most efficient situation would be all functional units running "flat out" all of the time.

Superscalar processors, on the other hand (shown in the middle diagram of Figure 5.30) fetch multiple instructions per cycle, and allow their internal functional units a degree of independence, where possible. This means that, for example, the internal ALU and MUL could operate simultaneously if tasked by two different instructions that were issued at the same time. As was explained in Section 5.4, this only works when the operands for neighboring instructions are independent, and where the instructions require different functional units. When there is a dependency of one instruction

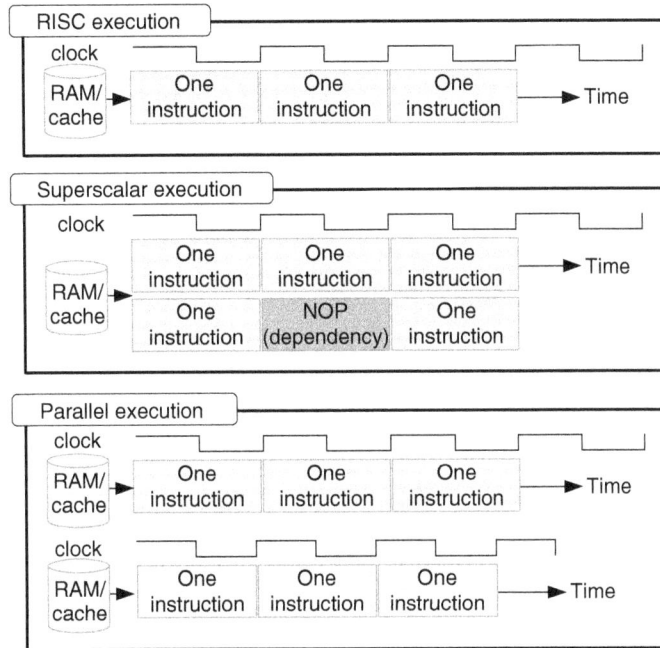

FIGURE 5.30 Single cycle execution in various architectures we have discussed up to now showing a simple RISC processor (top), a mythical RISC superscalar machine (middle), and parallel cores (bottom).

on another, the dependent instruction has to wait until the other instruction executes (this is a data dependency hazard as described in Section 5.2.4). Similarly, when two instructions both require the same functional unit, one instruction has to wait for the other one to execute first (another hazard, in the language of Chapter 5). The key feature in a superscalar architecture is that the dependencies (working out when an instruction is dependent upon a previous one, or when two instructions need to access the same functional unit) are determined at runtime; it is the job of the instruction fetch unit. Such dependencies obviously reduce overall execution speed, but even the process of checking for dependencies introduces complexities that limit performance.

If only there was a way of ensuring the neighboring instructions were already free of dependencies.

Parallel architectures (at the bottom of Figure 5.30) simply replicate the entire pipeline, fetch unit, and often memory block of a single pipeline CPU, and run them almost independently. Parallelism is popular, and effective in terms of raw performance in the real world (see Section 5.8.2) but simply replicates whatever inefficiencies might already exist in each single pipeline.

VLIW, by contrast, aims to reduce the inefficiencies at compile time. The performance gain is also achieved through execution in parallel, but in this case it is parallelism *within* the instruction word. Instead of two (or more) instructions being executed in parallel, it involves having very long instructions, each of which can contain a large number of parallel sub-instructions. The speedup is analogous to the way in which a parallel bus can clock

Instruction	Instruction	Instruction	→ Time
Instruction	Instruction	Instruction	
Instruction	NOP	Instruction	
Instruction	NOP	Instruction	
RAM/cache → Instruction	NOP	Instruction	
Instruction	Instruction	Instruction	
Instruction	Instruction	NOP	
Instruction	Instruction	NOP	
Instruction	Instruction	NOP	

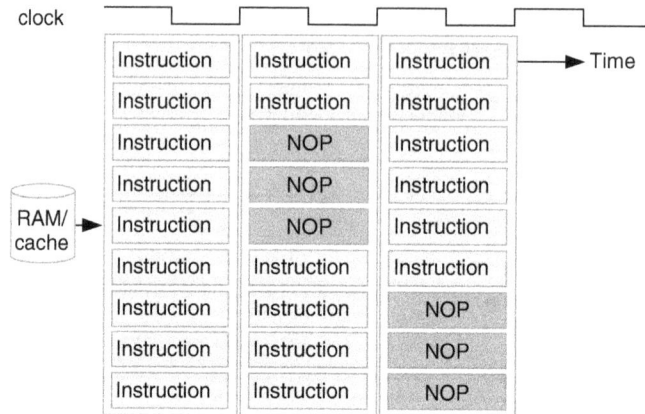

FIGURE 5.31   Unlike conventional architectures shown in Figure 5.30, the VLIW compiler packs as many sub-instructions into each instruction word as possible, based on its knowledge of dependencies and CPU resources.

transfer information faster than a serial bus, by transferring several related bits in a single clock cycle: and like the bus example, the trade-off is in terms of a greater instruction bandwidth required, and additional hardware resources.

5.10.2   The VLIW Rationale

Processor manufacturers have, for decades, tried to increase clock speeds year on year, yielding relentless processing gains. However, silicon speed increases have slowed as manufacturers approach the physical limits of the technology that they are working with. VLIW, like parallel processing, allows speed to remain constant, but enables the number of operations executed per clock cycle to increase, thus improving overall performance. The VLIW concept is illustrated in Figure 5.31. Since the exact sequencing, ordering, and arrangements of the operations (and combining these into a single instruction) is handled by the compiler, any runtime dependency checking is much simpler—limited mainly to conditionals—and therefore much more efficient. We will explore some of the disadvantages of this approach in Section 5.10.3, but first let us consider the advantages of VLIW.

Performing operations in parallel is not a new idea, and neither is VLIW. It really began as a concept with multimedia-based DSP processors from Texas Instruments in the late 1980s, resulting in some very fast processors from companies such as TI, Philips (in the Tri-Media range), and Mitsubishi (initially with the V30 processor).

Mainstream adoption of VLIW has been relatively slow, but the technique is likely to gradually become more popular over time simply because it helps to break the tight relationship between processing performance and clock speed. Outside of general computing, VLIW has found a niche in media processing—stream processing of audio and video in particular, and such content is continually increasing within processing systems.

VLIW architectures are sometimes known as EPIC (explicitly parallel instruction computing) and have been adopted by Intel in their IA-64-based machines. These are designed for high-performance servers—an indication of the strengths of the technique for high data throughput applications.

It is also possible to view VLIW as an extension of the RISC philosophy into parallel dimensions. It is inherently RISC-like in that individual sub-instructions are themselves simple and regular, and typically execute in a single cycle. Separate RISC-style sub-instructions are folded together by the compiler into a long instruction word that is to be executed in parallel by using:

- Independent CPU functional units (such as ADSP2181)

- Multiple copies of functional units (many DSPs and later Intel Pentiums)

- Pipelined functional units

An example output from a compiler for a multimedia task on a VLIW processor might look like the following:

	ALU 1	ALU 2	ALU 3	FPU 1	LOAD/STORE
Instruction 1	ADD	ADD	ADD	FMUL	NOP
Instruction 2	ADD	NOP	NOP	FMUL	STORE
Instruction 3	NOP	NOP	NOP	NOP	STORE

A superscalar machine doing a similar job would instead issue eight sequential instructions (ignoring the NOPs). They would *possibly* be parallelized to some degree depending on hardware flexibility and the current state of the system at the moment of execution.

VLIW instructions are commonly 1024 bits long, and directly control multiple hardware units, such as 16 ALUs, 4 FPUs, 4 branch units.

5.10.3 Difficulties with VLIW

As the discussion in the previous section has revealed, VLIW hardware is actually more regular, and simplified over and above an equivalent superscalar machine, but much of the complexity moves into the compiler.

To efficiently compile for VLIW machines, the compiler has to consider separate data flows—it rearranges the sequence of instructions in the user's program to improve instruction throughput, taking care where later instructions depend on the output from previous instructions. In other words, the compiler particularly needs to avoid pipeline hazards (such as those mentioned in Section 5.2).

Some of the other potential issues associated with VLIW code include

- *Poor code density*—sometimes it is not possible to make a program fully parallel, and in such cases VLIW code contains many NOPs "padding" out the instruction word.

- *Complex compilers are required*—this is simply a matter of transferring "difficulty" from a hardware domain to a software domain.

- *High-bandwidth memory needed*—on average, a VLIW processor will require more instruction bandwidth than other processors, such as superscalar, exacerbated by the padding of extra NOPs. The usual solution is for instruction memory to be 64, 128, or even 256 bits wide. This means more memory chips are needed, more PCB space to route buses, and more pins on the processor IC.

- *Hard to code in assembly language*—use of a high-level language (HLL) is almost an essential prerequisite for using VLIW processors.

The compiler complexity issue is one reason why VLIW has not generally been used in PC-architecture systems where backward code compatibility is required. If VLIW were to be adopted then compilers would need to be replaced by much more intelligent versions—object code would change and existing low-level tools would need to be replaced. By contrast, superscalar techniques are entirely compatible with legacy code. They require more complex instruction-handling hardware, but the compiler can remain simple.

Where companies, such as Mitsubishi and Philips have designed totally new architectures with no legacy code issues, they have been free to adopt VLIW, with some success.

5.10.4 Comparison with Superscalar

One question which arises, is how does VLIW differ from a superscalar architecture (Section 5.4), since both include multiple functional units and parallelism within the processor hardware. There are several differences, but most importantly the superscalar instruction fetch unit has to be able to *issue* instructions faster than individual execution units can process them, and instructions may have to wait to be processed. This is because the superscalar processor schedules itself what each instruction unit is doing, and which of the parallel execution units it is sent to, at run-time. By contrast the VLIW processor relies on the *compiler* to do the scheduling. It is the compiler that directs exactly what each execution unit is doing at every instant, where it gets its data from and where it writes it to. This kind of complex instruction reordering and scheduling is done at compile time rather than at run time, and therefore does not need a large and very complex CPU control unit that extreme superscalar systems will require. In VLIW, parallel instructions are issued, and executed, at a regular rate and the processor instruction handling hardware is less complex, and therefore potentially works faster.

5.11 Summary

While previous chapters concentrated mostly upon the foundations of computing, the functional units within a computer (and CPU) and the operation of these devices, this chapter has considered performance—mainly because this is one of the major drivers of innovation in the computing industry.

We looked at many types of speedup, from the traditional increasing clock speed, through the now well-established method of pipelining, CISC versus RISC, superscalar and other hardware accelerations such as zero-overhead looping and dedicated addressing hardware.

A large part of the chapter was devoted to issues related to pipelining, namely hazards and branch penalties, and how to alleviate these potential problems using delayed branching and/or branch prediction.

We have now completed our overview of CPU internal architecture (apart from some more esoteric methods in Chapter 12), and will next turn our attention to communicating with the CPU: getting information into and out of the system.

5.12 Problems

5.1   On some pipelined processors, a conditional branch can cause a stall or wasted cycle. The following code segment might stall a three-stage pipeline. Why?

```
MOV R0,R3          ;R0=R3
ORR R4,R3,R5       ;R4=R3 OR R5
AND R7,R6,R5       ;R6=R4 AND R5
ADDS R0,R1,R2      ;R0=R1+R2, set condition flags
BGT loop           ;Branch if the result is greater than 0
```

Note: *An "S" after the instruction means its result will set the condition codes. No "S" means that condition codes will not be set. Assume that every instruction completes in a single pipeline cycle in this processor.*

5.2   Reorder the code in question 5.1 to reduce the likelihood of a stall occurring.

5.3   If a delayed branch was available for the ARM, the BGT could be replaced by a BGTD in the code above. Rewrite the code in question 5.1 to use the delayed branch. (*Hint: You only need to move one instruction.*)

5.4   In an 8-bit RISC-style processor, starting from the initial conditions R0=0x0, R1=0x1 and R2=0xff, determine the state of the four condition flags after the following ARM-like instructions have completed. Assume that the instructions specified occur in sequence:

Instruction	N	Z	C	V
MOVS R3, #0x7f				
ADDS R4, R3, R1				
ANDS R5, R2, R0				
MOVS R5, R4, R4				
SUBS R5, R4, R1				
ORR R5, R4, R2				

5.5   Identify four hazards in the following segment of ARM-style assembler code which includes a delayed conditional branch:

```
i1    ADD R1, R2, R3
i2    NOTS R1, R2
i3    BEQD loop
i4    SUBS R4, R3, R2
i5    AND R5, R4, R1
i6    NOT R1, R2
```

5.6   Often, branches can cause pipeline stalls due to dependencies, code ordering, and pipeline hardware capabilities. Delayed branches can prevent such stalls. Name two other methods that can be used to improve branch performance.

5.7   Name three general methods of reducing or removing the effect of data hazards in a processor.

5.8   Draw a block diagram of hardware that can multiply any number by a value between 2 and 10, use data forwarding to apply a feedback path. The blocks you have available are

- Up to two single-bit shifters
- Up to two full adders

Ignore all control logic and storage registers.

5.9 Pipeline the design of the previous question. Use a single adder and a single shifter, again ignore control logic and registers.

5.10 Draw a reservation table for three pipelined multiplication examples from the previous question.

5.11 Identify the main mechanism for transferring data between a CPU and its co-processing unit, and state how this differs from a heterogeneous dual processing system.

5.12 List five typical features of RISC processors that differentiate them from their CISC predecessors.

5.13 What range of instructions per cycle (IPC) would be expected for a pure RISC processor? How would that differ for a perfect superscalar machine that can issue three instructions simultaneously?

5.14 A digital signal processor (DSP) implements simple zero-overhead loop hardware that has a loop counter, a start point address register and an end point address register. The hardware will detect when the program counter (PC) matches the end point address register, and if the loop counter is nonzero will reload the PC with the start point address. Identify which types of C loops can be catered for with this hardware:

```
a) for (loop = 0; loop <99; loop++){
        <do lots of calculations here>
   }

b) loop=99;
   do{
        <lots of calculations here>
   } while (loop-- >0)

c)   while(x + y != 23){
        <lots of calculations here>
   }
```

5.15 Calculate the parallel processing speedup possible when a program consisting of 1224 tasks (200 of them must be run sequentially but 1024 can be run in parallel) is executed in a 16-way homogeneous perfectly parallel machine. Each task requires 2 ms of CPU time to execute.

5.16 Referring to the pipeline speedup and efficiency calculations of Box 5.1, if one particular CPU pipeline design is found to have an efficiency of 68 percent and a speedup of 3.4, determine the number of stages in that pipeline.

5.17 To implement a digital audio delay, a processor has to continuously read in audio samples, delay them, and output them sometime later. For 16-bit audio at a sample rate of 8 kHz, how many samples must the wait be for a 1008 ms delay? Implement this on the ADSP2181 using a circular buffer, and write the pseudo code using the following instructions. Assume the buffer memory is empty at the beginning;

`<reg>=IO(audioport)`	to read in data to register
`IO(audioport)=<reg>`	to read out data from register
`<reg>=DM(I0,M0)`	to get data from memory location pointed to by I0, and pointer I0 is incremented by the value in M0 after the operation.
`DM(I0,M1)=<reg>`	stores value in register to memory location pointed to by I0, and after the operation I0=I0+M1
`I0=buffer_start`	sets I0 to point to a start of buffer in memory
`L0=buffer_end`	sets a circular buffer up (when I0 reaches L0, I0 is reset)
`M0=x`	sets address modifier M0 to contain a value of x
`M1=x`	sets address modifier M1 to contain a value of x
`B loop`	branches to the program label called *loop*

<reg> can be any register from the set AX0, AX1, AY0, or AX1

5.18   Identify the conditional flags that need to be set for the following conditional ARM instructions to be executed:

Instruction	Meaning	N	Z	C	V
`BEQ loop`	Branch if equal to zero				
`ADDLT R4, R9, R1`	Add if less than zero				
`ANDGE R1, R8, R0`	AND if greater than or equal to zero				
`BNE temp`	Branch if not zero				

5.19   Briefly explain when are shadow registers used, and what method do programmers use in this situation for processors that do not have shadow registers?

5.20   Trace the following code through a processor which has a global 2-bit branch predictor initialized to state "DT":

```
i1          MOV R8, #6          ; Load the value 9 into register R8
i2          MOV R5, #2          ; Load the value 3 into register R5
i3   lp1:   SUBS R8, R8, R5     ; R8=R8-R5
i4          BLE exit            ; branch if result <0
i5          BGT lp1             ; branch if result >0
```

CHAPTER 6

Externals

I n previous chapters the evolutionary, and very occasional revolutionary, heritage of microprocessors have been examined: including the drive for more capable devices with faster processing speeds, the concept of RISC, and the architectural or instruction set support for time-consuming programming and operating concepts.

In this chapter, to round off our studies of basic CPUs, we will examine some of the interactions between the core logic and the outside world, in terms of interfaces and buses, and something of particular relevance to many embedded systems—near real-time processing and interaction.

6.1   Interfacing Using a Bus

As mentioned previously, system-on-chip devices are available which integrate all components of a computer on a single IC. They include the CPU, surrounding logic, and can even include the memory. Such devices also provide the various internal and external buses that are required for a computer system. Parts of the computer that were historically implemented as external, off-chip devices, have now been integrated on chip. This means that the buses connecting those functional units to the CPU (or CPUs) may no longer be visible, but they are present nonetheless. As with all computers, buses connect the CPU and the peripheral handlers—although most, if not all, of those handlers and buses are now on-chip instead of off-chip. These system-on-chip or SoC processors are increasingly common, particularly for embedded applications.

As an example, consider the diagram in Figure 6.1 that describes a standard PC architecture from the late 1990s, showing the central CPU and various of the items clustered around it. This same architecture can be found implemented across a motherboard with 20 or so support chips, but more recently implemented within a single system-on-chip device. The same standard interfaces are present but are all within the same IC, shown in the hatched area.

Within ARM-based systems there are typically two standard buses—the AHB (ARM host bus) and the AMBA (advanced microcontroller bus architecture). Both can be found implemented within many ARM-based ICs from a variety of manufacturers, or equally implemented discretely on larger motherboards, such as the ARM integrator platform. The ARM buses have even become a de facto standard, being used to interface non-ARM processors, such as the SPARC-based ERC32 processor (used in satellites and space systems). Such standard buses, whether internal or external, help peripheral manufacturers—either separate IC vendors or internal logic block vendors—to produce standard items for incorporation in systems.

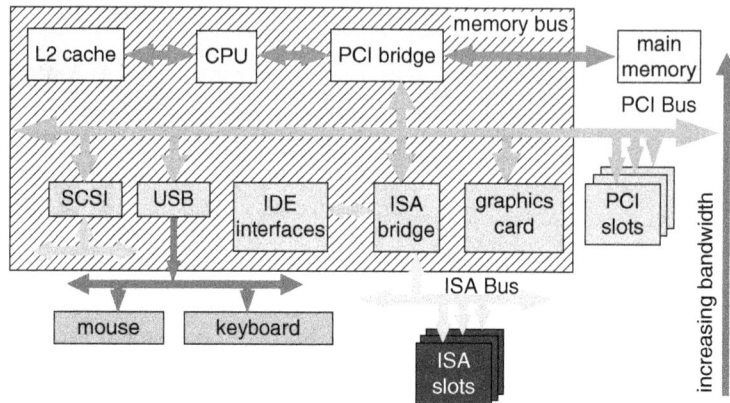

FIGURE 6.1   A block diagram of a fairly standard PC from the turn of the century.

Although it is expected that readers will have been introduced to the concept of a parallel bus previously, it will be reviewed briefly. Most important is the ability for the same physical resource—the data bus—to be shared by a number of devices for conveying information, either as input or output. A master device, usually the CPU, has the responsibility for controlling the parallel bus using control signals. Where two CPUs share the same bus, arbitration must be performed, either inside the CPUs themselves or using a separate external bus arbiter.

The master device uses bus control signals to tell other devices when to read from, or write to, the bus. Bus-compatible devices must ensure that whenever they are *not* writing to the bus, they do not drive the bus, that is, that their bus outputs are in a high-impedance state.

6.1.1   Bus Control Signals

Although there is some variation, bus control signals are typically as follows, where the lower case "n" indicates an active-low signal (meaning that a low voltage indicates an active state while a high voltage indicates an inactive state):

- *nOE and nRD:* Output enable/read enable, indicates that the master controller has allowed some device to write to the bus. The controller then selects a particular device either by its memory address or directly through activating its chip select signals.

- *nWE and nWR:* Write enable, indicates that the master controller has itself placed some value on the data bus, and that one or more other devices are to read this. Exactly which devices should read are those selected by their particular nOE/nRD signals becoming active.

- *RD/nWR:* Read not write. Any valid address or chip select occurring when this is high, indicates a read; any occurring when this is low, indicates a write.

- *nCS and nCE:* Chip enable/select is a one-per device signal indicating, when valid, which device is to "talk" to the bus. Originally, a separate address decoder chip would generate these, but most modern embedded processors internally generate chip selects.

In the days of dual-in-line through-hole chip packaging, there was such pressure on designers to minimize the number of pins on each IC that some strange multiplexed and hybrid parallel bus schemes were designed, with unusual bus control signals. However, the signals shown are most common among modern embedded processors and peripherals.

Other signals that may be associated with such buses include the nWAIT line, used by slower peripherals to cause a CPU that is accessing them to wait until they are ready before using the bus for other purposes. Also, there are bus ready, bus request, and bus grant lines, the latter two being reserved for buses which implement direct memory access (DMA).

6.1.2 Direct Memory Access

Direct memory access (DMA) allows two devices which share a bus to communicate with each other without the continuous intervention of a controlling CPU. Without DMA, a CPU instruction (or several) would be used to first read the external source device, then write to the external destination device. In a load-store architecture machine, this operation would also tie up an internal register for the duration of the transfer.

DMA requires a small amount of CPU intervention for setup, and then operates almost independently of the CPU. The source device delivers data to the destination device using the external bus, and does not require any CPU instructions per word delivered—excluding the initial set up of course—and does not occupy any CPU registers. While the transfer is progressing, the CPU is free to perform any other operations that may be required.

For systems with many devices sharing the same bus, there will be a number of DMA channels, each of which can be assigned different endpoints, and which have ordered priorities, such that if two DMA channels request operation simultaneously, the channel with highest priority will be granted use of the bus first. Box 6.1 examines the workings of the DMA system within one common ARM-based processor.

Box 6.1   DMA in a Commercial Processor

Let us consider a real example—the S3C2410, a popular system-on-chip ARM9 processor from Samsung. This has four channels of DMA, with the controller being located between internal and external buses and handling any combination of transitions between these.

Each of the four channels have five possible source triggers, with each channel being controlled by a three-state finite state machine. If we assume we have selected repetitive operation, and set up source and destination addresses correctly, then operation is as follows:

State 1: DMA controller waits for a DMA request. If seen, it transitions to state 2. DMA ACK and INT REQ are both inactive (0).

State 2: DMA ACK is set, and a counter is loaded to indicate the number of cycles to operate for (i.e., the amount of data to be transferred by that channel). Then it transitions to state 3.

State 3: Data is read from the source address and written to the destination address. This repeats, decrementing the counter, until it reaches zero, at which point it optionally interrupts the processor to indicate that it has finished the transfer. Upon finishing, it transitions back to state 1.

Although DMA improves processor efficiency in many designs, there are enhancements possible where performance is crucial. In fact, in some CPUs the DMA controller itself is intelligent enough to itself be a simple CPU. An example of this was the ARM-based Intel IXP425 network processor. This device contained a number of integrated peripherals such as USB, high-speed serial ports, and two Ethernet MACs (media access controller: a component of the Ethernet interface). The main processor clocked at 533 MHz, while three separate slave processors ran at 100 MHz and were dedicated to handling input/output on the system buses. These small RISC processors were designed to free the main ARM CPU from lengthy and inefficient bus handling and memory transfers. Normally one of the slave processors would be dedicated to running the MACs, making the processor very capable at performing network operations.

6.2   Parallel Bus Specifications

The bus transaction timing for the ARM9-based Samsung S3C2410 system-on-chip device is shown in Figure 6.2. This was chosen for an example first because Samsung had made an exceptional job of clarifying the timings and parameters, and matched each of the timings individually into a small number of control registers. With some CPUs, the situation is more complicated—with cycle calculations needed to be done by hand, combined and split parameters, and unusual behaviors being commonplace.

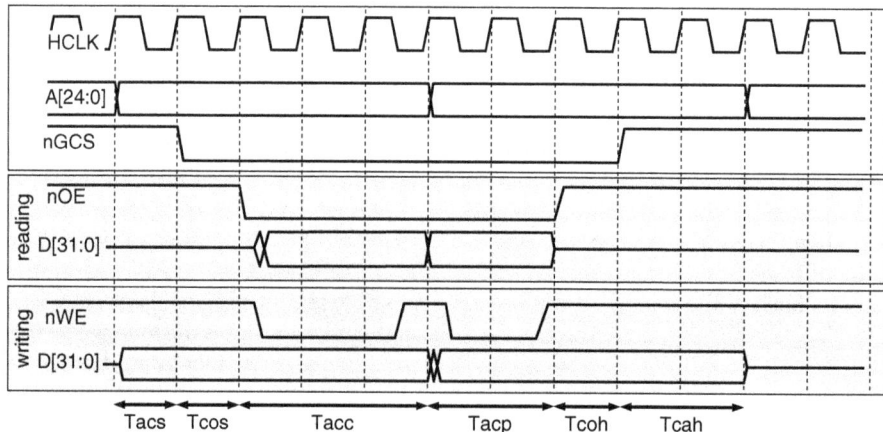

FIGURE 6.2   The SRAM bus transactions and timing diagram for the Samsung S3C2410 ARM9-based system-on-chip processor. The top section shows general control signals, the middle section indicates the signals for a read operation (during which nWE would remain high), and the bottom section indicates the signals for a write operation (during which nOE would in turn remain inactive—high).

The clock named HCLK is one of the main on-board clocks driving the memory interface, and other on-chip devices. It would typically be running at 100 MHz and is not available off chip—it is just shown in the diagram to provide a useful reference. The 25-bit external address bus and the nGCS chip select define an interface to an external device, probably a ROM or similar (including most external bus-interfaced peripheral devices which use the same interface). These signals are active during any bus transaction to that device. The bottom shaded boxes contain the read and write signals and behaviors, respectively.

The timing diagram shows several buses in a high-impedance (hi-Z) state, where the line is neither low nor high, but in between.

The timings shown apply for both reading and writing and are set up in the registers of the S3C2410 to control how it accesses external devices connected to that interface. Other peripheral devices would share the data, address, read and write lines, but bus timings are specified individually for each nGCS chip select. Thus fast and slow devices can coexist on the same physical bus, but do not share chip selects.

The table below gives the meanings of the timing signals shown, and their settings in the diagram:

Signal	Meaning	Setting shown
Tacs	Address set up time prior to nGCS active (0, 1, 2, or 4 cycles)	1 cycle
Tcos	Chip select set up time prior to nOE (0, 1, 2, or 4 cycles)	1 cycle
Tacc	Access cycle (1, 2, 3, 4, 6, 8, 10, or 14 cycles)	3 cycles
Tacp	Page mode access cycle (2, 3, 4, or 6, cycles)	2 cycles
Tcoh	Chip select hold time after nOE deactivates (0, 1, 2, or 4 cycles)	1 cycle
Tcah	Address hold time after nGCS deactivates (0, 1, 2, or 4 cycles)	2 cycles

Page mode is where a whole number of repetitive transactions are done in a quick burst, without accessing any other device in between. Box 6.2 discusses some examples of device connectivity using the bus, and possible settings shown above. Note that some devices, such as SDRAM, are connected very differently—and other devices will drive the nWAIT signal into the CPU which *tells* the CPU exactly how long to extend Tacc for (i.e., how long the device needs the CPU to wait for).

Box 6.2 Bus Settings for Peripheral Connectivity

Let us now identify a few device connection scenarios and see how we will handle them using the signals shown above, assuming a 100-MHz bus (i.e., a 10-ns cycle time).

Question: A fairly slow memory device that takes 120 ns to look up an internal address.

Answer: This means that being read or written to, the cycle has to extend over 120 ns. The relevant setting is Tacc, the access time, which would have to be set to 14 cycles, the next biggest after the 120 ns required.

Question: A peripheral where the chip select has to be activated at least 25 ns before the read or write signal.

Answer: In this case, the nGCS line has to go low before either nOE or nWE. The relevant setting is Tacs, and this would have to be set to 4 cycles, which is the next biggest setting after 25 ns.

Question: A peripheral that keeps driving the bus for 12 ns after it is read.

Answer: In this case, we need to make sure that nothing else can use the bus for at least 12 ns after a read to this device. The relevant setting is either Tcah or Tcoh or both (but most likely Tcoh). To be safe, we could set both to be 1 cycle, giving us a total of 20 ns. This is called the hold-off period.

Usually the data sheet of whatever peripheral you select will have a timing diagram from which it is possible to derive the required information. But if in doubt, select the longest and slowest values available as a starting point, and try to gradually reduce them while ensuring that the system still works reliably (and as an extra safety measure, make it slightly slower than the fastest settings which work for you—it might work at lab temperature, but once it's out in the cold or hot, or has aged a few years, it might no longer work at those fast settings.)

6.3 Standard Interfaces

Modern computers, whether embedded, desktop, or server, tend to use a limited set of very standard interface types. There is space in this book only to briefly highlight the more common interfaces and their characteristics.

These interfaces are classified according to their usage, whether this is low-speed data transfer, system control, or supporting mass storage devices. It should be borne in mind that ingenuity has bent many interfaces to uses different from those envisaged by the original designers.

6.3.1 System Control Interfaces

System control interfaces are those that control and set up various low-speed devices. They are typically pin and space efficient, and usually relatively low speed, but simple in structure.

- SPI, Serial Peripheral Interconnect, serial multi-drop addressed, 20 MHz
- IIC, Inter-IC Communications, serial multi-drop addressed, 1 MHz
- CAN, controller (or car) area network, serial multi-drop addressed, a few MHz

Other more recent variants now exist, such as Atmel's TWI (two-wire interface), Dallas Semiconductors 1-wire interface, and so on.

6.3.2 System Data Buses

Over the years there have been many attempts to introduce standard buses and parallel bus architectures. Many of those common parallel buses found in PC architecture systems are shown in the table below. It is useful to bear in mind that embedded systems tend to use different bus architectures—two very commonplace examples being the

AMBA (advanced microcontroller bus architecture) from ARM and GEC Plessey Semi-conductors (which later became part of Marconi Ltd and was finally swallowed by Mitel Semiconductors), as discussed in the S3C2410 example in Section 6.2, and the APB (ARM peripheral bus). Both AMBA and APB are found in a huge variety of system-on-chip and embedded processor arrangements. These fared far better than the IBM-introduced MCA (microchannel architecture) of the late 1980s, which despite being generally considered a fairly well-defined bus system, was dropped in favor of extended ISA (EISA).

Bus name	Width (bits)	Speed (MHz)	Data rate (MiB/s)
ISA (industry standard architecture) 8-bit	8	8	4
ISA 16-bit	16	8	8
EISA (extended ISA)	32	8.33	33.3
PCI (peripheral component interconnect) 32-bit	32	33	132
PCI 64-bit	64	33	264
AGP (advanced graphics port) 1x	64	66	266
AGP 8x	64	533	2100
VL-BUS	33	50	132
SCSI-I & II	8	5	40
Fast SCSI-II	8	10	80
Wide SCSI-II	16	10	60
Ultra SCSI-III	16	20	320
PCIexpress—per lane (up to 32) using LVDS[1]	1	2500	>500
RAMBUS (184-pin DRAM interface)	32	1066	4200
IDE[2]/ATA[3]	16[4]	66[4]	133[5]
SATA (serial ATA) using LVDS	1	1500	150
SATA-600	1	Unknown	600

[1]LVDS: low-voltage differential signaling.
[2]IDE: integrated drive electronics, corresponding to the first ATA implementation.
[3]ATA: advanced technology attachment, now renamed to parallel ATA or PATA to distinguish it from SATA.
[4]Assuming ATA-7 operation.
[5]133 MHz over 45 cm maximum length.

Although there are a vast number of bus systems (with those listed only being the more common ones), there is a fair degree of commonality since most use the same basic communications and arbitration strategies. There are several voltage and timing standards on offer.

Sometimes buses which are electrically identical have different names and uses differentiated by the actual communications protocol used on the bus. The OSI layered reference model (see Section 10.4) defines the low-level electrical, hardware, and timing parameters to be part of the physical layer, whereas the signaling protocol starts to be defined by the data link layer. An example of a physical layer interface is LVDS (low-voltage differential signaling) which is increasingly being used for high-speed serial buses in embedded computer systems.

We will examine two of the more common legacy buses here in a little more detail before we consider the physical LVDS layer used by SATA and other schemes.

6.3.2.1   ISA and Its Descendants

The industry standard architecture (ISA) bus was created by IBM in the early 1980s as a bus system for use within personal computers, in particular the IBM PC. The 8-bit bus was quickly expanded to 16 bits and then to 32 bits in the EISA version by 1988. Each new version was backward compatible with the previous ones.

As mentioned previously, IBM then attempted to move to the microchannel architecture (MCA), but since they did not release the full rights to this closed-standard bus, other computer vendors unsurprisingly preferred to stick with EISA. IBM then effectively backed down, dropping MCA, but two descendants of ISA, peripheral component interconnect (PCI) and VESA local bus, did incorporate some of the IBM MCA features.

As a bus, ISA and EISA performed reasonably well considering their age; however, they suffered from severe usability issues (see Box 6.3). These issues, coupled with relentless pressure to increase bus speeds, soon led to the definition and adoption of PCI in desktop systems.

Box 6.3   The Trouble with ISA

ISA, as a product of its time, was reasonable: It was designed for the 8-bit bus Intel 8088 processor, clocking at something like 4.77 MHz, and operating with 5 V logic. However, it inherited some severe hardware limitations and usability issues from these pioneering CPUs:

Hardware Limitations

The Intel 8086 and 8088 were built in a 40-pin dual in-line package (DIP) with 16-bit and 8-bit external data buses, respectively. Due to lack of pins, external buses were multiplexed, meaning that some physical pins were required to perform two functions. Even with this approach, there was only room for 20 address pins, thus only 1 Mbyte (2^{20}) of memory could only be accessed. Even more limiting was the fact that use of 16-bit address registers within the 8086 meant that memory could only be accessed in 64 kbyte (2^{16}) windows. Intel also provided two types of external access: memory accesses (using the 20-bit address bus) and I/O accesses (using 16 of the 20 address bits). Interestingly, the split between memory and address accesses is retained today in many systems—in contrast to the welcome simplicity of processors such as the ARM which have only memory-mapped external accesses.

Although the 8088 pins were buffered and demultiplexed before being connected to the ISA bus, the bus retained both the 20-bit address limitation and the separate I/O memory accesses (for which separate sets of control pins were provided). On a more positive note, the ISA bus did cater for four channels of DMA accesses quite nicely (Section 6.1.2).

Usability Issues

This is not particularly relevant within embedded computer systems, but helps to explain the replacement of ISA by PCI. Many PC users were faced with problems when installing ISA (and EISA) cards within their systems. Users would not only need to physically insert and screw down the cards but in most cases would have to inform the installation software what I/O port, DMA channel, and IRQ (interrupt request) lines the card is connected to, and this is not the sort of information that the average user would be able to provide. This was actually an improvement upon earlier devices where these settings were adjusted through changing tiny switches placed on the plug-in card itself.

Some installation software would scan the ISA bus looking for the installed card. Sometimes this worked, but at other times it would totally crash the system, as would a user entering incorrect details. Some PCs allowed the ISA slots to be swapped under BIOS control, or automatically at boot time. This meant that a card would work one day, but not the next.

Manufacturers, in exasperation, began to define a standard called "plug and play" or PnP for short. This would, in theory, allow a card to be inserted and simply work. The fact that the standard quickly became known as "plug and *pray*" is testament to the success of that strategy. Thankfully, the replacement of ISA/EISA by the PCI bus heralded a new era of simplification for users.

ISA spawned not only PCI and VESA local bus, but also the ATA standard, which itself led to IDE, enhanced IDE (EIDE), PATA, and SATA. In fact, it also led to the PC-card standard interface.[1] Despite being a 30-year-old standard, ISA can still be found in systems today where it is often referred to as a "legacy" bus.

6.3.2.2 PC/104

In embedded systems, perhaps the most enduring legacy of the ISA bus is in the **PC/104 standard** from the PC/104 consortium.[2]

The PC/104 standard mandates quite a small form factor printed circuit board size of 96 × 60 mm, which is ideal for many embedded systems. The board has, in its basic form, one connector on one edge that carries an 8-bit ISA bus. This 0.1-inch spacing connector has 64 pins arranged in two parallel rows. On the top side the connector presents a socket, while on the bottom side it presents long pins. This arrangement allows boards following the standard to be stacked, one on top of another. Normally, a second 40-pin connector, J2/P2, placed next to J1/P2, provides the ISA expansion to a 16-bit data bus.

The pin definitions for PC/104 are shown in Table 6.1. Rows A and B are the original ISA signals, encompassing the 8-bit data bus (SD0 to SD7) and the 20-bit address bus (SA0 to SA19) along with memory and I/O read and write (SMEMW*, SMEMR*, IOW*, IOR*), several IRQ pins and DMA signals (those beginning with "D"). The connector

---

[1] PC card was formerly known as a PCMCIA (personal computer memory card international association), although it is also known as "People Can't Memorize Computer Industry Acronyms" (see http://www.sucs.swan.ac.uk/~cmckenna/humour/computer/acronyms.html).

[2] http://www.pc104.org

Pin no.	J1/P1 row A	J1/P1 row B	J2/P2 row C1	J2/P2 row D1
0	–	–	GND	GND
1	IOCHCHK*	GND	SBHE*	MEMCS16*
2	SD7	RESETDRV	LA23	IOCS16*
3	SD6	+5V	LA22	IRQ10
4	SD5	IRQ9	LA21	IRQ11
5	SD4	−5V	LA20	IRQ12
6	SD3	DRQ2	LA19	IRQ15
7	SD2	−12V	LA18	IRQ14
8	SD1	ENDXFR*	LA17	DACK0*
9	SD0	+12V	MEMR*	DRQ0
10	IOCHRDY	key	MEMW*	DACK5*
11	AEN	SMEMW*	SD8	DRQ5
12	SA19	SMEMR*	SD9	DACK6*
13	SA18	IOW*	SD10	DRQ6
14	SA17	IOR*	SD11	DACK7*
15	SA16	DACK3*	SD12	DRQ7
16	SA15	DRQ3	SD13	+5V
17	SA14	DACK1*	SD14	MASTER*
18	SA13	DRQ1	SD15	GND
19	SA12	REFRESH*	key	GND
20	SA11	SYSCLK		
21	SA10	IRQ7		
22	SA9	IRQ6		
23	SA8	IRQ5		
24	SA7	IRQ4		
25	SA6	IRQ3		
26	SA5	DACK2*		
27	SA4	TC		
28	SA3	BALE		
29	SA2	+5V		
30	SA1	OSC		
31	SA0	GND		
32	GND	GND		

TABLE 6.1 The pin definitions of the PC/104 connector showing the two 2-row connectors J1/P1 and J2/P2. Active low signals are indicated with an asterisk "*," and the two keys shown indicate filled holes in the 0.1-inch connector.

specifies +5, −5, +12, and −12 V, along with ground (GND), although in practice often only +5 V is used unless items such as EIA232 and other line drivers are present.

The second connector, containing rows C1 and D1, provides a larger address range and expands the data bus to 16 bits (along with providing more DMA functionality). This is a parallel bus, and has all signals operating synchronous to SYSCLK.

6.3.2.3   PCI

Peripheral component interconnect (PCI) was a ground-up replacement for ISA/EISA, released in the early 1990s. It is probably the most common of the internal PC buses at the present time, although USB has emerged during recent years as the interface of choice for many peripherals that would once have been internal plug in cards for a PC. The much faster serial-based PCI express (PCIe) system is gradually replacing PCI in more recent systems.

PCI is similar to ISA in being synchronous, this time to a 33-MHz (or 66-MHz) clock, and like EISA is generally 32-bit, although 64-bit versions are available using a longer connector. The connector also differs depending upon the signaling voltage used—both 3.3 and 5 V versions are available, and these have different "notches" on the connector to prevent the wrong connector from being inserted (some "universal" cards have both notches and thus can plug into both systems). Like ISA, there are also +12 and −12 V pins, which are similarly not always utilized.

The PCI bus multiplexes the address and data pins, AD0 to AD31 (extending to AD63 in the 64-bit version), allowing for fast data transfer, and a large addressable memory space. There is a bus arbitration system defined for PCI which allows any connected device to request control of the bus, and the request to be granted by a central arbiter. A bus master is called an initiator, and a slave is called the target, with the bus master being the device that asserts the control signals. Practically this means that the voltages driving the PCI bus can come from any of the connected devices, and this is something that has a major implication on the integrity of electrical signals traversing the PCI bus. Therefore, PCI implements a very strict signal conditioning scheme for all connected devices.

Perhaps bearing in mind some of the usability issues associated with ISA and EISA, PCI devices must implement registers which are accessible over the bus to identify the device class, manufacturer, item numbers, and so on. More importantly, these registers define the device I/O addresses, interrupt details, and memory range.

6.3.2.4   LVDS

Low-voltage differential signaling (LVDS) is a very high-speed differential serial scheme relying on synchronized small-voltage swings to indicate data bits. Advocates of this standard have coined the slogan "gigabits at milliwatts," because LVDS can reach signaling speeds exceeding 2 Gbits per second.

Note that LVDS is not a bus protocol like ISA or PCI, it is simply a physical layer signaling scheme. LVDS, however, has increasingly been adopted by many of the bus standards that do exist. An example, which we shall discuss below, is PCI express.

In LVDS, each signal is transmitted over two wires, these are operated differentially, so that it is the difference between the voltage on the two wires, which indicates the presence of a logic "0" or logic "1." Differential transmission schemes are resistant to common-mode noise, that is noise that both wires experience (like power supply noise, and interference from nearby devices), and in fact LVDS can routinely cope with levels of common-mode noise that exceed the signaling voltage.

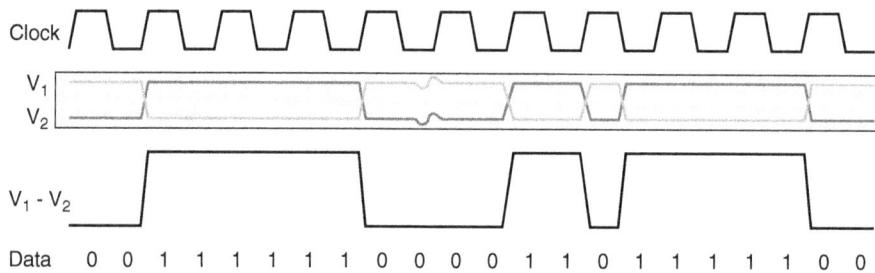

FIGURE 6.3   Illustration of low-voltage differential signaling (LVDS) showing two transmitted differential signals. At the receiver, the difference between these signals is calculated ($V_1 - V_2$), and used to determine the transmitted data at each clock cycle (shown at the bottom). Only the two signals (V_1 and V_2 on the grey background) are actually transmitted, although in practice the receiver and transmitter would both need to have accurate and synchronous timing information. A small amount of common-mode noise seen on both transmitted signals in the center of the plot, is completely removed through the differencing process.

This noise resistance means that lower voltage swings are necessary in LVDS connections. This, in turn, requires much less power to operate, allows faster signaling, and produces less electromagnetic interference (EMI). An illustration of the LVDS signaling scheme is shown in Figure 6.3, showing the differential nature of the system, and the rejection of common-mode noise.

Voltage swings in LVDS are typically around 0.25 to 0.3 V. Since switching (and data transmission) speed depends upon the time taken for a signal to change from one state to the next, with the very low-voltage swings of LVDS, switching can be extremely rapid. Power consumption in transmission systems also depends upon the square of the voltage, so a low-voltage signaling scheme like LVDS is significantly lower power than 3.3 or 5 V logic systems. Similarly, the low-voltage swings lead to low levels of EMI generated by LVDS.

Transmission using a differential pair means that as the voltage on one wire increases, the voltage on the other wire decreases. If we relate that to drive current, at any one time a transmitting device has to drive current into one wire, and out of the other wire. When the system is designed correctly, these current flows can actually be made to balance, something which contrasts very strongly to most switching schemes that experience supply current spikes every time a signal is switched. Supply current spikes translates to voltage fluctuations in the power supply voltage—something which can affect other circuitry in such systems.

LVDS receivers normally need to extract a clock signal from the differential data pairs, and this clock-recovery process implies that this is not a trivial interface to connect to (however, a separate clock signal—which could also be differential—can be transmitted alongside the data using another two wires, if required). Bus LVDS (BLVDS) is a variant of LVDS to allow for multiple devices to share the same physical differential wires.

PCI express (PCIe), as mentioned previously, gradually replaced PCI in desktop computer systems. PCIe systems usually specify how many lanes are available. For example, PCIe 1× has one lane, PCIe 4× has four lanes, and PCIe 32× has 32, with several intermediate steps being common. Each lane is actually one pair of LVDS transmitters and receivers (i.e., four electrical connections, two in each direction). Each lane operates at 2.5 GHz in the base standard.

The PCIe 1× connector is rather small, consisting of only 36 pins, and yet delivers a data rate at least 500 Mbits/s (after taking into account protocol overheads). The common PCIe 16× connector is similar in size to a parallel PCI connector (but is much, much faster of course).

6.3.3 I/O Buses

These buses shown below are typical communications buses, several being ones commonly found on PC-architecture systems (apart from USB which is discussed later).

Bus name	Type	Speed	Notes
EIA232 / RS232	Serial	115,200 bps	−12 and 0 V
EIA422 / RS422	32-device serial balanced multi-drop	Up to 10 Mbits/s	1 km at slow speed
EIA485 / RS485	As 422 but with multidrivers	Up to 10 Mbits/s	1 km at slow speed
DDC, Display Data Channel (monitor info)	Serial, data, clk, and gnd	Based on I^2C bus	
PS/2 (keyboard & mouse)	Serial, 6-pin miniDIN	Electrically the same as an AT interface	
IEEE1284 printer port	Parallel, 25-pin D	Up to 150 kbytes/s	Up to 8 m

EIA standards are ratified by the Electronic Industries Alliance (previously known as the Electronic Industries Association), which used the prefix "RS" to denote a recommended standard (i.e., proposed standards that had yet to be ratified). As an example, EIA232 was therefore known as RS232 before it became adopted as a standard. However, since it was implemented in almost every home and desktop computer for a generation with the prefix RS, this name has stuck. Perhaps there is a lesson here for the standards bodies, relating to the speed of their internal processes compared to the rate of adoption in the consumer market.

6.3.4 Peripheral Device Buses

Several common peripheral buses are now discussed. We will see that the trend in recent years has been toward simple plug-and-play serial-based buses. This is ironic as many older computer engineers will remember the pain of connecting printers to computers in the 1980s, when serial peripherals spelt trouble, and the only safe option was considered the parallel bus (also known as IEEE1284, and described in Section 6.3.3).

- USB1.2, universal serial bus, is a serial format originally envisaged for devices such as keyboard and mouse, but subsequently adopted for a wide variety of peripherals. USB1.2 is limited in distance to about 7 meters, and in speed to about 12 Mbits/s raw data rate. Being a serial bus, this bandwidth is shared with connected device, along with a significant control overhead for each. Perhaps the main driver for adoption of USB has been the fact that it can supply power to the peripheral, freeing up a separate power source and cable.

- USB2.0 appears to have been a response to the introduction of firewire (see below), and significantly improves on the speed of USB1.2—to 480 Mbits/s. In the gap between USB1.2 and USB2.0, firewire gained a strong foothold in the video market, becoming the de facto method of transferring video information to a computer.

- USB3.0 introduced in 2008 (although the latest and fastest variant, USB3.2, dates from 2017) increases data transfer rates to 5 Gbits/s and also supports true full duplex. All USB standards allow the connectors to transfer power to peripherals, with USB3.0 providing twice the power transfer capability of USB2.0. An improved connector, type-C was introduced along with the other improvements.

- Firewire, developed by Apple, and ratified as IEEE standard 1394 is another serial format, originally operating at 400 Mbits/s. IEEE1394b doubles the data rate to 800 Mbits/s, but maximum cable length is only 4.5 meters or so. Like USB, Firewire can also provide power to peripherals, but there does not appear to be a standard voltage or current rating across all providers.

- The PCMCIA, Personal Computer Memory Card International Association (mentioned briefly in Section 6.3.2), developed their card interface in the early 1990s based on the ATA or IDE interface. It is a parallel interface with many variants, but is potentially reasonably high speed. This has evolved into the compact flash (CF) interface.

- Multimedia card (MMC) is a serial interface adopted primarily for flash memory cards in cameras and portable audio players. This evolved into secure digital (SD and xD) that maintains the serial interface nature, but allows bits to be transferred in parallel. Sony memory stick is a proprietary alternative with similar specification (and similar shrinkage in package dimensions).

6.3.5 Interface to Networking Devices

These days networking has become ubiquitous: being off-line for any length of time feels like a bereavement. System-on-chip designers have not ignored this trend, and hardware blocks to handle networking are commonly integrated into modern embedded processors.

Typically, a media access controller (MAC) hardware block is integrated on chip while the physical layer driver (PHY) is not, mainly because of the analog drive and different voltage requirements of an Ethernet physical interface. It is, however, possible to purchase a combined MAC-PHY device, so eventually it is expected that a full MAC and PHY implementation will be possible within a system-on-chip. Current integrations are similar to that shown in Figure 6.4.

Considering that the majority of networked infrastructure is currently Ethernet, magnetics are shown in Figure 6.4 connected to the PHY to give a very common system arrangement. The interface between MAC and PHY is a media-independent interface (MII) indicating that communication is not confined simply to wired Ethernet, it could equally as well be through an optical interface that conforms to the MII standard, and possibly requiring a different PHY device. Wireless is another increasingly common alternative based around the same standards process (and is discussed separately in Section 6.6).

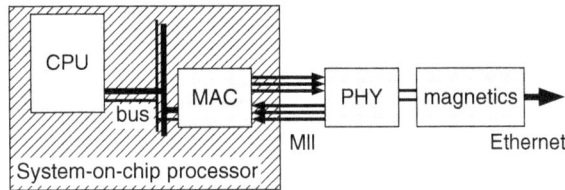

FIGURE 6.4   A block diagram of a network (Ethernet) data connection to a CPU through a media access controller (MAC).

6.4   Real-Time Issues

Remember the ancestor of today's computers: machines occupying an entire room engaged in abstract mathematical calculations, programmed with discrete switches, or punched cards, and delivering results minutes or even hours later? This is far removed from small devices embedded in a human body to adjust blood chemistry, or devices controlling the brake system in a family car. The latter examples are *hard* real-time systems of today. Hard in that they must respond to conditions within a certain time, and the consequences of not doing so are severe.

The former system has no real-time requirement, and its designers might think in terms of speeding up calculations so that they could go home earlier, but would probably not have envisaged a computer making a millisecond response to an external stimuli. This means that traditional computer architectures, and programming languages, did not evolve with real-time responses in mind.

Today, with many more embedded processors than PCs sitting on desktops (and many more PCs than room-sized mainframes), the computing world is increasingly running real time. The vast majority of embedded devices interact with the real world in a timely fashion and are thus real-time systems, either *hard* or *soft* (soft ones are where the consequence of missing a deadline is not catastrophic).

6.4.1   External Stimuli

External stimuli can take many forms, but are often derived from some form of sensor. Examples include an over-temperature sensor in a nuclear reactor, an accelerometer in a vehicle air-bag controller, a vacuum switch in an engine management system, or an optoelectronic gate around a slotted disc in an old-fashioned ball mouse. Each of these could be triggered at almost any time.

Other external stimuli might include data arriving over Ethernet, or data sent from PC via parallel port to a laser printer for printing. Both of these stimuli derive from computers themselves, but since they arrive at the destination at unpredictable times, they appear to be real-time stimuli to the destination processor.

6.4.2   Interrupts

Stimuli arriving at a real-time processor are almost always converted into standard forms to trigger a CPU. These interrupt signals are by convention active-low, attached to an interrupt pin (or possibly an on-chip signal converted to an active low input to the CPU core in the case of a system-on-chip processor).

Most processors have the ability to support many interrupt signals simultaneously. These signals will be prioritized so that when two or more are triggered together, the highest priority interrupt is serviced first.

Interrupts are discussed more completely in Section 6.5, but here it is only necessary to recognize that once an interrupt stimuli occurs, it takes a short amount of time for a CPU to notice this, then more time until the CPU can begin to service the interrupt, and finally even more time until the servicing has completed. Interrupt servicing is through an interrupt service routing (ISR)—which was introduced briefly back in Section 5.6.3 on shadow registers. When designing a real-time system, it is necessary to determine interrupt timings and relate them to the temporal scope of a task (as discussed below in Section 6.4.4).

6.4.3 Real-Time Definitions

Soft and *hard* deadlines were mentioned previously, and these are both real-time constraints, differentiated by the consequence of missing a required deadline. Missing a hard deadline would be catastrophic to the system, whereas missing a soft deadline is unfortunate but not a critical failure.

These terms can also relate to entire systems: a *hard real-time system* is one that includes some hard deadlines. If all deadlines are soft then it is a *soft real-time system*. When choosing an operating system, it is also possible to consider degree of "hardness": for example, uCos is capable of meeting hard deadlines whereas embedded Linux is often softer in its response and Microsoft's operating systems tend to be very soft—which is why they are generally avoided for "mission critical" real-time systems.

While Chapter 9 considers operating systems in general, it does not do so from the perspective of real-time operation, which will be our emphasis in this section.

A *task* is a piece of program code dedicated to handling one or more functions, perhaps tied up with a real-time input or output. In a multitasking *real-time operating system* (RTOS), there will be several tasks running concurrently, with each task having a priority associated with it.

Most systems are designed around interrupts or timers such that every time a particular interrupt occurs, one task will be triggered to handle it. Other tasks will trigger on expiration of a timer. Tasks can themselves be interrupt service routines, but generally they are separate code (in the interests of keeping the ISR as short as possible), so that when ISRs run, they release appropriate tasks using dedicated RTOS functions. These functions, include *semaphores*, *queues*, and *mailboxes* are used in specialized RTOS systems, but can often also be found in standard operating systems, and indeed we will briefly discuss semaphores again in Section 9.6.1.2.

Many tasks would spend most of their time sleeping, waiting to be woken up by an ISR or another task, but often a very low-priority background task runs to perform system-related functions and logging. This may also include adjusting prioritization of tasks yet to be run.

6.4.4 Temporal Scope

The *temporal scope* of a task is a set of five parameters that together describe its real-time requirements. This is a formalism that is very useful in systems with multiple tasks running, each of which have deadlines associated with them.

The following values define the temporal scope, and unless specified are all timed from the event which is supposed to trigger the task:

Minimum delay before task should start	Usually 0, but occasionally specified.
Maximum delay before task must start	Interrupts should be acknowledged as quickly as possible in principle, but this is a hard upper limit.
Maximum time for task processing	Elapsed time between the start and end of the task.
Task CPU time	This may be different to the parameter above since the task could be interrupted, prolonging the time taken but not CPU time.
Maximum task completion time	Elapsed time between the trigger event and the task being completed.

Temporal scope can mostly be determined through analysis of system requirements, although finding the CPU time can only be done either by counting the number of instructions in the task, or through OS tools designed to measure processor cycles. A note on CPU timings—remember that sometimes conditional loops might be longer or shorter depending on the data being processed, and this should be taken into consideration. The CPU time specified is the maximum with all loops being as long as they can possibly be ... it therefore stands to reason that writing compact task code is important.

A task diagram is shown in Figure 6.5, which lists the various tasks that available, and shows which of these occupies the CPU at any particular time. The vertical lines indicate points at which the scheduler has been run and is able to switch between tasks if required. The scheduler is often itself implemented in a system task, and chooses which user task occupies the CPU at which time. Depending on RTOS type, the scheduler will be invoked in different ways—cooperatively through calls in the software itself, at fixed time intervals, or at task despatch points. Task despatch points are usually incorporated in library functions that perform OS-level tasks, sometimes as simple as printf() or similar, but almost always at FIFO, queue, mailbox, and semaphore-related operations. Sometimes a combination of methods is used to call invoke scheduler.

In the task diagram shown in Figure 6.5, the first time the scheduler was invoked (at the first vertical line), task 1 was executing. The scheduler did not switch to another task in this instance, but task 1 continued. The reason may be because task 1 had the highest priority of the three user tasks shown. Conventionally, that is why task 1 is shown at the top!

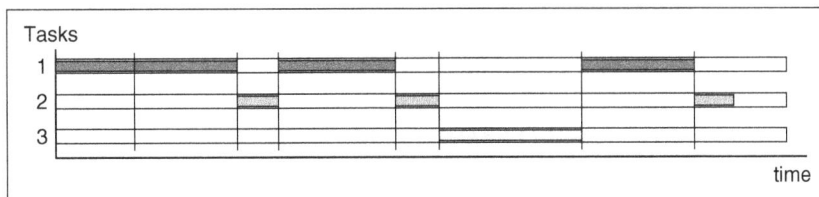

FIGURE 6.5   A scheduler diagram for three tasks, executing on a single CPU.

Task 2 appears to be about the same length each time, indicating that it is probably doing the same work each time it runs.

This brings us to a brief consideration of how a scheduler decides between tasks. First of all, tasks are allocated a priority. Top may be the scheduler, and bottom is the *idle task*, that gets executed when nothing else wants to run. In embedded systems, this might handle the low-priority I/O such as printout of debugging information, or flashing an activity LED (as an aside, the use of the low-priority task to print debugging information is very common—but will not help when debugging a total crash, because no debug information will be visible from the task which crashed, since if the task was running, by definition, the idle task will not get a chance to run).

A table within the scheduler keeps track of all tasks, and maintains the state of each: *running*, *runnable*, or *sleeping*. There will only be one running task at each particular instant, but many tasks could be runnable (indicating that they are waiting a chance to run). Sleeping tasks are those that are halted, perhaps temporarily waiting for a semaphore, or for some data to enter a queue or mailbox. Some methods of ordering scheduling priorities are shown in Box 6.4.

Box 6.4   Scheduling Priorities

Given a number of tasks in a real-time system, a designer is faced with the difficulty of deciding how to assign priorities to tasks to ensure that they can be scheduled. This is extremely important—some choices may result in a system that cannot meet the required deadlines (not schedulable), whereas a small change could make the system work. Some common formalized priority orderings are shown below, all require knowledge of temporal scope of tasks in the system:

Deadline monotonic scheduling: Tasks with the tightest deadlines have higher priority.

Rate monotonic scheduling: Tasks that trigger more often have higher priority.

Earliest deadline first scheduling: This is a dynamic scheme that has knowledge of when a deadline will occur—an assigns priority to whichever task must complete earliest.

Others include **most important first**, **ad hoc**, **round robin** and numerous hybrid schemes (most of which claim to be better than all others). We will later look at scheduling in general operating systems again in Section 9.7.1, and consider the example of the Linux scheduler.

We will revisit the topics of multitasking and scheduling in Chapter 9 from the perspective of the Linux operating system, and look a little more closely at this from a software perspective, rather than the hardware perspective we have used above.

6.4.5   Hardware Architecture Support for Real Time

This is a book about computer systems in general, not only real-time systems, thus it is more important to consider the hardware implications of running a real-time system on a processor than it is to discuss the real-time implications themselves. Let us review again the steps taken when a real-time event occurs:

1. The event causes an interrupt signal to the processor.
2. The processor "notices" the interrupt.

3. The processor may need a little time to finish what it is currently doing, then branches to an interrupt vector and from there to the address of whatever ISR is registered against that interrupt.

4. The processor switches from what it is currently executing into an interrupt service routine.

5. The ISR acknowledges the interrupt and "unlocks" a task to handle the event.

6. Any higher priority tasks waiting get executed first.

7. Finally, context switches to the task assigned to deal with the event.

8. The task handles the event.

Each of these eight steps (expanded in more detail in Section 6.5.2) potentially takes some time, and thereby slows down the real-time response of the system.

Hardware support for interrupts (explored further in Section 6.5) can significantly improve response time; however, the OS functions needed to service the task, particularly switching from previously running code into ISR, and then between tasks, are time consuming and can also be accelerated.

First, shadow registers (Section 5.6.3) speed up the changing of *context* from one piece of code to another. The ARM implements several sets of shadow registers, one of which, called *supervisor*, is dedicated to underlying OS code, such as the scheduler, so that running this does not entail a time-consuming context save and restore process.

Other CPUs take the approach further, implementing several register banks, each of which is allocated to a separate task. With this, switching between tasks is easy—no context save or restore is required, simply a switch to the correct register bank and then jump to the correct code location.

Hardware FIFOs and stacks can be used to implement mailboxes and queues efficiently, to communicate between tasks (the alternative is software to move data around a block of memory). These are generally less flexible because of their fixed size, but can be extremely quick.

It is theoretically feasible to implement a hardware scheduler, although this does not seem to be have been adopted by computer architects. Perhaps the highest performance hardware support for scheduling would be dual (or more) processors which can support hyper-threading or similar. In this instance there is the ability for two tasks to be running in each time instant rather than just one! This is an example of MIMD processing (see Section 2.1.1) that has been adopted in some of the latest processors from Intel in their Core devices. Other manufacturers are sure to follow (refer to Section 5.8.1 for more details of MIMD and dual cores).

6.5   Interrupts and Interrupt Handling

This section discusses interrupts, their overheads, and considers ways of servicing these quickly. Previously, the use of shadow registers for interrupt service routines (ISR) was covered in Section 5.6.3, and so this particular efficiency improvement will not be discussed again here.

6.5.1 The Importance of Interrupts

Interrupts and their handling is one of the most important topics in computer architecture and embedded software engineering. With the degree of interaction between computers and the real world increasing, and becoming more critical through the profusion of embedded computer deployments, it is the humble interrupt that is tasked with most of the burden. This burden includes ensuring that a processor responds when necessary, and as quickly as necessary to real-time events.

Real-time events were discussed previously, but here it is simply necessary to remember three important timings associated with an interrupt:

1. The interrupt detection time—how long after the event occurs that the CPU "notices" and can begin to take action.

2. The interrupt response time—how long after the event occurs that the CPU has "serviced" the event, that is, worst-case timing before the appropriate action has been taken.

3. The minimum interrupt period—the earliest time after one interrupt that the same interrupt can occur again. If not regular, then take the minimum allowable.

6.5.2 The Interrupt Process

Exactly what happens after an interrupt line asserts is important to understand, since these events have a huge impact on the system architecture as will be discussed. As already briefly described in Section 6.4.5, a table of the process is given below:

1. An external event causes an interrupt signal to the processor.

2. The processor "notices" that the interrupt has occurred.

3. The processor first finishes what it is currently doing, then branches to an interrupt vector and from there to the address of whatever ISR is registered to handle that interrupt.

4. The processor switches from the currently executing code, and branches to the appropriate interrupt service routine.

5. The ISR acknowledges the interrupt and ends. It will have "unlocked" any tasks pending on the interrupt event.

6. Any higher priority tasks waiting get executed first.

7. Finally, context switches to the task assigned to deal with the event.

8. The task handles the event.

We will look more closely at each of the first five steps in the following subsections since these are strongly influenced by architectural issues.

6.5.2.1 An Interrupt Event Signals the Processor

The interrupt signal to the CPU is, by convention, normally active-low, and can be edge triggered or level triggered. An edge-triggered interrupt signals to the CPU by the act of changing state. The processor then responds to this edge as soon as it can—even though the interrupt line may have reset itself in the meantime. Something like a key press might generate this type of interrupt (it should not matter how long the key is held down for, the processor will respond in the same way).

A level-triggered interrupt will be physically similar—but the processor samples this at predefined times to see what its state is, perhaps once per clock cycle. Once such a signal occurs, it needs to be asserted for a certain length of time before the processor "notices" it, and this time may be configurable. For example, it should be asserted for three consecutive sample times to be genuine, rather than a noise spike.

Once an interrupt signal is latched, whether or not the physical interrupt line deactivates again, the internal trigger remains set waiting. Eventually, some code in the processor will get around to servicing that interrupt. The question is, what happens if the interrupt line toggles again before the previous one has been serviced? As always, the answer depends on exactly which processor is being considered, but in general the second interrupt will be ignored: the internal "interrupt has happened" flag was set, and cannot be reset until it is cleared in software (in the ISR).

However, there have been several processors in the past that have been capable of queuing interrupt signals (especially processors which tended to be fairly slow about responding to interrupts). Queuing interrupt signals sounds like a fine idea, but it significantly complicates real-time handling, and is therefore not usually considered these days as a potential hardware solution. The best solution being to handle whatever interrupts occur as quickly as possible.

6.5.2.2 *The CPU Finishes What It Is Doing*

Modern processors cannot be interrupted in the middle of performing an instruction—they have to wait for that instruction to finish first. In the past, with CISC processors taking many cycles to perform some instructions, this was hugely detrimental to interrupt response time. For example, the Digital Equipment Corporation VAX computers are said to have had an instruction that took over 1 ms to complete, which is a long time to wait for an interrupt to be serviced (put in an audio context this means that a sample rate of 1 kHz would have been the maximum that could be supported by individual interrupts; far less than the 48 kHz and 44.1 kHz of today's MP3 players).

Attempts were made to allow sub-instruction interruption for processors using microcode, but this became horrendously complicated and was not popular. Real-time systems designers breathed a sigh of relief with the advent of RISC processors (Section 3.2.6) with their one-instruction-per-clock-cycle design rationale. This means that, in theory, the longest time taken for an instruction to complete is one instruction clock cycle, which tends to be very short on RISC processors. This would mean that the same short time is all that it takes between an interrupt being "noticed" and the branch to interrupt vectors.

In practice, this RISC concept is adhered to less strongly by some designers. The ARM, for example, has a multicycle register load or store instruction which is really useful for fast data moves or for context save and restore, but which does take up to 16 cycles to complete. So the worst case wait for the interrupt to hit the interrupt vectors is therefore 16 cycles (unless the programmer avoids using the multicycle load/store command).

One more thing to note is the effect of pipelines. With pipelined instructions, although one instruction enters the pipeline in each instruction cycle, it takes n cycles to actually complete an instruction, where n is the length of the pipeline. Without complex dedicated hardware support, a shadow register system will have to wait for the current instruction to flow through the pipeline, and store any result, before the jump to ISR can occur. Pipelines are great for very fast instruction throughput, but can be slower to respond to interrupts.

6.5.2.3   Branching to an Interrupt Service Routine

The traditional method of handling interrupts is that once one occurs, the PC is loaded with a preset value, thus causing the CPU to jump to a special place. Typically there is one of these special places for each type of CPU interrupt in the system. These places in memory are called interrupt vectors.

In the ARM, the interrupt vectors begin at address 0 in memory. Address 0 is called the *reset vector*—it is where the CPU starts at power up or after reset. Each event and interrupt in the CPU follows in order. What is stored in this vector table is simply a branch instruction to the handler for that event. For the reset vector this will be a branch to something like __start. For IRQ1 it will be to the ISR designated to handle IRQ1 (the use of double underscore is common when translating between C language and assembler).

Here is a typical interrupt vector table for an ARM program:

```
B       __start
B       _undefined_instruction
B       _software_interrupt
B       _prefetch_abort
B       _data_abort
B       _not_used
B       _irq
B       _fiq
```

Figure 6.6 illustrates use of the interrupt vector table to handle an interrupt occurring during execution of a routine.

It can be seen that execution starts at the initial reset vector, which branches to the start of the code that is to be run on the processor (B __start). This code progresses as normal (indicated by the dark solid arrows on the left-hand side) until an interrupt occurs during the SUB instruction. This instruction completes and then the processor jumps to the interrupt vector associated with that interrupt, which in this case is the IRQ interrupt. We can assume that, although it is not shown in Figure 6.6, there was a switch to shadow registers during this process. The IRQ interrupt vector contains a branch to the relevant interrupt service routine, which in this case is called ISR1. This services the interrupt and, once complete, returns to the instruction following the one in which the original interrupt occurred. Again, although it is not shown, it is assumed that a switch back from the shadow register set to the main set is performed during this return from interrupt. In some processors this happens automatically, but in others a different return instruction is required (e.g., the TMS320C50 has a RET to return from subroutine and a RETI to return from interrupt, which automatically POPs the shadow registers when it returns). It is fairly obvious in this case that the machine uses shadow registers—this is indicated by the fact that the ISR and the main code both use the same register names without any explicit context save and restore being performed in the ISR.

One more thing to note here is that the interrupts that are not used in the vector table are populated with NOP instructions, which would mean that if such an interrupt

FIGURE 6.6 An illustration of the calling of an interrupt handler via a branch to the interrupt vector table. In this diagram, execution from power-on involves an initial branching to location____start. The interrupt occurs during the second SUB instruction in the *handler* subroutine, and so the branch to service the interrupt happens before the ADDS instruction. Normal operation is shown with the solid arrows, control flow during the interrupt handling is shown with dotted arrows.

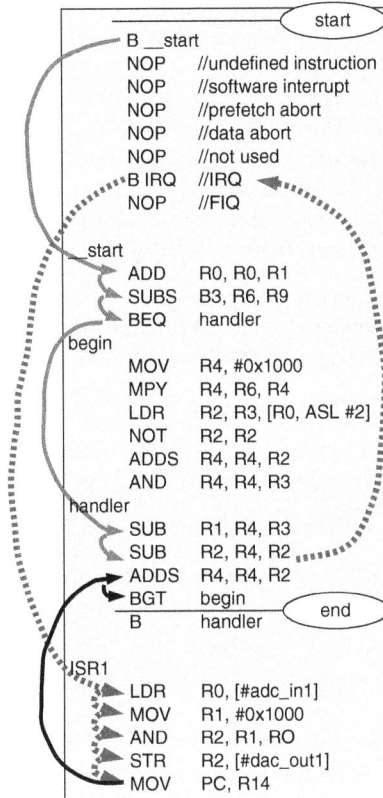

```
                                      start
            B __start
            NOP         //undefined instruction
            NOP         //software interrupt
            NOP         //prefetch abort
            NOP         //data abort
            NOP         //not used
            B IRQ       //IRQ
            NOP         //FIQ

        __start
            ADD     R0, R0, R1
            SUBS    B3, R6, R9
            BEQ     handler
        begin
            MOV     R4, #0x1000
            MPY     R4, R6, R4
            LDR     R2, R3, [R0, ASL #2]
            NOT     R2, R2
            ADDS    R4, R4, R2
            AND     R4, R4, R3
        handler
            SUB     R1, R4, R3
            SUB     R2, R4, R2
            ADDS    R4, R4, R2
            BGT     begin
            B       handler               end

        ISR1
            LDR     R0, [#adc_in1]
            MOV     R1, #0x1000
            AND     R2, R1, RO
            STR     R2, [#dac_out1]
            MOV     PC, R14
```

occurred, the NOP would execute, then the next NOP, and so on until something happened. For example, if there was a data abort event (caused by some sort of memory error) then control would branch there, do the NOP, then the next NOP and finally reach the branch to ISR. So IRQ1 would be executed—even though no IRQ interrupt had occurred. It is thus always better to put interrupt service routines for all interrupts whether they are used or not—and trap them displaying an error so that if the worst does happen and such an interrupt occurs, this error will at least be noted.

The interesting case of interrupt timing within the ARM processor is explored in Box 6.5.

Box 6.5   ARM Interrupt Timing Calculation

The ARM has two external interrupt sources; the standard interrupt (IRQ) and the fast interrupt (FIQ), with the FIQ having higher priority. The shadow register sets provide six usable shadow registers for the FIQ and only one for the IRQ. Assume we need to use four registers, and each register load to/from memory takes 2 cycles—because we have a 16-bit external bus, but instructions are 32 bits.

The IRQ interrupt vector is midway in the interrupt vector table, whereas the FIQ vector is at the end (*this means that no jump is needed for FIQ* from the vector table if the interrupt code is simply inserted at *this location onward*).

The longest instruction on the ARM7 is a multiple load of 16 registers from sequential memory locations, taking 20 clock cycles. It can take up to three cycles to latch an interrupt and two cycles are needed for every branch. There is one operation with higher priority than both FIQ and IRQ (and that is an SDRAM refresh operation). Assume that this takes 25 cycles to complete, and that the fictitious processor is clocked at 66 MHz.

We can now determine how long it will take to service an IRQ, and an FIQ.

Counting in cycles, timed from when the IRQ becoming active:

1. Time to recognize interrupt: 3 cycles.
2. Worst case current instruction must finish first: 20 cycles.
3. In case SDRAM is being refreshed, wait for that: 25 cycles.

At this point, the CPU is ready to respond.

4. Branch from current location to read line in vector table: 2 cycles.
5. Act on instruction in table: branch to ISR: 2 cycles.

Now we are within the interrupt service routine (ISR).

6. Context save 3 registers (we need 4, 1 is shadowed): $2 \times 3 = 6$ cycles.
7. Execute first instruction to respond to interrupt: 2 cycles.

Total (in instruction cycles): 60 cycles
Total (66 MHz processor cycle is approximated to 15 ns): 0.9 µs

One microsecond is considered relatively fast in terms of CPU response times, and indeed interrupt response time is one of the main advantages of the ARM architecture.

Now consider the case of the FIQ. In this instance there are two main differences. One being the fact that more registers are shadowed, and the other being that the FIQ code is resident at the interrupt vector, rather than one jump away. So the differences between FIQ and IRQ (above) are

5. No need to branch to ISR: −2 cycles.
6. FIQ has 6 shadow registers, so no context saves needed: −6 cycles.

Total (in instruction cycles): 52 cycles.
Total (66 MHz processor cycle is approximated to 15 ns): 0.78 µs.

Could we do anything to improve this still further (without over clocking!)? Yes, we could avoid the 20-cycle longest instruction in our code, or change memory technology. Avoiding multiple load/save instructions, and removing the SDRAM refresh cycle too, we can achieve 0.2 µs.

I should note here that ARM7-based processors do not normally use SDRAM, but those based on ARM9 and beyond do tend to.

6.5.2.4 *Interrupt Redirection*

One more point remains to be explained with regard to the interrupt vector table and that is in cases where the lower part of memory is mapped to non-volatile ROM since it contains a bootloader, and the upper part of the memory map contains RAM. Without some mechanism to alter the interrupt vector table, it means that whatever code is loaded into RAM cannot take advantage of the interrupt vectors. This is not at all useful to code in RAM that wants to use an interrupt.

There is thus often a mechanism in hardware to remap the interrupt vectors to another address in memory (see Box 6.6 shows an example of this from an ARM processor). This would mean that, on initial reset, a bootloader is executed which then loads some program and runs it. This program would cause the interrupt vector table to be remapped into RAM, into an address range that it occupies itself, can thus write to, and within which it places vectors for whatever interrupts it requires. We will examine booting and bootloaders more in Section 9.5.

Where an RTOS is used, there may be a second layer of vectorization: all interrupts trigger an appropriate ISR within the OS code itself, but external functions can register themselves with the OS to say that they should be called upon certain events. When such registered events occur, interrupts happen as normal, but the ISR is within the OS, and must initiate a further branch out to the registered interrupt handler. This mechanism can provide a handy way to implement shared interrupts on a processor, or system-on-chip, that does not support hardware interrupt sharing. In this case, it is the responsibility of the OS to decide exactly which of the shared interrupts had occurred and then branch to the relevant handler code. The usual way of interrupt sharing, in hardware, is covered in Section 6.5.4.

Box 6.6   Memory Remapping during Boot

Some processors get around the problem of needing to execute two branches to get to an ISR using a slightly different method. In the ARM-based Intel IXP425 XScale processors for example, on initial powerup, flash memory or ROM is mapped into memory address space 0 and upward, intended for storing boot code. A register inside the CPU allows the boot code memory to be mapped upward in memory, following which SDRAM is mapped at address 0 and upward.

Thus the bootloader simply needs to ensure that a program is loaded which contains its own interrupt vectors, and that these are located at the lowest address in RAM. Then the bootloader issues the remap command.

Unfortunately, it is not necessarily that easy since the bootloader itself is executing from an address in ROM, and when the remap occurs, the bootloader code will disappear. In other words, if the program counter (PC) is at address 0x00000104 executing the remap instruction, by the time the PC is incremented to point at the next instruction at 0x00000108 (steps of 4 bytes since each instruction is 32 bits), the instruction will not be there, it will have been remapped into a higher address space.

There is an easy, but tricky solution to this. See if you can think of it before I explain it…

We would avoid the problem if, after remap, exactly the same code is at exactly the same addresses as it was before. In practice this means saving a copy of the bootloader code to RAM at its higher address before the remapping occurs, and this is the approach used by many XScale bootloaders, such as U-Boot.

Another solution is to split the bootloader into two parts, or stages. The first stage exists to copy the second stage to a RAM address that does not get affected by the remapping. Then this first stage jumps to the second stage that performs the remapping and, being safely out of the way, is unaffected by it. The more sedate booting process for more typical embedded systems is described in Section 9.5.

6.5.3   Advanced Interrupt Handling

With the standard interrupt handling procedure in mind, it is instructive to examine one mechanism for efficiency of the process, and that is to preload the interrupt branch address into a register.

Consider the usual situation: when a particular interrupt occurs, the processor will jump to a given location in the interrupt vector table. This will contain a single instruction (or sometimes two) that normally commands the CPU to branch to another address where the relevant ISR resides. The process thus requires two sequential branches, and since Section 5.2 identified the branch instruction as one which is often inefficient in a pipelined machine, this solution is not particularly good.

Thinking about this, it seems that the CPU has to know where to branch in the interrupt vector table for each event. The vector addresses thus need to be stored within the processor—within some sort of register—and copied to the program counter (PC) when the trigger event occurs. Simply making the vector address register writable allows the vector address corresponding to a particular event to be changed. It is then possible to directly load the ISR start address into this vector address register. This would mean that when an event occurs, the processor can branch directly to the ISR without going through the vector table—and this applies to shared interrupts as well as dedicated ones.

The cost of this approach is a set of writable registers (which occupy more silicon than read-only locations), and a slightly more complex interrupt controller.

6.5.4   Sharing Interrupts

Many computer systems these days implement interrupt sharing. This was initially a consequence of limitation on the number of pins on the ICs used for hardware interrupts, and the limited register sizes inside such CPUs to control interrupts. A very small number of physical CPU interrupts would thus be shared by many separate interrupts. For example, the ARM has two separate interrupts: An interrupt request (IRQ) and a fast interrupt request (FIQ) but a typical ARM-based system-on-chip embedded processor may have up to 32 interrupt sources that share the IRQ and FIQ lines.

Upon a shared interrupt occurring, the ISR started in response would then need to read a register identifying which of the shared interrupts had been triggered, and finally trigger the correct code to respond to this. The triggering might be through using RTOS constructs, or by issuing a software interrupt (SWI). Sometimes, one huge ISR would service many shared interrupts.

Interrupt sharing requires an interrupt controller—either a separate IC dedicated to handling interrupts, or more commonly today an advanced interrupt controller (AIC) block within a system-on-chip embedded processor. An example is shown in Figure 6.7. In this example, it can be seen that the CPU itself has only a single interrupt line, and

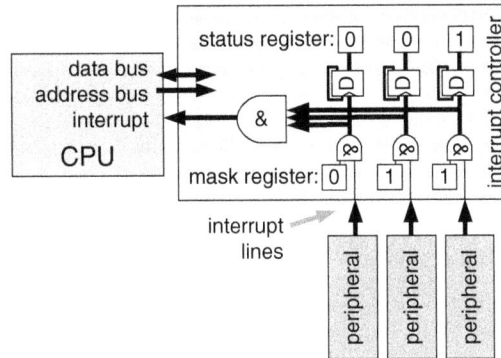

FIGURE 6.7 A block diagram of interrupt sharing hardware that might be used within an interrupt control block implemented in a system-on-chip processor.

this is shared among three peripherals. A CPU-writable register inside the interrupt controller can mask out any of the shared interrupt line, but those that are not masked can cause the CPU interrupt to be triggered.

When the CPU interrupt is triggered, the CPU is able to read the status register to determine which of the shared interrupt lines caused the event. Usually the act of reading this status register will clear the register ready for the next interrupt event (logic for which is not shown).

6.5.5 Reentrant Code

Although an interrupt that is asserted for long enough to trigger an interrupt response and then de-asserts will be physically ready to re-assert, it does not mean that the same interrupt can be triggered again immediately. Although it varies on a processor-by-processor basis, most devices, when servicing one interrupt, do NOT allow that same interrupt to be activated until the ISR has finished (i.e., do not allow *reentrant* interrupts). A second interrupt event will either be ignored while the first one is being serviced, or will cause a re-trigger as soon as the ISR has completed.

Some more advanced processors allow a high-priority event to interrupt a lower priority ISR, which requires hardware support through either separate shadow registers for each ISR, or a careful context save and restore when it occurs.

6.5.6 Software Interrupts

Software interrupts (SWIs) are methods for low-level software to interrupt higher level code, that are typically reserved for operating system (OS) intervention in task-level code handling in an RTOS. In the ARM processor, issuing an SWI command:

```
SWI 0x123456
```

will trigger a switch to a shadow register set. In this case, the processor will also enter supervisor mode (whereas normal programs operate in user mode). Supervisor mode on the ARM is privileged in that it can allow low-level settings to be altered that would be impossible in user mode, and supervisor mode is accompanied by jumping to the third entry in the interrupt vector table, at address 8 (refer to the ARM table shown in Section 6.5.2).

SWIs a type of processor trap, are useful for debugging. One way of breakpointing on a line of software is to replace the instruction with an SWI. Once that instruction is reached, the processor will be interrupted, jump to the SWI vector and on to an SWI service routine.

Inside the SWI service routine, the conditions of the registers (of user mode) and memory would be communicated to the debug software. The debug software would then wait for commands from the user.

6.6 Embedded Wireless Connectivity

This book considers computer devices primarily from the perspective of embedded system, and as such we need to recognize that embedded computing is increasingly networked, and embedded computers are increasingly connected in some way to the Internet. Therefore, in this section, we will very briefly consider wireless technology as it relates to embedded computer systems. However, we will leave discussion of the main wireless technologies in use today until Section 6.6 (since that topic is part of the larger networking topic, the subject of Chapter 11). We will summarize the main factors in wireless, then discuss the interfacing technologies used, before exploring several related issues.

6.6.1 Wireless Technology

Although wireless engineers have many classifications for wireless technology in terms of RF frequency band, channel bandwidth, power, modulation, and so on, for our purposes an embedded engineer would primarily consider different issues:

- *Connectivity to the CPU*—especially whether this is serial or parallel, as will be discussed in Section 6.6.2.

- *Data format*—is data sent in bits, bytes/characters, words, or packets? This relates to the connectivity, but also whether some standard form of data interchange is used, such as USB or IP (Internet protocol) packets.

- *Data rate*, typically measured in bits per second (and note the figure quoted by manufacturers is often before overheads such as packetization, headers, error control, and so on are included, so the rate available for application use may be significantly lower). Of course, this is important to match to the application, but for real-time use remember that data rate does not necessarily relate to latency. A system sending several megabits per second may respond to a single event slower than a system sending only several kilobits per second.

- *Form factor*, including physical size, number, and size of antennae. Lower frequency devices usually require a larger antenna.

- *Range*—also related to power, there will be limits imposed by regulatory authorities (often 0.25 W, and almost always before 1 W, depending upon frequency band and use).

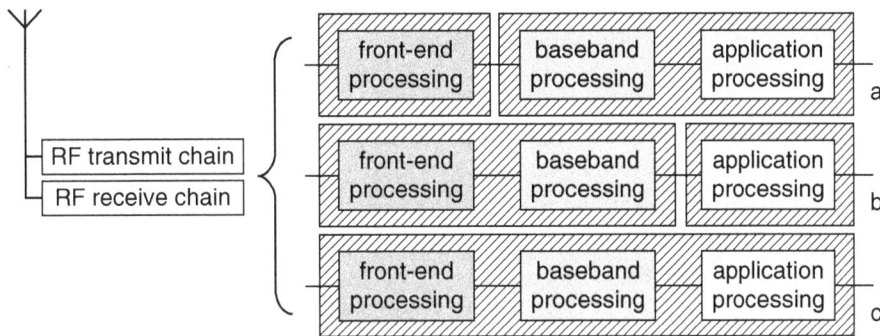

FIGURE 6.8 A block diagram of three alternative wireless processing schemes for an embedded computer showing two computational devices for the wireless processing plus the one embedded applications process. These devices either handle processing needs separately or (*a*) the baseband processing is performed together with the applications processing, (*b*) an add-on device providing wireless functionality to an application processor, and (*c*) an all-in-one wireless and applications processing device.

- *Power consumption*, again related to power, range, and data rate.
- *Error handling*—are communications "guaranteed" to be error free, or does the system need to take care of errors itself. This issue is covered more in Section 6.6.3.
- *CPU overhead* is another important factor to consider.

When a designer is given the task of providing wireless functionality for an embedded system, these considerations will need to be resolved, and some trade-off point reached between them.

Many wireless standards exist, and many are suitable for embedded systems so. In this section, we will consider the major issues that would allow a designer to sensibly analyze and evaluate the choices.

First of all, Figure 6.8 shows a block diagram relating the connection of a wireless solution to an application processor. The application processor is the CPU in the system being connected, and typically this is the only CPU in that application system.

Quite clearly, wireless communication usually requires a fair degree of signal processing, and most of the wireless solutions that embedded systems designers would be considering these days are digital in implementation. Front-end processing (which could be analog but is becoming more likely to be digital) is the very front-end signal conditioning performed on received, and to-be-transmitted wireless symbols. This data may require processing at a MHz or GHz rate, almost always at a multiple of the bit rate. Baseband processing, by contrast, is slower protocol-level computation, such as packet handling, packet error checking, tracking retries and resends.

When systems designers do not choose to comply with a standard and instead define their own wireless scheme, the baseband processing will probably be performed inside the application processor (as in Figure 6.8*a*). It may even be possible to use just a single computational device for *all* processing (Figure 6.8*c*). While this is also technically possible for simple standard protocols, many involve either a protocol overhead that is too great to be included in the application processor or is not freely available to embedded systems designers in source code form. Thus baseband processing needs to be undertaken separately, either in a separate device, or together with the front-end processing (as in Figure 6.8*b*).

Probably the greatest reason to separate the wireless processing out from the application processor is to not reinvent the wheel. It is truthfully quite difficult to create a reliable wireless communication system. Where a proven working solution is available off-the-shelf, this is a compelling reason to use it!

6.6.2   Wireless Interfacing

Since we had divided CPU buses into serial and parallel in Section 6.1, we can provide the same classification for wireless functionality. Although the actual data conveyed over air may be serial, parallel, or some combination, at heart every wireless device needs to interface to a CPU either by a serial interface, or by a parallel interface.

Simple and slow wireless interfaces tend to be serial connections: if we provide serial data at one end of a wireless link, we can receive it at the other end. If error control is provided in the link, then the received data can be assumed to be (relatively) error free. Otherwise, error checks should probably be added to the application code. Wireless USB standards also fall into this serial category.

Internet Protocol (IP)–based schemes, such as IEEE802.11 (WiFi) and IEEE802.16 (WiMAX), are block based. Entire packets of data are handled by the protocols. So wireless solutions for these standards interface to a CPU using a parallel bus, and often use DMA (see Section 6.1.2) to transfer data, to improve efficiency. In fact, this is much like the interfacing of a standard Ethernet device (Section 6.3.5).

6.6.3   Issues Relating to Wireless

At best, adding wireless functionality to a system simply allows another connection to be made to that system at will. Of course, wireless will obviously impact system power requirements, and so on; however, there are other issues that should be considered.

The first of these issues has been touched upon previously: CPU overhead. Obviously when the protocol handling is being undertaken in the application processor, a potentially significant proportion of the processing time will be occupied on that (and as always, consider the worst case—perhaps when every packet is received in error). However, even in the case where a separate device handles all of the wireless processing and protocol handling, even then an application processor doing nothing but streaming data in and out, may require many CPU cycles to handle the wireless traffic.

When error handling is considered, the issue is that behavior needs to be identified and coded in the case of the many different error types that are possible. Of course, this is just as true with wired Ethernet; however, wired Ethernet normally suffers at the extremes— no errors at all, or no packets at all. Wireless, by contrast, normally operates well within these endpoints.

A further issue is that of security—with wired connections, it is fairly easy to know what is connected (just follow the wires…); however, a wireless connection is invisible. Designers should note that it is not always the correct recipient who received data, and who replies. With the rapid growth of computer technology in embedded systems, many more people are entrusting their livelihoods and finances to such systems, and some observers feel that security considerations have moved more slowly than the technological advancement in this field.

Finally, the very wireless signals themselves permeate the free space around a transmit antenna. There are many cases where these signals can couple back into the system that is generating them, to become significant sources of electrical noise on the buses and wires within that system. This issue is known as *electromagnetic interference* or EMI, and has been recognized in recent years as a very important contributor to systems unreliability.

There are two main impacts on computer systems designers. The first is that any system which is being designed is a potential source of EMI. Different bus designs cause different levels of EMI. For example, an ISA bus will cause more interference than an LDVS bus by virtue of the larger voltage swings and unbalanced nature of the ISA bus. Memory technology also varies considerably as a cause of EMI. This EMI, created by a computer system, can affect the systems around it (some of us remember early home computers such as the Sinclair ZX Spectrum which, when turned on, would cause so much EMI that nearby FM radios would stop working), and can affect other parts of the system. The second issue is that embedded system designers probably ought to design their systems so that they can work, even when placed in the vicinity of a vintage ZX Spectrum: how to design such systems is not really a computer architecture issue, so will not be covered here, but is well covered in many books and papers on circuit design and PCB layout.

6.7   Summary

While having a wonderful calculating machine (CPU) is a good start to the building of a computer, it absolutely relies upon being provided with data, and communicating its output in some way. It is a common axiom in computing that useless input data will generally lead to useless output data. However, this axiom does not just apply to the quality of data, but also to the quantity and timeliness also.

In this chapter, we have considered computer interfacing, specifically using buses, both internal and external to convey that information. All computers, of whatever form, from room-sized mainframes to tiny medical diagnostic computers embedded in a pill, require buses to communicate. While there are a large number of standard buses available, more are being invented all the time (and there is nothing to stop an engineer from constructing his or her own bus design).

In this chapter, we tied our consideration of buses with the related discussion of real-time issues that are so important in many of today's human-centric embedded systems, and a separate consideration of wireless technology for embedded computational devices.

With this, we conclude much of our investigation into computer architecture. In the following chapter, we begin to put much of the techniques we have learnt into practice!

6.8   Problems

6.1   An embedded 40-MHz CISC CPU has a slowest instruction (a divide) that takes 100 clock cycles to complete. The fastest instruction (a branch) only requires two clock cycles. There are two interrupt pins for high (HIQ) and low (LIQ) priority interrupts.

Once an interrupt pin is asserted, four clock cycles are needed to recognize this fact and begin to initiate a branch to the interrupt vector table. Assume no other interrupts are enabled, and note that an interrupt must wait for the current instruction to complete before being serviced.

    a) Calculate the worst-case HIQ interrupt response time, timed from pin assertion until initiating a branch to the ISR contained in the interrupt vector table.

    b) The HIQ IRS requires 10 ms to complete execution (measured worst-case from when the HIQ pin is asserted). What is the worst-case LIQ response time?

6.2 The CPU in question 6.1 contains 16 general-purpose registers. Describe what hardware techniques could be used in the CPU design to improve ISR performance in terms of context save and restore (to reduce the time taken for an ISR to complete).

6.3 Comment on the following four techniques in terms of their effect on interrupt response times:

    1. Virtual memory

    2. A stack-based processor

    3. An RISC design (instead of CISC)

    4. A longer CPU pipeline

6.4 Determine the likely real-time requirements of the following systems and decide whether each real-time input or output is hard or soft:

    1. A portable MP3 player

    2. The anti-lock braking system on a family car

    3. A fire-alarm control and display panel

    4. A desktop PC

6.5 Draw a bus transaction diagram for a flash memory device connected to a 100-MHz processor. The flash memory datasheet specifies the following information:
40 ns access time
20 ns hold-off time
20 ns address select time

6.6 A real-time embedded system monitors the temperature in a pressure vessel, and if this exceeds a certain value, it must flash a warning light at 1 Hz and open a pressure relief valve. The system reads the temperature every 100 ms over a serial line, and takes around 10 ms to decode the serial received data into a temperature reading. In the worst case, the temperature can spike rapidly within 150 ms to levels that can cause an explosion.

If the three input and output signals (serial temperature input, pulsed warning light output, and pressure relief valve control) are each handled by separate tasks, determine the temporal scope of each of these, and classify them by degree of hardness.

6.7 Consider the PC104 interface and its pin definitions shown in Table 6.1. In an embedded system that implements the entire set of connections shown, how wide can the data bus be? When using the expansion connector J2/P2, the system has an extended address bus available. Calculate the maximum addressing space that this would allow, in Mibytes.

6.8   In the LVDS (low-voltage differential signaling) scheme, the voltage swings from representing a logic 0 to representing a logic 1 are much less than in other signaling formats. For example, a voltage difference of 12 V between logic 0 and logic 1 is common in EIA232 (RS232), whereas many LVDS drivers can only output a voltage difference of 0.25 V. Does that mean that EIA232 is likely to be a more reliable choice in systems experiencing high levels of electrical noise? Justify your answer.

6.9   Looking at the Ethernet driver in Section 6.3.5, try to decide what the software running on the CPU needs to do (in terms of read and write operations) in order to set up and then communicate data over a network.

6.10   A simple preemptive multitasking embedded computer executes three tasks, T1, T2, and T3, which are prioritized in that order (highest priority first). Task T1 requires 1 ms of CPU time, is triggered every 10 ms and must complete before it is triggered again. Task T2 requires 3 ms of CPU time, is triggered every 9 ms, and must complete within 8 ms of being triggered. Task T3 requires 1 ms of CPU time, is triggered every 6 ms, and must complete within 4 ms of being triggered.

Assuming that all tasks are triggered at time $t = 0$, draw a scheduler diagram for this system (similar to that shown in Figure 6.5), marking the time in ms along the x-axis, from time $t = 0$ up to $t = 40$ ms.

Determine whether, in the time interval shown, all tasks meet their respective deadlines.

6.11   Repeat question 6.10 with the difference that the tasks are now ordered using rate monotonic scheduling. Does this change make any difference in terms of tasks meeting their deadlines over the first $t = 40$ ms of operation?

6.12   A consumer electronics device requires a small, low-power, and medium-speed CPU controller. Discuss whether a parallel-connected data memory storage system, or a series-connected data memory storage system would be more appropriate.

6.13   If the system of question 6.12 was "souped up" so that performance and speed became more important than size and power consumption, would that affect the choice of bus you would choose?

6.14   Figure 6.9 shows the timing diagram for the Atmel AT29LV512 512 Kibit flash memory device. The timing parameters shown have the following values from the Atmel datasheet:

Parameter	Meaning	Minimum	Maximum
t_{ACC}	Access time (address valid to output delay)	–	120 ns
t_{CE}	nCE to output delay	–	120 ns
t_{OE}	nOE to output delay	0 ns	50 ns
t_{DF}	when nCE or nOE* de-assert to output Hi-Z	0 ns	30 ns
t_{OH}	output hold from address, nCE or nOE†	0 ns	–

*From whichever one was de-asserted first.

†From whichever one was de-asserted or changed first.

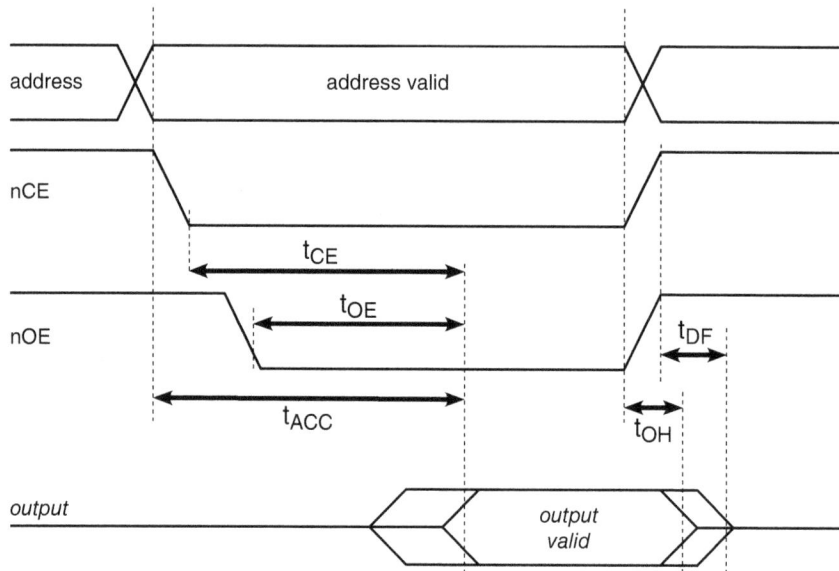

FIGURE 6.9 The read cycle of the Atmel AT29LV512 flash memory device (this waveform was drawn from inspection of the Atmel AT29LV512 datasheet).

Any values that are not given are assumed to be unimportant. Also, remember that this timing diagram is from the perspective of the flash memory device when being read from something external—presumably a CPU. It shows the timings that the CPU reads *must* comply with for the flash memory device to work correctly.

For this question, determine how to set up the S3C2410 parallel interface timing registers so that it could access a parallel connected Atmel AT29LV512 device correctly. This will require careful reading of Section 6.2 (and also Box 6.2), and the additional information that the HCLK signal (and hence the entire bus clock) is running at 100 MHz, and that the Atmel chip enable signal, nCE is connected to the S3C2410 nGCS signal.

The following table identifies the settings that need to be found (note, we ignore the page mode access cycle in this instance):

Signal	Meaning	No. of cycles
Tacs	Address set up time prior to nGCS active (0, 1, 2, or 4 cycles)	
Tcos	Chip select set up time prior to nOE (0, 1, 2, or 4 cycles)	
Tacc	Access cycle (1, 2, 3, 4, 6, 8, 10, or 14 cycles)	
Tcoh	Chip select hold time after nOE deactivates (0, 1, 2, or 4 cycles)	
Tcah	Address hold time after nGCS deactivates (0, 1, 2, or 4 cycles)	

6.15 Determine the worst-case duration of the single-word read transaction in question 6.14, and repeat the calculation for a more modern flash memory device that has a 55-ns access time and $t_{CE} = 55$ ns.

6.16 The Atmel AT25DF041A is a 4-Mibit serial flash device, using an SPI interface that runs up to 70 MHz.

To read a single byte from a selected AT25DF device requires that a controller CPU first outputs a read command (which is the byte 0x0B), followed by a 24-bit address, followed by a dummy byte. Each of these fields is clocked out serially, at up to 70 MHz from the serial output pin. Without stopping the clock, the Atmel device will then output the byte stored at that address, serially, over the next eight clock cycles for the CPU to read.

Determine how many clock cycles, in total, this "read byte" transaction is, and thus the minimum length of time taken to read a single byte from this device. From this simple calculation, how many times faster was the AT29LV512 single location read of question 6.14?

Note: It must be mentioned we are not being particularly fair in either instance. First of all, both devices are more efficient when reading a string of memory locations; the SPI device particularly so. Second, the SPI device has a faster read command available which we did not use—by commanding a read using command byte 0x03 instead of 0x0B it would not have been necessary to insert the dummy byte between the final address bit and the first output bit, although this mode is only specified for clock frequencies up to 33 MHz.

6.17 Match the following applications (a to d) to an appropriate bus technology, taking account of issues such as bandwidth, latency, power consumption, external/internal computer communication, number of wires, noise immunity, distance, and so on.

a) A device which is to be connected to an embedded computer for a disabled user to open and close a sliding window, and which has a single LED to warn when the window is open.

b) A graphics output device to be built into a powerful embedded computer which streams video data from a CPU at 1.8 Gibits/s.

c) An industrial automation computer needs to connect to a sensor located 500 m away across an electrically noisy factory (where wireless devices will not work due to interference). The sensor returns temperature data at just a few 10s of Kibits per second.

d) An FPGA coprocessor needs to be built into an x86 processor system to stream vast amounts of data as quickly as possible.

e) A small embedded industrial PC needs a peripheral card that can connect to a set of 20 analog-to-digital converters (ADCs) with a combined data rate of about 6 Mibytes/s.

For these five applications, there are five available bus technologies to choose from, one per application:

- AGP 4x
- USB 1.1
- PC/104 (16-bit ISA)
- 16x PCIe (16 lane PCI express)
- EIA422

6.18   What are five of the timings that can describe the temporal scope of a task in a real-time system?

6.19   Identify the general sequence of operations that occurs when an interrupt occurs in most embedded-sized CPUs.

6.20   Describe the hardware necessary to implement interrupt sharing for a processor such as the ARM that has only a single general-purpose interrupt signal (IRQ—if we ignore the fast FIQ), and note any additional overhead that this may impose on the software of the interrupt service routine.

CHAPTER 7

Practical Embedded CPUs

7.1 Introduction

Computer architecture has now been an academic discipline for several decades. In that time it has been taught to generations of computer engineering and science students. The teaching has often reflected much of the state of the hardware available during the decade prior to the time of teaching. A decade gap between course updates was fine when mainframe computers were the norm, but became a little troublesome as personal computers, and then smartphones became the cutting edge of computer technology.

In the early 1990s, the author himself fondly remembers being taught the 8086, 6502, and Z80, and yet he owned an early (but blazingly-fast for that time) ARM-powered desktop machine. Strangely the gap between what was taught and what industry is currently using also meant that students destined to work in the growing embedded systems industry, or exploding consumer electronics industry, were still being taught techniques and technology more suitable for mainframe computers.

This book has different aims—mainframe-only techniques are covered in passing, but techniques of interest to embedded systems engineers and programmers are covered in depth. The focus is on practicalities and encouraging the translation of the knowledge gained into real-world experience.

Up to this point, the book has primarily been foundational and theoretical. However, in this and the following chapter, we plunge boldly into practicalities: we enter the real world of embedded computer systems. We analyze what needs to be done to make computers work in that world, and in so doing cover several gaps that exist between the theory and reality of embedded computer architecture.

7.2 Microprocessors Are Core Plus More

One of the more popular microprocessors to be found in numerous embedded systems around the home and workplace is the 32-bit S3C2410 from Samsung, which we have mentioned before. Let us turn our attention to this little ARM9 device for a moment, examining the following list of device features:

- 1.8v/2.0v ARM9 processor core, running at up to 200 MHz
- 16 KiB instruction and 16 KiB data cache
- Internal MMU (memory management unit)

- Memory controller for external SDRAM
- Color LCD (liquid crystal display) controller
- Four-channel DMAs with external request pins
- Three-channel UART (universal asynchronous receiver/transmitter), with support for IrDA1.0, 16-Byte Tx FIFO, and 16-Byte Rx FIFO
- Two-channel channel SPI (serial peripheral interface)
- One-channel multi-master IIC bus driver and controller
- SD (secure digital) and MMC (multimedia card) interfaces
- Two-port USB host plus one-port USB device (version 1.1)
- Four-channel PWM (pulse width modulation) timers
- Internal timer
- Watchdog timer
- 117-bit general-purpose I/O ports
- 24-channel external interrupt sources
- Power control, with states for normal, slow, idle, and power-off modes
- Eight-channel 10-bit ADC (analog-to-digital converter) and touch-screen interface
- Real-time clock with calendar function
- On-chip clock generator

The S3C2410 is an excellent and feature-packed device, well suited for embedded systems, and consequently adopted by many industry developers over its lifetime. As we have seen in Section 6.1 such devices are sometimes called system on chip (SoC)[1] processors, to recognize the presence of so many peripheral components. The core at the heart of the system is the ARM processor, identical to that in many other ARM9-based devices.

Although Samsung probably does not reveal full internal details of the size and arrangements of the S3C2410 components in silicon, we can surmise that the largest part of the silicon IC is devoted to cache memory, the next part is the CPU core. Other large components would be the MMU, SDRAM memory handlers, and perhaps next is the ADCs.

In the early years of ICs, the CPU chip was just that, a single-chip CPU, which was itself an integration of many components that were previously separate. As time progressed, more and more functionality has been subsumed into some of these devices. For embedded systems, semiconductor companies realize that designers prefer to use fewer individual devices where possible, and hence the many on-chip features. Not all features will be needed in any one embedded system design, but conversely, any design will

---

[1] Smaller SoC systems are sometimes referred to as single-chip microprocessors or single-chip microcontrollers.

require at least some of the features mentioned. There are several practical implications of having such highly integrated SoC processors:

1. A reduced chip-count leads to reduced area, and usually reduced product cost.

2. When choosing an SoC, designers can draw up a "wish list" of features, and then try to find one device which matches this list as well as possible. Any item not integrated can still be incorporated externally.

3. Some hardware design is effectively subsumed into software (in that designer questions would be "how can I use this on-chip peripheral" rather than "how can I implement this function in hardware").

4. Occasionally, limitations in the on-chip features can constrain the functionality of products. It is easier to change an externally implemented feature than it is to change one which is included on-chip.

5. Designers now have to wade through CPU data "sheets" that can exceed 1000 pages in length (and often hide critically important details in a footnote on page 991).

6. Some functions cannot coexist. For example, a feature list might proudly proclaim both IIC and UART support, but neglect to mention that the device will support only one of these at a time (either due to insufficient multiplexed device pins, or insufficient internal serial hardware).

Even mainstream processors tend to devote more silicon area to cache than they do to normal CPU functionality, since cache memory is seen as an excellent method to improve processor performance. Consider as an example the 64-bit VIA Isaiah architecture (also known as the VIA Nano), a relatively recent x86 compatible processor, shown in Figure 7.1. It can be seen that the largest area on silicon is devoted to cache memory.

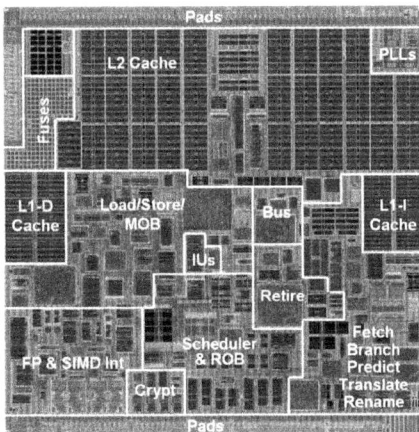

(a) Die photograph, showing functional area blocks overlaid.

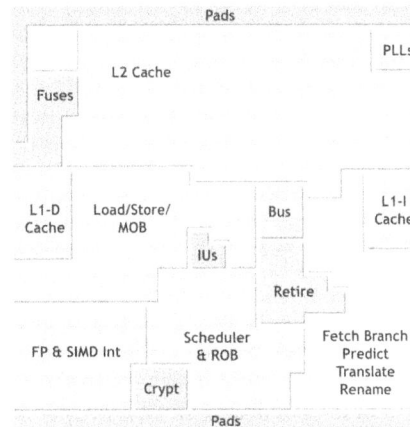

(b) Block diagram of the functional areas fit within the silicon area.

FIGURE 7.1   The VIA Isaiah Architecture, a lower-power x86-style CPU, particularly suited for mobile computing applications such as notebook computers, showing the internal arrangement of the device on the silicon. (Courtesy of VIA Gallery, Hsintien, Taiwan.)

FIGURE 7.2 AMD Phenom™ quad-core processor die photograph, provided courtesy of AMD. Note the horizontal and vertical lines of symmetry dividing the silicon into four distinct cores. The nonsymmetrical strips along the top and bottom of the device are the interface to DDR RAM and 2 MiB shared L3 cache, respectively. The central vertically oriented rectangle hosts the main bus bridging system for connecting the four cores together, while right and left sides host physical interfaces.

There are also separate blocks for clock generation (phase-locked loops—PLLs), very fast floating point (FP), SIMD architecture (specifically, the SSE-3 extensions that the device supports, as discussed in Section 4.7.4, which also explains why they are co-located with the FPU). Other interesting blocks are a section devoted to cryptographic processes, reorder buffer (ROB) for out-of-order execution, extensive branch prediction and retirement hardware at the end of the pipeline, reported to be more than 10 stages in length. There are also two 64-bit integer units (IUs) and three load/store units with memory reorder buffer (MOB). Pads along the top and bottom are used to "wire" the silicon to the lead frame within an IC package. This device, constructed on a 65 nm process, has a 64-kiB L1 cache and 1-MiB L2 cache and uses around 94 million transistors. For reference, compare this to a leading desktop/server CPU, the 450 million transistor quad-core Phenom device from AMD (which also includes 2 MiB of L3 cache) in Figure 7.2.

7.3 Required Functionality

In many systems, there are features which are "nice to have" and features that are essential. Judging between the two for SoC processors really depends upon the application that they are being considered for. For example, one system may require a serial port, another may require SPI.

For this reason also, SoC manufacturers do not quite agree upon a definitive list of "must have" peripherals, and this variety is a good thing for those of us searching for devices to embed within our designs. In fact, the situation is normally consumer device driven: a large company selling millions of systems is likely to be able to convince a semiconductor manufacturer to incorporate exactly what they require, whereas the pleas of a small independent designer to include a particular peripheral are likely to fall upon deaf ears.

However, there are one or two peripheral components that can be considered essential in almost any design, and will be found in the majority of SoC processors:

1. *Reset* circuitry, as explored further in Section 7.11.1, is necessary to ensure that any device starts with registers and state in a predictable, known condition.

2. *Clock* circuitry is needed to distribute a global clock to all parts of a synchronous design. Often, a phase-locked loop (PLL) or delay-locked loop (DLL) will be used to condition the oscillations generated by an external crystal, and to adjust the frequency.

3. *IO* (input/output) drivers to connect to external pins, driving sufficient current to toggle voltages on wires connected to the device. These also have some responsibility for protecting the delicate electronics inside an IC from static charges, shorts and voltage spikes picked up from off-chip sources. Many devices include GPIO—general-purpose IO, which is programmable in direction, drive characteristics, threshold and so on, discussed in Box 7.1.

4. *Bus* connections, again to the outside world, for connection to external memory, peripheral devices, and so on. Usually implemented as an array of IO drivers acting in concert.

5. *Memory* itself is required either on-chip or off-chip, and normally a combination of volatile storage for variables and stack, plus non-volatile storage of program code.

6. *Power management circuitry* required for power distribution throughout a device, turning off unused parts of a chip, and so on.

7. *Debug* circuitry, such as IEEE1149 JTAG is now considered a requirement rather than a nicety in most cases (we explore this in more detail later in Section 7.9.3).

Box 7.1   Configurable I/O Pins on the MSP430

The Texas Instruments MSP430 series of devices has, like many processors designed for embedded systems, great configurability in its I/O pins. As evidence, consider the pin definitions for one particular device, the MSP430F1611:

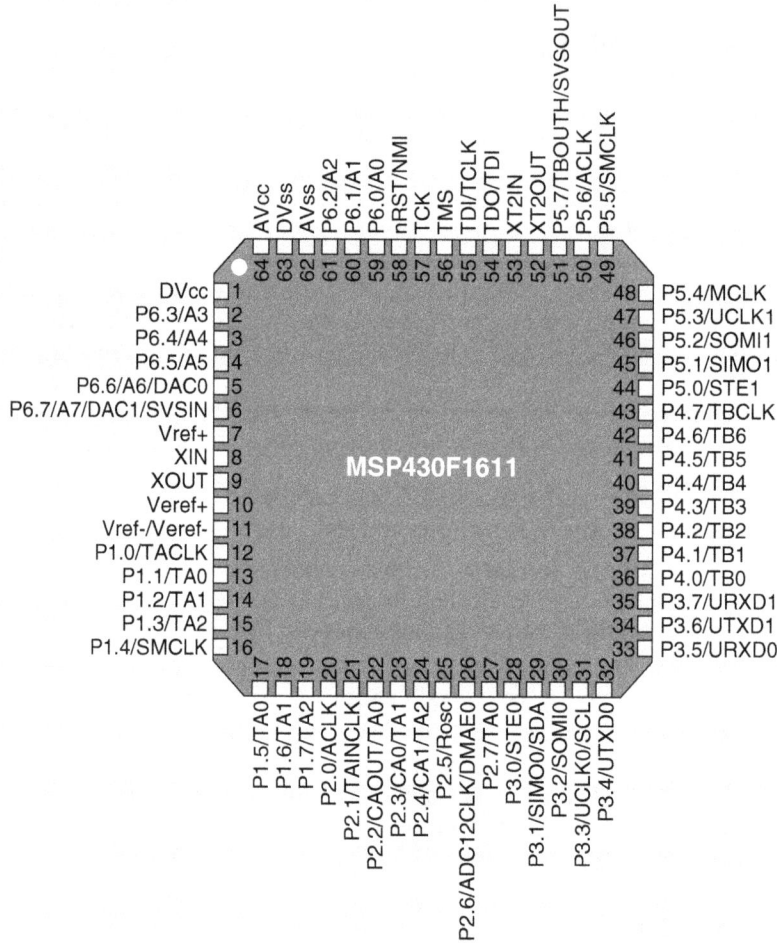

MSP430F1611

On this 64-pin package device, note that apart from the power and ground connections, voltage reference inputs, crystal oscillator connections, and two of the JTAG pins, every pin has multiple possible functions: 51 of the 64 pins are configurable.

As an example, refer to pin 5—this can serve as GPIO port 6 bit 6 (P6.6), as 12-bit ADC input channel 6, or as 12-bit DAC output channel 0, depending upon the particular configuration performed in software by the device programmer.

In Box 7.2 we will explore exactly how these pins can be configured.

Box 7.2   Pin Control on the MSP430

Box 7.1 had shown the pinout of a Texas Instruments MSP430F1611 to illustrate the ability for a single output pin to have many possible configurations. In fact, these pin configurations are under software control—so let us consider how this mechanism works.

The MSP430 has several pin control registers, arranged in 8-bit ports (such that P1.0 to P1.7 constitute port 1, P2.0 to P2.7 are port 2, and so on). Each port has eight I/O pins individually configurable for input or output, and which can individually be read from or written to, and in many cases can also be used as interrupt sources. Let us consider the registers for port 2:

Register P2DIR is an 8-bit direction register. Each bit in this register controls whether the corresponding pin is configured as an input or as an output. Writing a low value to a particular bit makes that pin an input. Writing a high value configures the pin as an output. For example, writing the value 0x83 would set P2.7, P2.1, and P2.0 as outputs, and the remaining pins as inputs.

Register P2IN is an 8-bit register with each bit reflecting the input value on the corresponding pin. So if this register is read, returning a value 0x09 then we know that the voltage on pins P2.3 and P2.0 is high, and the voltage on all other pins is low. Note that if we had configured P2.0 as an output and P2.3 as an input, then we would now know that P2.0 is currently outputting a logic high value, and some other device is providing a logic high input voltage to P2.3.

Register P2OUT is another 8-bit register, which determines the logic voltage output by each port pin that is currently configured in the output direction. Pins that are configured as inputs will ignore any value written to this register.

There remains one final configuration, and that is to choose between using those pins as a GPIO port, and connecting them to their alternative functions. For this, register P2SEL switches the pin between the GPIO port registers and the peripheral modules. Writing a logic low to each bit connects that pin to the GPIO register, and a logic high selects the peripheral function for that pin. For example, writing 0x81 to P2SEL will select the following functions:

Device pin	20	21	22	23	24	25	26	27
Function	ACLK	P2.1	P2.2	P2.3	P2.4	P2.5	P2.6	TA0

Two things should be noted at this point. The first is that the exact meaning of the peripheral function(s) is determined by the peripheral module, and its configuration as specified in the device data sheet. Where some pins have three meanings—one is always the GPIO port, and the other two belong to peripheral modules (and choosing between those is nothing to do with the pin select logic, and must be configured through the peripheral module).

The second point is that if a pin is configured for its peripheral function, the direction of the pin must be set appropriately (by writing to P2DIR). Some processors will do this automatically, but in the MSP430 it must be done by the programmer. So, for example, if one particular pin is defined as a serial port output, and has that function selected by a write to the P2SEL register, then the corresponding pin value in the P2DIR register should be set to logic 1, otherwise no output will occur.

Most devices also include one or more internal UART (universal asynchronous receiver/transmitter) or USART (universal synchronous/asynchronous receiver/transmitter), an internal real-time clock module (RTC), several timer-counter devices, internal cache memory, and so on.

It is interesting to compare the features of CPUs that have been designed to address different market segments, and we do that by comparing three example devices in Table 7.1. Each of the devices tabulated is characteristic of its class, is in popular use, and is of relevance to the embedded architect. The single-chip microprocessor, a Texas Instruments MSP430F1612, is an exceptionally low-power device (in lowest power modes it can literally be run from the electricity generated by two lemons), and has a wide range of low-level peripherals built into the system, with the emphasis being on ensuring a single-chip solution for those who choose this device for their designs. Hence, there is no provision for external memory. The Samsung S3C2410, by contrast, is a reasonably feature-rich ARM9-based SoC that is powerful enough for application in a simple smartphone. It has an SDRAM interface and extensive SRAM, ROM, and flash capabilities on its parallel bus (which we had seen exemplified in Section 6.2), but also a wide range of external feature interfaces—particularly communications and interconnection-based ones. Finally, the VIA Nano, which we had also met previously, in Section 7.2, is presented. In some ways this is a standard PC processor, although redesigned to be highly power efficient, and much smaller than typical desktop processors. Thus, it is a promising choice for an embedded system that requires an x86-style processor. This device concentrates on being excellent at computation: the emphasis is on performance at lower power. The many peripherals available in the other two devices are absent, although another add-on chip (also available from VIA) can provide most of these, and more functionality.

We will now examine a few of these "must have" CPU requirements in a little more detail, namely clocking, power control, and memory. Later (in Section 7.11) we will look at device resetting, and in particular, consider watchdog timers, reset supervisors, and brownout detectors.

7.4 Clocking

When looking at control of a CPU in Section 3.2.4, we had considered the important role of a system clock in controlling micro-operations. In fact, we had not emphasized the importance of clocking enough: apart from the very rare asynchronous processors (which we will encounter later in Section 12.3), all processors, most peripherals, buses and memory devices rely upon clock-synchronous signals for correct operation.

Clocking is particularly important around CPU blocks containing only combinational logic, such as an ALU. If a clock edge controls the input to an ALU, then the same clock edge cannot be used to capture the output from the ALU (since it takes a certain time for the ALU to do anything). It is necessary to either use a later clock edge, or use a two-phase clock (two asymmetrical clocks that are nonoverlapping and whose edges are separated by the maximum combinational logic delay in the clocked system).

In practice, it is often more convenient to use a single clock, but perform different functions on different edges of the waveform. An example of this is shown in Figure 7.3, where an ALU is operated using different edges of a clock. Starting with the first falling edge these operations are to (i) drive the single bus from R0, on the first rising edge to (ii) latch this value into the first ALU register, and de-assert the bus driver. Following from this, (iii) and (iv) repeat the procedure for R1 into the second ALU register. Having now received stable inputs, some time is required for the ALU signals to propagate through to a result in step (v). Step (vi) then loads this result into register R0.

	Single-chip micro	SoC CPU	x86 CPU
	TI MSP430F1612	Samsung S3C2410	VIA Nano
Clock speed	8 MHz	266 MHz	1.8 GHz
Power	< 1 mW	330 mW	5 to 25 W
Package	64-pin LQFN/P	272-pin FGBA	479-pin BGA
Internal cache	None	16 KiB I + 16 KiB D	128 KiB L1 + 1 MiB L2
Internal RAM	5 KiB	None	None
Internal flash	55 KiB	None	None
Internal width	16-bit	32-bit	64-bit
External data bus	None	32-bit	64-bit
External address bus	None	27-bit	Unknown
Memory support	None	ROM to SDRAM	DDR-2 RAM
ALU	1	1	2
FPU	No	No	Yes
SIMD	No	No	SSE-3
Multiply	16 bits	32 bits	up to 128 bits
ADCs	12-bit	8 × 10 bits	None
DACs	2 × 12 bits	None	None
RTC	No	Yes	No
PWM	No	4	No
GPIO	48 pins	117-pins	None
USARTs	2	3	No
I2C	Yes	Yes	No
SPI	2	2	No
USB	No	2 host, 1 device	No
WDT	Yes	Yes	No
Brownout detector	Yes	No	No
Timer	2	1	Yes
JTAG	Yes	Yes	Unknown

TABLE 7.1 Example devices from three classes of microprocessor: a single-chip microcontroller, system-on-chip microprocessor and a PC CPU, compared in terms of built-in features. Note that the Texas Instruments MSP430 family is available in up to 171 model variants at the time of writing, each having significantly different features and abilities—family devices can clock up to 25 MHz, contain up to 16 KiB of RAM and 256 KiB of flash, and add or drop a wide selection of peripherals. By contrast both the Samsung and VIA parts have, at most, a small handful of model variants.

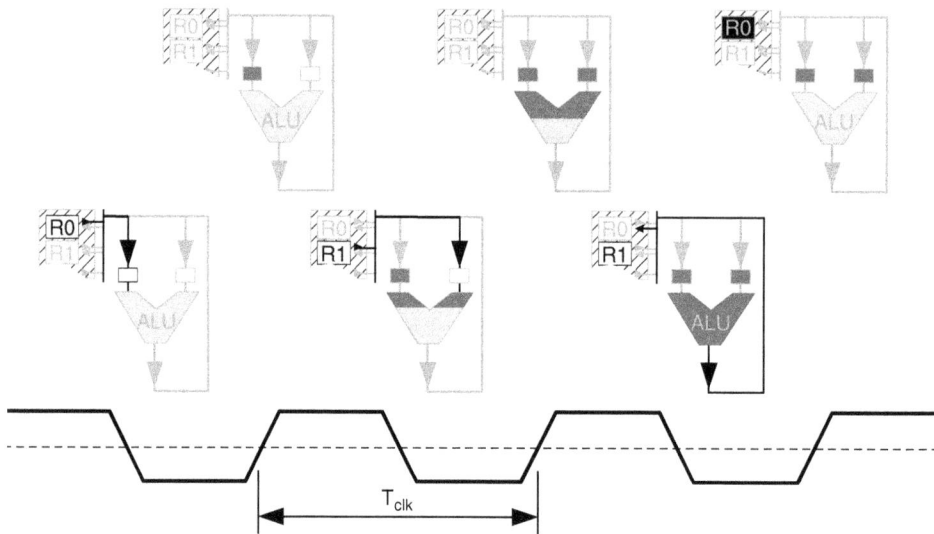

FIGURE 7.3 An example of different gates and latches driving an ALU synchronous to a single phase CPU clock, similar to the cycle-by-cycle timing diagram of Figure 3.3. The operation being performed is `R0 = R0 + R1`.

Figure 7.3 also shows the main clock signal at the bottom of the plot, operating at frequency $F_{clk} = 1/T_{clk}$. The operations fed from this clock, on ether the rising or the falling edge, are performed when the clock crosses some threshold voltage (shown as a dashed line). Note that the edges of this clock are not entirely vertical—there is both a rise time and a fall time associated with the clock. In fact, the point at which the clock crosses the threshold each cycle will vary slightly due to electrical noise, circuit capacitance, inductance, temperature, and so on. This is termed "jitter."

Jitter is also caused by the threshold voltage varying (or more often the threshold staying the same, but the clock voltage varying slowly with time), and causes the value of T_{clk} to change from one cycle to the next. Obviously, if a clock rate had been chosen so that it gave just enough time for a signal to propagate through an ALU, then any major jitter would cause the clock cycle to shorten occasionally, and for the ALU result to consequentially not be ready in time. The result—erratic behavior.

Therefore, clock integrity is very important, and most systems are clocked slower than the fastest cycle time that they can accommodate. This also means that with a very stable clock and power supply such systems can actually operate faster than their rated frequency (which is one reason why CPU overclocking was so popular in certain PC circles for many years).

7.4.1 Clock Generation

These days, most CPUs, and virtually all SoC processors can generate an internal clock frequency from an externally connected crystal oscillator, with at most a couple of tiny external capacitors required.

In order to achieve clock generation, these modern devices contain phase-locked loop (PLL) circuitry to condition the raw oscillator clock input, and usually there will also be internal frequency divider and multiplier hardware to allow, for example, the Samsung S3C2410 to clock at 266 MHz using a 12-MHz external clock (in fact, clock divide registers allow a large number of operating frequencies to be generated from any one particular external crystal).

The similar technology of a DLL (delay-locked loop) is slightly less flexible, and slightly less accurate, but is simpler and cheaper to construct in silicon. Note also that there is usually a provision for an external oscillator signal to be fed directly into such CPUs, if such a frequency is already available.

Many systems these days also require a *real-time clock*, usually provided from a separate 32.768-kHz external crystal (and separate PLL). A 32.768-kHz crystal is a very inexpensive device, and can be quite tiny. Often, it is referred to as a watch crystal due to its prevalence in timing circuits: the reason being that the signal can be divided by 2^{15} to yield a 1-second timing pulse that can drive clock and calendar circuitry (referred to as a 1 pps, or 1 pulse per second, signal).

Although very accurate crystals can be sourced, such as oven-controlled crystal oscillators (OCXOs) used in RF circuits, most microprocessors use either a standard quartz crystal or even a ceramic resonator. These have accuracies of around 100 ppm (parts per million), which equates to 0.0001%. This would translate to less than 1 hour per year inaccuracy, at worst. More expensive parts can easily achieve 10 ppm, and OCXOs can achieve accuracies in the range of a small fraction of 1 ppm.

7.5   Clocks and Power

Reading CPU data sheets, one can often find the clock and power control subsystems sharing a chapter—and in many cases sharing system control registers too. There is a very good reason for this, based upon the fact that clocking is the direct cause of most power consumption within a CPU.

Examine, for a moment, where power gets consumed in modern CMOS (complementary metal-oxide semiconductor) systems. Without delving too deeply into semiconductor theory, let us briefly consider a simple gate, such as the NAND structure shown in Figure 7.4. The "complementary" name comes from the fact that the output is connected either directly to Vss or directly to Vdd through the transistors (one path is always on and another path is always off).

In a perfect world, the CMOS system would connect the output to Vss or Vdd with no resistance. However, we know that a 0-Ω resistance is impossible in the real world, and that there will be some wire resistance, some drain-source resistance, and so on. The consequence of this resistance is to restrict the flow of current from or to Vss/Vdd, and it thus takes some time to charge up the output (or load) capacitance. Once the gate is switched from one state to another, an electrical current flow is triggered, either charging up or emptying the output capacitance. As the charge level changes, the voltage across the capacitor either rises or falls. This is shown more clearly in Figure 7.5, where the CMOS gate is replaced with a perfect switch. Most important is the logic level output at the bottom of the graph: in a digital circuit such as a NAND gate, the time taken from when an event happens (such as a switch position changing) until the output logic level stabilizes, causes the propagation delay that we had first discussed way back when considering the carry-propagate adder in Section 2.4.2.

FIGURE 7.4
Complementary
metal oxide
semiconductor
(CMOS) gate
design for a
NAND circuit,
showing the MOS
transistors
connecting directly
to source and drain
voltages. A
gray-colored output
capacitor is also
shown to reflect the
capacitance of the
load that is to be
switched by the
NAND output.

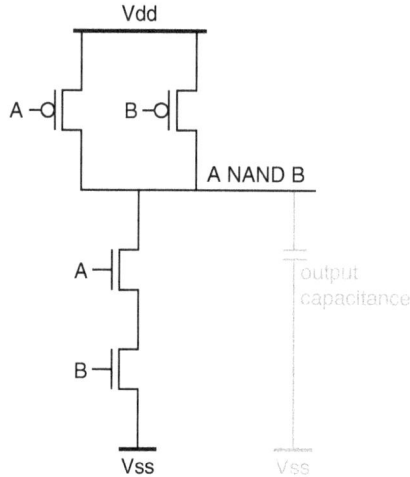

In fact, the situation is actually more complex than we have presented. There are parasitic capacitances within all of the silicon gates (not just on the output), parasitic resistance in each wire, gate connections, and so on, and even parasitic inductance in the wires and gates. These, in general, act to exacerbate the issue that we have observed for the load capacitance.

Having understood the basic issue of capacitance in the system, we can note two important consequences of this which we will examine further:

1. *Propagation delay* comes from the time taken to charge up, or discharge these capacitances, through the small resistance present in the wires and conductive tracks in silicon.

2. *Current flow* is caused by gate switching—since current must flow for the capacitors to charge or discharge.

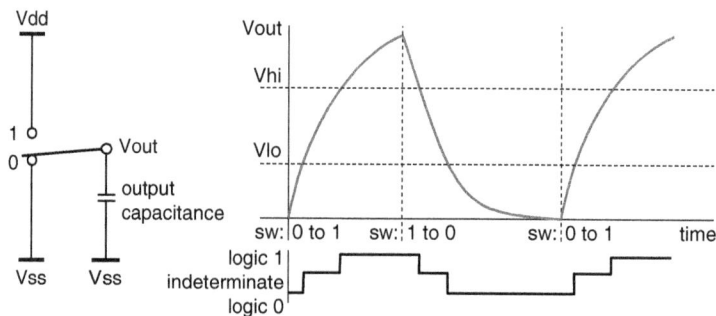

FIGURE 7.5   Switching voltages (left) into a capacitor requires time to charge and discharge, as shown by plotting the capacitor voltage over time as the switch position changes. Note the Vlo and Vhi thresholds for logic voltages, and the corresponding plot along the bottom showing the logic level of the voltage stored in the capacitor over time.

7.5.1 Propagation Delay

To reduce propagation delay (as described in the previous section), silicon designers can do several things: reduce the capacitance (primarily by making the gate smaller, since capacitance is proportional to the area of overlap of the gate structure etched into the silicon), reduce the voltage thresholds so that they are reached quicker or supply more current so the capacitor charges more quickly. Feature sizes of silicon gates have shrunk rapidly over the years, and may now be nearing the lower limit of possible sizes, but smaller sizes tend to mean higher resistance, which in turn restricts current flows, and so materials have changed to reduce semiconductor "on" resistance... Narrowing the voltage threshold limits has naturally been done as IC voltages have reduced from 5 V, through 3.3 to 1.8 V, 1.2 V or even less in some devices. However, reducing these thresholds makes such systems more susceptible to electrical noise.

Basically, silicon IC designers have squeezed their systems in all ways: carefully balanced trade-offs, and reduced propagation delays using all of the easily available means, and with many difficult ones too. This has led the year-on-year rise in device clock speeds from the 1950s up until 2010 or so. However, the difficulty associated with continuing this trend has prompted the widespread move into greater levels of parallelism—if you cannot go faster, then try and do more in parallel (Section 5.8.2).

Many of the techniques used to improve performance have increased the current flow (as we shall see in Section 7.5.2), and done so in a more restricted area as feature sizes have reduced. Since this current is flowing through parasitic resistors, it is expending energy in the form of heat: heat dissipation in a resistor is actually proportional to the square of the current multiplied by the resistance (and since the current is proportional to the voltage, the main reason why silicon manufacturers have been keen to reduce supply voltage is to reduce current flow, and thereby reduce power). Unfortunately, resistance is inversely proportional to area, and area has reduced due to feature size shrinkage, so resistance tends to increase. This is one area of significant trade-offs.

Overall, though, resistive losses have increased, and with clock frequencies having risen, so that switching is more often, the losses occur more often, and thus amount to a greater rate of power loss. This means that there is less time for heat caused by gate switching to dissipate, and thus temperatures to naturally increase. It is not uncommon for silicon junction temperatures to reach, or even exceed 125°C in a CPU.

Smaller feature sizes pack more hot junctions into a given volume, and shrinking IC packages mean that it can be more difficult to extract the heat from these, and thus fans, heat sinks, heat pipes, and so on are necessary to remove the heat.

While we will not consider fans and heat sinks further, we will look at methods of reducing power consumption in computers, something of particular relevance to embedded systems that are often required to operate on limited battery power.

7.5.2 The Trouble with Current

Static resistive loss in CMOS gates does consume some power (i.e., even when gates are not switching, there is a tiny current flow which consumes energy), but this is normally dwarfed by the current flow caused by gate switching.

The instantaneous inrush of current caused by a single MOS transistor switch is provided by a power supply circuit, often through a power plane or along wide power tracks on a PCB. Currents switched to ground are absorbed by a ground (GND) plane on a PCB. Unfortunately, the power tracks, power, and ground plane also each have a small resistance. When a very short but significant pulse of current caused by a gate switching passes through these resistors, there will be a compensatory voltage drop.

In reality, of course, there are hundreds of thousands of gates, all being switched at the same time, so the instantaneous current effect becomes multiplied. A sensitive oscilloscope, operating in differential mode, can detect the voltage drops, occurring in time with a system clock, quite easily, when connected between a power supply output and a device power pin. Good circuit design practice is to place external bypass capacitors close to device power and ground pins of digital logic. These act to couple much of the high-frequency noise found on a power supply pin directly to ground, but also act as power reservoirs to deliver the short sharp pulses of current that are synchronous to the system clock.

Switching currents can be very large, perhaps even hundreds of Amps for an x86-class device, but being so transient so that they last no more than a few nanoseconds. Another issue caused by this is electromagnetic interference (EMI—mentioned briefly in Section 6.6.3). Any time there is a movement of electrons, there is an associated movement in the electric field exerted by them, and in fact, circuit elements conveying pulses of current can very effectively act as antennae to radiate synchronous noise (or to pick it up from elsewhere).

7.5.3 Solutions for Clock Issues

Without reducing switching frequency, system voltage, or changing gate design, current flows will not alter radically, although techniques such as providing bypass and reservoir capacitors can, as we have seen in Section 7.5.2, alleviate some of the problems.

However, we can consider a number of solutions to the issue of clock-induced EMI. The first is to use multiple clocks, each slightly out of phase with each other. If there are four out-of-phase clocks, and a circuit is split so that roughly one quarter of the gates are clocked by each of the four clocks, then the peak current flows will drop by a factor of 4 (even though in total the same amount of current will be flowing overall).

Moving to a more steady current will significantly reduce EMI since electromagnetic radiation depends on voltage fluctuations: if we can approach DC, we solve all EMI issues.

Slightly more ambitious is the concept of a spread-spectrum clock. Essentially this is either periodically and randomly changing the clock frequency by small discrete steps, so that the energy radiated is spread over several frequency bands, or it is through deliberately introducing jitter into a system to prevent clock edges from "lining up" exactly.

EMI generated by power or signal lines as a result of current flows can also be counteracted by having a near-identical line running in parallel, and carrying an equal but opposite current flow. This is termed a balanced electrical circuit and is commonly used in LVDS signaling to reduce EMI.

7.5.4 Low-Power Design

If power consumed in a CPU relates primarily to clock frequency, then one good method to reduce power is to clock the CPU slower. In embedded systems, this is often possible by writing to clock scaling registers that are accessible in many microcontrollers and SoC processors. At certain times, processors may be "working hard," and at other times

```
start up  →  use     →  use          →  use          →  use
             ADC        serial port     PWM output      ADC
```

No power   Static power Dynamic power
control   control   control

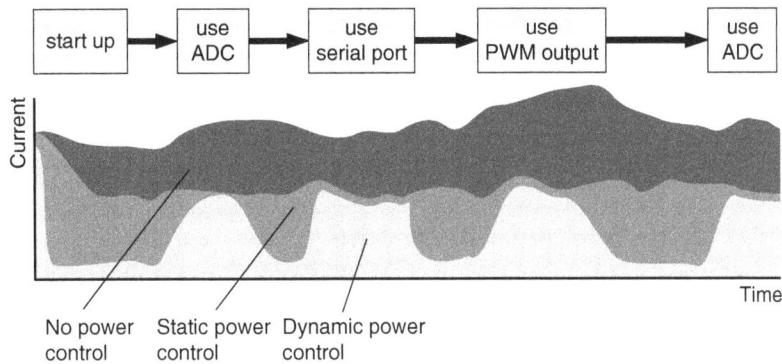

FIGURE 7.6   An illustration of power control within a CPU: a simple program operates several peripherals in turn (namely ADC, serial port, PWM, and then ADC again) and the current consumed by the device measured. Three scenarios are shown: no power control, static power control (when all other unused peripherals are turned off during start-up), and dynamic power control where all peripherals are turned off by default during start-up, and then are only enabled individually for the duration of their use. The area under each of the curves relates to the total energy consumed under the three scenarios.

may be mostly idle. Peak CPU clock speed, which is matched to the peak workload of a processor, does not need to be maintained at all times.

A simple method of scaling the clock in a real-time system that has many tasks operating, is to dedicate a single background task which runs at the lowest priority. An algorithm within the background task detects how much CPU time that task is occupying over a certain measurement period. If this becomes excessive, the system is evidently idle for most of the time and can scale back clock frequency. However, where the background task CPU time drops to zero, the system is working hard, and the clock frequency should be scaled up.

Most major CPU manufacturers, even those designing x86-class processors, now have variations of this system, which are essential for extending battery life in notebook computers.

Another method of reducing the power of a design is even simpler—turn off what is not being used. Surprisingly this idea took a while to become popular among IC designers, but now most processors designed for embedded systems contain power control registers, which can be used to de-power unused circuitry. Where these are used, most programmers simply enable the required blocks, and disable the others, during the start-up phase of their program. However, it is often better to control these dynamically.

The two methods of power control are illustrated in Figure 7.6, where the current consumption of an SoC processor is plotted as a program is executed, which uses a subset of available on-chip peripherals. Static power control reduces current by turning off all peripherals that will not be used, at the beginning of the program. Dynamic power control, by contrast, turns off all peripherals, and only turns them on when needed, and only for the duration of their use. In each case, the area under the graphs represents the total energy consumed—if this system was running from a battery, it would indicate the amount of battery power consumed in the three cases.

There are many other useful methods of power control in embedded systems. Consider the following unsorted list of hints and tips for embedded systems designers:

- There is no need to power an LED indicator continually since the human eye will still see a solid light even if it is, for example, turned on for 1 ms every 50 ms (and this will consume only 1/50 of the power). Other displays are similar.
- Use a combination of clock scaling and intelligent dynamic power control to achieve lowest power consumption.
- When waiting for an event in software, try to find a method of "sleeping," which can place most processors in a very low power mode, rather than using a *busy wait* loop which polls repetitively.
- Even if polling is necessary, consider entering a short sleep (which can be exited by a timer interrupt) wherever possible and where the CPU is idling.
- Fixed-point calculations are normally lower power than floating point calculations.
- On-chip memory is normally lower power than off-chip memory; therefore, wherever possible, use on-chip memory for frequently accessed variables.
- Data moves consume power (and time); therefore, it is a good idea to maximize operates on data structures in-place—that is, by passing a reference to them to operating functions, rather than passing a copy of the entire array.
- Block together operations that use higher power devices. For example, in the original iPod, the hard disc drive was a major consumer of battery power, so Apple designed a system with a large memory buffer. The system would read one or even two tracks from the hard disc into memory, then power down the hard disc while these tracks are replayed. Later, perhaps after a few minutes, the disc would be repowered to retrieve the next one or two tracks. In this way, the hard disc was powered for only very short times.

7.6   Memory

We have discussed memory many times in the previous chapters, and introduced several acronyms such as SDRAM, DDR, and so on. Let us now consider a few types of memory and their characteristics that might be relevant to computer architects, and to those building embedded computer systems. We shall begin with a short recap of computer memory history before looking in detail at ROM and then RAM technologies.

7.6.1   Early Computer Memory

It should be noted that in the early days of computing, there was not a single "memory," and in particular program storage and variable storage were seldom confused, or even considered in any way equivalent. It was only with the advent of von Neumann machines that program and data bytes began to share storage space.

Generally, the earliest programmable computers (such as those mentioned in Chapter 1) were either hard-coded through their wiring, or programmed with switches, and tended to use valves or delay lines for bit-level storage. Reprogramming such machines proved inflexible, as resetting wires (or even switches) every day to reprogram a system is time consuming and error prone. Punched cards (or tape) were quickly adopted for program storage—bearing in mind that these had been used effectively for more than 200 years to program looms for textile manufacture.

Data storage was accomplished through delay lines, sometimes with some quite interesting methods (such as cathode-ray tube delay lines, mercury delay lines, acoustic delay lines, and so on). These would hold a bit of information for a short time, allowing the computer to work on something else in the meantime: effectively the memory function in a simple digital calculator.

Later, magnetic core memory was invented, and magnetic storage was used for both variable storage and program storage, on tape. Magnetic discs were used for both and later evolved into both floppy and hard disc drives.

The greatest advance in memory technology came, as with many other areas, in the integration of circuits onto silicon. This provided rewritable memory storage for variables by the mid-1960s, and read-only memory for code during the same era. However, the higher cost per bit of silicon memory compared to magnetic storage has meant that, although silicon memory conquered most magnetic memory use in computers by the 1980s, the mass storage of data on hard disc drives has remained stubborn. It has only been very recently that hard disc-less computers have been considered viable for anything except the smallest of embedded systems.

Today, however, almost all embedded systems contain flash memory, and several brands of sub-notebook computer are similarly going solid state: these should in theory be lower power, less susceptible to shock damage and more reliable than their cousins which incorporate hard disc drives.

There is now little to differentiate memory for program code and for data: any of the devices discussed below in this chapter are capable of storing and handling bytes of both types; however, certain characteristics of access for each type of data can match the capabilities of memory types, so we shall consider these in turn.

7.6.2   ROM: Read-Only Memory

Read-only memory is not a technology, but rather a method of access: data stored in ROM can be read but not written to by the computer. This means that the data is non-volatile and unchanging, a characteristic that is well suited to program code, but could also be useful occasionally for data if that remains constant (e.g., digital filter coefficients or a start-up image for an MP3 player display).

At its basic level, a semiconductor ROM is simply a look-up table implemented in silicon. Given an address input, it selects a gate located "at" that address, which then outputs its content onto a data wire, with one data wire for each bit. This is shown in Figure 7.7, where a 4-byte ROM is illustrated, although the actual arrangement within ROM devices in use currently is a little more sophisticated than shown.

Some ROM devices are (despite their name) writable; however, the name indicates that the predominant action is reading, and that writing is either not possible when in situ, or is inconvenient. Consider the following varieties of ROM technology:

- A basic, or mask ROM IC is a simple silicon device having address bus, chip select input, read signal input, power and ground pins, and (when selected by both chip select and the read signal going low) will output the content of the currently selected memory location onto the data bus.

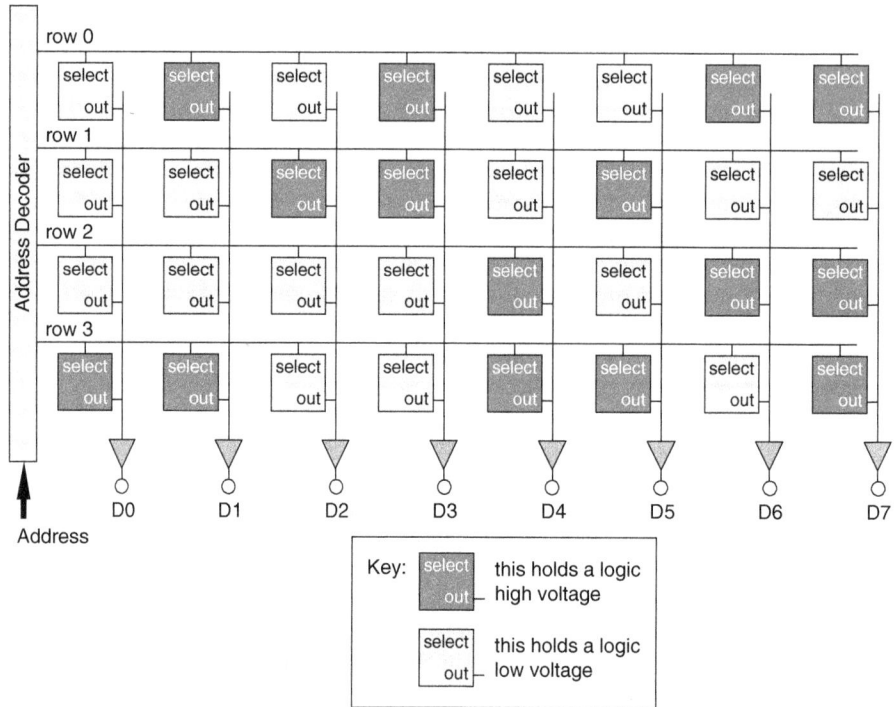

FIGURE 7.7 A simplified diagram of a ROM, showing a matrix of logic cells addressed by row, and feeding an 8-bit data output. If the dark cells are assumed to contain logic 1 and the light cells logic 0, and output their content when selected, then an address input that selected row 1 would cause a data output of `00110100`b, or 0x34 in hexadecimal. For correct operation, only one row should be selected at any one time.

- An EPROM device—an erasable programmable ROM (PROM) has a small silica "window" on the top of the device, through which the IC can be seen. By shining ultraviolet light through this for around 10 minutes, the data stored in the device can be erased.[2] The device can then be programmed by applying a high voltage to the data pins as the address is selected. This step is performed in a dedicated EPROM programming machine, which also means that EPROMs are usually socketed, so they can be removed and reinserted at will. When a device is manufactured without the silica window, it becomes a non-erasable EPROM (which is simply a PROM). Some silicon fuse-based ROMs are also available— in these, the high-voltage inputs blow fuse wires in the silicon to turn on or turn off connections.

- As an advancement on the EPROM, the E^2PROM or *EEPROM* is an electrically erasable PROM and (depending upon the manufacturer) can be synonymous with *flash memory*. These devices require a programming power supply of approximately 12 V to erase, and reprogram their memory contents; however,

---

[2] Daylight will also erase the device, but it takes a lot longer—therefore, engineers always had to remember to stick a label over the window if they wanted their program to last for more than a few days or weeks.

FIGURE 7.8 A diagram giving the pinout, of a popular (though rather old) electrically erasable and programmable read-only memory (EEPROM), showing 11 address pins, addressing 16 Kibits of memory (as 2048 bytes, hence the eight data lines). Chip select (nCE), write enable (nWE) and read/output enable (nOE) are also visible, as are GND and Vcc connections. This device, the 2816A can be written to more than 10,000 times, and can last for 10 years.

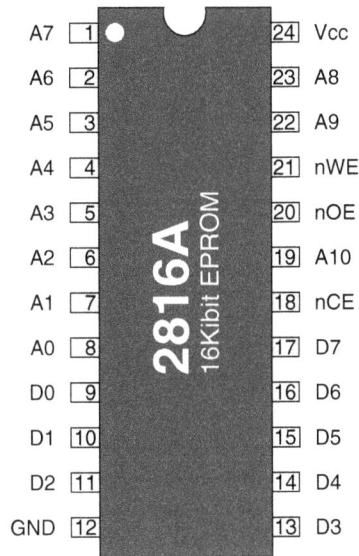

	2816A 16Kibit EPROM	
A7 [1]		[24] Vcc
A6 [2]		[23] A8
A5 [3]		[22] A9
A4 [4]		[21] nWE
A3 [5]		[20] nOE
A2 [6]		[19] A10
A1 [7]		[18] nCE
A0 [8]		[17] D7
D0 [9]		[16] D6
D1 [10]		[15] D5
D2 [11]		[14] D4
GND [12]		[13] D3

many modern devices can generate this 12 V internally from a 3.3- or 5-V power supply. Due to the technology used, these have a finite lifetime, normally specified in terms of data retention and erase cycles, which are typically over 10 years and 1000 to 10,000 times. The engineer choosing these devices should note that, while data read times are quick, and do not change over time, as the devices age, both the erase time and the reprogramming times can lengthen significantly. Figure 7.8 shows the pin arrangements of one of these devices, with a parallel address bus and a parallel data bus. The nWE pin (active-low write enable) is a give-away, indicating that this device can be written to. A true EPROM would look similar, and even have the same pin connections, apart from this one (which would probably be marked "NC" to denote "no connection").

There are actually two types of flash memory technology: NAND flash and NOR flash, as explained in Box 7.3.

Serial flash, shown in Figure 7.9, also contains flash memory, but in this case has a serial interface instead of a parallel interface. Having a 25-MHz serial bus, through which command words, address byte, and control signals must run, this is obviously significantly slower than the parallel-bus devices. Because of the nature of the addressing scheme in these devices, where a read/write address is specified (which takes some time to specify serially) followed by any number of reads or byte writes (which happen a lot faster), they particularly suit the storage of information which is to be read off sequentially—they are least efficient when randomly reading or writing individual bytes.

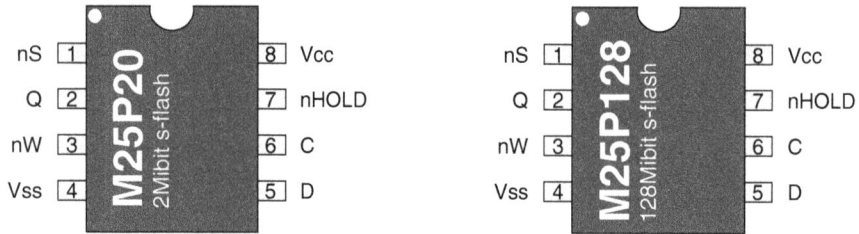

FIGURE 7.9 Serial flash devices, by using a serial interface, multiplex control, address and data on the same interface. Thus the size of memory array contained within the device on the right does not require extra dedicated address pins, despite containing 64 times as much data as the device on the left. Note also that this device is tiny—only 6 × 5 mm.

Most flash devices, whether parallel or serially accessed, are arranged internally into a number of *blocks* or pages. When the device is new, every byte within the device is initialized to 0xff; in other words, every bit stored within the device is initialized to "1." Any memory location can be addressed and read from, and each byte will return with the value 0xff.

Any location can also be programmed, and what happens is that each bit that is a "1" and is programmed with a "0" is then cleared to a "0." Any bit programmed with a "1" stays *unchanged*.

For example, starting with a byte 0xff, if a value 0xf3 is programmed, then that byte will become 0xf3. If the same location is programmed again with the value 0xa7, then the byte will become 0xf3 AND 0xa7, which is 0xa3 (since 1010 0111 AND 1111 0011 = 1010 0011). Clearly, if a byte is written to over and over again, it will eventually end up at 0x00. So developers working with flash memory can see that unerased portions of flash will be filled with 0xff.

Box 7.3   NAND and NOR Flash Memory

There are actually two different types of flash memory technology, named NAND and NOR flash after the gate structures used to implement them. One implementation of the technology is that NAND devices are block-based, high density, and low-cost, well suited to mass storage. They can replace the hard disc drives in embedded computers and are also suitable for storage of data in devices such as MP3 players.

A comparison of the two types of flash technology is given, as follows:

Feature	NOR	NAND
Capacity	Big	Bigger
Interface	Like SRAM	Block based
Access type	Random access	Sequential access
Erase cycles	Up to 100,000	Up to 1,000,000
Erase speed	Seconds	Milliseconds
Write speed	Slow	Fast
Read speed	Fast	Fast
Execute-in-place	Yes	No
Price	Higher	Lower

For embedded computer use, code storage, and so on, we will confine our discussion to NOR flash (which is the most likely that we will encounter, especially in parallel-connected devices).

When flash memory is erased, each byte needs to again be set to 0xff. In fact, the devices are erased block by block, so that once the erase command is issued, the entire block selected for erasure will become 0xff. It is also possible to lock memory blocks against erasure.

Reading parallel connected flash memory is accomplished in the same way as reading a ROM and follows a standard bus transaction as seen in Section 6.2. In essence, this means that a CPU connected to external flash can read it by (i) setting the address bus to the desired location; (ii) asserting the chip select signal, nCE; (iii) asserting output enable, nOE; (iv) allow the device some time to access the desired location, determine the contents, and output this on the data pins; then (v) reading the data bus before (vi) de-asserting all signals in turn.

A write to parallel flash is physically possible by performing much the same sequence of actions, except this time driving the data bus with the value to be written, and asserting write enable (nWE) instead of nOE. If this was performed on an SRAM chip (described in the following section), it would write to the desired address location; however, flash memory is slightly more complicated—it requires a specific command sequence to be written to the device to control it (and before any memory locations can be programmed). Some of these command sequences are shown for two typical flash memory devices, from Atmel and Intel, respectively, in the following table:

	Atmel AT29xxx		Intel 28F008SA	
	Data	Address	Data	Address
Program	0xaaaa	0x5555	0x10	<addr>
	0x5555	0x2aaa	<data>	<addr>
	0xa0a0	0x5555		
	<data>	<addr>		
Erase sector	0x00aa	0x5555	0x20	<addr>
	0x0055	0x2aaa	0xd0	<addr>
	0x0080	0x5555		
	0x00aa	0x5555		
	0x0055	0x2aaa		
	0x0050	<addr>		
Erase device	0x00aa	0x5555	Not supported	
	0x0055	0x2aaa		
	0x0080	0x5555		
	0x00aa	0x5555		
	0x0055	0x2aaa		
	0x0010	0x5555		

Thus, to program a word of value 0x1234 to address 0x1001 in the Atmel device would require four write cycles:

- Write 0xaaaa to address 0x5555.
- Write 0x5555 to address 0x2aaa.

- Write `0xa0a0` to address `0x5555`.
- Finally, the device is set to accept our data by writing `0x1234` to address `0x1001`.

For the Intel device, the sequence is somewhat shortened:

- Write `0x0010` to address `0x1001`.
- Write `0x1234` to address `0x1001`.

The reason for the complicated series of writes is to prevent spurious reprogramming of flash (which could happen when a CPU program operates incorrectly—it is not difficult to create a program that randomly writes data to different address locations). As a further protection mechanism, these devices carefully sense the voltage of the power supply and, if they detect undervoltage or significant fluctuations, will not allow a write to begin. Various status registers can be read from the devices (again by writing a series of commands to place the devices into "read status register mode" or equivalent, so that the following one or two read commands will return the status register contents). Another command is used to read back a manufacturer and device identifier, so a well-written program can determine the correct programming algorithm for the particular flash memory device that is connected.

Note that different manufacturers have different command sequences to control their parallel flash, although the two shown represent two main classes into which almost all other manufacturers fall (i.e., all other devices are handled similarly to these).

Flash memory is fundamentally a block-based technology—although individual words can be read and programmed as needed, it is entire blocks that get erased (and this is true in any flash-based technology such as compact flash (CF) cards, secure digital (SD) cards, memory sticks, and so on, even though this may not be noticeable to the user). The practical implication is that changing a single byte in one 64-KiB block of flash memory will usually require the following steps:

- Read the entire block from flash into RAM.
- Find the byte that needs to be changed in RAM and replace it with the new value.
- Issue the command sequence to erase the flash block.
- (Wait until the erase has completed)
- Issue the command sequence to begin writing, and then write the entire block back into flash.

Blocks may be quite large—the 64 KiB mentioned above is not uncommon, so flash memory is not a good choice for storing variables that change frequently.

From a programmer's perspective, it is useful to have different blocks dedicated to storing different types of information. In embedded systems, there are particular concerns over boot memory (and we will discuss this further in Section 7.8). A simple scheme is to place items that seldom need to be rewritten into one set of blocks, and items that may need to be rewritten more often (such as configuration settings) into another block.

As flash memory ages, it tends to slow down. Both erasing and programming bytes can become time consuming. Obviously it is better if a flash memory device does not slow

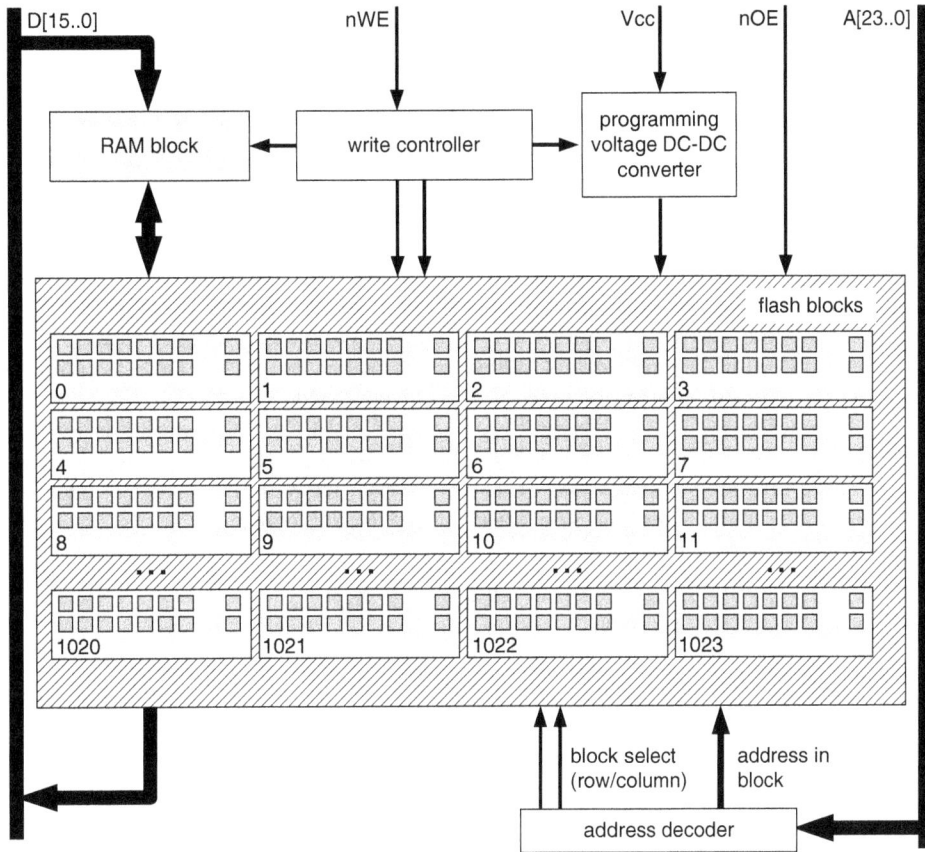

FIGURE 7.10   A block diagram of the internal structure of a flash memory device that contains a block-sized area of RAM for storage of programming data. Note the flash array consists of multiple identical blocks. This regular structure makes it very easy for a manufacturer to increase capacity by adding more rows of blocks to the device (and in practice there would probably be many more than four columns). Note the direction of arrows connecting to the data bus.

down a computer that it is attached to, and so the designers of flash memory have come up with some ingenious ways to tackle this problem. The block diagram in Figure 7.10 shows one such technique, that of incorporating a block-sized RAM area inside the device. Programmers wishing to write a block of memory to the device can first write the data, very quickly into the SRAM-based RAM block, then issue the programming command to cause the device to copy the entire RAM content into a flash memory block. Similarly, when only a single byte needs to be changed, the flash block can be internally copied into the RAM area, the programmer then adjusts the required byte, before issuing the command to erase, and then reprogram the desired flash block.

The flash memory structure shown in Figure 7.10 is also that which is used inside most serial flash devices—in the case of serial flash; however, the nOE, nWE, and other control signals are generated from a serial interface controller, rather than obtained directly from a parallel interface.

7.6.3 RAM: Random-Access Memory

The term "random-access memory" (RAM) is, like ROM, describing a method of access rather than a technology: it means that any memory location can be accessed (i.e., read from or written to) at will. We tend to take this ability for granted in computers these days, but the alternative is serial access, such as on magnetic tape and certain delay-based memories, in which data is available in the same order that it was written. The constraint of serial data access was not uncommon during the early years of computing.

Of course, there is another difference between serial access and random-access memory—the RAM is addressable and therefore requires an address to be specified to indicate the data locations that require access. For parallel bus memory, which is most common, this address specification is carried on a dedicated parallel address bus. Sometimes it is multiplexed with a data bus, and for serial memory devices, is conveyed over a serial connection (as in the serial flash device in Section 7.6.2).

In general, there are two technology classes of RAM: the first being static RAM (SRAM) and the second being dynamic RAM (DRAM). Although DRAM technology has mushroomed into several subclasses of its own, which we will briefly survey later, there are still some distinct differences between the two technologies:

SRAM	DRAM
Six transistors per bit	One transistor per bit
Lower density	Higher density
No refresh needed	Periodic refresh required
Large devices are expensive	Large devices are cheap
Higher power when active	Lower power when active

7.6.3.1 Static RAM

SRAM, although it is called "static," is still a volatile memory—when power is removed, stored data will be lost. The name static comes about because these memory cells will continually retain their state, as long as power is applied, without the need for the refreshing. Dynamic RAM, as we will see a little later, does require this periodic refreshing procedure.

SRAM tends to be fast, but because its logic cells are several times more complex than those of DRAM, it is more expensive, lower density, and consumes more electrical power during the process of reading and writing. Modern SRAM, however, can be lower power than DRAM when it is not being written to or read from—due to the refresh process in DRAM which must operate periodically even when the device is not being accessed.

SRAM is very similar in connectivity and use to ROM. Referring to the example of the pinout of the two SRAM devices shown in Figure 7.11, note the similarity to the EEPROM device of Figure 7.8, in terms of data connections, although the locations of specific pins may differ. Figure 7.11 actually shows two devices: a 16-Kibit and a 1-Mibit device. The former part has 11 address bus pins (since $2^{11} = 2048 \times 8$-bits $= 16,384$ bits), and the latter has six more, making a total of 17 (A[16..0], since $2^{17} = 131,072 \times 8$ bits $= 1024$ Kibits $= 1$ Mibit).

The internal structure of SRAM is very regular, cell like, and arranged similarly to ROM. Figure 7.12 presents a simplified block diagram of an internal SRAM matrix, showing logic cells that can be individually addressed in parallel (to form a connection to an 8-bit parallel bus in the figure), and apart from being selected on the basis of their address, can be read from or written to. Bidirectional buffers connect the external data bus

6116 16Kibit SRAM:

Left		Right	
A7	1	24	Vdd
A6	2	23	A8
A5	3	22	A9
A4	4	21	nWE
A3	5	20	nOE
A2	6	19	A10
A1	7	18	nCS
A0	8	17	I/O8
I/O1	9	16	I/O7
I/O2	10	15	I/O6
I/O3	11	14	I/O5
GND	12	13	I/O4

431000 1Mibit SRAM:

Left		Right	
NC	1	32	Vcc
A16	2	31	A15
A14	3	30	nCE2
A12	4	29	nWE
A7	5	28	A13
A6	6	27	A8
A5	7	26	A9
A4	8	25	A11
A3	9	24	nOE
A2	10	23	A10
A1	11	22	nCE1
A0	12	21	I/O8
I/O1	13	20	I/O7
I/O2	14	19	I/O6
I/O3	15	18	I/O5
GND	16	17	I/O4

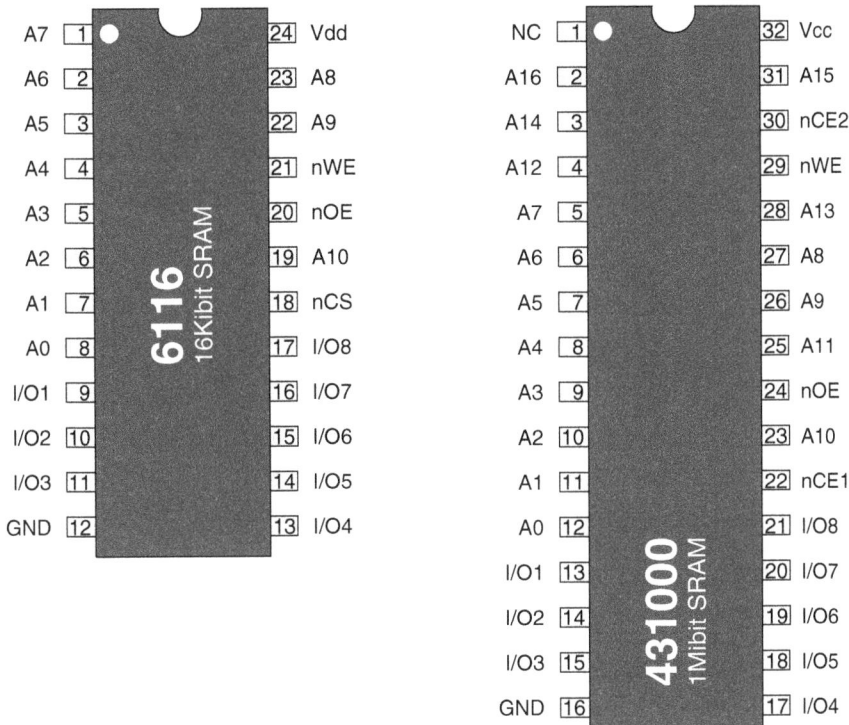

FIGURE 7.11 A pinout diagram for two early SRAM chips, the 16-Kibit 6116 and the 1-Mibit 431000. Note that both have the same 8-bit input/output port (usually connected to a data bus), and both have power supply, chip select (nCS) and read/write pins; however, the device on the right, containing 64 times as many bytes requires a further 6 address pins A11-A16 in order to access this.

to the internal data lines and are controlled in terms of directionality to avoid bus contentions with any other items, which may be connected to the same external data bus.

SRAM is used for cache memory and for the on-chip memory found in single-chip computers. It is also the external memory of choice for simple and small embedded microcontrollers, where memory sizes on the order of tens of kibibytes are sufficient (since at these low densities, the cost differential between DRAM and SRAM disappears, and because microcontrollers are simple, and lack support for DRAM).

7.6.3.2 Dynamic RAM

As we have mentioned previously, dynamic RAM is called dynamic because it is constantly in a state of change: The logic value of each cell is determined through the stored charge in a capacitor connected to the single transistor used per bit, and because the gates are "leaky," these capacitors are continually discharging. A *refresh* process reads each cell in turn and then "tops up" the stored charge in the capacitor appropriately. Any cell that is not refreshed will lose its charge within a few milliseconds.

The write process simply loads the required charge into the capacitor through the transistor (for a logic high), or discharges the capacitor (for a logic low). Interestingly, the

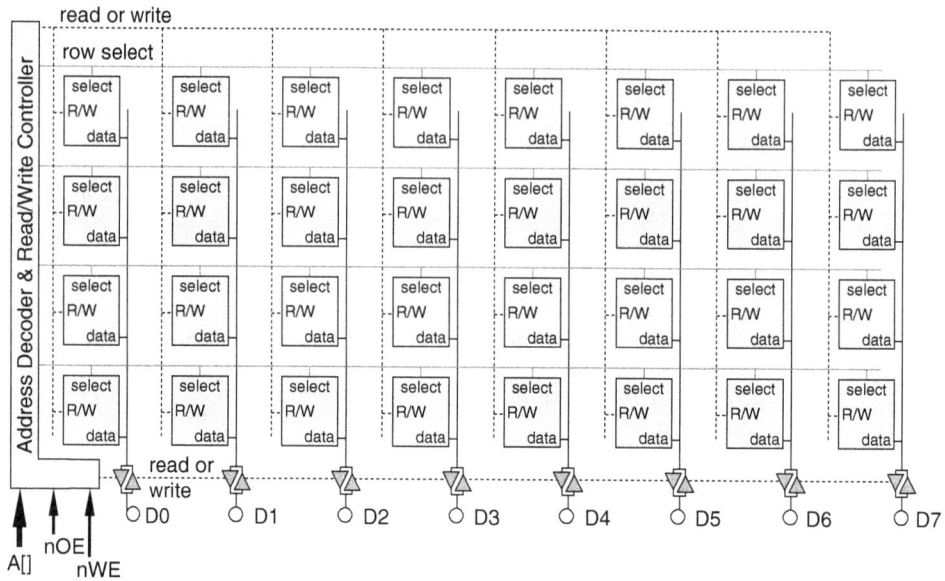

FIGURE 7.12 A simplified block diagram of the internal arrangement of an SRAM device showing an array of memory cells which can be read from and written to, controlled by an address decoder and read/write controller.

read process, through which the charge in a cell is determined, also refreshes that cell, so reading the entire device in a periodic fashion will refresh it.

Modern DRAM is highly integrated, and most microprocessors (or support ICs) that connect to DRAM will handle the refresh automatically—although may well require several configuration registers to be set up correctly. However, the refresh process takes a little time, which may be time that a CPU must spend waiting for its memory to become free. This can naturally impact CPU performance slightly.

DRAM has a very long history, beginning in the mid-1960s, and making several step change improvements along the way. Some of these more important development milestones are shown in Table 7.2, along with their approximate year of release, clock speed, and operating voltage.

DRAM differs from SRAM in its dynamic nature, requiring constant refresh. Since DRAM bit memory cells are a lot smaller than those of SRAM, DRAM is cheaper and is available in higher densities. However, DRAM is slower than SRAM, and the constant refreshing causes the devices to consume power even when they are not being read from or written to (although it must be remembered that SRAM consumes more power during accesses).

There is one other major difference between DRAM devices and SRAM devices, and that is in the addressing scheme for DRAM. Refer to the two early DRAM chip pinouts shown in Figure 7.13, for a 16-Kibit and 1-Mibit device, respectively. The first item to note is the several unusual signals named nWRITE, nRAS, nCAS, Din, and Dout, which we will discuss in a moment. The second item to note requires us to compare the DRAM pinouts to those of the SRAM shown previously in Figure 7.11. In both figures, the two

Name	In use from	Typical clock speed	Voltage
Basic DRAM	1966	–	5 V
Fast page mode (FPM)	1990	30 MHz	5 V
Extended data out (EDO)	1994	40 MHz	5 V
Synchronous DRAM (SDRAM)	1994*	40 MHz	3.3 V
Rambus DRAM (RDRAM)	1998	400 MHz	2.5 V
Double-data-rate (DDR) SDRAM	2000	266 MHz	2.5 V
DDR2 SDRAM	2003	533 MHz	1.8 V
DDR3 SDRAM	2007	800 MHz	1.5 V
DDR4 SDRAM	2014	1.6 GHz	1.05 V

*IBM had used synchronous DRAM much earlier than this, in isolated cases.
Note: RD and DDR RAM devices transfer data on both edges of the clock, so they operate at twice the speed of the rated clock frequency.

TABLE 7.2 Some of the more prominent landmark machines in the evolution of SDRAM technology.

devices have the same size memory content: in each figure the device on the right contains 64 times as much memory. For the SRAM case, the IC on the right has six more address pins than the one on the left. For the DRAM case, the IC on the right only has two more address pins than the one on the left. Since a 64 times increase in address space is an expansion of 2^6, this would normally require six extra pins. It seems that there is

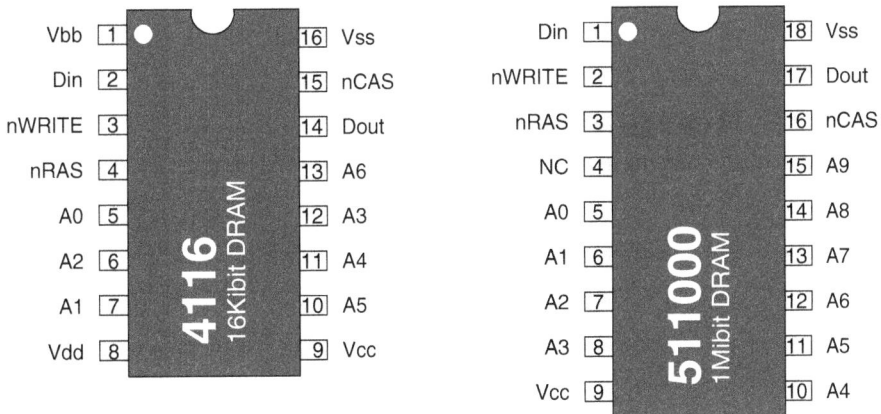

FIGURE 7.13 A pinout diagram for two early DRAM chips: the 16-Kibit 4116 and the 1-Mibit 511000. These devices both output a single data bit (and therefore eight of each would be connected in parallel when connected to an 8-bit data bus). Note that both share the same DRAM control signals, but the device on the right, despite containing 64 times as much data, has just three extra address pins (A7 to A9). Vbb, Vcc, Vdd, and Vss are various power supply pins.

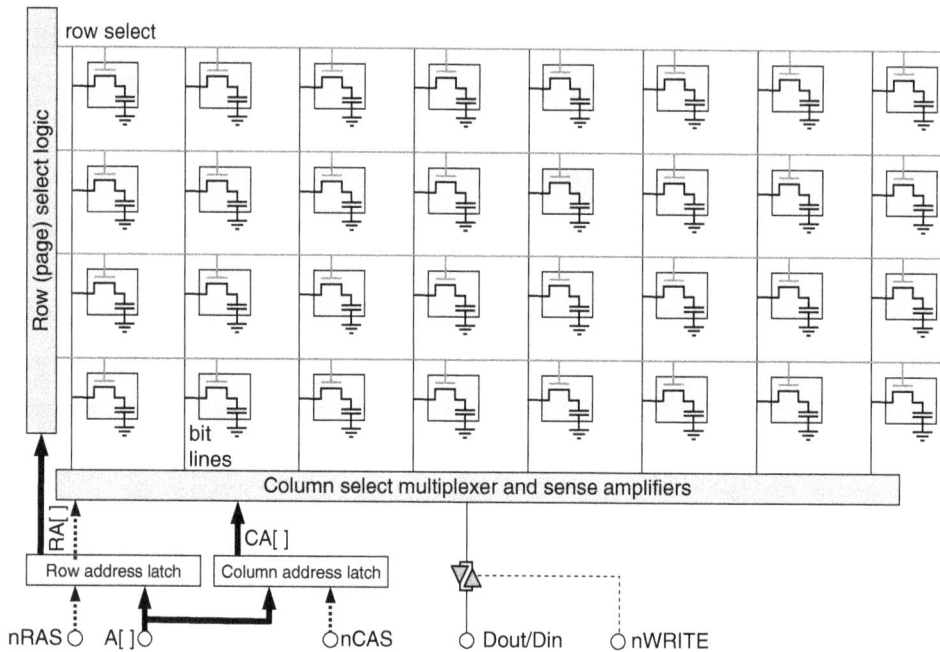

FIGURE 7.14 A diagram of the internal row/column select nature of the DRAM device, where a row latch and a column latch hold the row address and column address, respectively. Both row and column addresses are conveyed over the same address bus, identified through the row address strobe (nRAS) and column address strobe (nCAS) signals. The array shown connects to a single bit of a data bus.

more than that meets the eye inside the DRAM device. We will thus consider this a little further.

7.6.3.3 DRAM Addressing

First of all, let us note that DRAM devices are addressed by row (often called a page) and column. This is unlike the memory structures we presented previously in which only row addressing was performed. In fact, the devices with pinout shown in Figure 7.13 are 1-bit devices—in order to construct an 8-bit bus, eight of these would be required to operate in parallel, one per data bit. The Dout pin on these devices would be required to connect to data bus signals D0, D1, D2, D3, . . . respectively.

A clearer view of this row and column addressing can be seen by examining the internal structure of a DRAM device, as in Figure 7.14. Internal cells, each consisting of a transistor and a charge storage capacitor, are arranged in a rectangular fashion. The row address strobe (nRAS), when activated, will load the row address latch with the content of the address bus at that time. A demultiplexer maps the row address signal to a particular row (or page) of devices which are then selected to output their stored charge. The column address strobe (nCAS) then causes the column address latch to be loaded with the content of the address bus at that time. The column address determines which of the selected devices is chosen as the single bit output from the array.

Sense amplifiers, connected to each of the bit lines (columns), detect the charge stored in the capacitors for selected cells, and top it up. Thus, after selecting a particular page, if the charge is greater than a certain threshold on 1-bit line, the sense amplifier outputs a voltage to recharge the capacitor in the cell connected to that line. If the voltage was sensed to be lower than the threshold, then the sense amplifier does not output that voltage.

Actually, the sense amplifiers are triggered after the nRAS signal has selected a row, and this recharging process is entirely automatic. The practical implication is that the "refreshing" process in DRAM does not need to involve the column addresses—all that is required is for each row to be selected in turn (but as mentioned, most CPUs that support DRAM or SDRAM will perform this automatically). For DRAM that typically needs to be refreshed every 64 ms, each row will have to be selected sequentially within that time.

Of course, many DRAM devices are not single bit devices, but store bytes or words of data. In that case, the basic DRAM design is replicated on-chip several times. An example is shown in Figure 7.15 of an 8-bit bus-connected DRAM device, although this is very low density, being only a 256-bit memory. Since the device shown has 8 columns and 4 rows per bit, the row address would consist of 2 bits, and the column address would consist of 3 bits.

A 16-Kibit-sized device, such as the 4116 device shown in Figure 7.13, would perhaps have 128 rows and 128 columns (since $128 \times 128 = 16,384$), and thus require 7 address lines ($2^7 = 128$) to set up the address of the cell to be accessed. In fact, the steps required by a bus-connected CPU to read a single bit from this device, starting from the device being inactive (i.e., nRAS, nCAS, nWRITE are inactive; logic high) are as follows:

1. Output the required row on the address bus.
2. Assert nRAS (take it from logic high to logic low, causing the row address latch to capture the row address from the address bus).
3. Output the required column on the address bus.
4. Assert nCAS, to latch in the column address.
5. The device will, after some time, output the content of the addressed memory cell on to the connected wire of the data bus, which can then be read by the CPU.
6. De-assert nCAS, stop driving the address bus.
7. De-assert nRAS.

Of course, there are some very strict timings to observe when accessing the DRAM device in this way, or when performing a write. Clearly, with two address writes per memory access, this is significantly slower than a device that does not use row/column addressing, like an SRAM. This observation is true, but is tolerated for cost and density reasons: as seen in Figure 7.13, moving from a 16-Kibit to a 1-Mibit DRAM device requires just three more address lines, but in SRAM (Figure 7.11), this would require six extra address lines. For larger memory densities, this advantage in pin-count that DRAM has is very significant.

So instead of increasing pin-count, designers found more intelligent ways of using the row/column addressing scheme. For example, sequential reads from the same row do not require the nRAS signal to be activated (after all, reads from the same row all

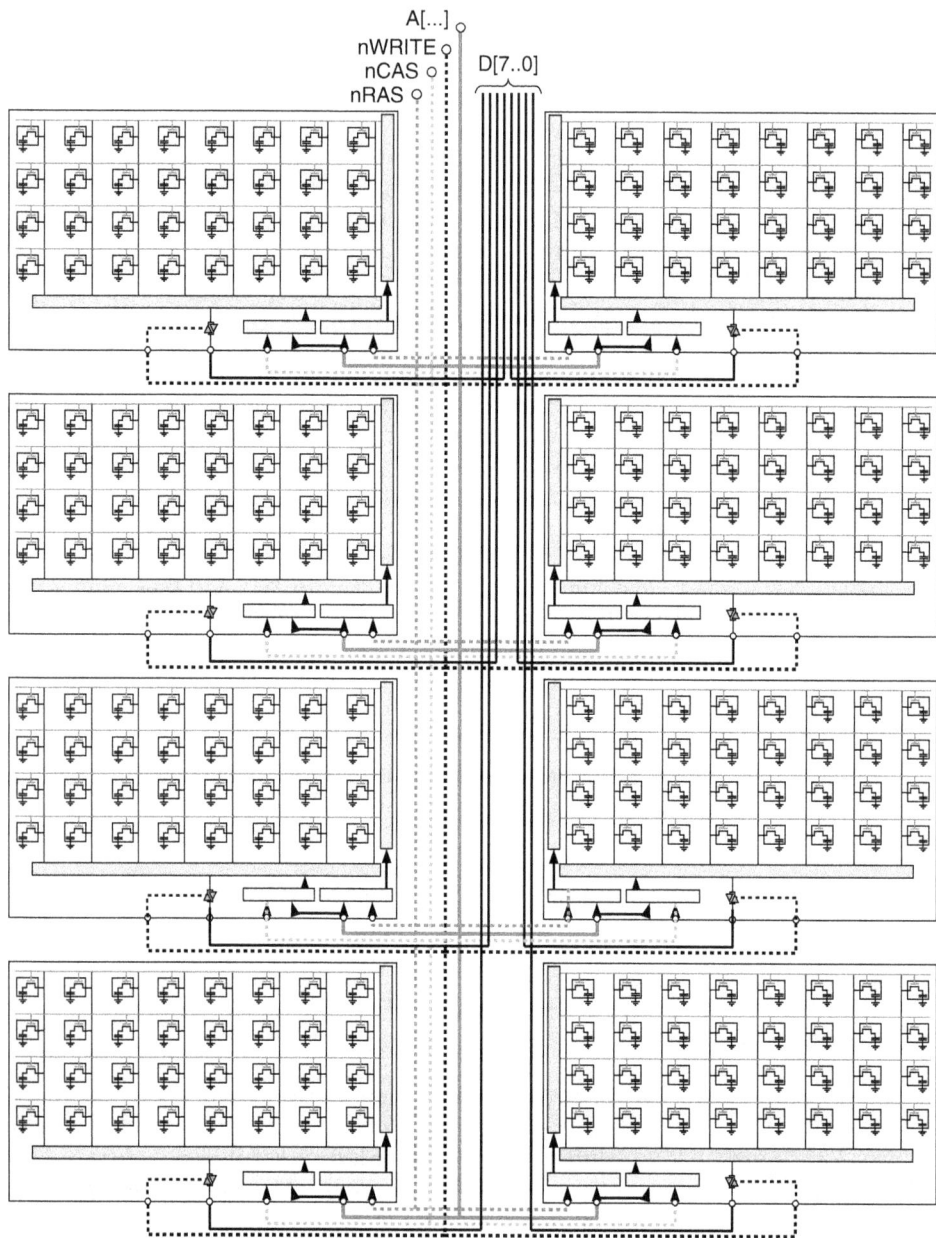

FIGURE 7.15 The basic single-bit DRAM array of Figure 7.14 has been replicated eight times to form a DRAM device that connects to an 8-bit bus. All control and addressing signals are common to each of the internal blocks (and in practice the blocks may all share a single common row and a single common column address latch).

have the same row address), and read-write or write-read combinations can similarly be simplified.

In fact, there have been many advances of these kinds, some of which we had listed previously in Table 7.2. The first innovation was the method of reading many locations from a page without re-asserting nRAS. This technique is called *fast page mode*.

DRAM was also adapted for use in video cards, becoming *video RAM* (VRAM), characterized by having two data ports for reading from the memory array. One port (the one connected to the main CPU) allows the processor to read from and write to this memory. A second part connected to video DACs (digital-to-analog converters) was read-only and allowed the data contained in the array to be accessed and read out pixel-by-pixel for display on a screen.

Moving back to general DRAM, *extended data out* (EDO) variants used an internal latch to store page data, so that this could be output and read by a CPU even as the CPU was beginning the process of reading the next page. This is, in fact, a form of pipelining and was improved further by blocking multiple reads so that they occurred together (up to four at a time in *burst mode* EDO DRAM). In multichip memory modules especially, clever use of interleaved memory banks also allowed reads to be staggered across banks, to further speed up access.

Up to now, each of the DRAM variants mentioned has been asynchronous, although controlled in a synchronous fashion by a CPU. In fact, it became obvious that squeezing any further performance out of these devices required them to have "knowledge" of the bus clock, and hence *synchronous DRAM* or SDRAM was invented. Being synchronous allows the devices to prefetch data ready for the next clock cycle, to more thoroughly pipeline, and to interleave internally.

The major performance improvements to SDRAM have been in increasing clock frequency, and allowing data to be transferred on both edges of the memory clock (i.e., instead of one word being transferred each clock cycle, two words can be transferred—one on the falling edge of the clock and one on the rising edge). This is termed "double data rate" or DDR SDRAM.

7.7 Pages and Overlays

Although we have only just looked at real memory devices in Section 7.6, it was way back in Section 4.3 that we introduced memory management using an MMU. MMU configuration is generally considered to be a fairly complex topic (the author can relate from first-hand experience that teaching and writing about it is nowhere near as tricky as having to actually configure a real MMU, in low-level assembly language, on a project with tight deadlines).

In most MMU-enabled systems, pages of memory are swapped in and out to external mass storage, typically provided by hard discs. The memory management system keeps track of which pages are actually resident in memory at any one time, and which are on disc, and loads or saves pages as required.

The MMU that controls this is actually the result of a long process of invention and evolutionary improvements, but stepping back now several generations, we can consider life without an MMU. This is not simply a thought-experiment; it is precisely the situation in many very modern embedded processors which have limited on-chip memory—designers very frequently run out of RAM in such devices.

Let us consider a real situation where software engineers are developing control code for an embedded processor that resides in a mobile radio. Nearing the end of their development, they total up the size of the code they have written for that CPU and it requires the following amounts of memory:

- Runtime memory: 18 Kibytes when executing from RAM

- Storage size of code: 15 KiB of ROM (read-only memory)

It happens that the processor has only 16 KiB of internal RAM, which is obviously insufficient to both hold the program code and execute it. If on-chip or parallel external ROM was available in the system then the program could be executed directly from this ROM (but with any read-write code sections located in RAM—and in most cases the "ROM" would actually be flash memory). However, let us suppose that in this case the only flash memory available is a 1-MiB device connected over a 25-MHz SPI (serial peripheral interface) serial port.

Unfortunately, this is far too slow for code to be executed directly from it.

In fact, designers measured the timing characteristics of the system as it was. From power on, the device took 5 ms to transfer the program code from flash memory to RAM before the program would start ($15 \times 1024 \times 8$ bits$/25 \times 10^6$ seconds).

Ignoring the obvious solutions of making the code more efficient or providing more RAM, designers were forced to use overlays to get the system to fit. These followed the principle that not all of the software was in use at any one time—in fact several sections were mutually exclusive. For example, the radio contained software that allowed it to operate in a legacy mode. This mode was selectable during power-up, such that it would either operate in "normal mode" or "legacy mode," but never both simultaneously. Bearing this in mind there is no reason why both parts of the code should reside in RAM together, much better to simply load whichever one is required.

Designers therefore split the operating code into two separate executables—or overlays: one for "legacy mode" and one for "normal mode." This appeared inefficient at first since the two modes shared quite a few functions, and these functions now had to be provided twice—once for each overlay. Also, an extra start-up code chooser was required to switch between the two overlays (in fact, to choose which overlay to use, load it, and then execute it). So did this provide a solution? Let us consider further.

Examining the memory situation, the code sizes were now as follows:

- Runtime memory in "normal mode": 12 KiB

- Runtime memory in "legacy mode": 10 KiB

- Storage size of code for overlay chooser: 1 KiB of ROM

- Storage size of code for "normal mode": 10 KiB of ROM

- Storage size of code for "legacy mode": 9 KiB of ROM

Total flash memory occupied had now become: $1 + 10 + 9 = 20$ KiB (compared to 15 previously); however, with 1 MiB of flash memory in total, this increase in size was not a concern.

But what about start-up speed? Several engineers were concerned that this approach would make the radio slow to start-up. However, tests showed that start-up time was actually faster.

For normal mode it required 3.6 ms to transfer the total 11 KiB of data, ignoring the few instructions of the selection code which may require just a couple of microseconds only ($11 \times 1024 \times 8$ bits$/25 \times 10^6$ seconds).

In legacy mode, the start-up time was even less: 3.3 ms ($10 \times 1024 \times 8$ bits$/25 \times 10^6$ seconds).

In summary, both start-up time and run-time RAM requirements improved through the use of overlays, at the expense of more software to get working, and more ROM space needed.

Without an MMU to handle memory management, options for expanding code beyond the RAM limitations are fairly straightforward: either write an overlay loader which could be as simple as a chooser between two executables or a more complicated device where overlays themselves contain code which chooses and loads the next overlay.

However, for modern embedded processors there is another choice: use of an advanced operating system (OS) that mimics, in part, the functionality provided by an MMU. One such prominent example is uCLinux (this is Linux for processors lacking an MMU), which allows a wide range of standard compiled Linux code to execute—including flash filing systems, execute in place (XIP) drivers, and so on.

One final point is that the overlay approach is finding a new lease of life with FPGA technology. These field reprogrammable devices can totally change their firmware functionality upon reprogramming, and as their name implies, they can be reprogrammed just about anywhere (even in a field...). A current hot topic applying this concept is software-defined radio (SDR). An SDR is a digital radio designed using common hardware, but able to load one of several decoding architectures to match whatever transmission scheme is being used at the frequency of interest. A front-end chooser monitors the wireless signals on the current frequency, decides what sort of modulation is in use within them, and then loads the correct firmware into the FPGA to demodulate and decode these signals. With such techniques likely to find a place in mobile phones over the next few years, it seems that overlay techniques are here to stay.

7.8 Memory in Embedded Systems

Most computer architecture textbooks describe memory subsystems for large computers, and some even cover shared memory for parallel processing machines (just as we have done), but they neglect to extend their discussion downward in dimension to embedded systems.

Embedded systems tend to use memory in a different way to desktop computers, and although embedded systems do come in all shapes and sizes for all manner of application, the majority of modern systems would contain flash memory in place of the hard disc in use within larger systems (as is reflected in the memory pyramid of Section 3.2.2).

At this point we shall examine a typical embedded system, built around an ARM9, and running embedded Linux. The arrangement we will reveal is actually quite typical of such systems and forms the majority class of such medium-sized embedded computers. We could also form a class of small systems, ones with up to 100 Kibits of RAM, which would have a monolithic real-time operating system (one which includes OS, application

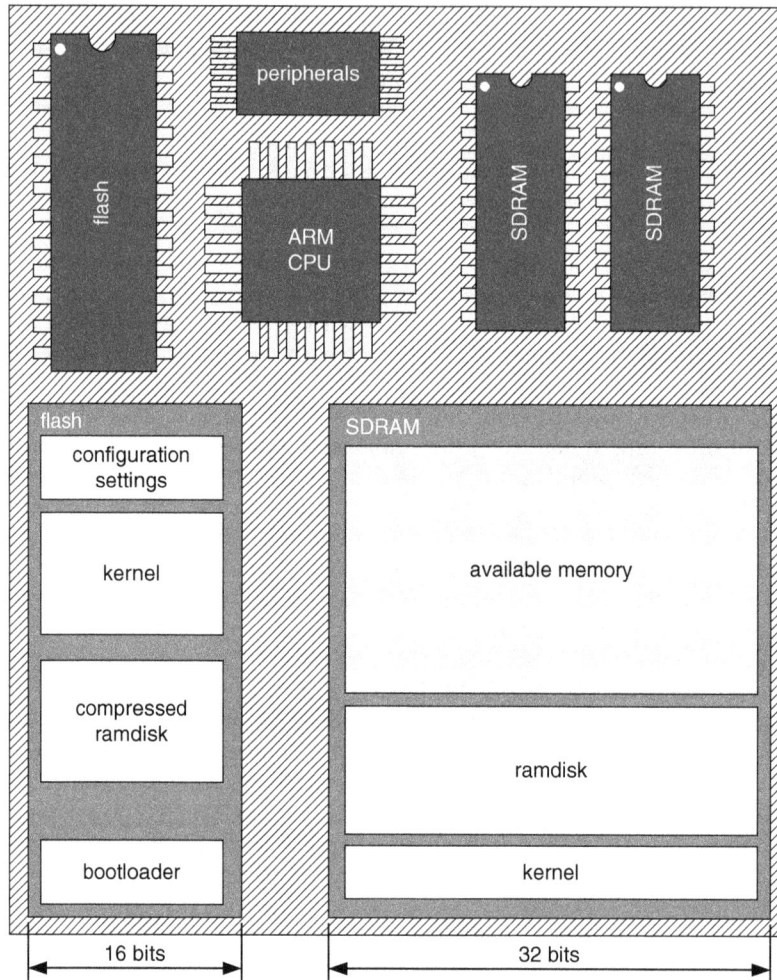

Figure 7.16 A block diagram showing the memory arrangement for an example ARM-based embedded system, with the memory content for both flash and SDRAM shown, during normal system operation.

code and start-up code in a single executable block), and larger PC-style systems which use smaller x86 processors, and are basically cut down low-power PCs.

For the medium-sized system shown in Figure 7.16 (which actually exists, and contains a Samsung S3C2410), non-volatile program code is stored in flash memory, and volatile running code plus data is contained in SDRAM. The flash memory device is 16 bits wide, and the SDRAM 32 bits wide (by using two 16-bit wide SDRAM devices in this case).

The lower part of Figure 7.16 shows the content of each type of memory during execution; however, we will look at memory content during three stages of operation.

7.8.1 Booting from Non-Volatile Memory

During power off, only the content of flash memory is preserved: SDRAM is essentially blank. When the ARM processor turns on and reset is de-asserted, the CPU begins to load instructions, and thus execute a program from address `0x0000 0000`. In this case, as in many embedded systems, flash memory is located at this position in the memory map. Thus the first instructions in flash get executed immediately after reset.

At this point the CPU is executing directly from flash. This important *bootloader* code needs to perform tasks such as resetting the processor state, turning off its watchdog timer (Section 7.11) and setting up SDRAM. This is one reason why most embedded developers need to learn about SDRAM: we need to configure it in order to progress beyond this point in the bootloader.

There are many freely available bootloaders to choose from, such as the popular U-boot; however, it is not uncommon for designers to write their own custom-designed boot code to perform the functionality they require. Some of the things a bootloader can be expected to do are as follows:

- Power on self-test (POST).
- Set up memory, particularly SDRAM.
- Set up CPU registers such as clock dividers, power control registers, MMU, cache memory.
- Write a message to serial port, LCD screen, or similar.
- Optionally wait for user intervention (such as "press any key to enter boot menu or wait 5 seconds to continue").
- Load kernel and/or ramdisk from flash to SDRAM.
- Run executable code (e.g., kernel) by jumping to its start address.
- Test memory.
- Erase blocks of flash memory.
- Download new kernel or ramdisk to SDRAM.
- Program a kernel or ramdisk from SDRAM into flash memory.

In the case of the system under consideration, there are three items loaded into flash memory. The first, located at the "bottom" of flash, beginning at address `0x0000 0000`, is the bootloader code. The next item is a compressed ramdisk, and the final item is a kernel.

The embedded Linux operating system is partitioned so that the ramdisk (which takes the place of the hard disc found in a desktop system) contains applications software and data, whereas the kernel contains the basic core of the operating system. The ramdisk is actually a filing system, which contains various files, some of them executable, all of which are compressed using gzip into a large compressed file, typically on the order or 1 or 2 MiB in size.

The kernel—the basic OS core—contains all of the system-level functionality, in-built drivers, low-level access routines, and so on. This code is designed to be unchanging, even when the ramdisk might be updated as new application code is developed. It is the kernel that the bootloader executes to begin running embedded Linux. However, first the kernel and ramdisk must be located in the correct place in memory, which is called booting up. This process is primarily a matter of software and is closely related to the operating system, so we will discuss it further in Chapter 9 (see in particular Section 9.5).

The explanation above is true of medium-sized and larger embedded systems. However, there are much smaller and simpler devices available. These tend to be less configurable and generally boot directly into their operating code. As an example, Box 7.4 describes the memory arrangement and start-up process for a small embedded processor, the MSP430x1 from Texas Instruments. This does not have an external data or address bus, so everything happens internally. There is no ramdisk, no separate boot program and kernel, and no decompression needed. It simply runs its internal software immediately from power-up (allowing it to have a much faster start-up time than any system which has a complex boot process).

Box 7.4   Memory Map in the MSP430

The MSP430 is a typical small and low-power microcontroller with a large amount of internal functionality, most of which is implemented using on-chip memory-mapped peripherals.

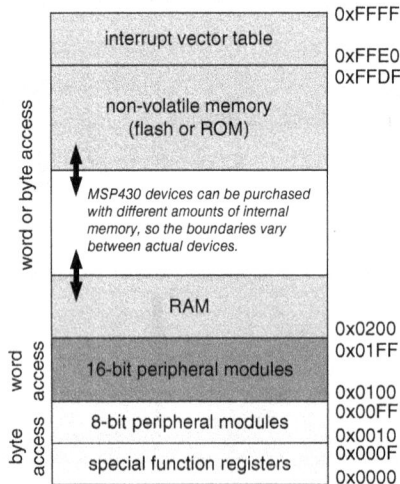

word or byte access		
	interrupt vector table	0xFFFF / 0xFFE0
		0xFFDF
	non-volatile memory (flash or ROM)	
	MSP430 devices can be purchased with different amounts of internal memory, so the boundaries vary between actual devices.	
	RAM	
word access	16-bit peripheral modules	0x0200 / 0x01FF
		0x0100
byte access	8-bit peripheral modules	0x00FF / 0x0010
	special function registers	0x000F / 0x0000

The memory map of the MSP430x1xxx series processors is shown in the diagram above, starting from address zero at the bottom. Note first that different parts of the memory map have different widths—some of it is 8-bit wide and some is 16-bit wide. Since both types of peripheral are available in those devices, Texas Instruments have segregated them into different areas of the memory map depending upon their width of access. For example, the setup and data registers of an 8-bit peripheral would lie between address 0x10 and 0xFF.

Special function registers, at the bottom of the map, control the entire system, processor, and so on (such as power and clock control). The interrupt vector table in this processor is actually located at the *top* of memory and that means when the device is reset, it will begin executing the code in that area. For this reason, non-volatile memory (flash or ROM) which will contain boot code is located at the top of the memory map to overlap this area. RAM is placed lower down.

Interestingly, there are a very wide variety of MSP devices available from Texas Instruments, each with a different selection of features and peripherals, and also varying widely in the amount of flash/ROM and RAM provided. Among all these parts, the memory maps are the same, apart from the upper boundary of RAM and the lower boundary of ROM, which depend upon exactly how much is present in the devices.

7.8.2 Other Memory

Many devices having a parallel interface can be added to the memory map of a CPU. This includes both external devices such as memory, Ethernet chips, and hard disc interfaces, but also internal devices such as many of the internal peripherals within an SoC processor.

However, there is one other common entity that is memory mapped, and that is the system and peripheral module control registers. These were identified clearly in the MSP430 memory map in Box 7.4 (at the bottom of the memory map, starting with special function registers, and continuing with the peripheral control registers). In fact, if you refer back, for a moment, to the description of pin control system on the MSP430 in Box 7.2, you will see several of the MSP430 registers named in our description.

All of these registers, and many more, are specified in the MSP430 data sheet, and all are memory mapped, which means that they occupy specific addresses in the memory map of the processor. For the registers mentioned in Box 7.2, these can be found at the following addresses in memory:

Name	Address
P2DIR	0x02A
P2IN	0x028
P2OUT	0x029
P2SEL	0x02E

Thus when the programmer writes to the given addresses, or reads from them, this action will alter or query the registers.

For the registers we are interested in, referring back to the memory map for a moment, we can see they lie within the "8-bit peripheral modules" section, which is what we should expect since the ports (and hence the registers controlling them) are 8 bits wide.

In C programming language code, the safest way to read from and write to these registers would probably be as follows:

```
unsigned char read_result;
void *addr;
read_result = *((volatile unsigned char *) addr); //to read
*((volatile unsigned char *) addr) = 0xFF;         //to write
```

The use of the `volatile` keyword is interesting. Let us examine why it is required.

Many compilers will detect a write-after-write within a program and simply delete the first write to improve efficiency. For example, if a program were to save something to memory location X and then save something to the same location a few clock cycles later, without reading from location X in between, then the first write is clearly a waste of time—whatever was written the first time would just be overwritten later.

This may be true when writing to RAM. However, there are some instances where we legitimately need to write-after-write to the same memory address: such as a flash memory programming algorithm, or when the location we are writing to is actually a memory-mapped register.

A case in point is the data output register of a serial port. A programmer wishing to serially output two bytes would first set up the serial port, and then write one byte after another to the memory-mapped serial transmit register.

The `volatile` keyword tells the compiler that the memory that is being written to is "volatile" that, it needs to be refreshed. The compiler will then ensure that the write-after-write removal does not take place, and the operation proceeds just as it has been coded.

It is not just the write-after-write cases that a compiler will detect—often compilers will detect read-after-read and optimize these to a single read if possible. Read-after-read does legitimately happen in code; in fact, the compiler will often deliberately insert these as part of the addition of spill code (see Section 3.4.4). However, the compilers' interpretation when a programmer writes a read-after-read is often that it was unintentional.

Of course, as we have seen, read-after-read can be just as necessary as write-after-write. For example, in reading serial data from a serial port input register. Or in polling a serial port status register to detect when the transmit buffer is empty. In each of these case, just as in the write-after-write case, the `volatile` keyword is used to tell the compiler that the read-after-read is deliberate.

The small code snipped above used `volatile` as a cast. It could equally as well have defined a volatile variable type:

```
volatile unsigned char * pointer;
```

7.9   Test and Verification

Test and verification need to be covered in any chapter purporting to address practical issues in computing (and particularly embedded computing). This is primarily because the performance improvements have made processors more and more complex and large over time. This has had the effect of making processor design and manufacture far more difficult, and introduced the need for test-support and failure-control mechanisms to be added to the devices themselves.

7.9.1   IC Design and Manufacture Problems

It is no longer possible for a single design engineer to understand and check an entire modern processor as it was in the 1970s. Although good teamwork and excellent design tools have largely taken the place of manual checking, it is easily possible for errors to be incorporated in the design of an IC. In fact, it is almost impossible to find a processor with no hardware design errors when first released. The vast majority are small inconveniences that can be fixed with a software workaround (e.g., "always put a NOP after a mode change if the serial port is operating"). Others are more serious.

One high-profile design error was the Intel FDIV bug in the 80486 processor—only detected once the CPUs were sold and installed in thousands, if not millions, of computers. This was literally a one in a million error that could remain undetected for months but nevertheless cost the company dearly in economic and public relations terms.

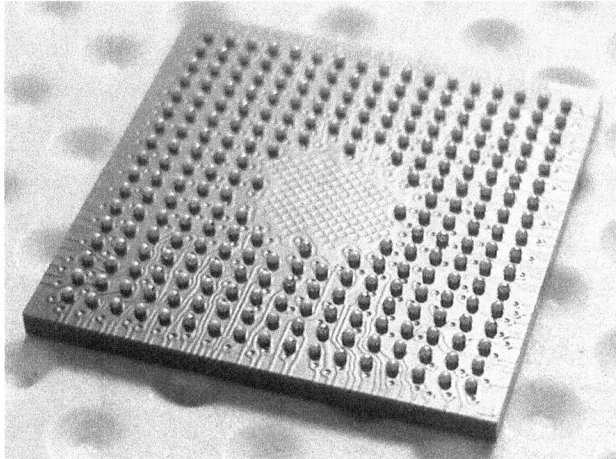

FIGURE 7.17   Underside of a modern ARM processor showing the grid array of solder balls (device size is 14 × 14mm. (Photograph taken by the author.)

Manufacturing faults are far more common. A glance at the printed circuit board (PCB) in a modern top-end PC compared to one built in the 1980s would reveal the gradual and relentless integration of separate components into silicon, but also that those silicon devices that are present on the circuit board today tend to be large with many pins (or rather balls). Figure 7.17 shows a photograph of the BGA (ball grid array) on the underside of an ARM processor. It clearly shows the small balls of solder that melt, when heated in a soldering oven, to connect the underside of the IC to corresponding pads on the surface of a PCB.

While the BGA is a very compact and efficient method for connecting an IC to a PCB, it is not at all debugging and repair friendly: with previous generations of IC packaging, it was possible to probe or test connections which were clustered around the outside of a device, and visible from above). The BGA, by contrast, hides all connections underneath itself—virtually the only way to check each connection physically is by taking an X-ray of the part after it has been placed on the PCB. An example of the ability of X-rays to "see through" a package is shown in Figure 7.18, where the internal detail of the IC, as well as PCB features below it, are visible.

As more design functionality is incorporated into single silicon devices, such as the SuperIO chip in a desktop PC, it is of course necessary to integrate whatever external interfaces that the original devices used, and this is one of the major drivers influencing device size. For example, to connect to storage media such as hard discs, CD-ROMs, floppy drives, and so on. However, these interfaces use multiple pin connections and were designed, in some cases, almost 30 years ago. It is no coincidence that the more recent interfacing techniques of USB (universal serial bus), Firewire, and SATA (serial ATA—Advanced Technology Attachment) use significantly fewer pins, and consequently much of the I/O (input/output) connections are present solely to support legacy interfaces rather than their modern counterparts. It seems likely that in time the older ISA, EISA,

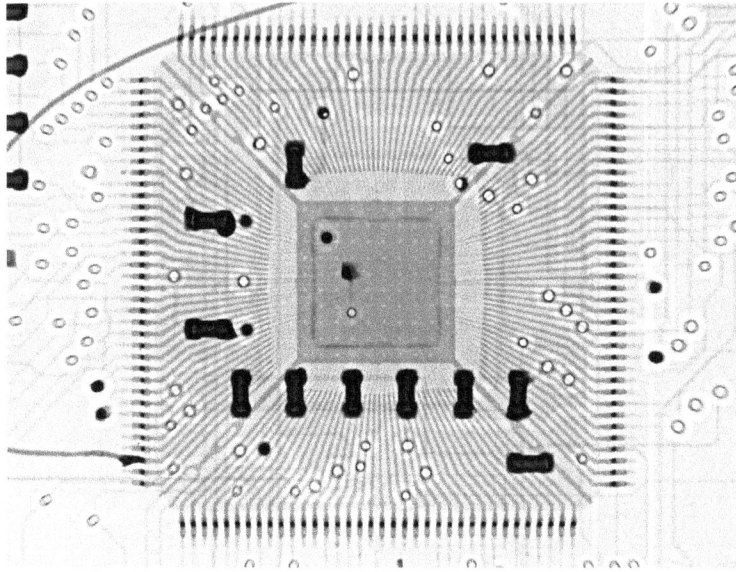

FIGURE 7.18 An X-ray photograph of an IC (in this case a quad flat pack package) showing visibility through the package of the IC, revealing the lead-frame and the silicon inside.

IDE, SCSI, and floppy disk buses will disappear, allowing a much smaller SuperIO[3] chip footprint.

Getting back to test and verification, there are two main issues with devices having a large number of I/O connections.

First is that when an IC is manufactured, it generally needs to be tested. Some manufacturers are content with batch testing only, but others prefer a zero tolerance to failures, and so will test every device made. These tests need to cover the two main manufacturing steps of making the silicon chip, and then attaching the legs/pins/balls to it.

Second, there is a small but finite probability of each solder joint not working, and therefore circuit board manufacturing failure rate is roughly proportional to the number of pins being soldered. Big devices with many I/O connections are thus more problematic, and there needs to be a way of verifying whether these soldered connections have been made correctly.

For the sake of clarity, we separate these techniques into two classes and discuss a number of solutions in the following subsections that apply to computer processors:

1. Device manufacture test—ways and means of ensuring that an IC works correctly before it leaves the semiconductor foundry, for this see:
   Section 7.9.2—BIST (built-in self-test)
   Section 7.9.3—JTAG (Joint Test Action Group)

---

[3]The SuperIO chip is the name given to the big IC that sits on a PC motherboard to provide much of the glue logic and functionality of the systems that must surround the CPU for the system to work—for example, the memory drivers, USB interface, parallel port, serial port, and so on.

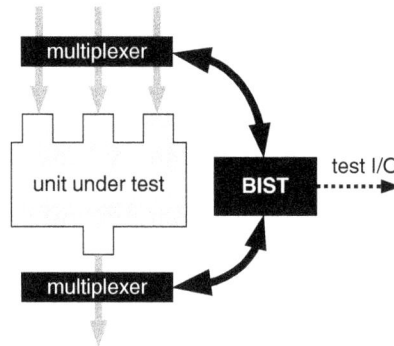

FIGURE 7.19 A built-in self-test (BIST) unit can isolate the input and output signals for a device under test—allowing the output to be verified correct for the given input conditions.

2. Runtime test and monitoring—ways to ensure that the final manufactured system is working correctly (in addition to the two mentioned above which are also useful in this way), see:
Section 7.10—EDAC (error detection and correction).
Section 7.11—Watchdog timers and brownout detectors.

7.9.2 Built-In Self-Test

Built-in self-test (BIST) is a device-specific on-chip hardware resource that is specifically designed to assist with the testing of internal device functionality.

This may be used, for example, by a testing machine as soon as a silicon wafer leaves the silicon etching production line, or perhaps as soon as the individual IC has been packaged and is ready to be shipped to customers. Sometimes customers will also be provided with the ability to access an internal BIST unit to aid in their own design verification.

The requirement of a BIST unit is that it can in some way isolate the part of the IC under test, feed known values and conditions into that part, and then check that the output from that part is correct. This is shown diagrammatically in Figure 7.19 where the multiplexers route data to/from the BIST unit when in test mode.

BIST may also involve an internal program within a CPU that can exercise various peripheral units. In this case it is usually required that there is some way of validating that the peripheral unit has functioned correctly, such as through a loop-back. This can be accomplished by a BIST unit as in the diagram of Figure 7.20, where multiplexers will feed back the analog output signals to the external input port when in test mode.

Feedback of external signals means that a manufacturer can generate a test sequence, output it through the analog output drivers (e.g., the EIA232 serial port which includes a negative voltage signaling level), then through the analog inputs thus validating the serial port hardware, the output driver or buffer, and the input detector.

This method of on-chip testing is certainly convenient, and easily capable of testing almost all logic, and many analog elements of an IC, but it comes at a cost in silicon area and complexity. There are three components to this cost:

1. The BIST unit itself.
2. Each unit and I/O port to be tested require a multiplexer or similar switch.
3. A switch and data connection from the BIST unit to each multiplexer.

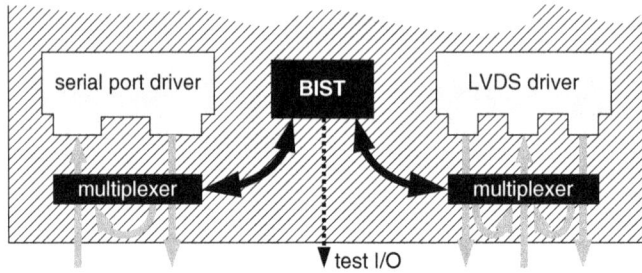

FIGURE 7.20   A built-in self-test (BIST) unit could be used to test or set input/output signals traveling between the external pins of a unit and its internal logic.

The BIST unit is not overly complex, and it scales readily to larger designs. For most logic entities, the addition of the input and output multiplexers does not significantly increase the amount of logic in the design. However, it is the data and switch connections from the BIST to each area of the device under test that become troublesome. These may have to operate at the same rate as the data paths they test, and can require bunches of parallel wires that connect to input and output buses. These wires (or metal/polysilicon tracks in a silicon IC) must run from all extremities of a device to a centralized BIST. Such routing makes designing an IC extremely difficult, and adds significantly to the cost. Decentralizing the BIST circuitry into a few, or many, smaller units can help, but the problem still remains that as IC design complexity increases, the overall BIST complexity also increases.

One method of decoupling this scaling is through the use of a serial "scan path" where the connections between the multiplexers are serial links, and the multiplexers themselves are simply parallel/serial registers. This is illustrated in Figure 7.21.

It can be seen that a single chain connects between the scan path control unit and all of the test points. This is called a scanchain, and its length is determined by the total number of bits in all of the serial/parallel registers around the chain. The chain consists of clock,

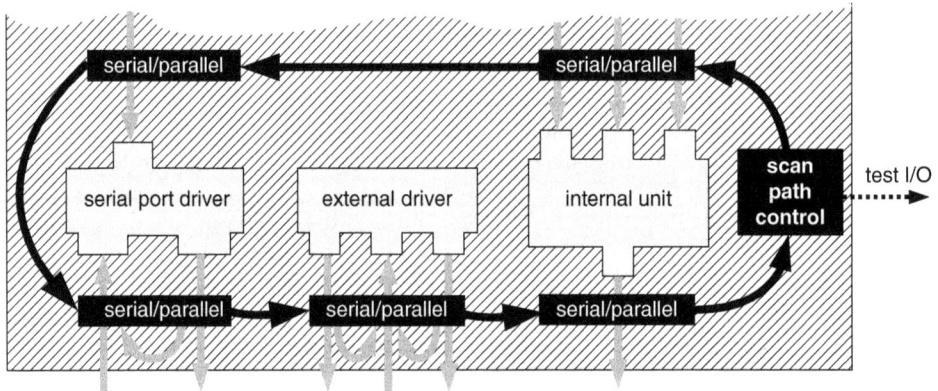

FIGURE 7.21   A scan path daisy-chains units to be tested, allowing the capability to isolate parts of a design using serial-to-parallel converter logic.

data, and control line, and is in essence a high-speed serial bus. Most importantly, this is much easier to route around an IC design, and the BIST unit (or scan path controller) can be located at the periphery of the chip rather than being central.

7.9.3 JTAG

The Joint Test Action Group were an IEEE work group that developed what is now IEEE standard 1149, as a common test control unit for scanchain control in various logic devices. Originally IEEE1149 applied to boundary scan testing, which is a subset of the scan path in that it connects to the external input and output of a device, rather than the I/O of internal units.

In fact, JTAG-compliant test units have gathered substantial extra functionality over the years, and they now commonly include various internal access functions in addition to the boundary scan path. In many cases, JTAG is the method used to enable hardware debugger access to a target processor, although some very modern processors have both a test JTAG unit and an in-circuit emulator (ICE) JTAG unit, the latter being used for debugging.

JTAG defines a standard test interface comprising the following external signals on the device that implements it:

1. TCK (test clock)
2. TMS (test mode select)
3. TDI (test data input)
4. TDO (test data output)
5. TRST (test reset—optional).

For those JTAG units implementing ICE functionality, there are generally four, or perhaps eight other input/output signals that can comprise a high-speed bus to transfer test data rapidly.

Moving back to "pure" JTAG, the hardware implemented in the unit for something like the ARM processor is shown in Figure 7.22.

The JTAG circuitry—which is definitely not shown to scale in Figure 7.22—is confined to the bottom half of the picture underneath the CPU internal logic, and as a boundary scan connecting to all the inputs and outputs of this block. Using the five JTAG pins, all of the input, output, and bidirectional pins connecting to the CPU internal logic can be queried and (where appropriate) adjusted.

JTAG is useful for many things, one example is in tracking down connectivity issues and soldering faults (see Box 7.5). Another very common application (which was unlikely to have been in the minds of the original designers) is the programming of boot code into flash memory in an embedded system, discussed in Box 7.6.

Box 7.5   Using JTAG for Finding a Soldering Fault

Imagine you have a newly made computer motherboard back from the factory. All looks correct: it does not have an over-current fault, the reset and clock signals are fine, but the board simply does not work. Perhaps there is a soldering fault?

Using JTAG, connected to the central CPU device, a test technician could set known values into the pins of the device, and then go around the PCB with a

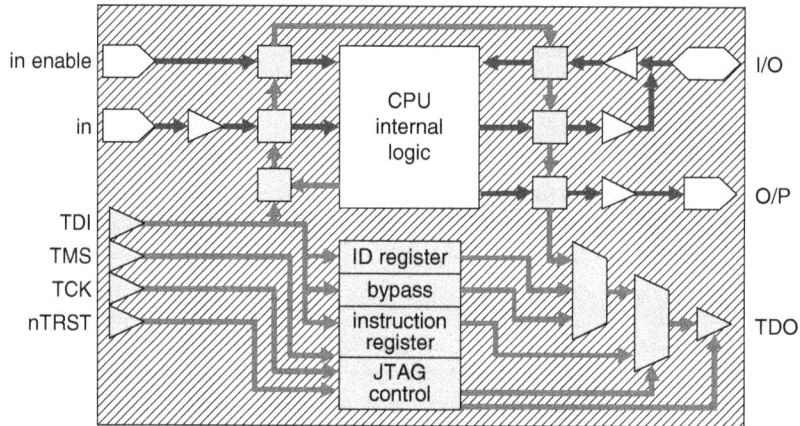

Figure 7.22 A block diagram of the main JTAG-related registers and examples of the serial data register interconnections in the ARM processor.

multimeter checking that these signals are correct. Perhaps set the address bus to 0xAAAA (which is binary pattern 1010101010101010) which will reveal whether any of those pins are shorted together, then set to 0x5555 (which is binary pattern 0101010101010101) so every pin now changes state and will reveal any pins that cannot drive either high or low correctly. It is important to measure both states because some signals on a PCB will float high if not driven, and some will float low...

Later the same technician may set various test points on the PCB to known values, and then use JTAG to read back the state of all input pins on the CPU. Then change the known value to something else (such as the inverted signals), and repeat.

In this way, all input, output and bidirectional signals on the CPU can be checked. If one pin or ball on the CPU is not soldered correctly, this will show up as the signal not able to be driven by the CPU, or as a CPU input being incorrect.

Good as this method is, it has its limitations. First is the pass/fail nature of the test which can tell if a solder joint is working, but not how *good* it is (which could help highlight potential future failures). Second, there are several pins which cannot be tested—power supply pins, analog I/O pins, and typically phase-locked loop input pins. Third, it is very slow since each pin has to be tested sequentially.

JTAG control is implemented as a simple state machine. Data is clocked in on the TDI pin on the falling edge of TCK. The TMS pin is used to select and change mode. Several modes exist, which typically include BYPASS which bypasses the scanchain so that whatever data is clocked in on TDI simply comes straight out on TDO. IDCODE clocks the contents of the ID register out to identify the manufacturer and the device. EXTEST and INTEST both clock data through the scanchain and exist to support testing of external and internal connectivity respectively.

A manufacturer may implement several alternative scanchains inside a device. One example is where integrated flash memory inside the same IC as the CPU has a separate

scanchain to service it independently of the main CPU (but using the same physical JTAG interface).

Typical scanchains are several hundred bits long. For example, the Samsung S3C2410 ARM9 processor has 272 BGA balls, but 427 bits in the scanchain. Each bit position in the scanchain corresponds to one of:

- Input pin.
- Output pin.
- Bidirectional pin.
- Control pin.
- Reserved or hidden.

Usually, output and bidirectional pins (or groups of similar pins) have a control bit associated with them that determines whether the output buffer is turned on or not. These control bits can be active high or active low—and this information along with everything else needed to control the JTAG of a particular device is stored in a boundary scan data (BSD, or BSD logic: BSDL) file, including scanchain length, command register length, the actual command words themselves, and the scanchain mapping of which bit relates to which pin or function.

Box 7.6   Using JTAG for Booting a CPU

Most ARM-based processors that do not contain internal flash memory will start to execute from address 0 following reset. This address relates to chip select 0 (*nCS0* to indicate it is active low) which is generally wired up to external flash memory.

This external flash would therefore contain a bootloader, which is the first small program run by the CPU after reset or power up, and which launches the main application or operating system—perhaps mobile Linux for a PDA (personal digital assistant), or SymbianOS for a fingerprint reader.

Before the 1990s boot code would be in an EPROM (erasable programmable read-only memory) that was typically socketed. It was simply a matter of inserting a programmed EPROM device, turning on the power, and the system would work. Today, EPROM has been superseded by flash memory which is reprogrammable, and a ROM socket is regarded as too expensive and too large to include in most manufactured electronics.

Every new device straight off the production line would have empty flash. There thus needs to be a step of placing the boot code inside the flash memory.

This can easily be accomplished with a JTAG-based programmer. This is driven from an external PC, connected to the CPU JTAG controller. It takes control of the CPU interface pins that connect to the flash memory, and then it drives the flash memory in such a way as to program in the boot code. As far as the flash memory is concerned, it does not know that this is controlled from an external PC: it simply sees the CPU controlling it in the normal way.

The external PC, working through the JTAG to the CPU, and then controlling the CPU interface, uses this to command the external flash device to erase itself, and then byte by byte, programs boot code into flash from address 0 onward.

Finally, it should be noted that since the JTAG standard is implemented as a serial connection, there is nothing to prevent a single JTAG interface from servicing several separate devices in a long daisy chain. An external test controller can then address and handle each one as required through a single JTAG interface.

JTAG is thus very hardware and resource efficient, and has become increasingly popular on CPUs, FPGAs (field programmable gate arrays), graphics chips, network controllers and configuration devices, etc. Anyone who can remember the difficulties in debugging and "running up" new digital hardware in the days before the adoption of JTAG would probably agree with the author that this technology, although quite simple, has revolutionized the ability of computer designers to get their prototype designs working.

7.10   Error Detection and Correction

Errors creep into digital systems in a number of ways apart from through incorrect programming. Poor system design may cause noise to corrupt digital lines, voltage droop may occur on power lines (also called brownout, described in Section 7.11.1), clock jitter (see Section 7.4) may cause a digital signal to be sampled at an incorrect time, and electromagnetic interference from other devices may corrupt signals.

One less commonly discussed cause is through cosmic radiation: so-called SEUs (Single Event Upsets) whereby a cosmic ray triggers a random bit-flip in an electronic device. Since the earth's atmosphere attenuates cosmic and solar radiation, SEUs become more prevalent with altitude. Consumer electronics at the altitude of a Galileo or Global Positioning Satellite (around 20,000 km) would be totally unusable, while at a low earth orbit altitude (500 km) may suffer several events per day. On a high mountain perhaps one or two a month, and on the ground, possibly a few per year. This does not sound like a cause for concern, but then imagine designing a computer to be used in an air traffic control system, for a nuclear reactor control room, or a premature baby life-support system.

Fortunately, well-established techniques exist to handle such errors, and this is an active research field in space science. Common techniques range from a NASA-like decision to run five separate computers in parallel and then "majority vote" on the decisions by each one to, at the simpler extreme, the use of parity on a memory bus.

In times gone by, well-engineered UNIX workstations by such giants as DEC (Digital Equipment Corporation), SUN Microsystems, and IBM were designed to accept parity memory. Parity memory stored 9 bits for every byte in memory—or 36 bits for a 32-bit data bus. One extra bit was provided for each stored byte to act as a parity check on the byte contents:

7	6	5	4	3	2	1	0	P
1	0	1	1	0	1	0	1	1
1	0	1	0	0	1	0	1	0

The P bit is a 1 if there are an odd number of 1's in the byte, and a 0 otherwise. It is therefore possible to detect if a single bit error occurs due to an SEU (for example), since the parity bit will then be wrong when compared to the contents of the byte. This applies even if the parity bit is the one affected by the SEU.

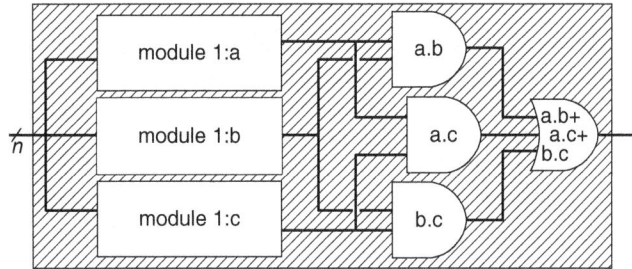

FIGURE 7.23 An example of triple module redundancy (TMR) where one processing module is repeated three times. A simple output circuit performs majority voting. For example, at a bit level if the three modules output 0, 0, 1, respectively, then the final output is 0 and we assume that the module which output the 1 was incorrect. Similarly, if the three modules output 1, 0, 1, respectively, then the final output will be 1 and we would assume that the middle module was in error. Note that the signals do not necessarily need to be bits, but could be larger items of data.

While this works well, 2 bits in error cannot be detected with a single parity bit. Even more unfortunate is the fact that although it is possible to know that an error has occurred, the scheme does not give any indication of *which* bit is in error, and so the error cannot be corrected.

More capable error detection systems utilize methods such as Hamming codes and Reed-Solomon encoding. One increasingly popular and relatively modern technique is the powerful Turbo Code, often used for satellite communications. Details of these techniques are outside the scope of this book, except to note that all the methods increase the amount of data that must be handled, and in return, improve the ability to recover corrupted data. In fact there is a multiway trade-off among:

- Encoding complexity: *how many MIPs to encode a data stream*
- Decoding complexity: *how many MIPs to decode a data stream*
- Coding overhead: *how many extra bits must be added to the data*
- Correction ability: *how many bits in error can be corrected*
- Detection ability: *how many bits in error can be detected*

It is possible to trade off each of these, and each scheme has its own particular characteristics. In addition, schemes are based on a unit of data that might range from a single byte (with a repetition code) to several kilobytes or more (turbo codes). This has the practical consideration that some schemes will output corrected data after a few bits have been processed, whereas with other schemes it may be necessary to wait until a large block of data has been processed before anything can be decoded. Some examples are

- *Triple redundancy:* Sometimes called a repetition code, each bit of data is repeated three times, so the coding overhead is 300%, and for that, one error can be corrected in every 3 bits. Encoding and decoding are extremely easy. An example of triple module redundancy (TMR), achieved by performing a "majority vote" on outputs of three (or more) modules, is shown in Figure 7.23. The signals being voted on do not necessarily have to be bits, but could be bytes, words, or even larger blocks of data. Voting can be performed individually for each bit or for the entire item of output data. One example of this is in NASA's space shuttle having five IBM flight computers. Four of these run identical code, and all feed a majority voter, while the fifth runs software which performs the same tasks, but was developed and written separately (and thus should not be susceptible to a software error common to the remaining computers).

- *Hamming codes:* A very popular family of codes, with a common choice being the (7, 4) code that adds 3 parity bits to each 4 bits of data. This can correct all single-bit errors per block, and additionally detect 2-bit errors per block. Encoding and decoding are both relatively trivial—requiring simple modulo-2 arithmetic[4] on matrices of 1s and 0s. Coding overhead is 75% for the (7, 4) code which is explored in Boxes 7.7 and 7.8. Note that many other variants of Hamming code exist having different overhead, detection, and correction characteristics.

- *Reed-Solomon (RS):* A block-based code characterized by relatively low encoding complexity, but higher decoding complexity. RS is actually a family of possible codes arranged in terms of block size—with the correction and detection ability set by the size of block being handled. One common code is RS(255,223) which works in coded blocks of 255 bytes. The coded block contains 223 data bytes and 32 parity bytes and can correct up to 16 bytes in error per 223 byte block. Note that these bytes may each have multiple errors so it is possible to correct significantly more than 16-bit errors at times. For RS(255,223), coding overhead is 32 in 223, or 14%.

Some CPUs (such as the European Space Agency version of the SPARC processor, called ERC32—also freely available as the Leon soft core) embed EDAC capabilities within themselves, but others rely on an external EDAC unit such as that shown in Figure 7.24.

In Figure 7.24, the data bus within, and connected to, the CPU is not EDAC protected, but an external EDAC device adds error-correcting codes to every memory word written out by the CPU and checks every word read into the CPU from memory. On detecting an unrecoverable error, an interrupt is triggered to inform the CPU; otherwise, recoverable errors are naturally corrected automatically without intervention required by the CPU.

Note that not all error-correcting codes are quick enough to sit between a CPU and memory. For example, Reed-Solomon codewords require relatively long to decode and would not be possible in such a scenario without causing the CPU to pause every time it reads an erroneous, but correctable, data word. Hamming, by contrast, is quick, and commonly used for this (see the example of Hamming coding in Boxes 7.7 and 7.8).

In summary, some computer systems are required to be highly reliable, and these are likely to require some form of error detection and correction, either internally, or on the

---

[4]Modulo-2 means counting with 0s and 1s, and that any value greater than 1 should be represented by the remainder of that value when divided by 2. Thus modulo-2 values are 0 for even numbers and 1 for odd numbers, for example, $3 = 1$ (mod 2) and $26 = 0$ (mod 2). Similarly, any number in modulo-n is the remainder of that number divided by n.

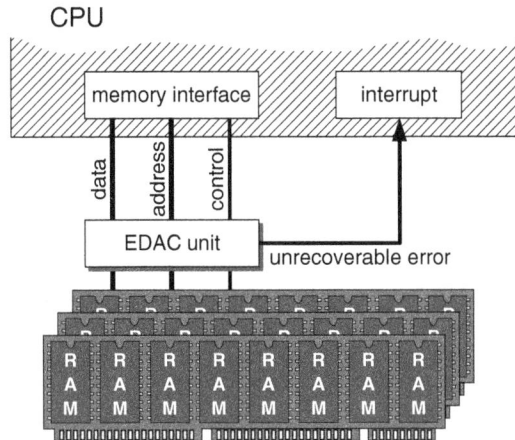

FIGURE 7.24   An error detection and correction (EDAC) unit located between a CPU's memory interface and external memory.

external buses that are more susceptible to noise. Similarly, high-density memory, which is more susceptible to SEU errors, may need to be protected with an EDAC unit.

Box 7.7   Hamming (7, 4) Encoding Example

For a 4-bit data word to be transmitted consisting of bits b_0, b_1, b_2, b_3, we can define four parity bits p_0 to p_3 using modulo-2 arithmetic:

$$p_0 = b_1 + b_2 + b_3$$
$$p_1 = b_0 + b_2 + b_3$$
$$p_2 = b_0 + b_1 + b_3$$
$$p_3 = b_0 + b_1 + b_2$$

The 7-bit word that actually gets transmitted is made up from the four original bits plus any three of the parity bits, such as the following:

b_0	b_1	b_2	b_3	p_0	p_1	p_2

When this 7-bit word is received, it is easy to recalculate the 3 parity bits and determine whether they are correct. If so, it means that the data has either been received correctly, or there was more than a single bit in error. If an error is detected then we can determine (assuming it is only a single bit in error) exactly which bit was affected: for example, if p_1 and p_2 are found to be incorrect, but p_0 is correct, then the data bit common to both must be suspect—in this case either b_0 or b_3. However, b_3 is used to calculate p_0 which was correct; thus, the error must be in b_0 alone.

It is more common to use matrices for Hamming (and most other) coding examples—see Box 7.8

Box 7.8   **Hamming (7, 4) Encoding Example Using Matrices**

In practice, Hamming encoding, verification, and correction are performed using linear algebra (matrices), defined using a generator matrix, **G**, and a parity-check matrix, **H** defined by Hamming:

$$
\mathbf{G} = \begin{bmatrix} 1&1&0&1 \\ 1&0&1&1 \\ 1&0&0&0 \\ 0&1&1&1 \\ 0&1&0&0 \\ 0&0&1&0 \\ 0&0&0&1 \end{bmatrix}
\qquad
\mathbf{H} = \begin{bmatrix} 1&0&1&0&1&0&1 \\ 0&1&1&0&0&1&1 \\ 0&0&0&1&1&1&1 \end{bmatrix}
$$

Let us test this out for an example 4-bit data vector, **d** (1101) which first needs to be multiplied by the generator matrix, **G** to form the 7-bit transmitted codeword:

$$
\mathbf{x} = \mathbf{Gd} = \begin{bmatrix} 1&1&0&1 \\ 1&0&1&1 \\ 1&0&0&0 \\ 0&1&1&1 \\ 0&1&0&0 \\ 0&0&1&0 \\ 0&0&0&1 \end{bmatrix} \begin{bmatrix} 1 \\ 1 \\ 0 \\ 1 \end{bmatrix} = \begin{bmatrix} 3 \\ 2 \\ 1 \\ 2 \\ 1 \\ 0 \\ 1 \end{bmatrix} \text{ modulo 2 } => \begin{bmatrix} 1 \\ 0 \\ 1 \\ 0 \\ 1 \\ 0 \\ 1 \end{bmatrix}
$$

So the transmitted data 1010101 represents the original data 1101. Now assume a single bit error, so we receive something different: **y**=1000101. Let us see how to use matrix **H** to check the received word:

$$
\mathbf{Hy} = \begin{bmatrix} 1&0&1&0&1&0&1 \\ 0&1&1&0&0&1&1 \\ 0&0&0&1&1&1&1 \end{bmatrix} \begin{bmatrix} 1 \\ 0 \\ 0 \\ 0 \\ 1 \\ 0 \\ 1 \end{bmatrix} = \begin{bmatrix} 3 \\ 1 \\ 2 \end{bmatrix} \text{ modulo 2 } => \begin{bmatrix} 1 \\ 1 \\ 0 \end{bmatrix}
$$

Looking back at the parity-check matrix, **H**, we see that the pattern [110] is found in column 3, which tells us that bit three of **y** was received in error. Comparing **x** and **y**, we see that is indeed the case. Toggling the indicated bit 3 re-creates the original message.

7.11   Watchdog Timers and Reset Supervision

While EDAC, apart from straightforward parity checking, is rare in ground-based computers, both watchdog timers and brownout detectors (Section 7.11.1), are extremely common and are often implemented inside a dedicated CPU support IC.

A watchdog to a processor is like a pacemaker to a human heart: a watchdog need to be reassured constantly that a processor is executing its code correctly. If a certain period expires without such a reassurance, the watchdog will assume that the processor has "hung" and will assert the reset line—just like a pacemaker delivering a small electric shock to a heart that has stopped beating.

From the programmers' perspective, a processor has to write to, or read from, a watchdog timer (WDT) repeatedly, within a time-out period. It can write or read as often as it likes, but failure to read or write at least once within the specified period, will cause the reset.

Internal WDTs usually allow the programmer to specify the time-out period by writing to an internal configuration register, which is usually memory-mapped. The devices are constructed as countdown timers fed by a divided-down system clock, with the divide ratio also being configurable in many cases. On system reset, the value in the watchdog count configuration register will be loaded into a hardware counter. Once out of reset, this counter is decremented by the clock. A comparator determines when it reaches zero, in which case the reset signal is asserted. Any time the CPU reads from, or writes to, the WDT, the counter is reloaded with the value in the count configuration register.

An external watchdog timer can be constructed from a capacitor, resistor, and a comparator. This works in a similar way to the external reset circuit (see below in Section 7.11.1), although the CPU continually "writes" a logic high to the capacitor to keep it charged up.

Typically the watchdog time-out period is a few hundred milliseconds, or perhaps a few seconds—anything too short would mean too many wasted CPU cycles as the code periodically accesses the WDT. Servicing this is best accomplished inside some periodic low-level code such as an operating system (OS) timer process that is executed every 100 ms. If this stops, we can assume that the OS has crashed, and the result will be the watchdog resetting the processor. The watchdog thereby ensures that the OS remains operational; otherwise, it will reset the CPU and restart the OS code cleanly.

7.11.1 Reset Supervisors and Brownout Detectors

Many veterans of the computer industry will remember the "big red switch" on the early IBM PCs and the prominent reset buttons sported by the machines. The prominence of these conveniences was probably a reflection of the reliability of the operating systems running on the machines, namely MS-DOS (Microsoft Disk Operating System) and Microsoft Windows. While MS-DOS is, thankfully, no longer with us in a meaningful way, windows unfortunately remains—although is not normally used in "mission critical" applications where reliability is paramount.

Embedded systems, by contrast, are unlikely to sport large reset switches and often have "soft" rather than "hard" power switches (i.e., those that are under software control, rather than ones that physically interrupt the power to the systems). It follows that embedded systems need to be more reliable, especially those in physically remote locations; for example, it would not be particularly useful to have a "big red switch" on the side of the Mars Rover.

In their quest to improve system reliability, embedded systems thus tend to make extensive use of watchdog timers (explained previously in Section 7.11), and also have supervisory circuits for power and reset.

Reset circuitry, usually driven by an external reset input, is important in ensuring that a device begins its operation in a known state. The lack of a clean reset signal has been

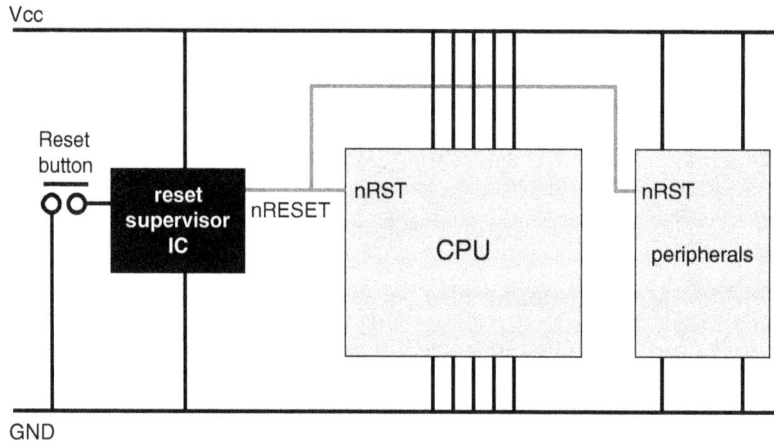

FIGURE 7.25 A reset supervisory IC will connect between Vcc and ground (GND) and generate an active low nRESET signal for a CPU and any peripherals that require it. By convention the reset signal is active low to ensure that devices are in reset when first powered on. If a reset button is required in a design, this is also supported as an input to the reset supervisor.

the cause of many system failures, whether in CPU, SoC, FPGA, or discrete hardware systems.

An external reset controller device, or supervisory IC, shown in Figure 7.25, normally "asserts" a reset signal as soon as power is applied to a system. Some time later, the reset signal is de-asserted, allowing the device to operate from a known starting position. A few SoC processors contain all of this reset logic and timing internally. Other devices may allow a designer to simply wire the reset pin to a capacitor connected to GND and a resistor connected to Vcc, but note that this is dangerous in most case, so beware.[5]

A *brownout* is a voltage drop on a power rail.[6] Since CPUs are only specified to operate within a very narrow range of power rail voltages, these droops can cause malfunction when they occur. External reset chips will assert the reset line if the power drops completely (i.e., once the power restores, they will hold the CPU in reset for at least the length of time specified by the manufacturer, before de-asserting the reset). However, only reset chips with a brownout detector will do the same whenever the voltage goes outside of the specified operating range.

---

[5] The reason for the danger is in the way the reset is triggered. As power is applied to the system, the voltage across the capacitor will initially be zero, meaning that the reset pin is held low. As the capacitor slowly charges up through the Vcc-connected resistor, the voltage will rise, until it reaches a threshold on the reset input pin, which then interprets it as a logic high, taking the device out of reset. Unfortunately, however, there is almost always electrical noise in any system, causing small fluctuations in voltage which, as the rising capacitor voltage passes the reset pin threshold, causes the device to rapidly toggle into and out of reset. The effect is often to "scramble" the reset action, prompting most manufacturers to specify a minimum time that their device should be held in reset for.

[6] A "brownout" is like a "blackout" but a little less severe. Perhaps we can follow the color analogies further and refer to a power surge as a "whiteout."

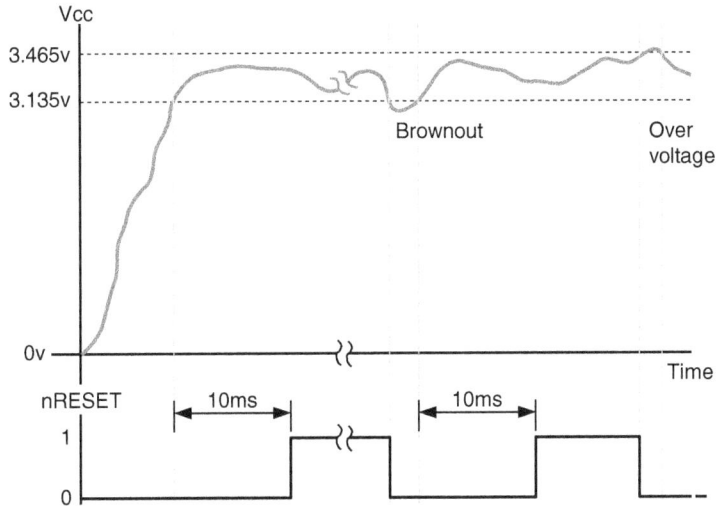

FIGURE 7.26   An illustration of a reset supervisory chip holding a device in reset (shown as the state of the nRESET signal on the lower axis) as the power supply (shown on the upper axis) rises to the nominal 3.3 V level. After some time of normal operation, a brownout occurs when the power rail voltage drops. The supervisory chip consequentially resets the processor cleanly until after the voltage has again risen to the nominal level. Later, the beginning of an overvoltage condition occurs, which will be handled similarly.

In addition, some brownout detecting reset chips can give an immediate power fail interrupt to a processor. This could allow the processor a few milliseconds to take action before the power fails totally, and thus power down cleanly. The process of reset supervision and brownout detection is illustrated in Figure 7.26, where the voltage of the Vcc power supply to a processor is plotted over time. The operating voltage of this device is 3.3 V ± 5 percent, and thus a reset supervisory system has been configured to detect any excursion of the Vcc voltage outside this range, and if detected, trigger a reset condition. The reset condition is held for 10 ms in each case (in reality this would be set to comfortably exceed the minimum time specified by the processor manufacturer, which is normally significantly less than 10 ms). The brownout device will be connected and used in the same way as the standard reset supervisory IC that was shown in Figure 7.25.

7.12   Reverse Engineering

The consumer of our embedded technology developments sees desirable and breathtaking new products, but for the developers these are often the culmination of a long, arduous, and expensive design process. Of course, any pioneering inventor of new embedded systems can expect there to be some competition in time, which may improve upon their original design; however, companies may often rely upon the first few months of sales in an uncrowded market to recoup large up-front design and manufacturing costs. Usually, competitor products will have a similar costing to the pioneer products—since these would have incurred similar development expense.

However, the economics changes substantially when a competitor cheaply and rapidly reverse engineers[7] a pioneering design. Their development costs are largely replaced by reverse engineering costs and, if we assume that these are significantly less, then the competitor would easily be able to undercut the pioneer device in price. The effects are twofold: first, the market lead of the pioneer company has been curtailed and second, their market share will reduce due to the lower pricing of the competitor product. The assumption that the reverse engineering (RE) process can be shorter, and less expensive than a full prototype-development project is borne out in the evidence of commercial examples of product piracy. The larger the differential between up-front development cost and RE cost, the greater the risk to a pioneering company and the greater reward to a nefarious competitor intent on pirating their products. The differential is greatest when a truly revolutionary product is introduced which is simple to reverse engineer.

Of course, it should be noted that reverse engineering to understand how something works is a well-established engineering approach; it is even a valid research area and something that many engineers love to do. However, design piracy through reverse engineering is a real concern in the embedded industry, and one which has led to some computer architecture-related challenges and responses which we will discuss.

First, however, it is useful to briefly survey the RE process itself, since this is the activity which prompts the later discussion.

7.12.1   The Reverse Engineering Process

In this section, we will work from the viewpoint of an offending company intent on RE an unprotected embedded system. The intention being to examine the difficulty, specialized equipment, and effort needed for each step, to allow determination of the cost structure of the process, and how this relates to the architecture of the system under "attack."

The RE process involves both top-down and bottom-up analysis of a system. The hierarchy of information which describes an embedded system is shown diagrammatically in Figure 7.27, where the system itself can be seen to potentially comprise different sub-assemblies each of which contain a module or modules of one or more printed circuit boards (PCBs). Top down means beginning with overall system functionality, and working down, partitioning the design as the process progresses, in order to elucidate more and more of the design functionality as a lower level is approached. Bottom up would most usually include identifying critical devices early and then inferring information from them. An example would be finding a known CPU on one PCB and thus inferring that much of the "intelligence" within the system is concentrated within that module.

Top-down RE of embedded systems typically involves several analytical steps. Although in practice a particular RE attack may not necessarily involve each step, or be in a particular sequence, a logical taxonomy of RE stages would be as follows:

    A. *Functionality*—understanding what the system does

    B. *Physical structure analysis:*

        B.1. Electromechanical arrangement

        B.2. Enclosure design

---

[7]"Reverse engineering" is normally defined as a process involving the analysis and understanding of the functionality, architecture, and technology of a device, then representing these in some manner which allows reuse or duplication of the original product, it's architecture or technology.

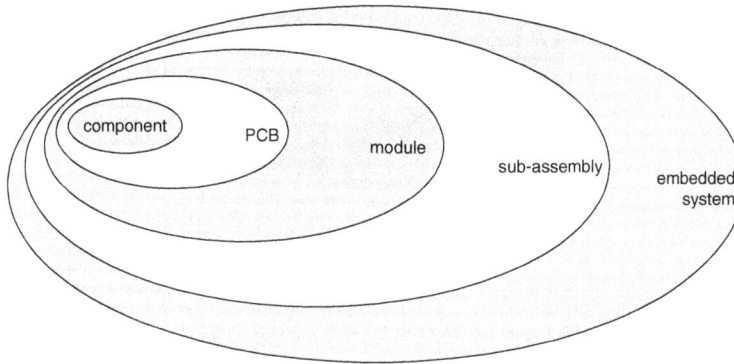

FIGURE 7.27   The hierarchy of information to be revealed when reverse engineering an embedded system comprises, from the outside in, analysis of the system as a whole, one or more subassemblies (including wiring looms), modules (and their fixtures), comprising one or more PCBs (including daughter-boards, plug-in cards, and so on), down to the individual components mounted on the PCBs or located elsewhere within the system.

    B.3.   Printed circuit board layout

    B.4.   Wiring looms and connectors

    B.5.   Assembly instructions

C. *Bill of materials:*

    C.1.   Active electronic components

    C.2.   Passive electronic components

    C.3.   Interconnect wires and connectors

    C.4.   Mechanical items

D. *System architecture:*

    D.1.   Functional blocks and their interfaces

    D.2.   Connectivity

E. *Detailed physical layout:*

    E.1.   Placement of individual components

    E.2.   Electrical connectivity between components

    E.3.   Impedance controlled and location-aware orientation

F. *Schematic of electrical connectivity*—system circuit diagram

G. *Object/executable code*—including

    G.1.   Isolation of code processors

    G.2.   Isolation of firmware code for reconfigurable logic

H. *Software*—analysis of the software within the object files, including embedded code, bootloader, firmware plus ASIC reverse engineering.

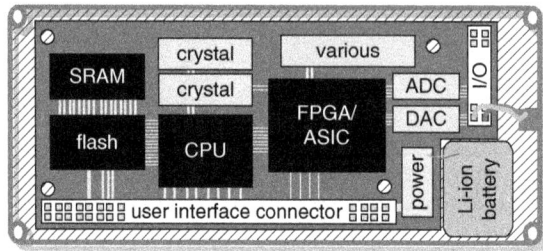

FIGURE 7.28   Block diagram of example embedded system under analysis, showing two active ICs (loosely identified as CPU and FPGA/ASIC), two memory elements (volatile being SRAM and non-volatile being flash), plus several connectors, power circuitry, crystals, and interfacing devices.

In order to highlight the process, each RE stage will now be discussed in relation to an unprotected/unhardened embedded system with a very generic system level diagram as shown in Figure 7.28. This consists of a large integrated circuit (IC) connected to volatile memory (SRAM in this case), non-volatile memory (flash), a field-programmable gate array (FPGA), a user interface of some kind, connectors, and some devices to interface with the outside world, generically termed analog-to-digital converter (ADC) and digital-to-analog converter (DAC) in this instance. Specific systems may differ, but as a general class, embedded systems often comprise a CPU booted from flash, executing out of SRAM (both of which are increasingly likely to be internal to the IC), connected to discrete or programmed logic (FPGA, programmable logic device, and so on), or ASIC. A user interface of some kind typically interfaces to the outside, analog, world. Larger systems would tend to use DRAM, SDRAM, or even hard disc storage. More integrated systems tend to incorporate a CPU soft core within an FPGA or ASIC.

For the example system of Figure 7.28, each stage in the RE process is now discussed, noting the assumption that the system has not been protected or deliberately hardened in any way.

7.12.1.1   Functionality

An RE team would normally receive several of the units to be reverse engineered. The process would begin by consulting user documentation, repair manuals, product briefs, and so on. At minimum, a list of functionality is required to double-check that subsequent analysis has revealed sufficient hardware and software to provide each of the functions identified.

This is relatively simple work and can be augmented by searching the Internet for information on newsgroups, blogs, hacking sites, and so on. Knowing the manufacturer and any OEM, postings by individuals from these email domains can be tracked and correlated.

7.12.1.2   Physical Structure Analysis

Disassembly may be as simple as removing a few screws to open a box, or as complex as having to work through layers of micromachinery. In many cases, documentation of a complex disassembly process is the key to determining the corresponding manufacturing assembly process. Considerable time and effort may have been devoted to manufacturability issues by the designers, and as such there is likely to be implicit value in understanding this. This information may also be found ready documented in a service manual.

Order and location of removed parts should be recorded—perhaps most easily accomplished by having a photographic or video record of the process. Ideally by one member of the team would be dedicated to documenting the process. Any observations and insights of those performing disassembly should also be noted at this stage. While detailed mechanical drawings for replicating enclosures, internal structures, wiring diagrams, and so on can be obtained from static analysis of the parts, assembly drawings need to be made through disassembly followed by reassembly. The analysis of the physical structure is unlikely to be an expensive component of the RE; however, understanding the reasons behind unusual mechanical arrangements and structures may at first be nonobvious and require brainstorming.

7.12.1.3 *Bill of Materials*

A bill of materials (BOM), listing of all components used in the design, can be as simple as counting the number of screws, resistors, and so on. However, it may transpire that some components are difficult to identify, particularly semi-custom ICs and devices in highly miniaturized packages (which do not have sufficient surface area for identification marks). If simplified codes are shown, these may follow the standardized formats from JEDEC, JIS, or Pro-Electron. Discrete parts may have to be removed to be tested and their characteristics painstakingly matched to known devices; however, with tolerances of 5% or more being common, it may be necessary to remove and test parts from a number of systems before an accurate determination can be made.

Some parts can be copied as is, a genuine field of academic research, especially in cases where obsolete parts need to be re-created. In general, solid models from physical measurement can be combined with materials analysis to completely describe many parts including structural items, fixings, and passive components.

PCB silk screen markings often provide useful clues to the identity of tiny unmarked parts (e.g., Z12 may well be a Zener diode and L101 an inductor). ICs with unusual or missing markings are more troublesome, especially where the manufacturer is not identified. Sometimes, system-on-chip processing cores are more likely to be identified with certain manufacturers than others. If a fabrication process can be identified, this can often be related to other known deliveries from that process.

Otherwise, subsequent system analysis (such as location and size of data bus, address bus, control signals, and power connections) can aid in identifying parts that are not immediately obvious.

Most embedded systems will incorporate off-the-shelf parts, and even provide helpful silk screen annotations, kindly assisting in allowing an inexpensive RE process. The most common difficulty appears to be associated with custom silicon devices, either from an OEM or in-house development of large-scale integrated (LSI) devices. While OEM silicon is often undocumented, it can sometimes be traceable online in Chinese, Korean, or Japanese documents. In addition, in-house LSI devices may be offered for sale by a commercialization arm of the parent company, in which case a feature list will be published somewhere—but a nondisclosure agreement (NDA) may be required before a full data sheet can be viewed.

Clearly the identification of major ICs during a cursory inspection process is preferable, but even if identification is not immediate, the process does not end there. Detailed and costly analysis can be performed to identify exact inputs and outputs and from there infer internal functionality. This may include examining voltage levels (such as CPU core voltage), clock frequencies, bus connectivity, decoupling arrangements, and so on. More destructively, the device casing can be opened and painstakingly analyzed silicon layer-by-silicon layer. IC reverse engineering will be discussed further in Section 7.12.2.2.

7.12.1.4   System Architecture

System architecture analysis reveals a rough block diagram of connectivity and subsystems responsible for various items of functionality: This involves understanding the partitioning of a design among the modules, boards, and devices within the system. Another important aspect to determine at this stage is the identification of power or ground planes and the power distribution to areas of the system. Equally necessary is an identification of bus connectivity within the system. The presence of a debug port or IEEE1149 JTAG (see Section 7.9.3) interface can be very significant in assisting the RE process, so any indication of this is an important find. Clues may include a set of five test points with pull up resistors located close to the CPU.

In most systems, circuit continuity tests and visual inspection of parts and their arrangements can be performed, and conclusions subsequently drawn. For example, it is likely in an embedded system that the same CPU data bus connects both parallel flash memory and SRAM. Continuity tests in conjunction with device data sheets easily reveal this type of arrangement. Such tests are complicated with modern packages such as ball-grid arrays (BGAs), but are still possible with some difficulty. Power pin locations, often predictable in advance, are easily tested for. For most embedded systems, this type of analysis is simple and inexpensive, but as we shall see in Section 7.13.1, can be deliberately complicated by designers.

Despite the ease of continuity testing most systems, the full architecture sometimes only becomes evident when a schematic has been re-created and organized.

7.12.2   Detailed Physical Layout

Where silk screen layout annotations are not present, photography of component placement and orientation can reveal required placement information for both outer layers. Next, all components are removed and drilled hole positions noted. As a quick check, the locations of holes on top and bottom can be compared: If these are identical then there are no blind vias and it is unlikely, although not impossible, that there are any buried vias.

The next stage is PCB delamination (peeling apart layer by layer) with photography of each layer from a constant reference position. This can be used to build a photographically correct layer stack-up. From this point it is relatively simple to copy the PCB; however, the composition and thickness of copper and each PCB layer is also required. In practice, this can be found by examining a section of the PCB where copper is present on every layer (many PCBs have a test coupon area designed for this purpose, since manufacturing process changes can affect copper thickness in particular, which in turn affect system performance and thus may need to be tested for).

Normally, a multilayer area is cut-out from the PCB test coupon and placed end-on into a hockey-puck shaped mold, which is then filled with epoxy. When set, a lens grinding

Name	Composition	Thickness (mm)
L1 signal	1/4 oz copper foil	0.0176
prepreg	7628 x2	0.3551
L2 signal	1/4 oz copper	0.0177
laminate	FR4	0.91
L3 signal	1/4 oz copper	0.0177
prepreg	7628 x2	0.3543
L4 signal	1/4 oz copper foil	0.0176
Total		**1.69**

TABLE 7.3 Example layer characteristics for four-layer PCB.

machine can be used to prepare an end-on section of the PCB for examination under a measuring microscope. Copper and layer thicknesses can simply be read off.

For large circuit boards, cut-outs from several areas on a PCB might have to be examined since variations may exist in the copper etching bath during manufacture (e.g., the edge of a PCB nearer to a top corner of the bath would have etched differently to an area nearer to the bottom center, and in either case local copper coverage density would similarly affect etching).

A growing number of embedded systems require track impedance control for high speed or RF-related signals. In this case, the exact characteristics of the PCB are important, including dielectric constant, the prepreg weave thickness, and resin type. Overall impedance can be determined through time-domain reflectometry or use of a network analyzer. The prepreg type and characteristic can be found through microscopy, and some determination of resin type can be made by looking at the overall figures.

An example of the information required for re-creation of electrically equivalent PCBs, apart from the photographically correct layer stack-up, is shown in Table 7.3.

X-rays may also be a viable method of extracting layout information, and can even provide useful information on the internals of unknown ICs. As an example, Figure 7.18 showed a low-magnification X-ray of an FPGA device mounted on an PCB within which electrical tracks, decoupling capacitors (on the underside of the PCB) and the internal lead-frame of the FPGA mounted on the top side of the board can be seen clearly. The solid circles are test points, whereas the hollow circles are vias interconnecting tracks on different PCB layers. The hair-like line across the top left is a thin wire soldered to one of the pins on the IC.

Although some specialized equipment (such as measuring microscope and reflectometer) may be required for the physical layout analysis stage, unless impedance control is involved, copying a PCB layout and stackup is neither difficult nor expensive.

7.12.2.1 *Schematic of Electrical Connectivity*

Electrical connectivity is most commonly represented as a netlist. This specifies the electrical connectivity between various nodes, and usually also specified devices connected to those nodes. The netlist itself does not take account of actual physical positioning, but only the connectivity relationship between nodes, although in real systems the physical positioning itself can be important (perhaps keep-out areas to reduce interference, or for safety reasons when high voltages may be present). The nodes are normally the pads and holes to which components are connected, and the connections are normally either wires or PCB tracks.

A netlist needs can be generated from the connectivity check, by inspection of X-rays photographs or photographs of a delaminated PCB. This is a time consuming and error-prone business, but is at least simple to verify by means such as (a) testing for expected continuity on the original board, (b) by referring to expected connectivity found on device data sheets, (c) by searching for hanging vertices and unexpected shorts, such as a two-pin component with only one pin connected, or both pins of a two-pin component commoned (meaning that they are shorted together).

Once a netlist has been found, and the devices are identified within a BOM, the next step would be to re-create a schematic diagram to represent the system. Netlist to schematic generation is an established research area, and there are even commercial tools available to assist in the process. However, in reality most RE attempts will involve a complete redrawing of a schematic directly from the revealed information. A forward netlist generation of this schematic can be compared to the deduced system netlist as a check of correctness.

Note also, that a BOM and known schematic allows for the possibility of using simulation tools to assist in the verification of both BOM and netlist accuracy.

7.12.2.2 Stored Program

Where multiple programmable devices are used (such as CPU and FPGA), the simplest electrical arrangement would be for each to have individual flash memory storage devices (with parallel and serial connections for CPU and FPGA, respectively). However, normally all non-volatile (NV) program storage within a system is clustered into a single device for cost reasons. In modern embedded systems, this device is often flash memory—serial connected if possible, otherwise parallel connected.

Items of storage within the NV memory could include separate boot code and operating code for a CPU, system configuration settings, FPGA configuration data, or other system-specific items. In this subsection, we consider methods of determining the memory location of stored programs, with a view to enabling individual extraction of these (in subsequent sections we discuss RE of the firmware/software programs themselves).

Mask-programmed gate arrays, non-volatile PLDs, and of course ASICs require no external NV devices, having their configuration stored internally. In some cases it is possible to isolate a programmable device and read out its internal configuration code. In cases where such readout is not possible, or device security measures are in force, the device will need to be either subject to extensive black box analysis, or examined internally. The latter can be accomplished through dissolving its plastic case and/or carefully grinding through layers of silicon, reading the state of each stored bit with an electron microscope, or a reflected laser.

Undoubtedly, stored program devices with security settings in place are far more troublesome and expensive to RE than the majority systems containing a single non-volatile storage block. The example system here falls into the majority category, where the CPU is responsible for programming the FPGA, and both in turn derive their code from the flash memory.

7.12.2.3 Software

Software obtained from a memory dump can easily be copied as is. Any changes may range from simple adjustments such as rewriting the content of strings to change a manufacturers name, serial number, and version codes. Executable code pieces can, with care, also be cut and pasted.

In contrast to embedded systems hardware RE, software RE of all scales is a very well researched field. At the benign end of the scale, software RE is a useful means to achieve the potential reuse of object-oriented code, whereas at the nefarious extreme, it is applied in the circumvention of copy protection schemes, leading to software piracy and theft. There is no indication that these conclusions are confined to software only. It is also the experience of the author that embedded system cloning and design theft are more prevalent in some regions than others. This may be due to such attitude differences, or more likely to variations in legal protection against design theft.

Software plays an increasingly important role in embedded systems, and although it is advisable for manufacturers to consider software RE and software security, in general it is a subset of overall software RE and protection.

However, an important subset of embedded system-specific software reverse engineering remains to be discussed. This includes the embedded operating system, bootloader, and NV memory storage arrangement of software in a typical embedded system. Consider a typical embedded system such as that discussed previously in Figure 7.28. A generic real-time operating system running on that hardware may contain a monolithic block of boot, OS, and applications code stored in flash; however, the growing use of embedded Linux and Android in embedded systems would usually present a different picture. Such embedded systems normally contain separate instances in memory of the following items:

- Boot code
- Operating system
- File system on non-volatile storage device
- System configuration settings
- FPGA configuration data

NV memory content can easily be extracted either by removing the device and dumping its content (static analysis) or by tapping off bus signals with a logic analyzer during operation (dynamic analysis). The logic analyzer method can give useful clues regarding context—for example, memory read signals detected immediately following powerup are likely to constitute boot code. However, this method obviously only reveals the content of memory addresses that are accessed during the analysis—in effect the current trace of execution/access: determining the entirety of stored code in this manner would be next to impossible in most real systems. It would require operating the system in every possible operating mode with every possible combination and timing of input signals in order to guarantee 100 percent code coverage. Nevertheless, a combination of both techniques is a powerful analytical tool.

Address and data bus lines are commonly jumbled on dense PCBs to aid in routing (see Box 7.9 for an explanation of this). This needs to be borne in mind with both methods, thus complicating the analysis slightly.

Static flash memory analysis first needs to determine the extent, boundaries, and identity of different storage areas. Where delimiters of erased flash are present (namely long strings of `0xFFFF` or `0xFF` ending on a block boundary), then this process is trivial. Otherwise, boot code is likely to begin with a vector table and is most likely to reside at the lowest address in flash, or in a specific boot block. An FPGA programming image will be approximately of the size specified in the FPGA data sheet or compressed using a standard algorithm (which if zip, gzip, or compress, will begin with a signature byte that can be searched for). A file system will be identifiable through its structure (and on a Linux desktop computer, the `file` command is available to rapidly identify the nature of many of these items once they are dumped to computer for analysis). The Linux kernel, along with other OS kernels, contain distinct signature code and may even contain readable strings (on a Linux desktop computer the `strings` command will scan through a block of code and display any strings that it finds).

The combination of static and dynamic analysis is very powerful and can provide significant information on memory content. For example, system configuration data may be stored anywhere in flash memory and may be difficult to identify by content alone. However, simply operating the device and changing a single configuration setting will cause a change in memory content. This can be identified by comparing content before and after, or by tracking the address of specific writes to flash memory with a logic analyzer.

In the extreme case, flash memory can be copied as is and replicated in a copied product. Overall, unless designers have specifically taken measures to protect their embedded system software, the process of reverse engineering NV memory to reveal stored programs, is not difficult.

Box 7.9 Bus Line Pin Swapping

For ICs such as quad-operational amplifiers that contain more than one amplifier per package, it usually does not matter which one gets used for any particular part of a circuit. So, during layout, even though the schematic would have connected individual amplifiers to different parts of the circuit, the designer is free to swap these to improve routing. This is a well-established technique.

In fact, the same can be true of memory devices: While we would naturally connect D0, D1, D2, D3, . . . on a CPU to D0, D1, D2, D3, . . . on a memory device, we are at liberty to swap the bit lines. In fact, we are also at liberty to swap address pins if we want (as long as the CPU always accesses memory with the same width—otherwise we can swap within individual bytes, but not between bytes). For example, consider the byte connection between a CPU and memory device:

CPU data pins	Memory data pins	Example bits
D0	D6	1
D1	D0	1
D2	D1	0
D3	D5	0
D4	D4	1
D5	D3	0
D6	D7	0
D7	D2	1

If this does not make sense, consider that as long as the CPU writes a byte B to location A, and receives the same byte B when reading back from location A, it will operate correctly. The exact way that byte B gets stored in memory is unimportant. The same is true of the address bus, when writing to SRAM:

CPU address pins	Memory address pins	Example bits
A0	A3	1
A1	A2	0
A2	A1	1
A3	A6	0
A4	A5	1
A5	A4	0
A6	A9	0
A7	A8	0
A8	A7	1
A9	A10	0
A10	A0	0

This works great with SRAM, but there are issues with flash memory. Remember the programming algorithms in Section 7.6.2? Well the flash expects to be receiving particular byte patterns, which means specific bits on specific pins. If the system designer has scrambled the data bus, then the programmer has to descramble the flash command words, and addresses to suit. For example, using the above scrambling scheme, if flash memory expects a byte 0x55 on address 0x0AA then the programmer would need to write byte 0x93 to address 0x115 (as shown in the tables above).

The type of bus scrambling shown here is very common as a means to solve tricky PCB routing problems. However, be very careful with SDRAM... some address pins are dedicated as column addresses, and some as row addresses (refer to Section 7.6.3.3). Furthermore, some SDRAM pins have other special meanings: for SDRAM in particular, which is actually programmed through a write state machine within the SDRAM controller, this is similar to the flash memory programming algorithms, except that it is *not under the programmers' control*, and so cannot be descrambled in software.

7.13 Preventing Reverse Engineering

Since RE cannot be prevented per se, the issue becomes an economic one: how can we maximize the RE cost experienced by competitors at minimal additional cost to ourselves. For determining this, first, the description of embedded systems RE from Section 7.12 will be drawn upon, related to an embedded context and then classified. First, mitigation methods are rated based upon their implementation complexity and cost, plus the economic impact of their implementation upon a RE-based attacker. We will first classify *all* methods of interest to embedded systems designers, before narrowing in on those with particular relevance to computer architecture.

To begin the classification, RE mitigation techniques are divided into categories of *passive methods* which are fixed at design time and *active methods* of resisting RE during an attack. The former tend to be structural changes that are less inexpensive to implement than the latter. We will explore both in turn.

Cost multipliers to the reverse engineers due to RE protection, come about through three major factors:

- Increased labor cost incurred as a result of greater time taken to RE the system.

- Increased labor cost due to higher levels of RE expertise required.

- Increased cost spent on purchase of specialized equipment required for the RE process.

In some cases, there will also be an increased BOM cost, if extra components are required.

Following the RE process of Section 7.12.1, the first level of protection can be applied to the functionality assessment: RE stage A. In this case, restricting the release of service manuals and documentation can reduce the degree of information available to an RE team. Manufacturers should control, monitor, and ideally limit information inadvertently provided by employees, especially when posting online. This will undoubtedly increase the time and effort needed to RE a system.

Stage B, the physical structure analysis can be made marginally more difficult through the use of tamper-proof fittings for enclosures such as torx and custom screws shapes which would require purchase of specialist equipment. One-way screws and adhesively bonded enclosures work similarly. Fully potting the space around a PCB provides another level of protection. At minimal cost, the primary detraction to the use of these methods comes from any requirement for product serviceability, which would normally necessitate ease of access.

Wiring which is not color-coded may complicate the manufacturing and servicing process, but will cause even greater difficulty and delay to an RE team working on a heavily wired system.

Unusual, custom and anonymous parts complicate the RE of a systems BOM in stage C. However, passive devices (stage C.2) can easily be removed and tested in isolation. A missing silk screen causes some difficulty in manufacturing and servicing, but limits the information provided to the RE team for stages C.3, E.1, E.2, and F. However, by far the most effective method of preventing BOM RE is through the use of custom silicon (or silicon that is not available for sale to the RE team). Reverse engineers in stage C.1 confronted by a large unmarked IC surrounded by minimal passive components, no silk screen and with no further information would face a very difficult and expensive RE process indeed. The need to identify and/or replicate custom silicon adds significant expense to the RE process, but also at great up-front cost, so may only be economical for large production runs.

For best security JTAG (Section 7.9.3) and other debug ports should be eliminated from semi-custom silicon, and not routed from standard parts to connectors or test pads, and certainly not labeled *TDI, TDO, TMS, TCK*. For device packages with exposed pins, these can still easily be accessed, so BGA (ball grid array) devices are preferred. But even for

BGA devices, unrouted JTAG pins can often be accessed by controlled depth drilling through the PCB from the opposite side, meaning that back-to-back BGA placement is most secure (such as a BGA processor on one side of a PCB with a BGA flash memory device directly underneath on the other side). The disadvantage here is that manufacturing cost increases by having double-sided component placement. Double-sided BGA placement is yet one step more expensive, but still there is no guarantee since it is possible, but extremely difficult, to remove a BGA device, reform the solder balls, and then refit this into a carrier which is soldered to the PCB. The intermediate signals through the carrier can then be made available for analysis.

Back-to-back BGA packaging generally necessitates blind and/or buried vias, which can increase PCB manufacturing costs (rule of thumb: by 10%), complicate the layout process, and significantly impact on any hardware debugging or modifications needed. It does, however, result in a very compact PCB which might itself be a useful product feature. Similarly, the number of PCB layers would often need to increase to accommodate back-to-back placement, therefore also increasing the RE cost to perform delamination and layer-by-layer analysis. Use of X-ray analysis to reveal layout details for stages E.2 and E.3 is difficult in multilayer PCB designs, but it can be complicated further by the useful practice of filling all available space on all layers with power plane fills. These can even be crosshatched on internal layers to mask individual tracking details on other layers on an X-ray photograph.

Electrical connectivity, stage E.2, can be difficult to ascertain when devices are operated in an unusual fashion such as jumbled address and data buses. Wiring unused pins to unused pins can add nothing to manufacturing cost, but complicate the RE process.

7.13.1 Passive Obfuscation of Stored Programs

There is much that can be done structurally to obfuscate the stored code in the flash memory of an embedded system, thus complicating RE stages G.1 and G.2. We will not consider that further since it is an active research area; however, there are some architectural aspects we can work on.

First, and as mentioned previously, the gaps between code sections (of unerased flash) can very easily be filled with random numbers or dummy code such that detection of separate memory areas is nontrivial. Apart from initial boot code, other sections of flash can also be encrypted if execute-from-flash is not required. This will cause difficulties in analyzing an image of flash contents; however, the unencrypted boot code may well be small and simple enough to trace and disassemble, revealing an unencrypted entry point to the system, hence the security of such encryption is questionable.

Scattering code, data, and configuration sections throughout flash memory will cause some programming difficulty but is primarily another means of protecting against stored program analysis. If an FPGA image is stored in flash, simple methods of obfuscating this apart from encryption include performing an exclusive OR operation on every data byte with some other area of flash, or storing a custom compressed FPGA image (not gzip, zip, or similar which have identifiable signatures).

A summary of various of the discussed RE mitigation methods are shown in Table 7.4, where the design cost, effectiveness at increasing RE cost, and manufacturing impact are identified using a five-point subjective scaling for the example embedded system.

	Design cost	RE cost	Manufacturing cost
Tamper-proof screws	2	0	1
Bonded case	1	1	1
Potting	1	1	2
No silk screen	1	1	1
Erased component identifiers	1	1	2
Use of BGA packages	1	3	3
Inner layer routing only	2	2	3
Blind or buried vias	2	2	4
Bus signal jumbling	1	1	0
ASIC signal router	5	3	2
FPGA signal router	2	2	2
No debug port	1	1	2
Random padding of unused memory	2	2	0

TABLE 7.4 Passive methods of increasing hardware reverse engineering cost rated on several criteria, 5 = most, 0 = least.

7.13.2 Programmable Logic Families

SRAM-based FPGAs normally require a configuration bitstream to be provided from an external device—such as a serial flash configurator, or provided by a microprocessor, such as the case in the example system. Since this bitstream can be accessed physically with little difficulty, this firmware can always be copied by tapping off and replicating the bitstream.

EEPROM-based programmable logic devices (PLDs), the otherwise obsolete EPROM versions, and newer flash-based products, are more secure since the configuration program resides internally and does not need to be transferred to the device following reset. Note that some flash-containing devices actually encapsulate two silicon dies in one chip—a memory die and a logic die, and thus are less secure since the configuration bitstream can always be tapped once the encapsulation is removed. In general, devices that are configured immediately following reset are those which contain non-volatile memory cells distributed around the silicon, and those that become configured several milliseconds after exiting reset are those in which a configuration bitstream may be accessible. In either case, many devices, including those from Altera and Xilinx, provide security settings which may prevent readout of program bitstream from a configured device. Use of this feature is highly recommended.

In regular cell-structure devices, including the mask-programmed gate array (MPGA), the location of memory configuration elements is known, determined by the manufacturer for all devices in that class. Using the methods of Section 7.12.2.2, this configuration data, and thus the original "program" can be retrieved—although this requires sophisticated technology.

A full-custom ASIC can be reverse-engineered by analyzing silicon layer-by-layer (similar to the PCB delamination, but with layers revealed through careful grinding), but even this technique can be complicated through countermeasures such as inserting mesh overlay layers. Antifuse FPGAs are generally considered to be the most secure of the standard programmable logic devices, due to the location of fuses buried deep below layers of silicon routing, rather than being exposed near the surface.

It is not impossible to RE systems incorporating ASICs or secured antifuse FPGAs, but this does require significant levels of expertise, requires the use of expensive specialized equipment, and is time consuming.

7.13.3 Active RE Mitigation

Many of the passive electronic methods given in Section 7.13.1 have active variants. Electrical connectivity can be confused by using spare inputs and outputs from processors to route signals which are not timing critical but which are functionally critical.

While jumbled address and data buses are more difficult to reverse engineer, dynamically jumbled buses provide one further level of complication, but add the cost of incorporating active devices to perform the jumbling/de-jumbling.

ASICs are probably the ultimate tool in mitigating RE attempts, but even the humble FPGA can be quite effective. In either case, IP cores implemented within logic are not easy to identify or isolate, and if using external memory, can access this in literally any way necessary—whether linearly, nonlinearly, or using some form of substitution or encryption. A CPU core which is completely custom, and without any public documentation, adds another layer of security through not revealing any details of its instruction set architecture. Furthermore, the instruction set could be deliberately changed in every implementation among several product versions to prevent repeated RE of the cores program. This would be an inexpensive software/firmware-only protection.

7.13.4 Active RE Mitigation Classification

The basic forms of RE mitigation can be subdivided in two dimensions. First is into methods of active confusion, hiding, or obfuscation, and the second is into temporal or spatial methods of achieving this. Any real system may employ a combination of these methods to maximum effect.

Information hiding uses existing resources in ways that attempt to conceal information from an attacker. This may involve combining code and data through concealing operating software within data arrays such as start-up boot images, or by sharing information across data reads in a nonobvious fashion. It may also include operating electronics at marginal voltage levels, relying upon unusual signaling or data handling schemes.

Obfuscation, normally a passive method (such as swapping the names of labels and functions within code, or jumbling the PCB silk screen annotations), can also be active in arrangements such as those that change bus connectivity, or device pin usage (for example multiplexing an interrupt input pin with a signal output function). This again uses existing resources in ways specifically designed to complicate the RE process by misdirecting the RE team.

Protection by confusion adds resources specifically to deliberately mislead or confuse an attacking RE team. This could include large pseudo-random data transfers, out-of-order code reading, and so on. There may be signal interconnections that employ current signaling but overlay a randomly modulated voltage signal upon the wire, or perhaps a meaningful signal driving a redundant signal wire. In a dynamic sense, this may include mode changes on tamper-detection or even a more extreme response of device erasure on tamper detection.

	Fixed timing	Dynamic timing
Information hiding	0	2
Obfuscation	1	3
Deliberate confusion	4	5

TABLE 7.5 Relative strength of active protection methods, 5 = most, 0 = least.

Spatial methods are those which operate at a placement or connectivity level, such as scrambling bus order depending upon memory address, turning on or off signal path routing devices in a nonobvious fashion or similar.

Temporal methods confuse through altering the sequence and/or timing of events. One example would be a boot loader that deliberately executes only a subset of fetched instructions. Another would be a memory management device able to prefetch code pages from memory and access these in nonlinear fashion, especially if these are out-of-sequence with respect to device operation.

The combination of these classifications is shown in Table 7.5 where their relative strength is categorized.

In terms of costs, dynamic methods are likely to cost more to develop, debug and test, and also increase both manufacturing and probably servicing costs, than fixed timing methods. Both information hiding and obfuscation could well be of similar development cost—mostly adding to NRE; however, deliberate confusion methods will undoubtedly cost more to develop than either hiding or obfuscation, and will add to manufacturing cost.

What is clear is that custom silicon, implementing active confusion and protection means, provides the greatest degree of protection. A DMC concerned by the costs involved in creating a full-custom ASIC for security purposes, could develop a generic security ASIC which can be used across a range of products. For the reverse engineers, the active protection methods in each category, particularly the dynamic timing cases, will require employing a highly skilled and flexible RE team. This team will require access to specialized equipment. For example, marginally operating timing signals may require analysis by high-speed digital oscilloscopes with very low capacitance active probes that do not load the signal lines, or even use of a superconducting quantum interference device (SQUID). A multichannel vector signal analyzer may be required for some of the more unusual signaling schemes. This kind of equipment is extremely costly.

7.14 Soft Core Processors

A soft core (or soft processor) is a CPU design that is written in a logic description language that allows it to be synthesized within a programmable logic device. Typically, a high-level language such as Verilog or VHDL[8] is used, and the end product synthesized on a field programmable gate array (FPGA).

---

[8]VHDL stands for "VHSIC hardware description language," where VHSIC refers to a "very high-speed integrated circuit."

This differs from the position of most processor manufacturers, who tend to create low-level designs that are specific to the semiconductor manufacturing process of their semiconductor fabrication partners. This happens mainly due to the need to squeeze maximum performance from the silicon that is being worked on. Sometimes, there are both custom, and soft core designs available for a particular processor, for example, the ARM. In such cases, the soft core design will usually provide inferior performance (slower, higher power), but be more flexible in where it can be used.

There are very many soft core processors available, many of them freely available,[9] although few could compare in efficiency, speed, or cost when implemented in FPGA, to dedicated microprocessors.

Other possibilities are the use of a commercial core—the main FPGA vendors each have such cores—and designing your own core. We will consider the anatomy of soft cores, then each of the three main possibilities of obtaining a core.

7.14.1 Microprocessors Are More Than Cores

A soft core processor, implemented on an FPGA, is a block of logic that can operate as a CPU. At its simplest, this block of logic, when reset and fed with a clock, will load in data, and process it as specified by a program. The program could reside internally within the FPGA, or could reside in external memory, either RAM or flash, as in most embedded systems.

This arrangement is fine; however, microprocessors are more than just cores. Refer back to the features available in the popular Samsung S3C2410 ARM-based microprocessor, discussed in Section 7.2. A long list of internal features and peripherals was presented, including the following more major ones:

- 16-KiB instruction and 16-KiB data cache plus internal MMU
- Memory controller for external SDRAM
- Color LCD controller
- Many serial ports, UARTs, SPI, IrDA, USB, IIC, etc.
- SD (secure digital) and MMC (multimedia card) interfaces
- An eight-channel 10-bit ADC (analog-to-digital converter) and touch-screen interface
- Real-time clock with calendar function

Clearly the processor core itself (which incidentally is the one item that was not listed in Samsung's own documentation) makes up only a small part of the IC named an S3C2410 which is purchased and included in an embedded system.

To clarify further, if an engineer somehow managed to obtain an ARM processor core written in a high-level HDL, and loaded this into an FPGA, she would not have a fully functioning microprocessor. Furthermore, this would be unlikely to operate at anything

---

[9] Refer to the project collection in `www.opencores.org` for free processor and other "IP" cores, where IP refers to "intellectual property."

approaching the S3C2410's 200 MHz in an FPGA (even in an FPGA advertised as supporting a 1-GHz clock speed).

The extra effort required to implement all of the other peripherals and interfaces on FPGA would be excessive and then the final result would be slower, more power hungry and far more expensive than an off-the-shelf ARM.

So given such disadvantages, why would anyone consider using a soft core?

7.14.2 The Advantages of Soft Core Processors

There are probably hundreds of millions of systems worldwide powered by soft cores, and although that is less than possibly hundreds of billions of ARM devices that have been built, there must be some good reasons why soft cores are sometimes selected. Let us consider a few of those good reasons under headings of performance, availability, and efficiency.

7.14.2.1 Performance

Performance should clearly be on the side of standard microprocessors, since we mentioned that soft cores are usually slower than dedicated devices. While that is true, remember that there are some performance issues that are more important than clock speed:

- Parallel systems allow multiple processors, or processor cores, to be implemented and run in parallel. It is quite easy to include several, or even many, soft cores inside a single CPU and thus create a parallel system. As always, learning how to use these multiple cores effectively is a task not to be overlooked.

- The CISC approach was known for creating custom instructions required by programmers. RISC, by contrast, eliminates the more complex or less common instructions and concentrates on making the most common instructions faster (so that the complex CISC instructions can be performed by multiple simple RISC instructions). However, in an embedded system, where code is often small and unchanging, it is quite possible that a different set of instructions would be chosen to be implemented. For example, in a system performing many division calculations and no logic operations, the optimal RISC processor may have a divider, but very few logic instructions. Where code is known, and fixed in advance, there is something to be said for custom-designing an instruction set specifically for the purpose of executing that code quickly.

- Even where the instruction set is not modified to suit a particular piece of code, it is always possible to add on a dedicated functional unit, or coprocessor, to a given core inside an FPGA. In the example above, we could opt to add a division unit to a standard core. Off-the-shelf parts cannot be modified in this way, although some do have external coprocessor interfaces.

- Soft cores supplied in VHDL or Verilog usually do not contain sophisticated buses and are without memory (sometimes even without a cache). The designer who uses these in an FPGA thus has to build buses and memory around them. While this fact appears to be a disadvantage, it is quite possible to use this as an advantage by creating a dedicated bus that matches the application. By contrast, an off-the-shelf standard part may choose to implement a bus scheme that does not match the application perfectly.

7.14.2.2   *Availability*

This has two meanings in the context of a soft core. The first relates to how easy it is to procure and use a device, and the second relates to ensuring that a processor works correctly when needed. We will cover both meanings:

- It is the bane of product designers (including the author, in an earlier life) to standardize on a CPU in their hardware design, to work toward a product release, and then days from launch to receive a notification from the CPU vendor that the device they are using is now EOL (end of life). This requires a very fundamental redesign of both software and hardware. While such a situation is unlikely to occur for designers selling mass-market products, it is all too common for small and medium embedded systems companies. With this in mind, consider the attraction of having *your own* CPU design: It is yours to keep forever and can never be dropped by a cost-cutting semiconductor vendor. You can program this, reuse code, reuse hardware, extend, and modify at will, in as many designs as you wish. Although it is synthesized in an FPGA, and the specific FPGA may go EOL, you can simply switch to another FPGA and the same code, same processor, will run there—perhaps even a little quicker.

- Similar issues are felt by designers in countries outside Europe and North America. New CPUs take time to become available in those markets, and stocks are usually slow or difficult to access. Again, for a company wishing to purchase several tens of thousands of devices, this is usually not a problem, but for small and medium embedded companies, it can be. In Singapore, for example, it was almost impossible for the author to purchase anything less than about 100 devices, something which effectively discourages prototyping. Thankfully, the FPGA vendors are a little more considerate to smaller companies and individuals.

- Availability in an electronic system means ensuring that the system is working correctly, and is working when you need it. Good design is the key to ensuring reliability, but sometimes, in order to ensure that a CPU is working and available, it is necessary to replicate it. Thus two CPUs can be better than one. In fact, three are better than two, and so on. A soft core can be replicated and parallelized as often as necessary, consuming just FPGA resources and power when turned on. A replicated dedicated processor means, for a start, twice as many ICs, also twice the cost.

7.14.2.3   *Efficiency*

Efficiency can be measured in terms of factors like power, cost, space, and so on. It turns out that there are arguments for each of these for soft cores; however, all relate to the same basic reasoning:

- The impressive list of S3C2410 features in Section 7.14.1 is hard for any designer to replicate in a custom soft core design. However, are all of these features really necessary? The answer is "yes" when designing a one-size-fits-all solution that is to be used by almost everyone. However, in individual cases only a small subset of these features would probably be necessary and therefore the answer is probably a "no." Soft cores only tend to include those features, interfaces, and peripherals that are absolutely necessary. They do not waste silicon space (or FPGA cells) on unused functionality in the way that a standard part may well do, and because of this will, at times, be more efficient than their standard cousins.

- Glue logic is the name given to those devices connecting microprocessors and other parts together. This includes items like inverters, AND gates, and so on. Sometimes, a large requirement for glue logic would be fulfilled by using a small FPGA. Given that glue logic is so ubiquitous, and is almost everywhere, replacing a standard microprocessor with an FPGA-implemented soft core can also allow the designer to fold all of the glue logic into the same FPGA. Sometimes the result will be reduced PCB space, lower manufacturing cost and so on over the dedicated CPU design.

7.14.2.4 Human Factors

The human factor is often overlooked by engineers and computer scientists[10]; however, there is as much importance in considering the human as there is in any technological reason. One can witness how upset and irrational some engineers can be when faced with the elimination of their ideas in a group design session. Some human factor reasons for considering soft cores might include the following:

- It's fun to develop your own computer! Well-motivated design engineers are efficient and hardworking. Motivation comes, in part, from doing something interesting, and building a custom soft core is something most engineers consider very interesting—something that most managers might not realize.

- Ownership of a design, while running the risk of the irrational behavior mentioned above, is another great motivator for engineers, and aids in the pursuit of design perfection.

- When embarking on a new embedded design project, there is usually a time to consider which embedded processor should power the new project. The "degree of fit" will be determined of various devices to the design requirements, and the best fit chosen, at least in theory (this process may well trigger more of that irrational behavior as various parties push their own agendas). However, something that is less often considered is the "learning curve" required to retrain engineers to use a new microprocessor. Sometimes the need to switch to a totally new device will incur months of delay while designers familiarize themselves with new devices, or may lengthen the design process through unanticipated beginners mistakes. It is often better to use a device that the team is familiar with, but is a less optimal fit. The use of soft cores can help here in that once a team is familiar with that soft core, it can be used in many successive designs. Small changes to the FPGA-implemented peripherals, functional units, and co-processors can be made to ensure that the core remains an optimal choice for new projects, and yet does not need to involve the team in lengthy retraining activities.

---

[10] Perhaps because some engineers and scientists rarely encounter humans.

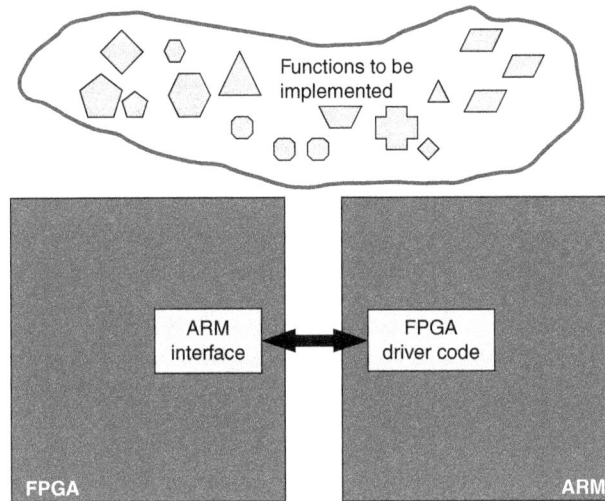

FIGURE 7.29   The process of designing an embedded system that contains both a CPU and an FPGA involves identifying the tasks to be performed, and allocating these to one of the two processing units. Of course, this presupposes that a working CPU-FPGA interface exists.

7.15   Hardware Software Codesign

Hardware software codesign is the term given to the process of designing a system that contains both hardware and software. It is particularly relevant to embedded systems, since such systems normally entail custom hardware and custom software.

When writing software for a desktop PC, programmers will normally expect that the hardware is error free and functions correctly. When designing a new PC, designers are able to run diagnostic software which has been proven correct and error free on working hardware (such as on the previous generation of PCs).

In embedded systems, the potential problem area is that both the hardware and the software are usually developed together—neither can be proven error free without the other, and so the process of debugging and making a working system is mired in problems that may lie in either domain (or even cross domains).[11]

Given a system containing an FPGA and a CPU, such as that shown in Figure 7.29, an embedded systems designer, knowing the requirements for the system being designed, must decide how to implement each of those requirements. Some will be implemented in software, some in hardware, and many will require a little of both. In general, software implementations are more flexible, easier to debug and change and easier to add features, whereas hardware implementation are higher performance and potentially lower power.

---

[11] There is a great tradition among hardware designers to blame programmers when something does not work correctly.
There is a great tradition among programmers to "blame the hardware" when code crashes.
This makes for an interesting development environment but is not particularly productive from a management perspective.

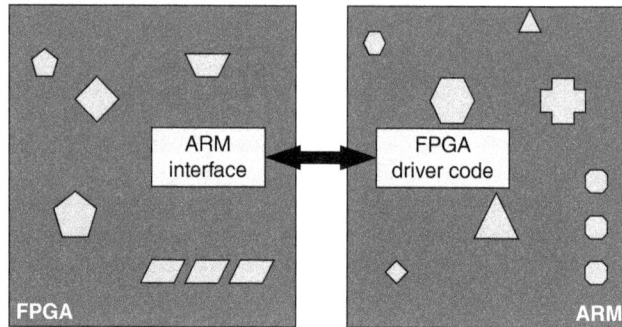

FIGURE 7.30 Tasks allocated to either a CPU or an FPGA as part of the hardware software partitioning step in embedded system design.

Some tasks are more naturally FPGA-oriented (e.g., bit-level manipulation, serial processing or parallelism) and some suited for high-level software on a CPU (e.g., control software, high-level protocols, textual manipulation, and so on). Knowing the size of FPGA, and MIPS/memory constraints in the processor will assist the designer in the partitioning process. There are many other issues that must be considered, and usually there will be an element of trade-off required. These issues include questions such as "Who will do the coding?" "How maintainable does the code need to be?" "Will the system need to be upgraded later?"

One particular area of concern could be in the connection between FPGA and CPU. This connection will have both bandwidth and latency constraints: It can only support a certain amount of data flow, and will naturally involve a small delay in message passing (an important consideration in real-time systems). Also, it would be normal for one device (usually the CPU) to be as master and the other a slave. Messages and data are initiated from the master, and so latency may well be different for messages in the two directions. Bandwidth could differ too. Most probably the two devices are not clock synchronous, and so any data that is streaming between the two may have to be buffered, possibly on both sides—adding to the data transfer latency.

The situation is exacerbated also when an FPGA (field programmable gate array) becomes available that could contain a soft core processor. This means a further decision needs to be made regarding whether tasks will be implemented in the CPU, in the FPGA as logic functions/state machine, or in the FPGA executed by a sort core processor.

Despite the difficulties, a partitioned design will eventually be agreed upon, such as that shown in Figure 7.30. Separate specifications for this system, including interface specifications, would then be drawn up and handed to the software team, and to the hardware (or firmware) team who would then go away and implement their parts of the system.

Some time later, *integration* would start—the process of fitting together the hardware and software designs, and (typically) discovering that the system does not work. At this point, the two teams tend to apportion some element of blame, before setting out on the long and hard process of getting their domains to "talk" to each other and work together.

Unfortunately, even when the system does finally work, it will seldom be an optimal solution because there is just too much human subjectivity involved in the original partitioning process, and the subsequent implementation.

Hardware software codesign has emerged relatively recently as a response to these design difficulties in systems that involve both hardware and software design. Codesign methodologies are implemented as a type of CAD tools, with the aim of simplifying the design process (to reduce time, money, and errors), optimizing the partitioning between hardware and software, and easing the integration process.

Hardware software codesign involves the following stages, assuming we are targeting a mixed FPGA/CPU system:

1. *Modelling*—in which some specification of what the system must do is created in a machine-readable format. This might be a formal design language (which we, thankfully, will not consider further here), or a simple program in "C" or MATLAB that emulates the output of the system to given input. This model will be used later to verify that the new system works correctly.

2. *Partitioning*—as mentioned above, and probably best performed by a human aided with information describing the system. Sometimes it is easy to split a system into different blocks, but usually there is at least some difficulty, and may require the original model to be rewritten slightly.

3. *Co-synthesis*—uses CAD tools to create a model of three items: the FPGA code, the C programming language code, and the interface between the two. FPGA code is synthesized in FPGA design tools, C code is compiled, and loaded into a processor emulator, and the interface between the two is often file-based.

4. *Co-simulation*—means running these three items together within the design tools. Ideally this would be in real time, but often it is thousands of times slower than the real hardware; however, it is bit-level accurate to an actual hardware implementation.

5. *Verification*—this means comparing the cosimulated system to the original model for veracity.

There are likely to be several iterations in this process: as errors are found (or more likely as opportunities for greater optimization are identified), slight changes to partitioning and design will be possible. A flowchart showing these stages is shown in Figure 7.31 where the importance of the system model is clear through the verification process that takes place at every stage in the design process.

The important fact is that *everything gets simulated together*: the hardware (usually FPGA), software, and interface between them can be developed using the design tools, and tested thoroughly in simulation. Problems can be identified and rectified early. When the system is finally working as modeled, it can be constructed in hardware and tested. At this point, it is hoped that the software and hardware will work perfectly together, so that the programmers and hardware developers can celebrate together.

7.16   Off-the-Shelf Cores

Earlier in this chapter, we have seen that many free processing cores are available for synthesis within FPGA. It is also possible to use commercial cores from several vendors, not least the main FPGA manufacturers, so let us for moment consider these offerings as they exist at the time of writing:

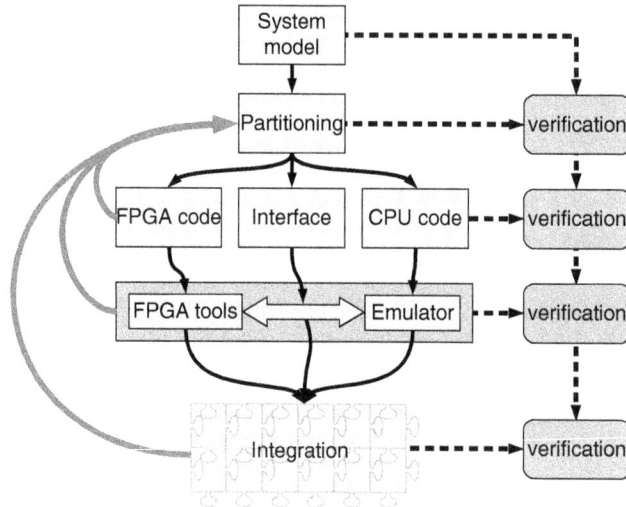

FIGURE 7.31 An illustration of the hardware software codesign process showing the sequential development flow, with verification performed at each step, and iteration back to previous steps in the case of errors detected.

- The *Altera Nios II* is a 32-bit RISC processor optimized for Altera FPGAs. In many ways it is seen as a response to the widely popular ARM processor, and builds upon the original Nios core. A single instruction set can be executed by the Nios II in one of many sized configurations ranging from 700 logic elements (LEs) upward. The largest, and fastest configuration, with a six-stage pipeline, separate data and instruction caches, dedicated multiplier, branch predictor unit, and even optional divider and MMU, is quite powerful.

  Most usefully, from an embedded computer architects' perspective, the core allows up to 256 custom instructions to access dedicated blocks of custom logic, and for dedicated hardware accelerates to be included within the pipeline. Nios II is supported by a variety of operating systems, including embedded Linux.

- The *Xilinx MicroBlaze* is also a 32-bit RISC processor, for use within Xilinx devices. It can have either a three- or five-stage pipeline and has many configurable options in terms of buses, functional units, MMU, and so on. The MicroBlaze has a Harvard architecture with configurable cache sizes. Hardware division, fast multiply and an IEEE754-compatible FPU are available, and like Nios II, MicroBlaze is also supported by several operating systems, again including embedded Linux.

- *Actel* joined the soft core party a little later, initially not having a response to the two larger vendors, but finally signed a significant agreement with *ARM* to ship an amazing *ARM7*-based soft core. The advantage of this approach is the very wide range of support and existing code base available for the ARM7; however, Actel is a much smaller vendor than either Altera or Xilinx, and target a different segment of the FPGA market. So while ARM is clear winner in off-the-shelf microprocessors, only time will tell whether this success is repeated within the FPGA soft core market.

- *Lattice*, the final contender in this market, also developed and released a 32-bit soft core RISC processor. The *LatticeMico32* uses less than 2000 look-up tables (LUTs) in a Lattice FPGA and, although not quite as configurable as the Xilinx and Altera offerings, nor quite as powerful, is small and fast. Various peripherals such as UART and bus interfaces are available and configurable. Furthermore, it is completely open, meaning that it can be used and modified anywhere—it does not need to be licensed when used and sold within a design.

Apart from these cores, there are a few companies specializing in the IP-cores market, selling their cores for use on any FPGA. Even ARM have released a small soft core ARM Cortex device. Clearly this field is active and of growing importance to embedded systems.

A final note of importance is to remember the fact that these cores do not exist in isolation. Yes, we have seen that they require synthesizing with an FPGA and require external buses, peripherals such as memory, clock signals, and other facilities in order to operate. However, they also need programs.

Software development for soft core processors is an integral part of ensuring that they can operate correctly within a design. Therefore, important issues to resolve are whether a toolchain is available (which is used to develop software), whether an operating system (OS) is available for that processor, and what types of debug tools are available.

A standard embedded toolchain, such as the GNU toolchain, incorporates several elements which include a C (and C++) compiler, assembler, and linker (e.g., gcc). There is often a need for library management tools, object file tools, a stripper (to remove debugging comments from within an object file in order to reduce its size), analytical tools, and so on. A debugger, such as GDB, is highly recommended for debugging, since it can execute, single step, breakpoint, watch point, and monitor running code. The GNU toolchain also contains software to allow running code to be profiled (i.e., to determine the amount of CPU time spent within each function, the program trace, and the number of looks executed).

An operating system, particularly a real-time operating system (RTOS), is often required in many developments. Unfortunately, it can be difficult writing, or porting, an OS to a new processor, and this is one major argument in favor of choosing a core that is already supported by a good OS such as embedded Linux. Despite this there are reasons to custom design a soft core, for example, when only small items of code, such as hand-written assembly language are used.

7.17 Summary

This chapter has considered many of the practical aspects of computing, such as memory technology, on-chip peripherals, clocking strategies, and the provision of reset signals. Embedded systems, in particular, often suffer from memory shortages, which can be alleviated through the use of memory pages and overlays (and we also examined the memory structure of a typical embedded system using the popular embedded Linux OS). Watchdog timers were described, as useful means of ensuring overall reliability in real-time and embedded systems, and for this aim we also discussed error detection and correction.

As CPUs have become faster and more complex over the years, manufacturing and development difficulties abound due to this complexity. This has highlighted the need for test and verification in such systems—so we split this into methods of provision during IC manufacture, system manufacture, and at runtime.

The issue of reverse engineering was then surveyed. This is a particularly relevant issue in many embedded systems, especially those within consumer devices, and as such we looked at how nefarious reverse engineering is performed, and with this in mind, surveyed methods to prevent this. Finally, we looked at the trade-offs between designs that use an off-the-shelf CPU or system-on-chip, and those that create bespoke solutions, perhaps using a soft core. We also reflected that while CPUs such as the ARM are dominant in embedded systems, due to their efficiency, flexibility and ease of use, there are occasions when a custom-designed soft core can provide important advantages.

7.18   Problems

7.1   Identify four factors that would argue for the use of system-on-chip (SoC) processors in an embedded system.

7.2   List the minimum set of control register settings necessary to implement programmable I/O pins on a microcontroller given that these are required to support the following functionality:

- Can be configured either as general-purpose input/output (GPIO) or as a dedicated output from an in-built peripheral device such as a UART.
- When in GPIO mode, can be configured as either an input or an output.
- Each pin can be individually read from, and written to.

7.3   Indicate whether you would expect a single-chip microcontroller or a quad-core high-speed server processor to devote a greater proportion of its silicon area to memory. Justify your answer by noting the primary use of that area in both machines.

7.4   Note a few of the approaches that semiconductor designers have taken to reducing propagation delay in CPUs over the past two or three decades.

7.5   What changes can be made to a computer system clocking strategy (or to the clock itself) to reduce the amount of electromagnetic interference (EMI) generated by that system?

7.6   What external devices, located close to the power pins of a CPU, can reduce the amount of EMI generated? Explain the mechanism that causes the EMI, and how these devices can reduce it.

7.7   Identify the most appropriate memory technologies, from those listed, for the following applications (a to d):

a) An MP3 player needs to access audio data from 8 Gibyte memory at a rate up to 350 Kibits per second. The data (your songs) should remain in memory even when the power is turned off.

b) The program memory within a small and simple embedded system is designed to do one thing, and one thing only. The manufacturer will build millions of these devices, which have no provision for reprogramming.

c) The 256-Mibyte system memory within an ARM9 embedded system, built to run an advanced embedded OS such as embedded Linux, in a PDA.

d) The 16-Mibyte non-volatile program memory in the above system—assuming that many of the OS routines remain in flash memory, and are executed directly from there.

e) A 4-Kibyte run-time memory to be connected to a medium-size microcontroller in a small embedded system.

The set of memory technologies (one to be used for each application) is as follows:

- Serial flash
- Parallel flash
- SDRAM
- SRAM
- ROM

7.8   Note seven common functions that can be found in an embedded system boot-loader such as u-Boot.

7.9   If a typical embedded system CPU, implemented in a BGA package, is mounted on the PCB of a prototype embedded system, and the designer suspects that a soldering fault is preventing the system from operating correctly, identify two methods by which the potential system problems can be identified.

7.10   A byte $0xF3$ is to be transmitted over a noisy wireless channel as two nibbles, each encoded using Hamming (7, 4). Referring to the method shown in Box 7.7, identify the two 7-bit transmit words in hexadecimal.

7.11   Repeat the Hamming encoding of question 7.10, this time transmitting byte $0xB7$ using the method of Box 7.8.

7.12   Identify the three main reasons why, although it is sometimes necessary to incorporate reverse engineering protection in an embedded system, it may lead to slightly reduced profitability to the manufacturer.

7.13   In what ways would a working JTAG connection to the CPU in an embedded system be usable by a reverse engineering team trying to determine:

a) The identity of that CPU.

b) Circuit connectivity and system schematic.

c) The content of system non-volatile (flash) memory.

7.14   Why do so many SoC microprocessors have 32.768 kHz crystals connected to them?

7.15   What is clock jitter, and how does this influence the determination of maximum clock speed that a processor is capable of?

7.16   If a byte `0xa7` is programmed to one location in parallel flash memory, and later another byte `0x9a` is programmed to the same location (without it being erased in between), what value would the location then contain?

7.17   EPROM memory devices have a small glass window which can be used to expose the silicon die to ultraviolet light in order to erase the memory array. Flash memory devices (and EEPROM), by contrast, can erase their memory electronically. Identify two major advantages that flash memory technology offers over the EPROM.

7.18   Imagine you are leading a small design team for a new embedded product: The hardware is ready, and the software engineers are putting the finishing touches to the system code. There is a huge amount of serial flash memory in the system, but only a small amount of SRAM available. Just weeks before product launch, the software team reveals that their runtime code cannot fit within the SRAM, and there is no way of reducing the code size. Without changing the hardware, suggest a method of memory handling that will provide a way around this problem.

7.19   A JTAG scanchain may be several hundred bits long. This chain can be serially clocked into a CPUs JTAG scan path to change the device behavior, or clocked out to read the device state. What are the meanings of some of the bit positions (i.e., what behavior can they change, and what state can they determine)?

7.20   How can triple module redundancy be used to determine the correct output of a calculation? Illustrate your answer by considering three supposedly identical blocks in a malfunctioning system that output bytes `0xB9`, `0x33`, and `0x2B`, respectively. If these were wired to a bitwise majority voter, what would the final corrected output byte from the system be?

CHAPTER 8

Programming

We know that a computer is, at heart, simply a number-crunching unit. Everything it does can be summed up by the word "processing," which really means handling numbers; moving them around and changing them. Those numbers might be pixels of different colors in an image, samples of sound in an audio recording, wind speeds in a meteorological map, share prices, Internet protocol addresses, characters on a webpage, the text of an email, or any other quantity within the digital world. Programming means "getting the computer to do what you want" with those numbers.

In this book so far, we have looked at how computers work in great detail. We have seen memory maps, the control unit, the ALU, and even discussed how computers boot once the power is turned on. However, many people today have an experience of a computer which is quite different from what we have described. Those people will power up their notebook computer or turn on their smartphone and be presented by a graphical user interface (GUI) that lets them run an Internet browser, listen to music, Skype their friends, update a spreadsheet, or edit photos. It basically just works—and works well.

As we will see soon in Chapter 9, a lot of this functionality is thanks to an operating system (OS). Booting is the process of powering up a CPU and getting it to run software. This was described briefly in Section 7.8.1, and will be discussed much more thoroughly when we get to OS booting in Section 9.5. In a single-sentence overview of the process, when the computer powers on, it executes the software located at a special place in memory—in ARM processors this *boot vector* is usually address 0x0000 0000—and the software found there is usually a bootloader (or BIOS on bigger machines). A chain of programs may then get run, eventually launching the operating system. On many of our devices the OS will in turn launch a window manager or GUI. Once this is running, the user may then choose to launch some user programs—for example, by clicking on a program icon, specifying a program name in a command window or simply speaking a command.

We will see in Chapter 9 that a computer does not need to have an operating system (many smaller embedded systems do not have any OS). These systems turn on and immediately run the program that their embedded programmer has developed, with a single program often handling everything. To make that work, the embedded developer simply makes sure that the wanted program is located so that its start is placed at the boot vector.[1]

---

[1] To be honest, it can be a little more complicated than described in reality: The user code will need to be relocatable, and must not rely on any OS functionality or libraries, so it is not an identical program to one that would be launched from within the OS. But the idea is simple.

In this chapter, we will be looking first at how a program is written, starting with a recap of how the program is run, and what is happening inside the computer to make that possible.

What we do *not* discuss is the important detail of program design, implementation and programming, the structure of programs, debugging, verification, and validation: These are all topics for software engineering or software development textbooks. However, we will look at compiled and interpreted languages, then end with a discussion of the UNIX programming model and the ubiquitous and powerful UNIX shell.

8.1 Running a Program

A compiled and linked executable is a ready-to-run piece of code. On larger computers which have an OS, executables normally reside on secondary or non-volatile storage like hard disc until needed. When they are needed—for example, when the user clicks on the icon representing that program, or executes the program by name in a command window—the executable code block will be moved to RAM and run from there.

Within embedded systems, or when there is no OS present, the code block would usually be moved into RAM during the boot process (by the bootloader), and then executed by the bootloader. However, in the smallest of space-efficient embedded systems, code is sometimes written that can be stored in fast flash memory, and executed directly from flash memory (called "execute-in-place"). In fact, this is what a first-stage bootloader is; a small block of code in non-volatile memory (at the boot vector) that gets executed when the CPU powers on. The first thing it does is usually to copy a second stage bootloader from non-volatile memory into RAM and then execute that.

8.1.1 What Does Executing Mean?

We know from discussions in Chapter 3 how a CPU executes code: It fetches the machine code instruction located in memory at the current program counter (PC) address, and then it does whatever that instruction tells it to do (i.e., it executes the instruction). After that, unless the current instruction tells it differently, the CPU will increment the PC to make it "point at" the next instruction in memory. The CPU then fetches that next instruction, and executes it. Programs are therefore designed to be executed step by step and line by line. Except that some instructions can cause the PC to change—these are branch, call or jump instructions (different processors have different terminologies). What they do is set the PC to some different value, overriding the default "fetch next instruction" behavior.

Figure 8.1 helps to explain how those instructions work, and how they enable us to execute different programs on a computer. Since the figure contains a lot of information, we will begin by explaining it. At heart, it shows a segment of an ARM memory map, expect that it is reversed from the way we usually present ARM memory maps. In this case address 0 is on top (usually we would draw a map with address 0 at the bottom). It uses ARM instructions for clarity, although many modern ARM processors may do things slightly differently in practice. The figure shows a block of memory spanning the address range 0x0000 0000 to 0x0002 012c (although we skip a big chunk in the middle for clarity). The addresses are written along the left, and the 32-bit instructions shown

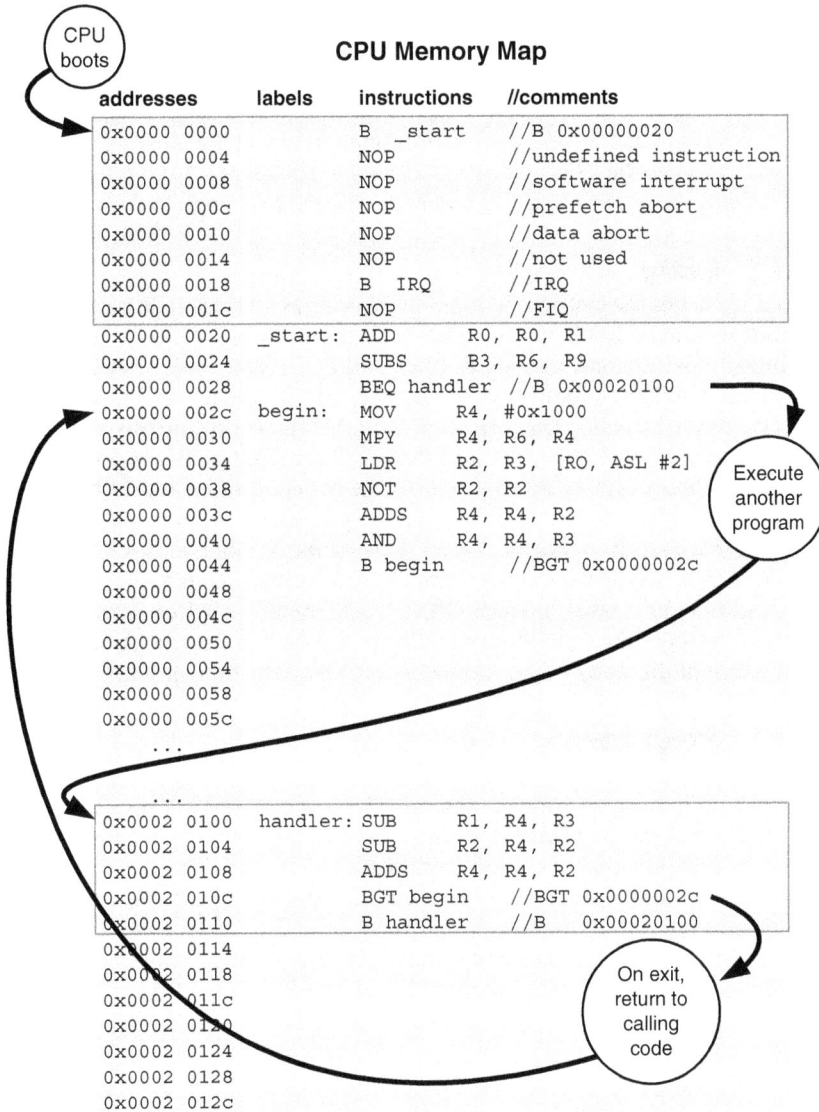

FIGURE 8.1   Illustration of how boot code is run when an ARM processor first powers up, and how this then executes code located elsewhere in memory.

to the right. Since each instruction is 32 bits in size, the addresses count upward in 32-bit steps, that is, `0x0000 0000`, `0x0000 0004`, `0x0000 0008`, and then `0x0000 000c`.

The assembly language displayed shows labels in the first column, the instruction in the second column, and comments for some instructions in the final column. Comments are explanatory text starting with `//` and running to the end of the line. The figure does not mention what type of memory this is, but since it covers the boot vector address, it is most likely non-volatile memory, probably some kind of flash memory.

8.1.1.1 Booting

When this CPU first turns on, the program counter is automatically reset to 0 by the hardware. The CPU fetches its first instruction from, as usual, the address the PC points to. Looking at the map in Figure 8.1, we can see that the instruction at address 0x0000 0000 is B _start, which means an unconditional branch to the label _start. The comment shows that, since the label _start is located at address 0x0000 0020, then this instruction is really B 0x0000 0020. It will cause the PC to change to the value 0x0000 0020.

8.1.1.2 Running

The CPU fetches the instruction that the PC points to, which in this case is ADD R0, R0, R1, meaning an addition of two registers (in this example, it is only necessary to examine the branch instructions, namely B, BEQ, and BGT). Once this instruction has been executed, the PC will automatically increment from 0x0000 0020 to 0x0000 0024 ready to fetch the next instruction (a subtraction, SUBS). After executing this, the program counter will increment to 0x0000 0028 and the BEQ instruction will be fetched. This is a conditional branch to address 0x0002 0100. Since it is conditional on "EQ" or "branch if equal to zero," the branch will only be taken if the previous condition code-setting instruction resulted in a zero (it is not necessary to review condition codes here, but a full set is shown in Box 3.3). If the current state of the condition code register (CCR) shows that the condition is not met, then the PC will get incremented as normal and the next instruction fetched, the MOV. However, if the CCR shows that the condition was met (i.e., if the result of the SUBS instruction was zero) then the branch will be taken.

8.1.1.3 Taking a Branch

Taking the branch means that the PC will be set to the value specified in the branch instruction. In this case it means setting PC=0x0002 0100 (although we should remind the reader that the ARM processor uses relative branching, so it is actually not setting the PC to a 32-bit value, it is simply adding an offset to the PC to get to the right place—this was explained previously in Section 3.3.2.3). Once the PC has been set, everything proceeds as normal for the following instruction: The CPU fetches whatever instruction the PC points to, and then executes that. The CPU is now executing code that is located in a very different section of memory.

8.1.1.4 Running a Program

Now we can see how to run a new program:

1. Load the required program into some place in memory.
2. Set the PC to the start address of the program.

Following that, the next instruction fetched and executed will be the first instruction of the new program. The PC then steps line by line from there, through the program, until it encounters another branch instruction, of course.

8.1.1.5 When a Program Ends

When the program ends, it is simply a matter of resetting the PC to another value *outside* of that program. In Figure 8.1, the instruction at address 0x0002 010c ends that separate block of code. It is another conditional branch. If its condition is met, the branch will "send" the PC back to address 0x0000 002c, which happens to be the instruction following the one that caused that block of code to be executed in the first place.

8.1.2   Other Things to Note

As usual, the above descriptions are a slight simplification of what happens in practice, although it does contain the most important elements. Apart from what was described, it is worth noting the following:

- Before executing another program, the calling process needs to save its state, and then restore that state when the program terminates—otherwise the program being called will change the registers and state of the machine in unpredictable ways, which could cause havoc with the calling process. The state includes various registers that contain pointers, condition code register, and so on. All of those need to first be stored onto a software stack in RAM. The stack is defined by a pointer (in the ARM it is the stack pointer, SP, which is register R13), and can pop and push stored registers, causing the stack memory to grow or contract as required. So before calling a program, the content of the current set of important registers is usually pushed onto the stack. When that program terminates and control passes back (i.e., the PC jumps back) to the calling process, the register contents are restored by popping them off the stack.

- When the program being executed ends, the PC needs to jump back to the calling process. In the ARM, this can be done through the link register, LR, which is register R14. To make this work, we use a "branch and link" instruction, BL in place of the "branch" instruction, B. The BL instruction causes the CPU to automatically store a copy of the return address to register R14 at the same time as it branches. Since the PC gets automatically incremented after each instruction fetch anyway, when a BL instruction is executed, the CPU has already calculated the return address (which is PC + 4). It is this return address that gets stored in R14. Having done that, the program being run is able to return to where it was called from, by changing the PC to the value stored in R14, that is, with the instruction MOV PC, R14 (which can also be written MOV PC, LR, or achieved with the instruction BX LR).

- Many programs that we want to execute will need to be passed information (parameters or arguments) before they are run. This information can be supplied to a program in two ways. The first is to store the information in a register, such as register R0, before calling a program. For this to work, the program will have been written to assume that R0 contains the required information, and the calling process will need to place that information into R0 beforehand. Sometimes, though, the required information will not fit into the 32-bit register R0. Either the data size is too large, or there is more than one item of data. In that case, the calling process places the information into a block of memory, that is, the stack, before the program is called. For this to work, the program will have been written to assume that the *stack* contains the required information. If you have followed this discussion so far, you will realize that these need to be two different programs, written with different assumptions. To prevent such confusion, most computers and OSs make use of some kinds of standards. These define exactly how such things work. They specify the requirements for the calling process, including stack and register usage, calling and return behavior, as well as the assumptions that individual programs must make when they execute.

- In the ARM processor, the ARM procedure call standard (APCS) has been used for a long time to set out these assumptions and requirements.[2]

Before the reader becomes overly concerned with setting up the link register, saving state and following the procedure call standard, it should be pointed out that these factors never affect most programmers directly. This is because it is the job of the compiler to ensure that compiled code works within a system, and is compliant with all standards. The only exceptions would be when a programmer is writing freestanding code for systems without an OS, when writing a bootloader, or when hand-crafting assembly language functions that need to be called from within a program. While these are fun and fascinating activities, only a minute fraction of programmers will ever undertake them.

From this point on, we will move to a higher level consideration of how to write and execute programs. First, through the use of a compiler, and then a brief overview of interpreted languages.

8.2 Writing a Program

In the early days of programmable computing, programs tended to be written by the same engineers who had designed and built the computers in the first place. Since they had intimate knowledge of the hardware, programs were tied very closely to that hardware. Those early computers were typically programmed in machine code, and later in assembly language—which was more convenient to write, more human-readable, and included a number of features designed to assist the programmer. A program called an *assembler* was used to convert the assembly language program into machine code. The illustration in Figure 8.1 used assembly language to illustrate how programs can be run (and how branching works). But since assembly language is the human-readable version of machine code, the information that is actually stored in memory would be the equivalent machine code. Hence, the actual memory map would look more like the following, where addresses are shown along the left, and 32-bit instructions shown, in sequence, to the right:

```
0x00000000 | 0xea000006   0xeafffffe   0xeafffffe   0xeafffffe
0x00000020 | 0xeafffffe   0xeafffffe   0xe51fff20   0xeafffffe
0x00000040 | 0xe3a01981   0xe3c11003   0xe1a0d001   0xeb0000bc
0x00000060 | 0xe3a00981   0xe321f0d3   0xe1a0d000   0xe2400010
0x00000100 | 0xe321f0d2   0xe1a0d000   0xe2400010   0xe321f0d1
```

---

[2] The function of the original APCS has been replaced by the ARM Architecture Procedure Call Standard (AAPCS), which includes both ARM and Thumb mode standards. This is joined by a newer specification called the Application Binary Interface (ABI), which specifies how code must be written to work in a given operating system. The embedded version, EABI, is used where no OS is present—such as in a bootloader or freestanding embedded system with no OS.

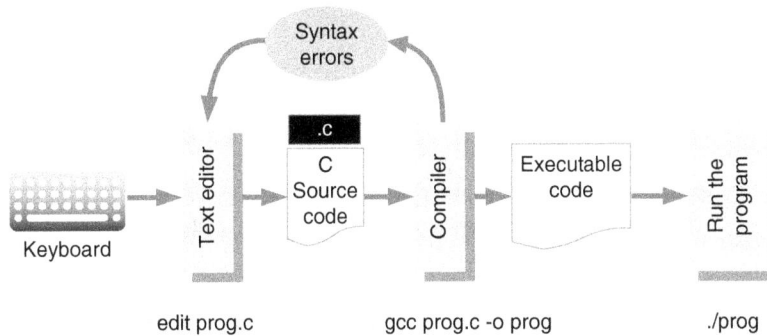

FIGURE 8.2 An illustration of entering, compiling, and executing a "C" language program on a computer. The appropriate commands for doing this under a typical Linux installation are shown in the grey boxes underneath.

As computer programming has become more specialized, and eventually became a discipline in its own right (computer science), so programming methods and languages have been revolutionized.

8.2.1 Compiled Languages

Compiled languages are written in a more human-readable form than assembly language, making use of a greater number of conveniences for the writer. These conveniences include the use of variable names, functions and function names, pointers, arithmetic, loops, conditionals, and a host of other structures and features.

Taking the "C" language as an example, on modern computers the programmer typically types a program into the computer using an editor. The code is saved as a ".c" file which is then compiled by a C compiler (and built-in linker) to produce an executable—a program that can be executed directly. This is illustrated in Figure 8.2, with examples for Linux-based compilation and execution at the command line using the gcc (GNU compiler collection) C compiler, of a program called `prog.c`. It is rare for a newly written program to compile immediately without errors, and so the process shows the compiler identifying syntax errors, which the programmer can then correct before recompiling.

8.2.1.1 Compiling and Linking

In fact, the process discussed above is a simplified view of compilation, which does not take into account the fact that most programs make use of library functions.

Libraries include many standard functions or operations that may be required by several programs on a computer, and are hence provided centrally alongside the OS for any program to use (instead of being rewritten in full inside a number of those programs). A collection of functions is gathered together into something called a shared library, with libraries for such groups of functionality as input and output, mathematics functions, string handling, cryptography, and hundreds of other library types.

Figure 8.3 illustrates a more complete example of compiling a program that includes references to library functions (it shows more detail than Figure 8.2 but is doing the same thing). The program shown is in "C," but a similar sequence is followed for most

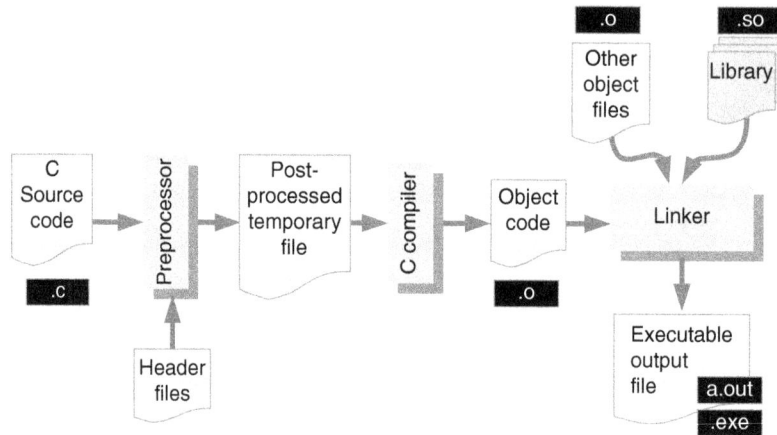

FIGURE 8.3 An illustration of a "C" language program making use of a library function, showing a preprocessor checking these against the library header files and ensuring the function call code is correct, the compiler creating an object file of unlinked machine code, and then the linker combining this code with the library function code (or combining a method to call the correct library function).

compiled languages. Working from the left to the right, a program has been written that makes use of a library function—something like `printf()` or `sqrt()`, from the standard input/output (stdio) or mathematics (math) libraries, respectively. In order to use them in the code, the programmer is required to include a reference to the appropriate *header* files in their program (e.g., `stdio.h` or `math.h`). These contain *function prototypes* that are like outline functions that help to ensure that the programmer calls the functions correctly (specifically, that the correct types of information are passed into the function and returned from them).

Assuming that the programmer has used the functions in their code, the first step in compilation is that the C preprocessor checks those functions against the prototypes in the header files, and assuming a match, creates a temporary file that replaces the library functions with some other code. The altered program is then compiled into *object code* by the compiler (which is usually a file ending in ".o"). Object code is a block of machine code with everything ready for execution except the memory references (things like branches, jumps, and calls), which are encoded in a special way that keeps the code flexible until it is "linked" by a *linker*. The linker combines the block of program code with other blocks, including blocks of code from the library, as well as a memory map which fixes where everything fits into memory. The output of the linker is an executable program; code that can be run (executed) directly.

Note that the library is usually setup to provide something called a ".so" or shared object file for the linker. There are two options for linking to this in general. The first is to copy the library object code responsible for the library functions into the program code (so the compiled program becomes bigger). The second is to link in a small piece of code that causes execution to jump to the correct part of the shared object code block, and execute the function code from there, before jumping back into the program. The latter method is much more space efficient when many programs use the same code, but is slightly slower due to the jumps.

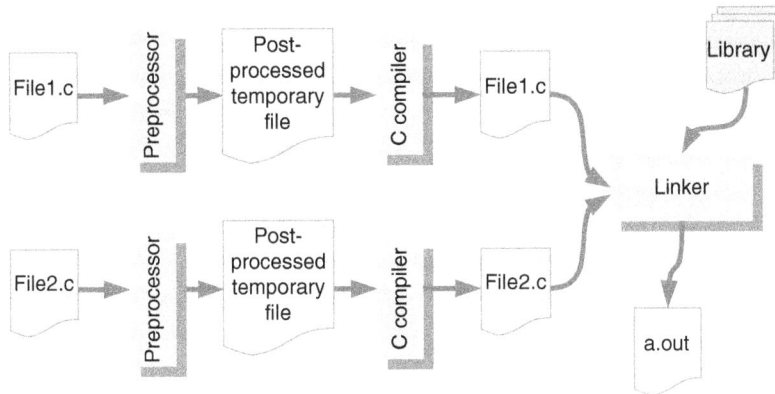

FIGURE 8.4   Most programs consist of multiple source files, which are illustrated here being combined by the linker into a single executable file called "a.out."

To see how this works, imagine writing programs that use the `printf()` library function. The library that contains the `printf()` code can be over 1 Mibyte in size. But if this function is used, without the benefit of a shared library, in 1000 programs, it would need to be copied 1000 times and linked into each separate program (i.e., 1000 Mibytes of identical code). But with the second linking method, a tiny piece of code—perhaps just a few tens of bytes—would instruct each program in how to call the `printf()` function inside the shared object file at runtime. In that case, instead of copying the 1-Mibyte library 1000 times, there would be only a single library in existence, and every program would call that library when needed. The calling code, even repeated 1000 times, may be only 0.1 Mibyte in size, that is, much more space efficient.

Now repeat this for the hundreds of library functions and shared object libraries, and it is clear why it is so efficient to share an object library.

The final aspect of compiling and linking that we had simplified earlier is that many programs consist of multiple source files. Splitting code into multiple source files is preferable when the program is large, is worked upon by multiple people, splits reusable and non-reusable parts, or splits by major functionality. During compilation, each separate ".c" program is compiled to ".o" object code, and then combined together with all other ".o" objects and library objects, to create the executable.[3] This is illustrated in Figure 8.4.

8.2.1.2 Cross-Compiling

Since compilers create executable code, they are processor- and machine-specific. This means that a compiler for an ARM processor will output ARM machine code, whereas an x86 compiler will output x86 machine code.

---

[3] In practice, many programmers utilize an excellent utility called make to control the compilation and linking process. make ensures that all files are kept up to date, are compiled and then linked correctly, combined with the correct siblings ".o" files, and linked in the correct sequence.

For embedded systems development in particular, a lot of ARM code is written by programmers on x86 hardware, both on desktop as well as notebook computers, under a variety of operating systems (e.g., MacOS-X, Ubuntu, Microsoft Windows, and so on). The system that the code is being written for may have a different OS (or no OS), as well as having a different processor, different memory map, and different peripherals to the machine that the programmer is using.

Cross-compiling means the process of building software on one computer that is destined to be executed on another, different, machine. Two important terms that are used when talking about cross-compiling are

- *Target*—the system that we are building software for
- *Host*—the system upon which the development is taking place

Both the target and the host are often identified in terms of the processor and hardware systems as well as their operating system. For example, we may be using an x86-linux host to write code for an arm-linux target.

A typical compiler for this, running on the x86 host, would be called `arm-linux-gcc` or similar. This is a program that runs on the x86 machine, but outputs compiled machine code for an ARM architecture computer. The ARM executable would be produced on the x86 host, and then needs to be downloaded or copied to the ARM target for testing. Depending upon the exact functionality of the ARM system, this could be over JTAG connection, serial port, USB, network, or programmed directly into on-board flash memory.

8.2.1.3   Hardware Emulation

It is also possible to execute code written for one architecture on the architecture of a second computer using an emulator. Emulation technology (which will be discussed briefly in Section 9.3.6.1) allows executables for one type of computer to be run on a different kind of machine. This is a significant and fascinating research area in its own right, but has a very practical application allied with cross-compiling. If, for example, an arm-linux program has been cross-compiled on an x86 host, then an ARM emulator can be used to execute the ARM architecture program directly on the x86 architecture computer.

This can be a very convenient method of testing the compiled code—certainly it can be faster than the alternative of downloading the executable over a serial link to the target hardware, or programming it into the flash memory, and then rebooting the target hardware for testing.

8.2.1.4   Programming Language Characteristics

Of the several hundred different types of compiled programming languages that have been developed, just a handful of languages make up the vast majority of code written today. These include C (as well as C++ and other variants) and more recently, Java. Older applications code written in FORTRAN (for numerical computation) and COBOL (for business applications) are still, surprisingly, found in niche areas.

Languages differ greatly in their ease of coding, how easy they are to debug, their robustness, memory efficiency, speed, security, flexibility and how error prone they are. Many languages have particular strengths in different areas. These include C for embedded systems, Java for graphical user interfaces, Ada for real-time and reliable systems, C++ for games and fast graphics, and a number of specialized languages for all kinds of application area.

What we have talked about up to now are compiled languages, where programs are written in source code and then compiled into machine code for execution. The alternative are interpreted languages, in which the source code is executed directly, step by step, on a program called an interpreter.

Before we look in more detail at interpreted languages, we should note that some languages are intermediate. Java is one example, where source code is compiled to Bytecode, which is a compact set of instructions that can be executed directly on a specialized Java processor (e.g., ARMs Jazelle of Section 4.8), or executed on an interpreter (e.g., a Java virtual machine, JVM, or Java runtime environment, JRE).

8.2.2 Interpreted Languages

As with compiled languages, there are all kinds of interpreted languages for different applications. These range from text processing, compiler design, mathematics processing, system booting, graphical user interfaces, database handling to shell processing (namely, the command and control of an operating system). Despite having a lot of variety, interpreted languages usually tend to be faster and simpler to write than compiled languages; they are often more convenient to use, and typically execute slower—although there are some very notable exceptions to each of those characteristics. For interpreted programming languages, the style and structure of the program, and the way of using the language vary widely. When choosing between different kinds of language, they provide different advantages in terms of factors such as

- Efficiency of operation
- Compactness of code
- Speed, throughput, or response time
- Ease of writing (or debugging)
- Development cost, time, or expertise
- Platform independence (the range of hardware the language is able to operate on)
- Availability (and cost) of support

8.2.2.1 Interpreter

Interpreted languages are written using a text editor, saved as a source file, and then are run within an *interpreter*, as shown in Figure 8.5. The interpreter is itself an executable program running on the computer (unlike compiled programs which are executed directly). So while compiled programs are compiled into machine code programs which can be executed, interpreted programs are saved in source form and then run within a separate executable program called an interpreter which "understands" their code. Within the interpreter, the program is loaded from file and then run step by step, executing each command in turn, following loops, branches and functions, and calling shared libraries as appropriate.

There is actually quite a lot of commonality between the idea of an interpreter and the ideas behind microcode (see Section 3.2.5). Microcode is effectively executed by a *hardware interpreter* that processes each language instruction (token) using multiple machine instructions. A single interpreted language instruction will often require several hundreds or even thousands of machine instructions to be spent in order to execute it.

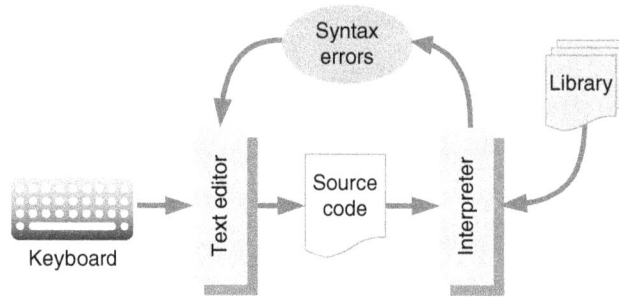

FIGURE 8.5   A program in an interpreted language is written with a text editor, saved as source code, and then executed directly within the interpreter.

This makes interpreted languages more "concentrated" in that a shorter source code program can often accomplish more than a similar length of source code with a compiled language.

Of course, in both cases we have mentioned library functions. Much of the complex operation within modern computers is handled by such functions. This includes much of the graphical user interface code, encryption code, document handling and printing, networking and communications operations, mathematical functions, text processing, video, graphics, and sound. With such rich functionality in shared libraries, some programs become little more than a set of calls to library functions that are glued together.

When interpreted programs largely consist of library calls, they may be just as efficient as compiled programs calling the same function, but are of course easier to write and modify (because there is no need for the compile-link step). However, in general, a sequence of simple operations or steps in a compiled language will be faster and more efficient to execute than the same sequence of operations in an interpreted program. Similarly, assembly language is even faster to execute (but not usually to write). It is for this reason that assembly language is a "very low level" language, compiled languages are typically "low level" and interpreted languages are "high level languages." The highest level language is often said to be our native human language. The idea is that we can one day "program" a computer by simply speaking to it. But that is well beyond the scope of this book.

8.2.2.2   Integrated Development Environment

Many programmers these days will write, execute, debug, and test their code within an integrated development environment (IDE). These can be used for both compiled and interpreted languages, and in some senses they blur the distinction between them, as far as the programmer is concerned.

Code written in an IDE is entered directly using a built-in text editor, then is compiled and executed or interpreted (depending on whether a compiled or an interpreted language is used) within the IDE. The execution/interpretation step can be done outside the IDE, but with most IDEs providing extensive debugging capabilities, there is little incentive for the programmer to do so—however, the final code that is developed, when finished, will almost always need to operate outside the IDE.

An IDE—which is a large application running on a computer—may be graphically intensive, may be endowed with built-in help functionality, and will probably significantly improve the speed and convenience of programming for novice users. While these are excellent features, they do have the potential to hide the low-level sequence of operations from the eyes of the programmer. It is therefore important for IDE users to understand that the previous sections (8.2.1 and 8.2.2) described what is actually happening underneath the IDE, although the programmer's view may be smoothly integrated and graphically rich.

8.2.2.3   *Other Interpreted Languages*

The growing predominance of Web-based interactions, such as Web browsers accessing websites or cloud computing-based connectivity, has led to large growth in Web-based interpreted languages. Hypertext markup language (HTML), Javascript, PHP (PHP: hypertext preprocessor), and a variety of other interpreted languages have been designed to operate across a link between a client and a server (see Chapter 10).

Depending upon the exact language, the interpretation (i.e., execution) may happen on the client's Web browser or on the webserver, with the final display being in the client's browser window. For example, when a Web browser connects to a remote webserver to access a webpage, the action may trigger a set of PHP code to be run on that webserver which generates a webpage containing HTML and Javascript. The PHP program being interpreted on the webserver is designed to create that output "document" and the webserver is thus partly an interpreter for PHP code.

The output document containing HTML and Javascript is sent to the client Web browser. The role of that Web browser is to interpret the HTML and Javascript code and display the result. The Web browser is therefore also an interpreter, and the final display in the browser window contains the text and graphics that the HTML and Javascript programs have produced.

This is just a single example of the millions of Web interactions happening daily. But interpreted languages are in widespread use in many other areas too, including database access, graphical user interface displays on smartphones, human-computer interaction, computer startup scripts, satellite operations scheduling, control systems (which span such diverse areas as industrial automation to home heating control), and a growing variety of other areas.

8.3   The UNIX Programming Model

UNIX-based operating systems, as well as some others, assume that the overall operating system is controlled by a console. In years gone by, this role might have physically been performed at a teletype terminal that the superuser, or system operator sat at, but today it is the control input for system startup and management. After the OS boots and the kernel runs, automated scripts will launch graphical user interfaces (GUIs), webservers, and other operating software that needs to be automatically launched. These start-up scripts could be entered manually, by a human operator, and if so, the commands would be typed into a *console.*

In modern operating systems, the console is reached by running a terminal program (this is true of Ubuntu Linux as well as Mac OS-X, where it is called `terminal`). For embedded systems, an external terminal (or more usually an external computer emulating a terminal) may need to be attached via a serial port, or using USB. On network-connected embedded systems, a program called Telnet (i.e., `telnet`), running on an externally networked computer, is often able to connect to the embedded system and open a terminal session, accessible on that external computer. A more secure alternative is the `slogin` program which uses secure socket layer (SSL).

For users of Microsoft Windows wishing to connect to an embedded system in this way, much of the functionality is available with a third-party program called `putty`. Microsoft Windows does not itself require a terminal, but has something similar which is often called the "DOS prompt" or the "Command window." Like the UNIX console, it is a type of command line interface.

8.3.1   The Shell

Command line interfaces, terminals, consoles, and so on were the primary human-computer interfaces of computers in the era before window managers and GUIs became popular (or were invented). As such, they needed to operate in a way that was both human-readable and meaningful to the computer. This required the use of an interpreter—a piece of software that interpreted human-readable commands into computer-actionable commands. In early home computers, the terminal directly booted into an interpreter for a language like BASIC, but scientific and commercial computers usually booted into a command shell (or DOS prompt).

The purpose of a shell is to give the operator (user) the ability to directly control the operating system, often at quite a low level. It is therefore potentially a very powerful interface. Experienced users of the shell can also control their computers much faster than any operator with a mouse and GUI could do.

Although different computers used different software, with varying command names and syntax, to implement their command shells, there was a great degree of convergence toward shells that combined features of being powerful, easy, and fast to use. Several shell programs were ported to different operating systems, and are now in common use. The following can be used on almost any modern computer (and that includes embedded Linux and Android systems, iOS, and Mac-OS):

- `bash`—Bourne again shell
- `kshell`—Korn Shell
- `zsh`—Z shell
- `csh`—C-shell
- `sh`—the most basic and common shell

Most contemporary UNIX-based systems implement `bash` as the default shell, although some still use `csh` and other alternatives. Embedded Linux computers often use something called `ash` or `lash` which are lightweight versions of the Bourne again shell, and are built into Busybox (see Section 9.5.3). Virtually all UNIX systems also include the basic shell, `/bin/sh`, as a backup system shell. As always in the UNIX world, users can opt for their own choice of shell—and it is easy to support different shells for as all different users of a system, even simultaneously. Nothing prevents two or even hundreds of shells running simultaneously, all providing command input to the OS.

Shell scripts are programs written in the shell language and saved to HDD. These often make use of other tools and utility programs in the path (i.e., command line utilities). Shell scripts can be incredibly powerful, and it is probably not an exaggeration to state that shell scripts hold the Internet together.

There are many types of shell (far more than the tiny list above), firstly because anyone can write a shell if they wish, but also they use different styles of language, and differ in their user-friendliness or how well they "suit" a particular user.

Experienced UNIX programmers will often use the shell for writing their programs, in whatever language they wish, compiling and linking them, executing and debugging them, and then automating them. After all, the final program, when fully working, will need to be launched from a shell—whether at boot time, by the user, or via a GUI.

To clarify, all programs in a UNIX-based computer are ultimately either part of the kernel, or are executed from the shell (or from something that was itself executed from the shell, or was executed from something that was executed from the shell, and so on).

Executing a program from the shell allows its priority to be changed (for multitasking), for it to be paused, placed in the background (not graphically, but executing "behind the scenes") or terminated at will. It also allows programs to receive input from various places, files, or programs and to send output to other files, programs, and places—in particular through the use of redirection.

8.3.2   Redirections and Data Flow

All executable programs being run in UNIX-based operating systems can receive input and produce output—as depicted in Figure 8.6. What input they receive and what happens to their output depends upon how and where they were originally executed.

For a program executed manually at a terminal (or terminal window) running a command shell (and assuming it was not executed as a background task), subsequent keyboard presses at that terminal (or terminal window) would be passed by the OS to the shell, and by the shell on to the program. The program would thus receive those key presses as its input. If the program prints or displays a message of text to the screen, that text is passed out of the program to the shell, and by the shell to the OS to display in the terminal (or terminal window). All of this is normal behavior of programs running in the shell and those two streams are called *standard input* and *standard output*, respectively.

In UNIX, compiled executables by default have the possibility of generating another stream of output text. This is called *standard error*, and is to allow printed error messages to be easily distinguished from normal printed output messages. Typically, for programs executed at the shell command line, both types of message would be printed at the shell (in the terminal window), but it is very easy for the shell programmer to disentangle the message streams if required. Figure 8.7 shows the three default streams. By convention they are numbered; under UNIX they would be 0, 1, and 2.

There are several methods of *file handling* available to programs written in most languages, but UNIX streams are among the most common. The standard input, output,

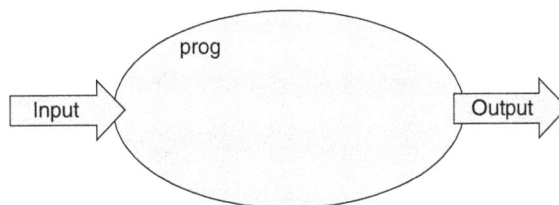

FIGURE 8.6   A program is executed. By default it has an input and an output data stream.

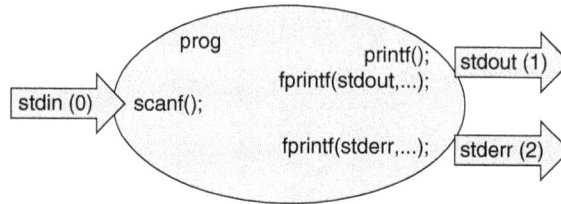

FIGURE 8.7 A program is executed. Input is on stream 0, standard output is on steam 1. Error messages are output on stream 2.

and error interfaces discussed above are examples of streams—ones that are present by default. However, the programmer can choose to open additional streams which read from or write to files on the filesystem (and would sequentially number 3, 4, 5, and so on).

One of the most powerful features of the UNIX shell is the ability to redirect the three default streams of programs launched within the shell. The command language is relatively simple. For example, if a compiled program is normally executed like this:

```
./prog
```

then the following four bash shell commands would redirect its output, its input, its error log, and both its input and output, respectively;

```
./prog > output_file
./prog < input_file
./prog 2> error_log
./prog < input_file > output_file 2> error_log
```

The two text files called output_file and error_log would be created if the program produced any output (i.e., if it printed any text or error messages) on those respective streams. If files with those names happened to already exist in the current directory then it would cause them to be overwritten. The text file input_file would need to exist when the second and final commands are run, otherwise an error (e.g., "No such file or directory") would be generated. Figure 8.8 graphically illustrates the final command redirection example.

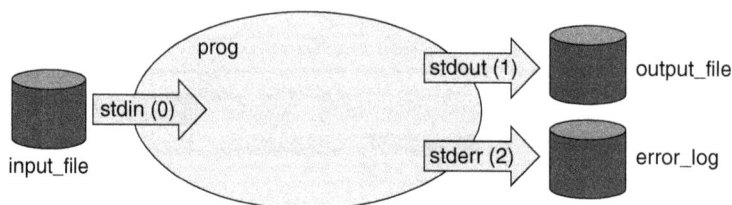

FIGURE 8.8 The power of the UNIX shell allows steams to very easily be redirected to and from files.

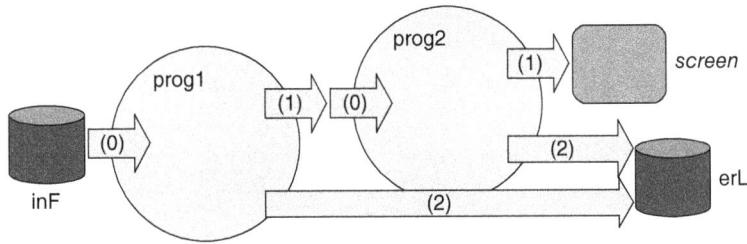

FIGURE 8.9   A program is executed—it has an input and an output data stream.

In fact, a shell such as bash is powerful enough to combine many possibilities together. This includes the ability to stream information from one program to another. One way of streaming between programs is to use the pipe character (|), as in the following command which is illustrated graphically in Figure 8.9:

```
./prog1  < inF  2> erL  |  ./prog2 2>> erL
```

Shell scripting is an extremely rich resource for system control and automation. In this section, we have only touched upon a very small section of its capabilities, but there is far more than this brief summary—including types of looping, conditional evaluation, file handling, error control, shell multitasking, mathematical and text processing, and so on.

8.3.3   Utility Software

It would be unfortunate to end our discussion of the UNIX programming model without mentioning several utilities that make the job of the programmer and user much easier. While we have already discussed the GNU compiler collection (gcc), briefly commented on the highly flexible make program, and mentioned the GNU debugger, gdb (Section 7.16)—which are all outstanding UNIX tools—there are three shell-based tool programs that are extremely commonly used within shell scripts:

- grep—The GNU regular expression parser (although the original grep was not the GNU version) allows for character-based data to be quickly and conveniently searched for using patterns. Regular expressions are standard ways of describing patterns; ways that are understood by many UNIX file handling and text processing programs. For example, grep ^z.{4} could be used to search a dictionary file for every occurrence of five-letter words starting with the letter "z."

- sed—The GNU serial editor allows streams of data to be quickly and conveniently modified based on regular expressions. It can easily modify or delete every occurrence of one or more specified sets of phrases or words in a file. For example sed s/me/you/g. would change every occurrence of "me" to "you" in a file or data stream.

- awk—Often thought to be an apt abbreviation of the word "awkward," this utility was actually named after its inventors A. Aho, P. Weinberger, and Brian Kernighan. It is the ultimate text analysis tool for multicolumn, multiline character data.

awk has an extremely powerful programming language used to control it (along with supporting regular expressions). For a simple example,

```
ls -l | awk '{ size=size+$5; print size}'
```

would print the running total size of all files in the current directory (since file size is reported in the fifth column of the output text that ls -l reports when it prints out details of all files in the current directory).

The software utilities and resources available on any UNIX system—even most embedded Linux and Android devices—are extraordinarily powerful. They are well documented, generally extremely efficient and reliable.

8.4   Summary

This chapter began with an overview of how a program is written, beginning with recapping how a program is executed on a computer—specifically what is happening inside the computer to make such a thing possible. We did not attempt to describe program design, implementation, or programming, since they are all topics for software development rather than computer systems. However, we did discuss the different characteristics and strengths of compiled and interpreted languages. We ended the chapter with a discussion of the UNIX programming model and the powerful UNIX shell.

In the following chapter, the operating system is discussed; the standard software layers that lie between user code on most computers and the underlying hardware. Most programmers like to write software without being too concerned about the vagaries and complexities of the particular computer that is currently executing the code—and it is the OS that, above all, frees the programmer from that responsibility.

8.5   Problems

8.1   Programs are generally executed line by line; however, some instructions will change this sequence by causing execution to jump to a new location in the program. Which CPU register is changed to enable a program to "follow a jump or a branch"?

8.2   For user programs located on non-volatile storage medium such as hard disc or DVD-ROM, are these executed directly from their storage media or will they be moved elsewhere first? If so, where will they be moved to?

8.3   What is the name of the program used to turn an assembly language program into machine code?

8.4   What is the name of the program used to turn most of a "C" language program into machine code?

8.5   When writing programs and cross-compiling them for an Android smartphone using integrated development environment (IDE) software installed on a MacBook computer, which device is the *target* and which is the *host*?

8.6   Undertake some research to identify which of the following languages are compiled and which are normally interpreted: Python, XML, FORTH, C++, Ada, Perl, and Lisp.

8.7   Describe, in a single sentence, the function and location of a bootloader on an embedded ARM computer.

8.8   If you write a program in a compiled language like "C" consisting of a number of different source files, what is the name of the program that joins the code from your source files into a final executable program?

8.9   Syntax errors are mistakes in the way program code is written whereas runtime errors are problems that occur during program execution. Which type of error is caught during compilation of a language like "C"?

8.10   A Web browser, running on a desktop computer, connects to a remote webserver to request a page of HTML which is then displayed on the screen. There are three main programs involved in this transaction—the Web browser software, the downloaded HTML code, and the webserver software. Identify for each program whether they are more likely to be compiled or interpreted software.

CHAPTER 9

Operating Systems

I n Chapter 8, we looked at programming a computer, noting that the aim of doing so is to make the computer do what we want it to do. This means that programming is a way of obtaining value from the hardware. Without a program, a computer is just a lump of silicon, metal, and plastic, but with a good enough program it comes alive to perform tasks as diverse as keeping a pacemaker running, piloting a cargo ship, plotting a course through the asteroid belt, running a virtual reality game, providing music at a concert, or teaching a baby to speak. It is amazing to consider that the same basic lump of hardware can accomplish so many different things, in the hands of a skillful programmer.

However, most of us know that there is something that lies between the program and the "metal," and that is the operating system (OS). Like the hardware itself, the OS remains a constant presence even while different programs allow the overall system to do different things.

9.1 What Is an Operating System?

A simple description of an OS is that it is low-level software that lies between the hardware and user programs, but like many simple definitions, this is not particularly helpful. In fact, we can view an OS in two quite different ways. The first is to look at how the OS became what it is today from its earliest beginnings, and the second is to look at the OS only as it is today. Strangely, we had the same choice when we were looking at the hardware, the CPU, in Chapter 1. We started with a brief overview of the evolution of computers and then we began to consider what is a modern computer. Both approaches have value. The evolutionary path is interesting because each step along the way was made as an intelligent response to a problem, and every new idea came about for a reason. Some ideas were good ones in the context of the time they were invented, but have since become obsolete, but quite a number of them have stood the test of time and can still be seen today. By looking at those ideas and why they were introduced, we can gain a deeper understanding of how they work. On the other hand, dissecting and explaining a modern device definitely allows us to take a shortcut!

This chapter presents a pragmatic combination of both viewpoints. We will begin with a firm description of the modern OS and its characteristics. Then we will explain these in more detail. Where it could add value or provide an interesting explanation, we will also delve into the historical reasons behind certain features.

9.2 Why Do We Need an Operating System?

Actually we do not need an operating system—any computer can run a program without having an OS, provided the program is written in such a way that it replaces the low-level OS functionality. However, the vast majority of computers larger than a wristwatch make use of an OS. The main reason is simply that it is a lot easier to make the computer do what you want, and do more of what you want, with the support of a well-written OS.

This is because modern operating systems almost all provide the following characteristics:

- They "abstract away" the hardware (so the programmer does not need to understand exactly what the hardware contains, and so that programs do not need to be rewritten when there are small changes in hardware).

- They provide consistent software interfaces to control hardware functions and allow for various forms of I/O.

- They allow each program to pretend that it is the only thing running on the computer, when in fact many programs might be sharing the computer.

- They make developing (and porting) code a lot quicker.

- Usually they will provide a lot of assistance when debugging.

- They can enforce and ensure system security.

- They can give a consistent means for the user or programmer to access and control the computer.

It is important to clarify one point, and that is related to the look and feel, appearance or user interface. In fact, one cannot see an OS. It is *not* the graphical user interface (GUI) that greets a user when they interact with a desktop computer or notebook. It is not the welcome screen on an Android smartphone or the display on an iPad.

Many people are confused into thinking that the computer display—the GUI or the windows, icons, mouse, and pointer (WIMP) interface—is the OS. It is not. That kind of interface is often termed a "window manager," and most operating systems for larger computers actually support a wide variety of window managers with different characteristics, features, and emphases. These allow the user to choose the kind of interface, without affecting the underlying operation, security, and features of the system.

While many OSs allow a choice of window manager, some are tied to a fixed GUI. Common examples of such restrictions include iOS, Microsoft Windows, and Android, which all have fixed window managers, although each of these do have limited look-and-feel configuration settings. Truly multiuser OSs, such as Linux and FreeBSD are far more flexible. These not only allow a choice of window managers, but are easily capable of running multiple different window managers simultaneously. This feature allows Linux (along with similar UNIX-based systems) to support multiple users on a single machine, and even allow each user to interact with the system through a completely different window manager of their own choice.

9.2.1   Operating System Characteristics

Common operating systems include Android (as used on the majority of smartphones, at the time of this writing), Linux (as used on the majority of Internet servers, large-scale cluster computers, and which also forms the core of Android), FreeBSD, MacOS, Microsoft Windows (which is still used on most desktop computers, particularly by novice users), uCOS, and VxWorks (both commonly used in small embedded systems or high-reliability systems).

Some OSs are suited to small and low-power CPUs, some to medium, and some to large systems. This is generally due to their functionality, support, and design emphasis, but also to market pressure and familiarity of developers. All OSs differ by their characteristics, in particular in terms of

- Responsivity—how quickly they react to external events.
- Efficiency—how much of the CPU time is occupied just by the OS.
- Flexibility—what they can be made to do, and on what kind of CPU (and by far the most flexible OS to date in this respect is Linux which works on almost every current CPU).
- Security—even when not running special security software, some OSs are inherently more secure than others.
- Reliability—the frequency of crashing, rebooting, or other errors.

The first four of these characteristics are explored in the following table for nine common OSs (out of hundreds of possible examples), where a verdict is given about the ability of the OS to deliver the desired characteristic. This is to be taken as an approximate indicator, since the tasks are not fully defined, and there is no numerical analysis given.

	Responsivity	Efficiency	Flexibility	Security
Android	Medium	Medium	Good	Medium
Linux	Medium	Good	Good	Good
FreeBSD	Medium	Good	Good	Good
MacOS	Medium	Medium	Good	Good
MS-DOS	Medium	Medium	Medium	Bad
MS Windows	Bad	Bad	Medium	Medium
uCOS	Good	Good	Good	Medium
VxWorks	Good	Good	Medium	Medium

9.2.1.1   Reliability

In terms of reliability, most observers would probably rank uCOS and VxWorks as the most reliable, followed by Linux and FreeBSD, MacOS, then MS-DOS, and finally MS Windows. Reliability is a different aspect to security (which we will consider next), and is often measured in mean time before failure (MTBF). Since there are different aspects to failure (such as catastrophic failure, or the opposite which is inconsequential failure), we know that not all failures are equal, so for very high-reliability systems we might also consider what happens when a failure does occur, and whether or not a system "fails safe."

Because of this, software engineers tend to prefer to consider how trustworthy or dependable a piece of software is instead. This includes whether it is (i) available when required, whether it (ii) operates correctly and whether (iii) it has any undesirable side effects (which might include unwanted information disclosure). Dependability covers system attributes of availability, reliability, safety, and security, and sometimes dependability might be more important than basic functionality, particularly when the cost of failure outweighs the cost of the system in the first place.

Availability is defined as the probability that a system will be working correctly at the time when it is needed. *Reliability* is defined as the probability that the system will continue to correctly deliver services as expected by the user over a given period of time. *Safety* involves a judgment as to how likely it is that the system will cause danger to people or its environment, and similarly, *security* is a judgment as to how well a system is able to resist deliberate attacks or accidental leakage of information.

A nuclear power station control computer might be designed with safety and reliability as the paramount aspects, whereas your internet banking software probably ranks security as its first attribute—and at least this should be more important than availability (i.e., it is much better that sometimes you cannot access your bank account when you want to, compared to having other people sometimes access your bank account when you do not want them to—this is ranking security above availability). Most large-scale software systems will need to consider the importance of these attributes, and this is also true of the OS.

9.2.2 Types of Operating Systems

It is tempting to class operating systems by the *size* of system they work with (either physical, memory, or complexity terms); however, this has become impossible with the spread of Linux which is currently by far the most important OS for very large-scale systems such as server farms or supercomputers, as well as the smallest wearables and mobile devices. It thus spans right across the "size" spectrum, failing to dominate only in the middle ground of traditional PCs and notebooks, which are the domain of MacOS and MS Windows. In fact, this is testament to the inherent scalability of UNIX-based operating systems.

More useful is being able to classify an OS in terms of it being a *single-user system* which provides just one virtual machine (VM), for one user or a *multiuser system*. Single-user systems include MS-DOS running on a PC (something you may find in a computer history museum perhaps). But some single-user systems support more than one VM. These allow multiple tasks, all for the same user, but running concurrently. Examples include Microsoft Windows (95, 98, ME, NT, 2000, XP, Vista, 7, 8, 10) but also the smaller embedded system OSs such as uCOS and VxWorks.

Multiuser systems, by contrast, provide multiple VMs for multiple users, each of which can have multiple concurrent tasks. Prominent examples are Linux, MacOS-X and FreeBSD, plus other UNIX versions These all have much more sophisticated security and protection mechanisms than single-user systems. They were designed, from the bottom-up, to cater for multiple users and concurrent tasks, and were thus created with multiuser security in mind—something that had to be almost "bolted on" as an afterthought to the single-user systems.

In fact, many of the most infamous security failures reported in recent years came from single-user systems being connected to the Internet (or to a wider network) where different users could attempt to gain access. This includes the Uranium enrichment centrifuge controllers in the Iranian nuclear program, the hacking of in-vehicle networks and the widescale ransomware penetration of the UK's National Health Service.

9.2.2.1 Embedded Devices

Embedded systems that are used to control a particular process or a single piece of equipment are single-user computer systems: for example, washing machines, simple TV/satellite receivers, engine management units, security alarms, and so on. These use relatively simple and fixed-function devices with CPUs that need only handle single tasks.

This is changing though. Embedded systems are increasingly being networked, and some of these devices have become multiuser, multitasking. At the time of this writing, it is becoming increasingly popular for kitchen appliances—such as kettles, fridge/freezers, microwaves and coffee makers—to be Internet enabled. What is unclear to the author is why anyone would deliberately allow a stranger in Shanghai, Nairobi, or Moscow to access (and potentially hack) their microwave, but such is the relentless march of Internet connectivity that a connected future for daily appliances seems inevitable.

Internet-connected appliances tend to run more secure operating systems like embedded versions of Linux, iOS, or Android (all of which are UNIX based), but at least one model of fridge/freezer contains a full MS Windows-based PC. Fortunately, UNIX-based OSs currently have over 90% share of the embedded market. With both Android and Apple's iOS being UNIX based, with Android being built on top of a Linux kernel, the OSs of choice for embedded systems designers worldwide is almost entirely dominated by UNIX-based systems.

9.2.2.2 Clusters

Clusters that are used for cloud-based computing and large-scale information systems infrastructure need to support multiprocessing inherently. These need OS-level support for running programs across multiple CPUs, and must have the highest level of multiuser security, and hence almost always run some version of UNIX.

The largest cluster computers in the world are surveyed every 6 months and tracked on a website named `www.top500.org`. This provides ranked lists of the 500 fastest supercomputers, with breakdowns by operating system, country, application area, processor technology, and so on. We will look later at some examples from this resource in Box 12.1, but just perusing the current list, we note the following breakdown of operating system share:

- Linux powers 488 of the systems.
- Other Unix powers 10 the systems.
- MS Windows powers 1 system.

And presumably one is unknown. However, the inference is clear—among the 500 most powerful and most expensive computer systems on the planet, where performance is the most important characteristic, designers have selected UNIX-based operating systems.

9.3   The Role of an Operating System

A modern OS actually handles many aspects of computing, but the two most important traditional functions are to provide:

1. A *resource manager* controls the hardware and peripherals, allowing them to be shared between different programs and users.

2. A *virtual machine* within which programs are executed as if they are the only thing running on that hardware.

9.3.1   Resource Management

Consider the resources that need to be shared between programs and users. These are both internal and external, and could change dynamically. Internal resources include the CPU (see Section 9.3.3), registers, memory, stack, network access, interrupts, storage (e.g., hard disc or solid state drive), and execution units like FPU or GPU. External resources include displays (there could be multiple), sound input/output, keyboards, buttons, mouse, touchpad, or touchscreen, as well as things that get plugged in to peripheral ports (like USB discs).

If every program just tried to use these without coordination then it is not difficult to imagine what might happen...utter chaos! For example, if two programs both waited for keyboard input at the same time—both waiting for some keys to be pressed—then how does the user know which program will be listening when she presses a key? The answer is that the operating system needs to give one program at a time the access to the keyboard, and make that fact obvious to the user.

In a system with a GUI, which program receives keypresses is usually indicated by one window having "focus" (which means coming to the forefront of the display). The user expects the program which currently has the focus will receive any keyboard input. Other programs without focus will be waiting for their turn. Similarly on the command line, the program currently running in the foreground will receive keyboard input while other programs will wait in the background. While handling this is not normally the job of the OS itself in a UNIX system (instead the OS allows it to be handled by another program called a *shell* or a window manager), it is an example of why such resource management is necessary—and this kind of arbitration must be done for every shared resource (including display, sound, mouse, and so on).

9.3.2   Virtual Machine

When a programmer writes some software, he or she does not want to be concerned about what other programs might be running on that computer at the same time. Neither do they want to be concerned about exactly what kind of hardware the programmer is running on, and have to write different code for each possible case. Instead they just want to go ahead and write the software needed to perform their task without wasting time being concerned with hardware.

In fact, they want to write their program as if it is the only thing executing on a perfectly standard, empty computer that has plenty of memory. They want their program to use memory, read and write files and consume CPU time without being affected by whatever other programs might currently be running, or whatever quirky hardware the current user might actually be using.

Luckily for modern programmers, they really do not need to be concerned with anything except their own software: The modern operating system uses something called a virtual machine concept to provide each new program with a virtual computer to run in. Each program gets its own virtual machine. When running it can "see" an empty computer, with plenty of unused memory and uninterrupted CPU time. Little does the program "know" that, behind the scenes, the OS is cleverly managing things so that multiple programs can be executed—each of which runs as if it has a dedicated machine to itself.

We will explore how this works and examine several other attributes of the modern OS in the subsequent few sections.

9.3.3   CPU Time

Within the virtual machine, a program can start and run, seemingly without interruptions, until it completes. However, behind the scenes the OS is doing quite a lot of management when a program executes. First, it has to create a virtual machine, which is kind of like an empty container, that it loads the program into. The container has attributes which include CPU time, a memory area, access to peripherals, security permissions, and a priority (which we will discuss later). The container is loaded into main memory by the OS which then allocates it some CPU time, which will cause the program to begin executing. In UNIX terminology, the container is called a process (and will be explained in more detail in Section 9.6). On a single CPU machine, with many other processes, they each get allocated small "slices" of CPU time, which is seldom enough for the entire program to complete. Instead, after a time which could range from a few microseconds up to a milliseconds or so, the OS will cause the process to be temporarily paused, its state (including memory, registers, stack, current operation) saved, and another process will be chosen to enjoy a small slice of CPU time.

From the program's perspective, it does not "know" about any pause. At various times it is either running or is in suspended animation, frozen in the middle of an operation. The OS will, hopefully, allow a paused process a little more CPU time in future, at which point it will resume as if nothing had happened in the meantime. But soon it will be paused again while the process repeats.

Over the course of executing even a short program, the process may have been temporarily paused hundreds, thousands or even millions of times. Its execution might be interspersed with hundreds or thousands of other processes, all of which are independent, and none of which interfere with each other. This is shown in Figure 9.1 in which three programs share a single CPU. The OS switches between the three programs so that each gets a share of CPU time. The important point is that the OS handles this so that *none of the programs has to interact with any of the other programs*. This kind of operating system is more properly known as a multitasking OS.[1]

The only aspects that remain to be described are how the OS does the switching, how it decides which process gets to execute at each switching point, and what happens to

---

[1] There are several kinds of task-switching OS, including cooperative multitasking, in which programs have to relinquish control rather than the OS deciding when to switch, but the majority of multitasking operating systems are as described above.

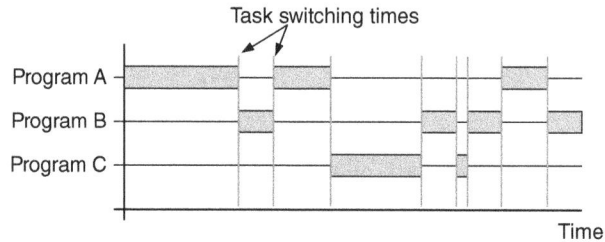

FIGURE 9.1 An example of an operating system switching between three separate programs, fairly allocating CPU time to each.

processes that are not currently executing. It is also worth noting that in reality it is tasks that get switched rather than programs—because multitasking programs can each be split into multiple tasks. All of this is described in Section 9.6.

One important aspect to mention at this point is that switching itself takes time, and hence contributes to inefficiencies (or OS overheads). More frequent switching allows a computer to be more responsive to events, but has a greater overhead. Less frequent switching is more efficient, but less responsive. For this reason, most multitasking OSs allow task switching to be configured depending upon the current application. In Linux, for example, this can be done on a task-by-task basis as well as through adjusting the overall task scheduler.[2]

It is not difficult on a UNIX-based OS to measure CPU execution time (e.g., using the system command called `time` or looking at the data reported by the `ps` utility, which we will meet again later), and in fact will report several items of information: First, the *elapsed time* which is like a stopwatch that starts when the program is first executed, and stops when the program completes. Second, is the *CPU time*, which is the sum of all the tiny durations during which the program is actually executed on a CPU. The difference between elapsed and CPU time tells us how long the process was paused for (i.e., in "suspended animation"). UNIX timing utilities will also indicate how much system time was consumed during execution, which is the sum of task switching times for that process. It is a measure of the overheads occupied by the OS.

9.3.4 Memory Management

In addition to providing CPU time to different programs (or tasks), the OS is responsible for providing memory to programs. Typically, a program is granted a certain amount of memory when it is first run, and the operating system provides a method for the programmer to request more blocks of memory (and release memory blocks when they have been finished with).

There is only a certain amount of memory available in each machine, but the virtual memory technique (see Section 4.3.1) allows the available memory space for programs to be larger than the physical amount of RAM in a computer. Using virtual memory

---

[2] On UNIX-based systems we normally consider each program or task as forming a separate *process*, and so we tend to use terms like "process switching" rather than "task switching" and "process management" rather than "task management." In this chapter the terms "task" and "process" will be used interchangeably and can be considered to be equivalent, except where noted.

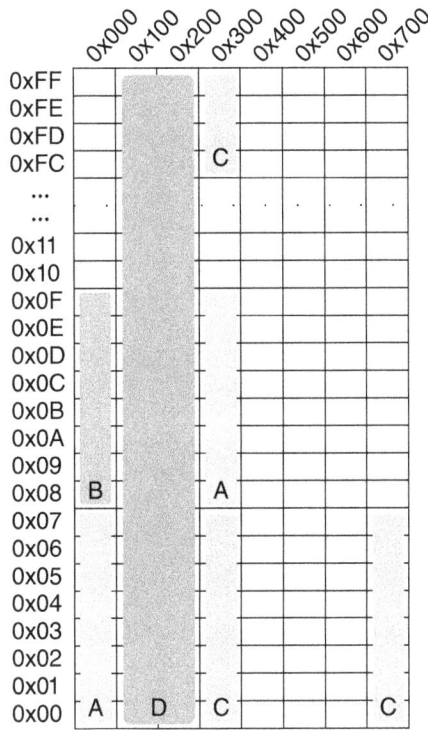

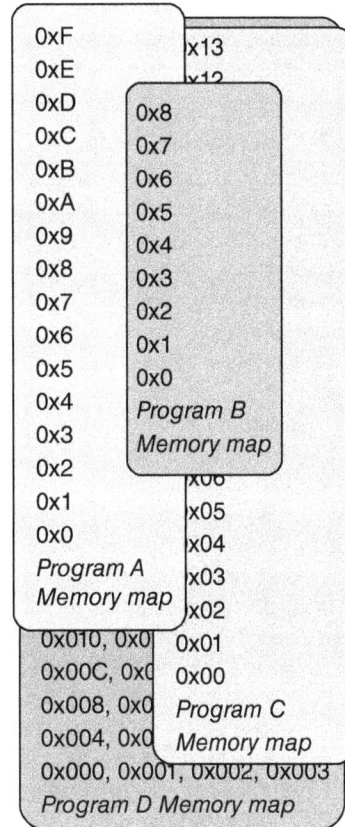

The physical memory map contains blocks for each virtual machine

Each virtual machine "sees" unique continuous memory spaces

FIGURE 9.2 A diagram showing how virtual machines each receive a block that they see is continuous, unique, and uncluttered, but these are distributed through physical memory by the operating system.

approaches, the OS can provide each virtual machine with sufficient memory, but can also provide a virtual machine with more memory than the real machine contains, allowing large programs to be run in small machines. All of this is part of the concept of providing virtual machines with what they need, irrespective of the actual physical constraints of the computer.

Without a virtual machine concept, each program would have to manage its own memory, and cooperate with whatever other programs are running at the time. This would need to be dynamic because the memory map changes whenever a new program was executed or a running program was terminated.

A view of the actual physical memory in a machine is shown in Figure 9.2, where four virtual machines each experience their own uncluttered and unique continuous virtual memory spaces, but these are in reality scattered across physical memory by the operating system. The actual mapping from physical to virtual memory has to be extremely fast, because memory access is one of the most important factors influencing program execution speed, and so the mapping is done in hardware. This is the job of the memory management unit, discussed from a hardware perspective back in Section 4.3.

9.3.5 Storage and Filing

Another function of the modern operating system is to provide persistent storage of information. If you think about storage for a moment, the important facts are that a piece of information gets stored somewhere, and can later be retrieved without error.

While storage media has different physical attributes, we can be fairly sure that the average programmer does not want to learn how to specify the exact location on the surface of a hard disc in which to store information, and does not want to learn exactly how flash memory hardware works in a solid state drive. Instead, they are able to simply pass the information to the OS and effectively tell it to "store this."

When saving information to a file, the program specifies the data to be saved and a human-readable filename and path (the path is the location—that is, the name of the device and in which directory/folder) to the OS. The OS then does the actual saving of information, including "talking to" the storage device. Sometimes the programmer will specify the name and location of the file in advance, but often this is something that the user can specify or overrule.

Similarly when reading a file, the programmer simply asks the OS to retrieve a file of information from a named device or location. When the required file is selected, the OS "talks to" the device and reads in the file. If you stop and think for a moment, this is much more programmer-friendly than having to specify the physical position, head, sector, and cylinder on the hard disc, then having to determine the storage protocol and test for errors.

In fact, the modern convenience is made possible both by the OS and by use of something called a filing system, which imposes a degree of structure and organization onto a storage medium like a hard disc. In order to do this, almost every OS provides a set of functions or APIs (application programming interfaces[3]) for file handling.

These are namely

- `int open(char *path, int flags)` specifies a filename and path which is to be opened (depending on flags) for reading, writing, or a combination of both. When writing, this will either create a new file that is specified by the name and path, or open a file of that name and path if it currently exists, in which case writing either appends the information to the end of the existing file or overwrites with new data, depending on the flags. The `open()` function returns an integer file descriptor, `fd`, or −1 if the function failed.

- `int read(int fd, void *data, int nbytes)` when given a file descriptor (such as that returned by the `open()` function above), reads the next `nbytes` bytes from the file, starting at the beginning, and places the information into memory specified at the memory address called `data`. It returns the number of bytes successfully read (which might be less than `nbytes` if the end

---

[3] An API or application programming interface provides a way for some user code to run pre-written software library functions by calling a defined function name and providing the required information. The API is the specification that shows how the function is called, and made use of, by other programs.

of file is reached) or -1 when there is no more data to be read or when there was an error with reading.

- `int write(int fd, void *data, int nbytes)` is the opposite, writing the specified number of bytes from the data buffer in memory to the file specified by the file descriptor.

- `close(int fd)` ends the process by closing the file specified by the file descriptor.

The above four functions are standard across languages, operating systems, computers, generations, decades, scales, and sizes of machine. From the programmers perspective, all that is needed is to specify a human readable filename, and keep track of the file descriptor. From within a program this could allow saving to a hard disc, across the network to cloud storage, to a USB drive, a solid-state drive (SSD), punched tape, or flash memory. The important feature is that the programmer does not need to know, or understand, how all of those work; the OS takes care of everything.

9.3.6 Protection and Error Handling

The final useful feature provided by a modern OS is that each virtual machine is protected against errors in the other virtual machines. This was not always the case. In the early days of computing, errors in a real machine would cause a catastrophic halt, bringing everything running in the computer to a complete standstill and potentially losing a lot of information. Since programs of any complexity *always* contain errors, this was understandably a concern, and thus hardware and OS designers devised means to lessen the impact of so-called catastrophic errors. One important method was the extension of the virtual machine concept so that errors within one virtual machine only affect that virtual machine and not others. Adding such protection meant that virtual machines were effectively protected from one another. This also meant that malicious programs running in one virtual machine are not supposed (in theory) to be able to affect other virtual machines—something with obviously positive security implications.

9.3.6.1 Emulation Technology

Apart from those main points discussed above, there are many other important aspects to virtual machines. The technique, at its most extreme, allows different kinds of machines to be run on hardware of another type. This has led to something called hardware emulation which we met in Chapter 8. This allows, for example, Android code to be developed, prototyped, and executed on a MacOS-X device. While the resulting virtual machine is likely to execute slowly, it is useful for developing embedded systems when programmers prefer to write their code on Mac or PC devices rather than directly on tiny embedded systems. Another extreme endpoint is the virtualization technology commonly used in cloud computing, where a PC "instance" can be run as a virtual machine on a cloud server and accessed by a remote user. The user does not know where in the world the underlying hardware running the machine is currently located—and indeed this location and device might change from one session to another, but when they access the cloud they are able to interface with the same virtual machine during each session. If desired, they could then even choose to run multiple copies of the same machine in parallel, providing a scalable and fault tolerant computing resource.

9.4   OS Structure

Computer designers and programmers do not tend to be the kind of people who pass up the opportunity to take unusual approaches when designing systems, so it should be no surprise that there has been a multitude of operating system designs down the years. Fortunately these can be classified into three basic structures:

- *Monolithic* in which the OS and operating code are combined into a single large program. This is very unusual in larger systems but common in single-purpose embedded systems which have limited memory availability. The design can be difficult to maintain or enhance, but is also quite secure through being very hard to reverse engineer or hack into, due to the degree of customization involved.

- *Layered* operating systems divide their functionality into modules that are layered between user programs and the underlying hardware.

- *Client-server* operating systems consist of modules that have roughly equal status, that communicate by sending messages and requests. The client-server OS is also referred to as a microkernel approach (microkernels are very small and compact items of OS code that are called when needed, in a kind of client-server arrangement).

Despite these neat categorizations, most modern operating systems take a pragmatic approach and are essentially a hybrid of different features.

Looking at the structure of operating systems, most can be divided into two or three main parts. The first is the core essential code named the *kernel*, the second are the associated programs that are needed alongside the OS for maintaining, controlling, and communicating with it. The third are device drivers or modules that are often provided as separate blocks of code (i.e., not part of the core OS kernel). The kernel is explained further in Box 9.1.

Box 9.1   More about the Kernel

The core layer of a layered system or the major part of a monolithic system is the kernel. This contains the most critical code. Since errors in this code are almost certain to crash the system, it should be as small, simple, and reliable as possible. it also needs to validate any requests, connections, or inputs, to prevent external processes from causing a crash.

In general, OS kernels provide the means for

- Process switching and control
- Mechanisms for communication and synchronization
- Interaction with the actual hardware (e.g., drivers)

In practice, certain other functions are needed for convenience or reliability, but the main considerations are size and reliability. Some operating systems split the kernel into two sections, an "upper" and a "lower" part. The upper part is always the same and provides a consistent interface for code and user programs to interact with whereas the lower part changes on different hardware. In Windows and some other operating systems like VxWorks, the lower part is referred to as a hardware abstraction layer (HAL). In Linux the functionality of the lower part is mainly provided by device drivers or kernel modules.

A very basic kernel contains three main components which are a first-level interrupt handler (FLIH), a low-level scheduler, and communications primitives. For modular systems, anything else is optional and can be added on as needed. This is the case with the Linux kernel. It is also popular (especially among programmers) to class anything else that deals with "hardware" as being part of the kernel, such as connectivity, memory management, and so on, and this meaning has grown over the years to allow "messy" hardware to be abstracted as far away from the programmer as possible!

The *first-level interrupt handler* is the piece of code that is executed whenever an interrupt occurs. Discussed more fully in Section 6.5, an interrupt or exception causes the processor to execute code at the corresponding interrupt vector, which is usually a branch instruction. This branch will jump directly to the FLIH which, when run, first saves the machine state (context) and then cause the interrupt to be serviced. It then makes the interrupt service routine runnable, giving it very high priority, and then calling the scheduler.

The *low-level scheduler* performs process start-up and switching. It allocates, initializes, and updates something called a process control block (PCB), which describes each process and is used for switching. Process control will be described further in Section 9.6, and scheduling will be discussed in Section 9.7.

9.4.1   Layered Operating Systems

The most prominent examples of layered operating systems would be Linux and Android (which is itself layered upon the Linux kernel). The diagram in Figure 9.3 shows how the operating system kernel is structured in layers between user programs (top) and hardware (bottom). Take the example of an EIA232 serial port device. The hardware is part of the "Console, serial" block at the bottom. This is serviced by a hardware driver in the kernel which is a "character device." A user program (at the top of the diagram) which wishes to use a serial port simply needs to interface to a block called "device control." In practice, the user program would open a special file using the standard open() function of Section 9.3.5 and then read or write data in the normal way using read() or write() functions. In Linux, several special character device file nodes exist in the file system which correspond to hardware interfaces. In a typical Linux system, the file named /dev/ttyS0 (along with many other files in the /dev directory) are used to access hardware devices. Thus, a user program that opened the file /dev/ttyS0, wrote several bytes of data and then closed the file would—assuming it has sufficient security privileges to write to hardware—have caused the serial port to output those several bytes of information.

The exact sequence of events starts with the user program calling the open() function. This is an OS system call, which prompts the OS to first determine if the user program is allowed to open such a file (i.e., does it have sufficient security permissions). To do this, it will look on the file system to examine the security flags for that file—every file in the Linux file system has security flags or permissions associated with it. It will also look at the type of file. In this case the file is a special type called a "character" file. A character file has two numbers associated with it called a major and a minor number that, together,

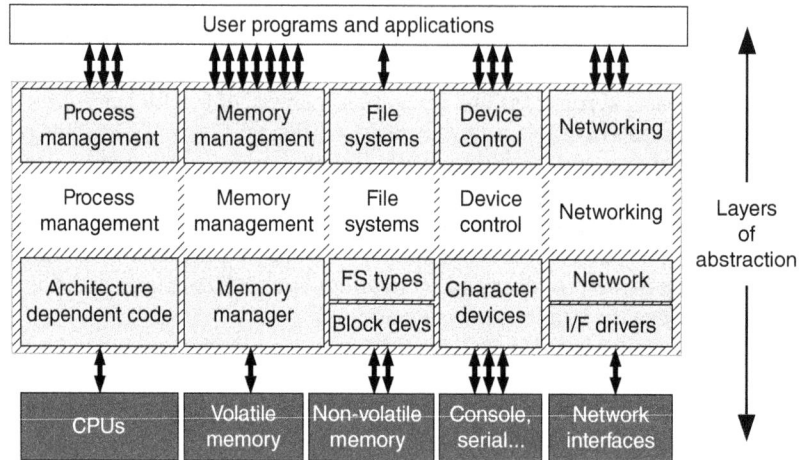

FIGURE 9.3   The layered nature of the Linux operating system, between user code on the top and hardware at the bottom, inspired by a similar diagram in Alessandro Rubini's famous book *Linux Device Drivers* published by O'Reilly Press.

identify which device driver handles that kind of file, and what it relates to (e.g., which serial port, if there are several). Now knowing that the user program requested to open a special file, it has permissions to open it, and knowing which device driver handles that file, the kernel will pass the open request to the device driver to handle.

Within each device driver there are functions to handle open, close, read, and write functionality (and often a lot more besides). The kernel simply passes any such user program file requests to the corresponding device driver functions. Each device driver has a part that handles the user program requests, and a part that handles the low-level hardware control. It thus spans the middle layer in Figure 9.3.

While this kind of functionality might not at first glance sound important, it has the considerable benefit of allowing anyone with the ability to open, read/write, and close a file, to control hardware. The exact brand of serial port, the complex internal workings, and the timings are all taken care of by the OS. Meanwhile this structure also allows device driver code to be very regular—one device driver is usually written in a very similar way to other ones.

The final point is that device drivers in Linux are usually written as loadable kernel modules. These are not normally resident within the kernel until they are required, at which point they are loaded in and plugged into the kernel. Some time after they have finished being used, they will be automatically unloaded (unplugged). This has the advantage of keeping the main part of the kernel small and efficient, while allowing functionality to be expanded and contracted as it becomes necessary.

9.4.2   Client-Server Operating Systems

In a client-server OS, the basic kernel is small, and provides only basic services, which include communications between modules. Much of the functionality is contained in modules that each handle one particular aspect of the system's functionality.

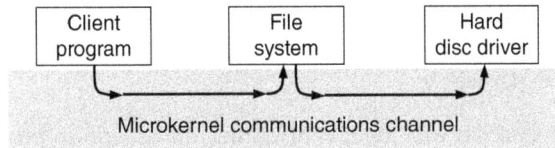

FIGURE 9.4 Microkernel operating system diagram showing the microkernel communications channel connecting a client program (user code) with a device driver, via the file system module.

Figure 9.4 shows how the microkernel (lower layer) allows communications between user programs and modules. This horizontal communication process is quite similar to the vertical communications in the Linux OS, but has the advantage of regularizing communications channels and the disadvantage of sharing communications channels, which has security implications.

Windows NT/2000/XP/Vista/7/8/10 have partly object-oriented modules but are essentially a client-server operating system. Unlike Linux, programmers are not encouraged to delve into the core Windows operating system and begin to rewrite parts of the OS code, so there is not the same level of public information available regarding the internal kernel structure. However, despite the different approach, different structure, and degree of openness, Windows still provides essentially the same open, read/write, and close functionality as the UNIX-like operating systems.

9.5 Booting

Although this is not strictly a function of an operating system, there is a close relationship between the OS and the sequence of events between powering a computer on and it being ready to execute programs. Computers, of whatever size, contain a mixture of non-volatile memory (e.g., flash memory or HDD) and volatile memory (some form of RAM). When powered off, the volatile memory is blank, and so when powering on the computer, everything needs to start by running some code in non-volatile memory. The process of going from power on to a fully running system is called *booting*. We have already examined how the code resident at the start-up, boot, or reset vector (depending on which processor is being used) gets executed automatically at powerup for an example ARM system in Section 7.8.1. This outlined the precise sequence of events that needed to occur in that particular embedded system, to launch the bootloader code. In the same section, we also discussed the role of the bootloader in such systems.

However, computers vary widely. The precise sequence followed depends on the type of CPU, the boot medium used, which bootloader is run and what OS will be used (if any). Not only is there wide variety between systems, but many systems have a large number of configurable options. In this section, we will consider matters from a software perspective for fairly typical embedded Linux systems. We will look separately at how such systems can boot from two different media—hard disc drive (HDD) and parallel flash memory (non-volatile memory, usually a variety of NOR-flash). From a software perspective a solid-state disc (SSD) works similarly to an HDD, even though it consists internally of flash memory (although in this case it is usually higher density NAND-flash), and so we consider HDD and SSD together, and parallel flash separately.

When any embedded Linux system boots up, from a software perspective, there are a few well-defined steps in the sequence no matter what medium is being used to start up from. These are namely

- A *bootloader* is executed, configuring the system to prepare for the kernel and ramdisk.
- The *kernel* is executed, decompresses itself (if compressed) and then sets up the system memory and hardware.
- The kernel looks for a *disc image* which contains the root file system. When found, this is mounted, and initialization code residing on there is executed.

The bootloader is a very low-level and basic system utility that configures hardware before finding and launching the kernel. We met a common bootloader in Section 7.8.1, where we also described the features and functions of bootloaders in general. In that section, the ARM was booting from parallel flash, and both kernel and ramdisk were compressed.

The Linux kernel is an executable called vmlinux, bzimage, zimage, or similar, that is usually *mostly* compressed using either the `gzip` (GNU zip), `bzip2` or `xz` programs. The only part that is not compressed is the very first few bytes of the kernel which contains a small decompressor. When the kernel is first executed, this small piece of code decompresses the remainder of the kernel into RAM, and then computes a CRC (cyclic redundancy check), which is a kind of checksum that is used to ensure there were no errors when decompressing the kernel. If an error is detected in the decompressed kernel, the kernel will simply print "CRC Error" to the console and then immediately halt at that point. If no error is detected, then the code immediately jumps to the newly decompressed kernel code and continues by executing it.

Next, the kernel will look for a disc image in a particular location in memory or on the HDD (depending upon its configuration and command line parameters), which it can mount to become the root file system. There are a large number of options and possibilities, but these generally divide into two classes of systems that boot from parallel flash and systems that boot from HDD/SSD.

We will examine what happens after booting in Section 9.5.3. Before that we will look at the different mechanisms that are normally used to boot in each of our two classes of system. In both cases, the systems start turned off and so main memory (some form of RAM) begins empty. We will follow both cases until they have mounted a ramdisk and are about to begin execution of user code.

9.5.1 Booting from Parallel Flash

The most common method of booting embedded Linux systems, this was the method described in Section 7.8.1. To recap briefly, and with reference to Figure 9.5, the following sequence occurs:

1. The system is turned off, RAM is empty, flash memory contains a bootloader, a compressed kernel and a compressed ramdisk.
2. Power is turned on. The CPU starts execution at the boot vector (on an ARM, this is usually address `0x0000 0000`, which either jumps to the bootloader, or is the start location of the bootloader).

1. Power off state   2. Ready to execute kernel   3. Linux running from ramdisk

Flash	RAM		Flash	RAM		USB/flash device is no longer mounted to the system	RAM
							system memory
ramdisk (bzip2)			ramdisk (bzip2)				ramdisk mounted
kernel (comp.)			kernel (comp.)	kernel (comp.)			
syslinux			syslinux				kernel
MBR			MBR				

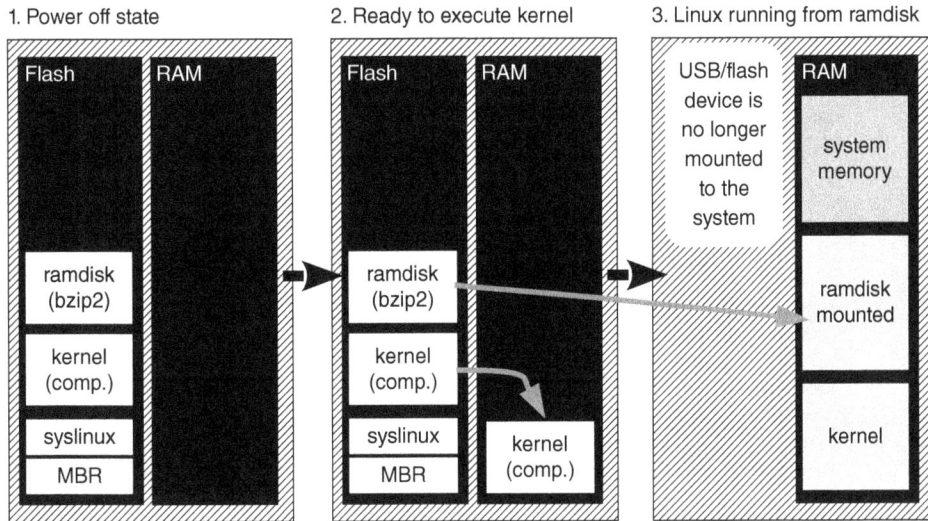

FIGURE 9.5   Diagram showing three separate stages during the process of booting an embedded Linux system from parallel flash memory. Booting from HDD/SSD is shown in Figure 9.6.

3. The system executes the bootloader which sets up hardware and configures the system memory (particularly SDRAM, which needs to follow an initialization process). It usually provides some way for the programmer to access configuration options.

4. The bootloader finds the compressed kernel in flash memory and copies it into RAM.

5. The compressed kernel is executed. It decompresses itself (moving to a different location in RAM in the process).

6. The kernel completes system setup and then looks for a compressed ramdisk in flash memory. This is decompressed into RAM.

7. The ramdisk is mounted as a root file system and the Linux `init` process begins.

The final stage, shown in Figure 9.5, is that everything is operating from RAM. The root file system is a ramdisk, resident in RAM, the kernel is running from RAM and the system memory for running code is also in RAM. Flash memory is typically not used once the system is booted and running.

This might seem wasteful; however, it is very pragmatic. Both kernel and ramdisk were stored compressed, and had to be decompressed to execute (decompression needs a read-write storage system, and the only space large enough will be in RAM). The root file system on Linux also has to be read-write, as does the kernel (e.g., to support loadable kernel modules). Despite these requirements, many alternative options are possible, including splitting the file system into read-write and read-only parts and potentially storing a kernel uncompressed and executing directly (probably difficult with Linux, but very easy with some of the smaller embedded OSs).

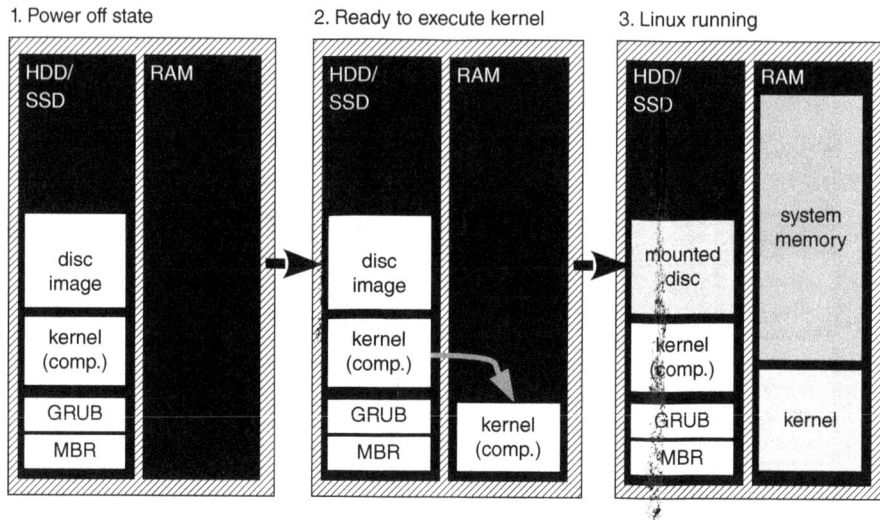

FIGURE 9.6 Diagram showing three separate stages during the process of booting an embedded Linux system from HDD, SSD, or even USB-connected flash storage. Booting from parallel flash memory is shown in Figure 9.5.

9.5.2 Booting from HDD/SSD

Booting from HDD is the most common method for larger Linux systems, deriving from standard desktop PCs. The process shown below, describing an Intel Atom-based embedded system, is almost identical to running Linux on a modern desktop computer. Exactly the same process is used for booting from an SSD, and even for booting from USB-connected storage (if the BIOS supports this). The process is shown in Figure 9.6 and ends up with a hard disc image being used as root file system while the kernel executes from RAM.

1. The system is turned off, RAM is empty, the HDD contains a bootloader, a compressed kernel and a disc image, probably on a separate HDD partition.

2. Power is turned on. The CPU starts execution at the boot vector which contains (or is redirected to) a BIOS (basic input output stream).

3. The system executes the BIOS which sets up hardware, configures the system memory and usually provides some way for the user to access the configuration options. The BIOS also configures the storage medium. This could be hard disc (or solid state drive), USB-connected flash memory, or even boot over a network. Typically, it boots from HDD, and thus needs to setup the disc controller first.

4. BIOS looks for the master boot record (MBR) on the selected storage medium. This is a specially organized section of the disc (or other medium) that instructs BIOS about which program to run from disc.

5. It could run literally anything from this point, but would typically execute a bootloader such as GRUB, LILO or syslinux. These small programs are similar to the bootloader we had seen for smaller embedded systems, but typically have much greater functionality in terms of media they support.

6. The bootloader executes, finds the compressed kernel on HDD, and copies it into RAM.

7. The compressed kernel is executed. It decompresses itself (moving to a different location in RAM in the process).

8. The kernel completes system setup and then looks for a root file system (RFS), which is typically located on a separate partition of the HDD.

9. The root file system is mounted and the Linux `init` process begins.

The boot system is so configurable that it can be made to do many things, but we have presented the most normal, straightforward way. Interestingly, the compressed kernel is often (but not always) found *inside* the RFS, so in step 6, the bootloader is actually reading from the RFS before it has been mounted.

9.5.3   What Happens Next

After booting, the kernel (plus all required kernel modules and processes) will have been launched and will be running normally. For Linux, a root file system *must* be present (the important role of the root file system will be explored more in Section 9.8.3.2). Larger operating systems generally require file systems, but some of the smaller OSs like VxWorks and uCos tend to package all functionality and operating software up into a single monolithic executable. This would take the place of the kernel, and would not need a separate RFS—making these systems easier to get working when developing new embedded systems, but significantly reducing flexibility since the entire OS must be rebuilt and reloaded whenever even small changes are made to operating code (with Linux, the kernel is built and, once working, remains unchanged while the code on the RFS is modified and developed).

So to recap, for embedded Linux, the situation after booting is;

• The kernel is loaded, decompressed into RAM and is running.

• A decompressed ramdisk or HDD partition has been mounted as the root file system [giving a root (/) directory and everything below it].

Next, the kernel searches for an `init` program to execute. In most embedded systems, `init` is part of Busybox—an amazing executable that incorporates almost all UNIX tool programs, utilities and low-level helper applications. The Busybox developers call it the "Swiss Army knife of embedded systems," and this is no exaggeration! On desktop and larger systems, `init` is a standalone program. Whichever way it is provided, it has only one job, and that is to read a configuration script that launches every other program. At this point the system is not yet multitasking in user space; so far it has not run *any* user code at all—that will happen once `init` begins executing its configuration script.

Since `init` is so important, the kernel has some built in backup strategies for situations where it cannot find a program called `init` in the right places on the file system. These include searching in different places and, as a final backup, launching the default system shell (`/bin/sh`). It has been known for a novice embedded systems developer to corrupt either `init` or its configuration script (and some of us have done this more than once), but we will assume here that everything is working correctly.

The usual name and location of the start-up configuration script is `/etc/inittab` (we will see later that a lot of configuration material in Linux is located in the `/etc` directory), and this is a text file that is supposed to be human readable. Once init has found /etc/inittab, it performs the actions specified in that file. Obviously the precise format of /etc/inittab depends on which init program is being used—and these differ between desktop and embedded systems, but a typical inittab for an embedded system running Busybox would be as follows:

```
::sysinit:/etc/init.d/rcS
#
# Put a getty on the serial line (for a terminal)
::respawn:/sbin/getty -L ttyS0 115200 vt100
#
# Set up a main console
::askfirst:/bin/sh
#
# Restarting the init process when needed
::restart:/sbin/init
#
# Before rebooting, run this
::ctrlaltdel:/sbin/reboot
```

Without needed to explain the format in detail, this listing shows separate sections for system initialization (the lines at the top) and things to do when rebooting (the final lines). Lines beginning with a hash (#) are comments. The script specifies a serial interface (called a teletype terminal or TTY) running over the first serial port /dev/ttyS0, a main console and a line that restarts the init process when necessary (i.e., if it crashes). The respawn command for the serial line means that the process is relaunched if it terminates or crashes and the askfirst command for the main console launches it only if the user "requests" it by pressing a key. Under the various circumstances shown, it calls additional programs such as /sbin/getty, /sbin/reboot, /bin/sh but the main task is to run the system start-up code /etc/init.d/rcS.

/etc/init.d/rcS is a standard shell script, written in the system's shell language that the developer can edit to change the start-up behavior of the system. Usually, the software developers will not be allowed to edit or change /etc/inittab directly but could definitely modify /etc/init.d/rcS to change the after-boot sequence and tasks. Such tasks include printing out a welcome message, setting up networking, and launching the main system programs.

On desktop machines, this will be where the GUI is finally launched from, where the system first starts to connect to the Internet, and will run the code necessary to allow different users to "log on" to the system. In software terms, it is the start point of "user space"—everything up to this point has been running from "kernel space."

9.6 Processes

We can think of a running OS as a set of activities that are all competing for CPU time. Some of those activities do work on behalf of the system to manage and maintain itself, whereas others work for the users of the system. Most (but not always all) of the system activities are part of the kernel, while the other activities are generally known as processes. We know from Section 9.3.3 that a process can be thought of as a container within which a program is running (or more properly, a "task" is running, because some programs are written so that they split into separate independent tasks). Most processes are "visible" to users who are interested to examine them, and can be viewed on most computer systems as described in Box 9.2.

Box 9.2   Finding Out about Processes

Most operating systems provide a means of viewing the activities of different processes on the system, including desktop systems as well as mobile systems running Android and iOS.

On a Windows computer, at time of writing, users can press Ctrl-Alt-Del and select "Processes" to view a list of what is currently running. Any UNIX-based system such as Linux, FreeBSD, and MacOS-X will provide similar information using the ps command (as well as several other ways, including through a GUI or, in Linux by examining the /proc directory.

Executing ps ax to list all processes in the Terminal my 2015 Apple laptop (running macOS Sierra version 10.12.4) gives about 370 lines of currently running process. The first 20 lines (ps ax | head -20) are as follows (where the longest lines have been truncated at "…" and two lines have been deleted to ensure my own security);

```
PID  TT  STAT     TIME COMMAND
  1  ??  Ss   65:00.11 /sbin/launchd
 54  ??  Ss    5:38.85 /usr/libexec/UserEventAgent (System)
 55  ??  Ss    2:10.75 /usr/sbin/syslogd
 57  ??  Ss    0:49.64 /System/Library/PrivateFrameworks/...
 58  ??  Ss    3:27.56 /usr/libexec/kextd
 59  ??  Ss    9:11.41 /System/Library/Frameworks/...
 61  ??  Ss    1:13.64 /opt/cisco/anyconnect/bin/vpnagentd ...
 62  ??  Ss    1:26.20 /System/Library/PrivateFrameworks/...
 66  ??  Ss    0:10.86 /System/Library/CoreServices/...
 67  ??  Ss    4:44.88 /usr/libexec/configd
 68  ??  Ss    2:15.04 /System/Library/CoreServices/...
 69  ??  Ss    0:22.23 /usr/libexec/mobileassetd
 76  ??  Ss   51:48.19 /usr/libexec/logd
 80  ??  Ss   42:34.00 /usr/libexec/airportd
 82  ??  SNs   0:07.67 /usr/libexec/warmd
 83  ??  Ss   22:28.61 /System/Library/Frameworks/...
 88  ??  Ss    0:02.55 /System/Library/CoreServices/...
```

In a Linux-based system, the first process in the list would be init, but otherwise the format is similar—and lots of system and library code is being executed.

9.6.1   Processes, Processors, and Concurrency

Any computer system contains one or more processors which the processes run on. The most obvious processor in the computer is usually the CPU, and there might be

A machine with one CPU

Running on the CPU

Runnable processes

Suspended processes

Time

A machine with two CPUs

Running on CPU A
Running on CPU B

Runnable processes

Suspended processes

Time

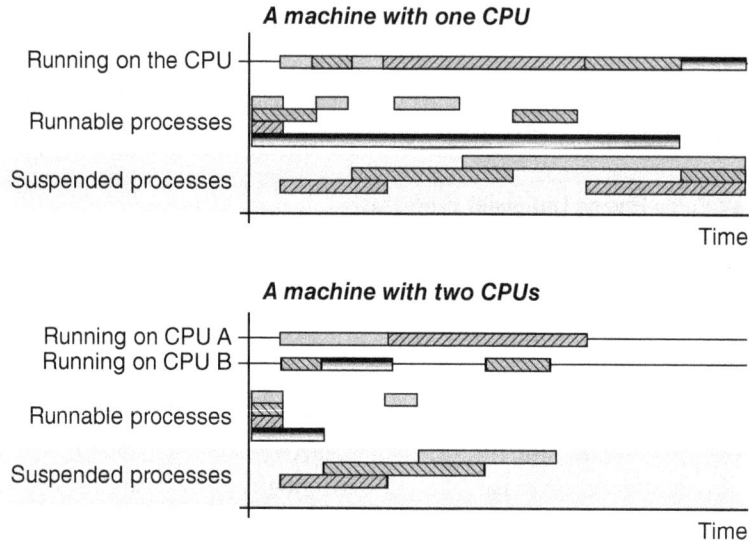

FIGURE 9.7 Two task diagrams showing roughly the same tasks being performed on single and dual core computers. In the former, tasks spend longer in the runnable state (waiting to execute). The actual CPU time occupied by each task in both cases is similar, but the elapsed time between first becoming runnable and task completion is very different.

more than one of those. A multicore machine may contain several CPUs, each of which can comprise several processor cores (each core is a separate processor, or nearly so—something we examine more closely in Section 5.8.1). Other processors include built-in GPUs and dedicated acceleration hardware (all discussed in Chapter 5). These days, devices ranging from an iPhone to a Playstation 4 are actually multiprocessor machines.

If, at any time, there are N processors, a maximum of N processes can be running simultaneously (although many more may be runnable, waiting for their chance to execute).

In its simplest form, concurrency is when several processes are active at the same time. On a single CPU machine, this is always just an illusion produced by the OS. Processes "think" that they are all running continuously, but in reality are being switched on and off very rapidly as we discovered in Section 9.3.3. In modern OSs, this kind of task switching is involuntary and unpredictable. The program cannot predict and can never really know when it is running and when it is waiting. In a multi-core machine, since there are several processors, it is possible that programs are experiencing *real* concurrency (not just the illusion), but processes are still switched on and off and will still be unaware of it. Figure 9.7 presents a diagram of a set of tasks being executed on single and dual core systems. In each case the tasks become runnable at the same times and take roughly the same amount of CPU time, but the dual core system, as expected, completes the set of tasks earlier—and probably does so with fewer task switching operations. Task switching is very fast, but it still requires valuable CPU time, so minimizing the amount of time spent on switching is generally a good way to improve efficiency.

To handle concurrency, processes and processors need to be able to synchronize themselves as well as communicate in some way (e.g., to share data), and these activities therefore need to be at the heart of any operating system. In computer science terms, when we talk about concurrency, we also need to mention two points.

9.6.1.1 *Mutual Exclusion*

When more than one process competes for a common, but un-shareable, resource (e.g., updating a single file, printing a page to a printer, or a single CPU), these processes have mutually exclusive access. Mutually exclusive means that at most one of them can have access, but never both at the same time.

Imagine what would happen if two programs trying to print a page to the printer were *not* mutually exclusive—you would get half a page of printout from one program, then half a page from the other program. It would be much better if the OS was able to enforce mutual exclusion: It could grant the first program exclusive access until that program had finished printing, and only then grant access to the second program for it to print.

Mutual exclusion is often abbreviated as "mutex" and is a type of exclusion that can be enforced by the kernel on external tasks (and indeed also within the kernel itself for internal OS operations on shared resources).

9.6.1.2 *Synchronization*

Mutual exclusion requires communications between tasks so that they can be synchronized, and such synchronization is generally provided by an OS. Apart from mutual exclusivity, cooperating tasks often need to sequence themselves so that one operation occurs before another, and so on. Process synchronization is the action where a process requires some other process to have reached a certain point in its execution, before it can itself continue. It is the activity of one process giving work to another, or one process setting something up for another process and waiting until it has been done before continuing.

There are many ways of synchronizing tasks. These include semaphores (signals), message passing and mailboxes. Semaphores were developed by Edsger Dijkstra (the most common form is thus known as a Dijkstra semaphore) and are simple building blocks, but these simple objects can be used to achieve very complex functionality in reality.

9.7 Scheduling

On any system that has multiple tasks or processes—which includes the vast majority of modern operating systems—there needs to be a way of deciding which task runs on the CPU at any particular time. To do this there needs to be a way of telling the CPU which task to execute, and then a way of switching so that another task gets executed instead. We looked at *task switching* in Section 9.3.3 but did not yet consider how the OS chooses which task to execute at a particular time. Box 9.2 examined a list of processes, displaying various items of information about each, but we had not yet linked this to task (process) switching. Meanwhile Box 9.1 looked at the OS kernel, mentioning its most important functions. This included a low-level scheduler, but we have yet to examine the operation of this.

In many ways, the scheduler is the part of the OS that controls task switching, and can therefore be likened to the beating heart of the OS.

9.7.1 The Scheduler

The low-level scheduler controls process start-up, switching, and termination, following a complex decision-making process to decide which process runs at which time. As we discussed very briefly back in Section 6.4.4, processes can be in one of three states: *running*, *runnable*, or *sleeping*. Sleeping processes are not ready to be executed, instead they are waiting for some event to happen or some resource to become free—generally they are waiting for a semaphore or signal from another process. Runnable processes are those that are ready for CPU time and can be executed right away if allowed the chance. The running processes are those lucky ones that are currently allocated CPU time and are thus actively executing at present. There cannot be more running processes than there are CPUs or cores in the machine.

We will begin looking at the role of the scheduler by examining process start-up, which can be considered a special case of switching, and then look at switching itself.

9.7.1.1 Process Start-Up

When starting up a new process, the scheduler first allocates it a data structure that will be used to keep track of and manage the process. This data structure, in the form of a *process control block* (PCB), is created by the scheduler in the empty state and then initialized. It will subsequently be used for scheduling as well as to provide the raw information shown in the process list, such as the one reproduced in Box 9.2.

One PCB, also sometimes known as a process descriptor, is maintained for the lifetime of each process. The information it contains includes

- Saved contents of hardware registers (register state)
- Owner identification
- Process name and number
- Privileges (there may be several sets)
- Current state (stopped/running/runnable)
- CPU time, elapsed time
- Resource usage (e.g., memory, open files)
- Resource limits
- Static priority level and dynamic priority level

The PCB is not accessible by the process itself; in fact, it is not directly accessible from user space at all, but is purely for use within the kernel. The scheduler reads this block, using the information for decision making, but is also responsible for updating it. On process start-up, the scheduler also needs to grant the new process a dedicated memory space in which to run (in fact this comprises several areas, including for code and working variables), and perform any memory copy that is needed to locate the process code within its allocated memory area.

Once the process is set up, the scheduler adds the newly initialized PCB to the list of runnable processes, and updates other system tables to account for the new process and its resources.

9.7.1.2 Process Switching — When to Switch

On a single-CPU machine, if some process is currently executing on that CPU then no other code can be run at the same time (there is only one CPU and nothing else can execute code). If the scheduler is itself not executing, it cannot stop or affect the current process. So unless something else intervenes or happens, the current process could continue executing on the CPU—potentially forever if the code contains an infinite loop.

What happens is either the current code calls a system-level function such as `write()` or `malloc()`, or a hardware timer in the system causes an interrupt to occur. System-level functions are handled by the kernel, and interrupts are also serviced by the kernel. In both events, some kernel code is executed, and the kernel takes the opportunity to execute the scheduler function before returning back to the calling process. The scheduler generally handles interrupts first, then handles system-level or internal kernel functions, and finally will allow one of the runnable processes to execute.

In normal operation of a computer, with multiple programs (processes) being run seemingly concurrently, the kernel scheduler will be executed more often than probably any other piece of code. It will run any time an interrupt occurs, including timer interrupts (such as a microsecond timer that interrupts every microsecond), and any time a system-level function is run—which includes all forms of input or output from the system, any file handling, or network operations.

Some early multitasking operating systems used a time slicing approach, triggered by a timer interrupt. Early multitasking versions of Microsoft Windows would interrupt every 100 ms to run the scheduler. While the timing was configurable, 100 ms is approximately the shortest latency that humans can perceive without the delay becoming annoying. So, for example, pressing a key and getting a response 100 ms later feels much better than getting a response 200 ms later, whereas a 10-ms delay is not really distinguishable from a 50-ms delay. The idea was, presumably, to improve human-computer interaction. Unfortunately the delay was too lengthy in practice to allow the computer any semblance of real-time operating capability, and too short to be efficient. The systems had a reputation for sluggishness, and time slicing was rapidly preempted by a combination of much faster timer and triggering at actions such as system-level function calls.

9.7.1.3 Process Switching — How to Switch

Possibly the most important part of process switching means choosing which process gets to use the processor next (could be more than one process on multi-core and multi-CPU systems). Switching does not always occur; sometimes the current process needs to continue running, and has high enough priority to do so, in which case it can simply resume where it left off. But when processes are switched, the scheduler first selects the next process to run, and then saves the context (state) of the current process into its PCB. Next it loads the context (state) of the selected process, making it the current process, before passing control to the current process to allow it to execute.

9.7.1.4 Process Switching — Selecting the Next Process

Different operating systems have their own ways of choosing which process to run next. This is related to the theory of scheduling real-time operating systems (see Section 6.4.4 and Box 6.4) but tends to be a lot more pragmatic in general-purpose computers. In most cases, apart from simple round-robin scheduling, the choice of which process to execute next is governed by the relative priority among the runnable processes.

Priority is itself formed from some combination of the following factors:

- Base priority (importance)
- Urgency (timing criteria)
- Previous resource usage
- Performance
- Resources being used

In Linux, whenever the scheduler runs, it chooses from the set of runnable tasks based on their *dynamic priority*, according to the current scheduling policy.

The Linux function `sched_setscheduler()` defines the scheduling policy and parameters for the specified processes (specified through their process identifiers or PIDs). Any user (with root access) can call this function in their programs to change how they are scheduled.

While more embedded and exotic versions of Linux define various alternatives, there are three basic scheduling policies in standard Linux, as follows:

- SCHED_OTHER is the default time-sharing preemptive scheduler. It uses a static process "nice" level and a counter that determines how long each process has been waiting, to define a dynamic priority (such that the dynamic priority increases the longer a process has been waiting).

- SCHED_FIFO is the first-in, first-out preemptive non-timesliced scheduler that will run a high-priority process to completion first. Then it will allow newly runnable processes to preempt lower priority processes.

- SCHED_RR is like the above SCHED_FIFO but includes round-robin capabilities so that high-priority processes do not directly run to completion, but are instead time-sliced so that other processes can run in the meantime.

9.7.1.5 Changing Process Priorities

In UNIX-based operating systems such as Linux, FreeBSD, and MacOS-X, the system user can make use of a command-line utility called `nice`. When used at the shell prompt, this can run a program and set its base (or static) priority to one of 40 levels ranging from −20 to 19. For example,

```
nice    ./myprog   -5
```

A similar program called `renice` can change the priority of a running process, specified by its PID, assuming of course that the current user has sufficient security privileges to change the priority of the selected process. Be aware that a "nice" program does not try to keep all of the CPU time for itself, so if we were to decrease the "niceness" of PID 232, for example, like this:

```
renice   -9   232
```

Then its priority will actually *increase* as it is being *less* nice than before, and it will be allocated more CPU time. By contrast, if we increase its niceness to +19, it will become a very nice program indeed, sharing more of its CPU time and executing much slower. Effectively, a more "nice" process is more willing to share CPU time than a less "nice" process.

For those who are keen to experiment with scheduling processes and priorities on their own computers, the `sched_setscheduler()` man (manual) page carries a timely warning:

> "As a non-blocking end-less loop in a process scheduled under SCHED_FIFO or SCHED_RR will block all processes with lower priority forever, a software developer should always keep available on the console a shell scheduled under a higher static priority than the tested application.
>
> This will allow an emergency kill of tested real-time applications that do not block or terminate as expected."

9.8   Storage and File Systems

In general computer architecture terms, the classical way to consider memory is in terms of *primary storage* which is fast and volatile. This is the main memory used for running programs to store variables into, and for a working stack. *Secondary storage* is slower, larger, and provides non-volatile semi-permanent memory. Cache memory is sometimes considered to be a third type of memory whose role (see Section 4.4) is in being fast, usually expensive (and hence small) storage used to reduce the average access time of larger slower memory. We could use a cache with primary or secondary memory or even use different levels of cache so that a very fast cache is used to speed up the average access time of a slower but larger cache, which itself speeds up average access time of slower memory storage. Cache is always made from volatile memory in practice.

Non-volatile memory retains its contents when the power is turned off. This could be programmable non-volatile memory (like flash, EEPROM, EPROM), or fixed non-programmable memory (like ROM and mask-programmed ROM). The earliest computers used wires and plugs or punched card for their non-volatile memory.

Volatile memory, by contrast, gets wiped whenever the power goes off. This type of memory includes SRAM and DRAM (including SDRAM, DDR, etc.).

Section 7.6 discussed memory, including all of those individual technologies mentioned above.

9.8.1   Secondary Storage

Having already discussed memory, and demonstrating the need for both volatile and non-volatile (NV) devices, we turn our attention to types of NV memory. Most embedded systems use flash memory for NV storage of program code and data (sometimes called *mass storage*), but in the PC world, the standard is currently the hard disc drive (HDD) or solid state drive (SSD).

We can characterize mass storage devices in several ways:

- Cost per Mibyte
- Power consumption (per Mibyte, to keep the data stored)
- Power consumption (per Mibyte, during accesses—often is higher during writing than reading)

- Data density (Mibytes per cubic centimeter)
- The type of interface (USB, ATA, FireWire, etc.)
- How large, heavy, or slim the storage device is
- Other physical aspects like it being fragile in use, or fragile to store
- The power supply requirements (generally 12, 5, or 3.3 V)
- Access time (usually faster to read than to write)
- Data retention time (e.g., 10 years)
- Write/erase cycles (e.g., 10,000, 100,000, or more)

At the time of this writing, for very large amounts of data storage (above a few tens of Tibytes), tape is still the best choice of medium. For between a half and a few Tibytes, the HDD rules, but below that size, flash memory is king. But the kingdom of flash is growing ever bigger year by year, and there are other technologies under development which promise either larger, faster, or lower power storage in future.

9.8.1.1 *Storage for Different Scales*

Large systems like desktop and notebook computers typically rely on HDDs for secondary storage, and therefore OS support is good for this. SSDs "look" like an HDD to the OS and to the programmer, with the same interface (and sometimes they even look physically the same too), but have flash memory inside. These are thus used interchangeably with HDDs. For the smallest embedded systems, where size and power are critical, flash memory is used, either parallel NAND flash or serial NOR flash.

Medium-sized intelligent systems fall into a transition area in which developers may prefer the convenience and capacity of an HDD but wish for the compactness and robustness of flash.

At the time of this writing, SSDs are becoming more popular than HDDs, particularly in notebook computers. Although they appear similar to the programmer, they do have different characteristics. Let us briefly compare HDD, SSD and standalone flash memory.

- *HDD*—very long data retention time, the largest capacity, have a standard interface (usually SATA), slow access time but a built-in cache speeds up average access time. Tend to be fragile in use and bulky.

- *SSD*—lower capacity than HDD, significantly more expensive, but physically smaller and lower power. Often with the same SATA interface, SSDs are typically slower to write large blocks of data but faster to read data than HDDs. At the time of this writing the data retention time of an SSD (how long it reliably stores data for) might be 10 years, but does depend upon how heavily it is used.

- *Flash*—used either singly or in simple arrays can have similar capacity to SSD, be smaller, less expensive, and lower power. Data reading speeds may be similar to SSDs for small blocks of data, but unless a separate memory cache is used, writing data would be slower (because SSDs have internal cache memory). The main disadvantage of individual flash memory devices is that they have many different types of serial or parallel interface that can vary between manufacturers.

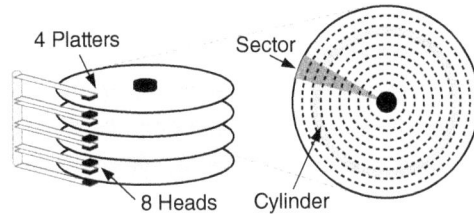

FIGURE 9.8   Rotating storage in the form of a hard disc drive, showing four platters, with a separate read/write head for each surface, and illustrating how sectors and cylinders (also known as tracks) are laid out magnetically on the surface of each platter. In reality there may be tens of platters with hundreds of cylinders and thousands of sectors.

9.8.1.2 More about HDDs

Although they are seldom used in embedded systems, and are being supplanted by SSDs in many other areas, most file systems were designed for use on HDDs or similar rotating storage devices. Even when used in embedded systems, file systems retain the terminology from their history, so it is worth taking a minor detour into understanding a little about this arcane technology.

With reference to Figure 9.8, the HDD contains a number of "plates" called platters that are coated in magnetic material. Both sides of every platter have dedicated electromagnetic "heads" that can rotate on a common spindle to read or write magnetic information anywhere on the surface of the disc. The track that the head skims over on each side of the platter is called a cylinder, and this ring is divided up into a number of sectors on which data is stored. The location of any item of information stored on the disc can therefore be specified by the head, the cylinder, and the sector. These addresses are abbreviated into C, H, and S "coordinates."

When the BIOS on a desktop or notebook computer first boots, it "asks" the HDD (or rather it asks the disc controller, a small CPU built into the HDD) to provide information about its internal capacity. The HDD should reply by giving the total number of cylinders, C, the total number of heads, H, and the number of sectors per cylinder, S.

This would be straightforward, except that some early BIOS manufactures wrote their BIOS software so that it limited the maximum number of cylinders it could support. When HDD manufacturers began making larger capacity discs, they had to start telling lies to the BIOS by reporting an incorrect geometry. As long as the total capacity $C \times H \times S$ was correct, things *usually* worked out fine.

In an attempt to get around this problem, the standard C/H/S geometry was superseded over a decade ago by something called logical block addressing (LBA), which simply numbered sectors from 0 up to however many billions there were. However, even today, if we use a flash-memory-based SSD or plug in a flash-based USB memory device, these will still report back a geometry to the OS in terms of C/H/S. None of these devices really contains any H (heads), C (cylinders), or S (sectors), but the legacy addressing system still continues today.

Any HDD (and most other types of secondary storage media) needs to be formatted before they can be used. Formatting organizes the structure of data on the disc into a logical arrangement through the use of a *file system* (see Section 9.8.2). Often, HDDs and other secondary storage media are *partitioned* before use. Partitioning means dividing the usable space into smaller areas. If they are partitioned, then a separate file system can be used for each partition (i.e., each partition can be formatted differently).

Partitioning might be performed as a means of organizing a disc, for example having separate areas for normal file storage and for backup data, or two separate OS code from user data. It might also be because different characteristics are needed within a system; for example, some partitions could be accessed as read-only to users, whereas some are read-write. Alternatively, some users may want a partition that is extremely fast at storing large amounts of data, whereas others may need a partition that has the highest reliability (which usually implies it is slow).

9.8.1.3 More about Booting

In Section 9.5.2, we examined how a system boots from HDD or SSD. We detailed how the BIOS reads the bootloader from the boot media by looking at the master boot record (MBR). This entire arrangement began when users could insert a different floppy disc (or plug in a different hard disc) to boot a computer. After first powering on, the computer would not "know" where the bootloader could be found, and so a standard developed. On discs (whether floppy or hard) that were specified as being bootable (i.e., able to be booted from and, hopefully, containing a valid bootloader), the first few sectors of the disc (HDD, floppy, SSD and now USB-connected flash memory device) should contain a master boot record (MBR). By convention the MBR is located at $C = 0, H = 0, S = 1$. This sector of data contains first a master partition table and second a tiny block of master boot code.

The master partition table lists up to four primary partitions on the device. Each partition is allocated a type and can be set as being "active" or "bootable" (which means the BIOS can boot from it). The master boot code is the software that the BIOS loads and executes to begin the boot process. It might contain the bootloader itself, but more normally simply contains a small loader that locates and runs the actual bootloader from elsewhere on the disc.

There are hundreds of alternative partition types, but the main ones used today, with their numeric IDs, are

ID	Type	ID	Type
0x0	Empty	0x82	Linux swap
0x1	FAT12	0x83	Linux
0x4	FAT16 <32M	0x85	Linux extended
0x5	Extended (CHS addressing)	0x8e	Linux LVM
0x6	FAT16		
0x7	HPFS/NTFS/extended FAT		
0xb	W95 FAT32		
0xc	W95 FAT32 (LBA)		
0xe	W95 FAT16 (LBA)		
0xf	W95 Ext'd (LBA)		

In Linux, the `fdisk` program can provide a full list, which is a lot longer than the few lines shown above.

Dual-booting a computer means selecting one of two (or more) operating systems when booting a computer. For example, allowing a MacBook to boot up into its native MacOS-X or choosing to boot it into Linux. A triple-boot example would be improving

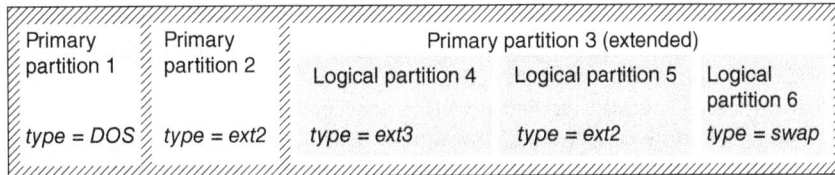

FIGURE 9.9   Example of a partition table containing three primary partitions, with the third one containing three logical partitions.

a PC running Windows so that it can be made to boot up into Linux or into FreeBSD. This is usually (but not only) achieved by having separate partitions for each operating system. Each partition is formatted according to the native file system of the respective operating system.

The MBR, in this case, launches a chooser program that allows the user to specify which OS to boot into. If Linux is chosen, then the boot process continues by executing the boot record at the start of the partition dedicated to Linux. If FreeBSD is chosen, then booting continues by executing the boot code on the FreeBSD partition. Once booted, the systems (particularly for Linux and FreeBSD) can easily access data on the other partitions, but they each have their own "native" space on the disc in the form of their own partitions.

9.8.1.4   Additional Partitions

We mentioned the four partition limit in the MBR. While the original developers probably assumed that it would be sufficient for anyone, it eventually became a bottleneck. As a response, one special primary partition-type number was set aside to indicate that the partition contains more partitions inside itself.

Thus, the partition ID 0x5 means "extended." This partition has its own separate partition table which further subdivides within that partition. An example is shown in Figure 9.9 which is probably from a dual-boot computer, having a DOS or Windows partition plus several native Linux partitions (ext2 and ext3) plus a Linux swap space, used for temporary data storage when running the kernel.

Readers may have noticed in the discussion above that each partition begins with its own boot record. In this case it is known as a volume boot record (VBR), although it operates just like the MBR except that it only describes the content within that partition, rather than within that disc. The boot record for extended partitions has a special name; this is called an extended boot record (EBR). MBR, VBR, and EBR are essentially very similar and perform pretty much the same function; they define the size of their container and what can be found inside.

9.8.2   Need for File Systems

The job of a file system (FS) is to allocate units of space on a storage device and organize files and directories. An FS needs to keep track of which parts of the media are allocated to which file and which are not currently being used, and be able to write data to a file and subsequently be able to locate and read it.

Any partitions that contain programs or data will first need to be formatted with some kind of file system, and it may not surprise the reader to discover that there are hundreds of possibilities to choose from. The UNIX standard, adopted by Linux, FreeBSD, MacOS-X, and other modern operating systems, is for files to be arranged in a single tree structure. The origin of the entire structure is called the root and is denoted in UNIX systems by the symbol /, which we will revisit in Section 9.8.3.2.

The most common Linux file systems—used across all scales of systems from tiny embedded medical diagnosis computers that can be swallowed to the largest supercomputers on the planet—are the range of **ext** (extended) file systems.

These FSs, which are very much intertwined with the history of UNIX and Linux, are explored more in Box 9.3, but in brief the second extended file system, **ext2** is usually considered the Linux default. It is fast and efficient, but not as reliable as **ext3** which adds journaling support (journaling, which will be discussed in Section 9.8.3.3, tends to increase reliability to events such as power failure during writing, at the expense of being slower when writing data). Some quite significant speed and scalability improvements were introduced with **ext4**, and this is also highly configurable (see Box 9.4 entitled *File System Configuration*).

Box 9.3   Practical File Systems—ext4

The original ext2, although still very much in use (especially in embedded systems), is increasingly being replaced by ext3, which adds journaling support. In most cases, on HDDs, ext2 and ext3 are interchangeable, but designers might choose ext2 for unchanging read-only disc partitions (where journaling is unimportant), and ext3 for partitions where crucial data is to be written.

The main cost involved in using ext3 is in speed and power efficiency. Hence the newest variant, ext4, has been designed to be faster and more efficient, while also introducing several new features:

- Supports for very large partitions (ext4 is a 64-bit FS)
- Allows programmers to preallocate space for files if they know how big they will be
- Doubles the ext3 limit of 32,000 files per directory
- Is much faster when checking the file system
- Timestamps files with nanosecond resolution (instead of seconds)

Plus three additional features which will be explained below:

- Supports delayed allocation
- Performs journal checksumming
- Uses something called "extents" instead of block-wise mapping

Some of the improvements are relevant only for large systems, but ext4 is highly configurable, and able to be made more suitable for resource constrained embedded systems (see Box 9.4 for an overview on its configurability).

Delayed allocation—simply means that the file system waits until it knows the file size before it chooses which blocks to use (the idea being that if it knows the exact size of file it can choose a space that is the perfect size, instead of making a guess that might end up wrong). This reduces overall fragmentation at the expense of a slight reduction in writing speed; however, the support for pre-allocating space (where possible) aims to alleviate some of this cost.

Journal checksumming—means that there is less effort required in order to validate that the journal is correct. When checking, it computes a checksum over the journal memory area and then compares the calculated checksum to the stored checksum on disc.

Extents—replaces the block mapping using in earlier ext FS implementations. An extent is a pointer to a start block, plus a length field to indicate how many blocks are used. So instead of storing a pointer to every block in the file (i.e., the old system), ext4 inodes contain up to 4 extents.

Barriers—are where the file system forces the order of data writes to confirm to the sequence in the journal. This is needed because hard discs with a built-in cache (i.e., all modern ones) often write the physical data to the disc surface in a different order to the sequence in which the OS passed it to the HDD. Barriers force in-order writing, so that the journal is always correct, at the expense of slightly reduced performance. No change to the FS is needed, but if a storage device does not reorder writes then it can be mounted *without* barriers by using `mount -o barrier=0`, in which case it should be slightly quicker at writing.

Some advice for system designers

A file system (FS) that is only going to be read and not written to (i.e., the boot code for an embedded system) can safely be ext2. But any FS that has frequent write operations can get corrupted and needs to be checked periodically. Journaling makes this more efficient—hence ext3 or ext4.

A final point to note is that you might not need to use any FS at all if you (i) manage data blocks yourself, (ii) assume fixed size data writes, and (iii) do not require storage to be compatible with any other system.

Box 9.4   File System Configuration—ext4

There are many configurable options for ext4 that can be set either at creation time (when using the `mke2fs` program to format the partition/disc), or after creation when the `tune2fs` program can tune the FS (as long as it is not currently mounted). A summary of the main options would include

- **No journal**—allows journaling support to be turned off. Useful for read-only partitions, although ext2 could equally be used in such cases.

- **inode size**—ext3 default is 128 bytes. By contrast, ext4 needs at least 256 bytes, but can be extended, which is useful when the system needs to store large numbers of files.

- **Default commit time**—also supported by earlier versions of ext, data is written to disc after 5s by default, but delayed allocation can increase that to 30s.

- **Block size**—when using flash memory, this should match, or be a multiple of, the natural erase block size of the flash device.

- **Bytes per inode**—should equal multiple block size bytes. Defines the number of files that can be stored for a given partition size.

- **Journal location**—can be stored in different locations, including on a separate device.

- **Checking frequency**—this tuning option specifies how often the FS needs to be checked, counted by the number of times it is mounted (i.e., used), or by the amount of elapsed time. It also supports the extremes of *every time* and *never*.

In addition, the file /etc/mke2fs.conf specifies defaults for making a new ext4 (or other ext) file system, and also gives some hints for different sized systems.

Checking the FS is done using the e2fsck tool, and so this needs to be configured by editing the file /etc/e2fsck.conf to specify (i) what should the system do when an unrecoverable FS error occurs, (ii) what the system should class as an "unrecoverable error," and (iii) what type of errors should e2fsck automatically fix (probably as many as possible).

Some further interesting file systems include **JFS** the Journaled FS from IBM, designed to be fast and stable, and **ReiserFS**, a file system developed by, and named after a colorful but dubious character around the turn of the millennium. ReiserFS was extremely advanced, and many of the so-called new improvements in more modern FSs were originally introduced in ResierFS.

In addition to many FSs for HDDs, some have been designed specifically for use with flash memory, assuming it can be directly accessed at a low level (i.e., not on a SSD which hides the flash nature of the media by making the device appear to be an HDD).

YAFFS, which stands for "yet another flash filing system," is a robust FS designed for NAND flash, handling flash blocks in a sensitive manner, including being aware of the maximum number of erase/write cycles for each block. **JFFS2** is the journaling flash filing system, which combines the wear-leveling properties of YAFFS with the journaling of JFS or ext3. **UBIFS**, the "unsorted block image FS" is another system that attempts to retain the advantages of file systems like ext3, but with wear-leveling and appreciation of the flash memory limitations.

9.8.3 What Are File Systems?

File systems contain user data (i.e., your files and programs) as well as metadata (structural information, indexes and attributes). Structural information and attributes include the name and size of the file, the time and date it was created, who created it, and who owns it (used for UNIX-style system security).

In the ext FS, the metadata starts with something called a superblock. This contains information about the file system type, its overall size, current status, and where to find other metadata structures on the disc.

The superblock is quite important—without this information all of the files could potentially be lost—and so several copies always exist. In Linux, we can find exactly how many copies and where they are located quite easily using the dumpe2fs command.

In this example, we are examining the partition called `sda6` which is the seventh partition (0 . . . 6) on the first SATA HDD (denoted by the "a" in `sda`):

```
$ sudo dumpe2fs /dev/sda6 | grep -i superblock
dumpe2fs 1.42 (29-Nov-2011)
  Primary superblock at 0, Group descriptors at 1-19
  Backup superblock at 32768, Group descriptors at 32769-32787
  Backup superblock at 98304, Group descriptors at 98305-98323
  Backup superblock at 163840, Group descriptors at 163841-163859
  Backup superblock at 229376, Group descriptors at 229377-229395
  Backup superblock at 294912, Group descriptors at 294913-294931
  Backup superblock at 819200, Group descriptors at 819201-819219
```

The superblock should be 1024 bytes long and the original is always at byte offset 1024 from the start of the partition (this is in block 0 in the above example).

File systems are divided into blocks which are arranged as block groups, in ext2 and beyond. Each file usually gets spread over several blocks, but they are normally all located within the same block group.

Some block groups contain copies of the superblock and they also contain a block group descriptor table, a block bitmap, an inode bitmap, an inode table and, of course, the actual data. The bitmaps are structures used to help speed up scanning the disc, and the inodes contain all data and metadata (they are described below). For recent ext2 systems, superblocks are stored at blocks 0 and 1 as well as powers of 3, 5, and 7 but this is configurable when the device is first formatted.

Fixed-size block groups are placed on the device when you format it, leading to a structure such as that shown in Figure 9.10. The figure shows a number of block groups on the disc (above) and zooms in on three block groups (below), some of which contain copies of the superblock.

To examine how this works in a little more detail, Table 9.1 shows a sample structure for a 20-Mibyte HDD formatted using ext2 with a 1-kiB block size (the raw data for this table was obtained from `www.nongnu.org`).

Something to note is that if the block size is greater than 1 Kibyte, the first superblock lies inside Block 0 instead of inside block 1 (because it is at a fixed location of 1024 byes, from the start, and that offset does not change irrespective of what block size has been configured).

9.8.3.1 Inodes

An inode is the basic data structure in a UNIX-style file system such as ext2. This is the data "container" that the file system is comprised of, containing literally everything that gets stored on the disc. In addition to the actual user data stored in the form of files, inodes contain information about each file (or about each file system object), including their type (e.g., executable, char or block special), permissions associated with the file (e.g., read, write, executable, or special—R, W, X, or S), the user ID of the owner, the group ID of the group that the file belongs to, the file size, the times of file access/change/modification/creation/deletion, links (either soft or hard), and several other pieces of even more complex information. Strangely, it does *not* hold the name of the file (see later). It is important to note that there is just one inode per file, On Linux, inode information is very

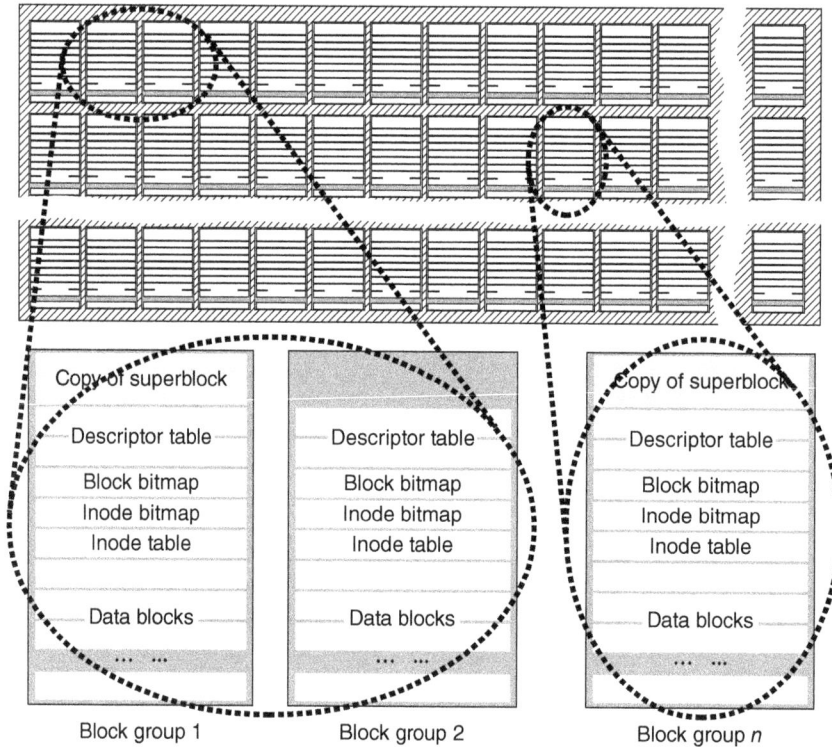

FIGURE 9.10   An illustration of block groups as arranged on storage media (top) showing the zoomed-in structure of three of the block groups below. Two of those block groups contain copies of the superblock, but all contain the usual descriptor tables for inodes.

easy to access. Here are two different ways, showing just the inode number (11535538), as well as further information for the second command:

```
$ ls -i /etc/passwd
11535538 /etc/passwd

$ stat /etc/passwd
File: '/etc/passwd'
Size:   1878        Blocks: 8   IO Block: 4096   regular file
Device: 806h/2054d  Inode:   11535538        Links: 1
Access: (0644/-rw-r--r--)  Uid: (0/root)   Gid: (0/root)
Access: 2012-11-21 15:53:22.359678777 +0800
Modify: 2012-09-21 13:25:36.939097771 +0800
Change: 2012-09-21 13:25:36.995097774 +0800
 Birth: -
```

The inode container, shown in Figure 9.11, holds either data, or information about data. In programming, a "container," is usually termed a "structure," and this is precisely

Address	Length	Description
0	512 bytes	Boot record
Byte 512	512 bytes	Boot record extras
Block Group 0 (blocks 1 to 8192)		
Byte 1024	1024 bytes	Superblock
Block 2	1 block	BG descriptor table
Block 3	1 block	Block bitmap
Block 4	1 block	Inode bitmap
Block 5	214 blocks	Inode table
Block 219	7974 blocks	Data blocks
Block Group 1 (blocks 8193 to 16384)		
Block 8193	1 block	Superblock backup
Block 8194	1 block	BG descr tbl b/up
Block 8195	1 block	Block bitmap
Block 8196	1 block	Inode bitmap
Block 8197	214 blocks	Inode table
Block 8411	7974 blocks	Data blocks
Block Group 2 (blocks 16385 to 24576)		
Block 16385	1 block	Block bitmap
Block 16386	1 block	Inode bitmap
Block 16387	214 blocks	Inode table
Block 16601	7976 blocks	Data blocks
Block Group 3 (blocks 24577 to …)		

TABLE 9.1 An example block structure for a 20-Mibyte HDD formatted using ext2 with 1-KiB block size.

what an inode is. It is an information structure stored on disc. Files start with an inode structure on one block, but by the end this could spread across several blocks.

The inode structure is arranged as a set of pointers to blocks. The first 12 blocks pointed to directly, following the inode, definition actually contain file data, but files are often larger than this (i.e., for 1 Kibyte blocks, this would be 12 Kibytes). In such cases, the 13th block contains a pointer that points to an indirect block. The indirect block itself contains a set of pointers which themselves point directly to other data blocks. If all of that is not enough then the 14th pointer points to a doubly indirect block. The doubly indirect block contains a set of pointers that point to a set of indirect blocks which, as we have described, each point directly to actual data blocks.

If the file is larger than that, then the 15th pointer points to a trebly indirect block. This points to another set of doubly indirect blocks which each point to a set of indirect blocks, which each point directly to a set of blocks. The process continues until the entire file can be stored. This is shown as a tree diagram in Figure 9.12.

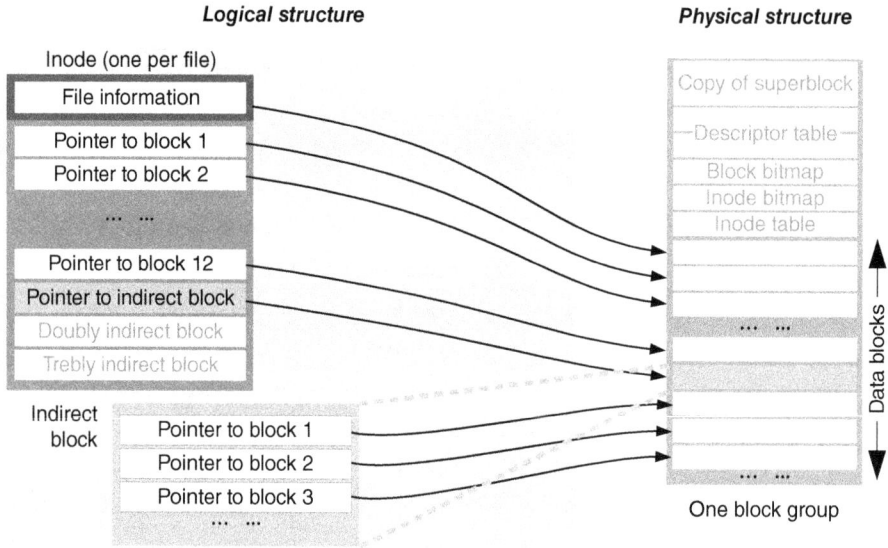

FIGURE 9.11   The logical inode structure (left) showing how this relates to the physical data on disc (right) by either pointing directly to data blocks or to indirect blocks that point to data blocks.

9.8.3.2   Directories and the Root File System

A directory (sometimes known as a "folder") appears to the user of a GUI-based computer to be a holder for files. They can drop files into a folder, rename them, delete them, and move them to another folder. However, the clustering of files is merely an illusion: Files are stored all over a disc, scattered throughout the filing system, and stored using

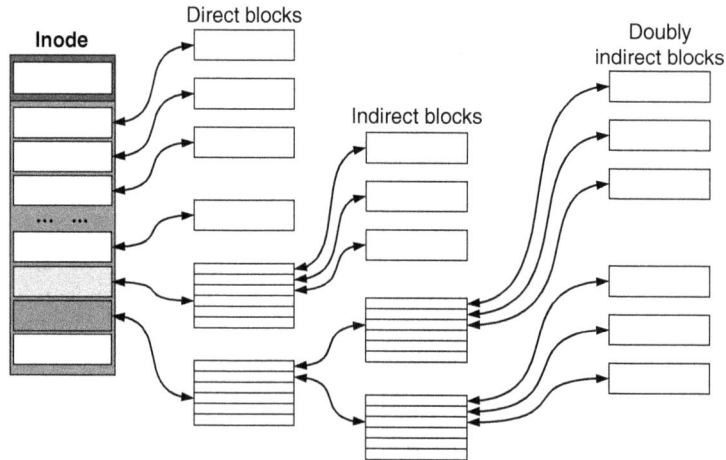

FIGURE 9.12   An inode structure (left) pointing directly to data blocks (three are shown), as well as pointing to indirect blocks and doubly indirect blocks. Doubly indirect blocks point to a set of indirect blocks while all indirect blocks all point directly to data blocks.

inodes as described in the previous section; they are not stored according to separate folders. The folders do not have any relevance to *where* on a disc they are stored. It is the role of a directory to impose the kind of order that allows a user to view groups of some files as being in one folder and groups of other files as being in a different folder.

At its most simple, a directory is just a data structure containing a list of files (or links). There is one data structure per folder, and that structure contains one entry per file in that folder. It is called a directory because it enables files to be searched for in a convenient way.

The directory information stored for each file is simply its inode number (remember there is only one inode specified per file), the *length* of the filename (i.e., number of characters in the filename), and the filename itself. The length is important because UNIX allows for some extremely long filenames, and it would be inefficient if every file was allocated space to hold the maximum filename length.

It probably will not surprise readers to find that the directory data structure is itself another inode object. For ext2, the root directory (/) is always fixed to inode number 2:

```
$ sudo stat /
  File: '/'
  Size: 4096    Blocks: 8    IO Block: 4096    directory
Device: 806h/2054d    Inode: 2         Links: 24
Access: (0755/drwxr-xr-x)  Uid: (0/root)   Gid: (0/root)
Access: 2012-11-22 12:46:26.405056381 +0800
Modify: 2012-10-02 15:21:11.571237232 +0800
Change: 2012-10-02 15:21:11.571237232 +0800
 Birth: -
```

When the operating system reads through a file system, perhaps to search for a file, it always starts in a fixed place with the root directory (/).

Inside the root directory, it can find the names, and the inode numbers for each subdirectory and file. To "look" inside a subdirectory to see what it contains is easy—find the inode for that subdirectory and read the inode data. This inode data is itself a directory data structure that contains a list of names of items in that directory, and their corresponding inodes.

As long as the OS knows where to start (namely with / on block 2), it is possible to find all files and directories on a machine by simply scanning through the inode links to each subdirectory. Everything eventually connects back to /.

This is illustrated in Figure 9.13 which shows part of a UNIX directory tree. For clarity, the diagram only shows one file, named `Lecture01.pdf`, and a few of the many branches on the system. The full path of that file (i.e., including its directory location) would be

/home/asian/docs/teaching/Lecture01.pdf

Every file that is accessible within that computer would appear on that tree somewhere. It is also possible for files to appear twice. UNIX supports both "hard" and "soft" symbolic links. Hard symbolic links allow the same inode to be listed in two or more directory data structures, whereas soft symbolic links list something different in the directory structure—in this case it lists a link to the *directory structure* of another directory. A hard link allows the file to be accessed directly whereas a soft link leads into the listing from another directory.

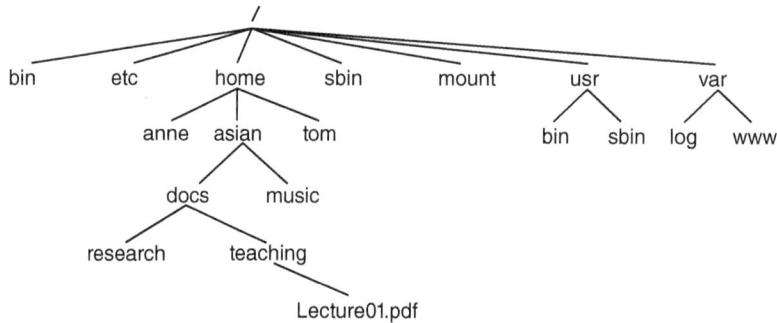

FIGURE 9.13   A simple ext2 directory structure from a Linux system, only showing a few of the important branches and subbranches.

When another storage device is connected to a Linux system, such as plugging in a USB storage device, this gets "mounted" into the directory tree. On a desktop or notebook machine, the mounting typically happens automatically, but on an embedded system, it has to be accomplished with a command, usually issued by the superuser (the user with full permissions to do almost anything on the system, known as having "root access"). On both types of system, the device should be unmounted before being removed, to prevent data loss. Figure 9.14 shows a Linux root directory tree that has had two external devices mounted onto it. The first device is a DVD-ROM device mounted at /media/dvd and the second is something mounted at /home/tom_in_nz, which is actually a network mount: It is a shared folder from a machine on the other side of the world.

Finally, we should note that Windows and MS-DOS do not have a single directory tree, but create separate directory trees for each device. This is shown in Figure 9.15 where two HDDs named C: and D: are present in a machine.

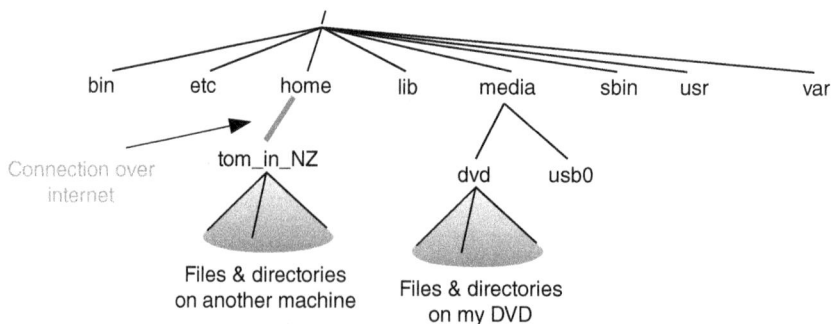

FIGURE 9.14   The single UNIX tree from a Linux system showing an ext2 HDD, a mounted DVD-ROM device and a directory mounted over the network (in this case one that is physically located in New Zealand, but appears as part of the directory tree on the local machine).

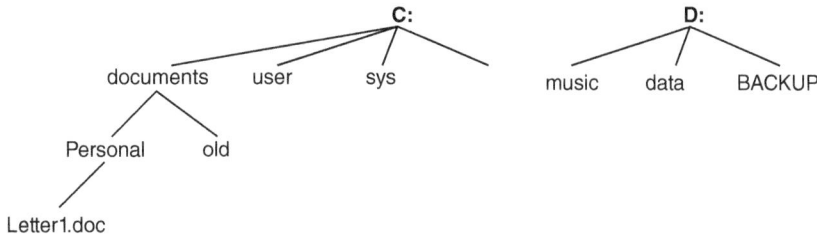

FIGURE 9.15 Windows and MS-DOS directory structure showing separate trees from two HDDs.

9.8.3.3 Journaling

Journaling is a very simple idea—if the FS maintains a record of everything it is about to do, and has done to the disc, then there should never be a situation when the file system is found in an uncertain (or unrecoverable) state. If power is restored after a failure, the system is either consistent with the record, or inconsistent with the record, but can be recovered with the help of the record.

The main positive use of journaling is in recovering a corrupt file system, something that usually is caused by a system crash or power failure during a data write operation. It is a fact of life that all file systems will get into trouble eventually, and hence they need to be checked periodically as well as maintained to prevent problems from growing. Older file systems may need to be de-fragmented periodically too.

Journaling is an attempt to reduce the additional maintenance caused by partial writes, to reduce the need for maintenance as well as make maintenance easier. It also simplifies and speeds up the file system checking process (which has to be done from time to time on all systems).

A journal is created which contains a step-by-step timestamped list of all the operations being performed on the file system. The journal is itself a file that usually resides on the same partition at the data. It will be written to any time a file change takes place and can be read by file system recovery and checking tools. The journal operations are shown in Figure 9.16 for a write operation. In this process, when the file system is given a block of data to write to the disc, it first writes a record to the journal stating what it is about to write, then it performs the actual write, and finally it records to the journal that the write has been completed.

If any problem occurs during writing, the journal can be checked to determine whether the write succeeded or not, and if not can repair the problem. In addition, the file system can be checked by following through the journal step-by-step during regular disc checking procedures.

9.8.4 Backup

Data within a computer can easily be lost due to software bugs or hardware malfunction, and even data saved onto an HDD is vulnerable if that device is damaged. For this reason, a backup system should be regarded as an essential part of any computer system, particularly when the data is important.

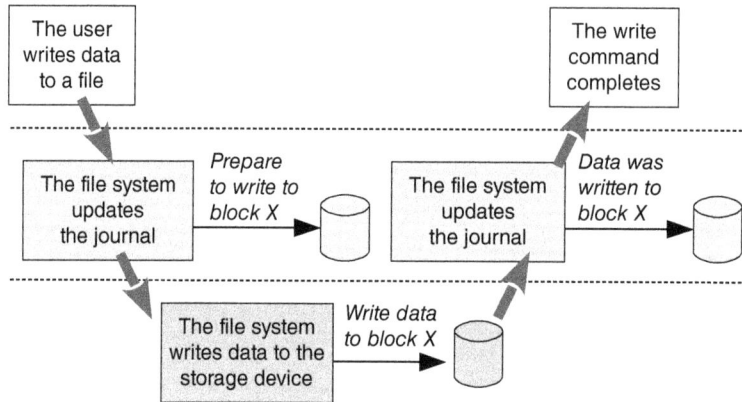

FIGURE 9.16 Illustration of journaling for a data write operation.

The requirement of any backup system is similar to any secondary storage system—to be able to restore part or all of the contents of the filing system to its state as it was at some previous time. For backups this includes

- Directory structure
- Names of files
- Content of files
- File metadata (timestamps, size, permissions, and so on)

The recovery operation should ideally be fast and convenient and allow for either full or selective recovery (i.e., either restore everything or recover just a single file). It should also be as space-efficient as possible (meaning that taking full copies of everything is not always the best solution).

A complete backup means that everything is replicated somewhere else. This takes a lot of storage space, and is slow, but is very easy to recover either fully or partially. For businesses and those with important data, the backups are best stored off-site (and under lock and key, because they are a potential security vulnerability), and backups should be made before any substantial or important changes to the files.

Assuming a complete and reliable backup has been made, subsequent backups could save time and storage space by only saving files that have changed since the last backup. This process is called *incremental backup*. Since there is less data to store, it is faster than a full backup, but cannot be used on its own to recover all of the files if the original full backup is damaged or missing.

A *differential backup* stores just the difference in the data between the last backup and the current state. In some systems, if a file existed before but became longer, only the additional part will need to be backed up. This might be particularly appropriate for databases—by being far more efficient than re-saving an entire database.

In all cases, backups are required above all to be

- Complete and correct
- Reliable
- Able to be restored
- Secure (e.g., protected against being accessed or modified)

Sometimes it is useful to deliberately add redundancy to the backup in the form of extra data that will help the system to reconstruct or restore lost data in the event of errors during restoration.

For many users, it is nice to be able to run backups of an active file system during use—meaning that work does not need to stop when either backing up or restoring data.

Backups do not need to be automated, and do not need complex or expensive software. Indeed, backing up UNIX-based systems can often best be done from the command line using utilities such as `scp` or `rsync`. Dedicated HDDs are also unnecessary because backups can easily be made to places such as cloud-based storage, dropbox, an external hard disc, tape drive, USB-attached storage, CD-ROM, or DVD-ROM.

When planning a backup strategy, apart from questions such as "how often," "when," "to what media," and "how to verify," it is important to consider the physical security of the backup data (including protecting it from theft), how susceptible it is to physical damage (e.g., fire/flood), what is the lifetime of the medium, its cost to purchase and to store, whether it is reusable, and even whether it is likely to become obsolete in the near future; many of us still have backup floppy discs, but with no hardware to read them!

9.9   Summary

This chapter has taken us a long way. Starting with the bare metal of a computer that has no software (as defined in Chapters 3 to 6), and knowing how software is written (as defined in Chapter 8), we have bridged the gap that allows software to run on a computer operating system and do something useful for the user.

We have seen that much of the character of a computer that we use is defined by the operating system, but also that the graphical user interface (GUI) is not the same thing as the OS. While many operating systems allow the user to choose among several types of GUI, some do not, and some have no GUI at all. Similarly, we noted that some computers—particularly those that are small, single function, and high reliability—make do without an OS. However, an OS can provide quite a few advantages to the programmer.

Apart from considering the role and structure of the OS, this chapter has considered system booting—for both embedded systems booting from flash memory, and systems booting from hard disc drive, how processes are initiated, switched, prioritized, and examined.

Finally, we looked at the importance of organizing data on storage media, which is the role of file systems. We considered the span from the low-level information written to the platter of a disc or the cells of flash memory, up through the file system structure, folders, and high-level tree-based file organization. We ended with a consideration of backup, and this should be a timely reminder to all readers to ensure that their important documents and files are frequently backed up.

9.10   Problems

9.1   Some of the following computer systems might be better implemented without an operating system. For each computer system, state which features (if any) mean an OS might be unnecessary:

1. A smart watch that has multiuser capabilities, and built-in wireless networking, but very limited battery capacity.

2. A toaster with multiple toast-cooking programs that are selected when bread is inserted, and which run to completion unless a "stop" button is pressed. There is no networking capability.

3. One of the slave processors clustered around a large and fast CPU to handle memory accesses, including DMA and network packet handling for the main processor.

4. A tablet computer used by a family, with which each member of the family can log in to their own account to access their emails and personal files remotely.

9.2   Of the operating systems mentioned in Section 9.2.1, determine which would be the best choice for the following applications. Justify your choice:

1. A system controlling a vent valve in a nuclear power station control room. The valve is required to operate within 100 ms of an over-heat sensor being triggered in order to prevent a melt-down. Both the sensor and the valve actuator connect to the CPU.

2. A network-attached storage device in a company. It is required to securely host important documents for several work teams.

9.3   Identify five items of information that would normally be stored inside a process control block (PCB) for a newly created process on a multitasking operating system.

9.4   The scheduler chose the new process in Question 9.3 to run on the CPU for a while but now the scheduler has chosen another process to replace it. Identify which items in the PCB the scheduling code will update when the new process is moved from "running" to the "runnable process" list.

9.5   As a programmer, you write a program which runs in user space on a modern multitasking operating system. Due to an error in your code, part of your program writes data to an address that is located in a block of memory that is used by another program. Explain what is likely to happen; will either of the programs crash or become corrupted?

9.6   An embedded CPU has a fixed boot vector address of $0x0000\ 0000$. You are told that it is connected to two types of memory—NAND flash and SDRAM—and that it has no internal non-volatile storage. Which type of memory do you think will be located in the lower region of the memory map (i.e., starting at address 0), and which will be located in a higher region of the memory map. Justify your choice.

9.7   The `init` program has become corrupted on an embedded Linux system. When the kernel finds that it cannot launch `init`, is it likely to stop immediately with an error, or will it attempt a different start-up method?

9.8   On a computer with two quad-core CPUs, what is the maximum number of processes that can execute concurrently?

9.9   When the Linux scheduler, running its default scheduling policy, chooses which of two runnable tasks to execute next, will it always select the task with the highest static priority? Justify your answer.

9.10   A user on a shared-access UNIX mainframe computer is annoyed that their large piece of scientific software is running slowly. After finding out that there are many other users also running scientific programs on the same machine, they attempt to change the priority of their software with `renice +20 101`. Assuming that their software has a PID of 101, is this command likely to improve the situation and speed up their code?

9.11   Identify which kind of storage media (tape, HDD, SSD, parallel flash memory) would be best suited to the following applications:

1. A light and portable notebook computer that is required to boot up quickly and be extremely power-efficient.
2. Archival of several TiBytes of video recordings that may possibly be needed in future.
3. Storage of the latest Hollywood movies that will be streamed by multiple users from a network-attached storage device across a university campus.
4. Non-volatile file storage in a smart wristwatch device.
5. A system that continually stores thousands of small files per second that only need to be kept for 1 minute each and then deleted.

9.12   Identify the address, in terms of C, H, and S, at which the master boot record (MBR) is conventionally stored on an HDD.

9.13   Specify five types of information that an inode, which identifies a file, will contain for an ext file system.

9.14   A computer with an ext3 file system is in the middle of writing a file to the HDD when the system power fails. When power is restored, what feature of the ext3 file system would improve the chances of the file being recovered?

9.15   A data center performs regular daily incremental backups of all of their machines. At the end of each day, automatic scripts look for newly created or newly changed files on each computer, and ensure that those files are backed up to a separate DVD-ROM. They do not undertake any other backup activity on a regular basis. After 1 year in operation, a catastrophic failure occurs on a computer. Although all of the backup DVD-ROMs are available, restoring files to that one computer is extremely troublesome. Explain the difficulty in restoring files, and suggest a method by which they could improve their backup strategy in future.

CHAPTER 10

Connectivity

Today it is difficult to imagine a world in which there is no Internet; a world with computers that cannot talk to one another; and a world without instant access to information such that the only methods of communicating with a distant friend are voice, telephone, or postal service. Most of us take the connectivity we enjoy for granted, few stop to think about it, and very few would stop to ask *why* we should be connected.

Yet from the early days of computing, engineers had designed more and more powerful computers that were essentially standalone calculating machines. Aside from experiments in the occasional research laboratory, for most of the history of computers, these devices have operated individually and independently.

From our post–World Wide Web standpoint, it is very natural to connect a computer to the world via the Internet or to a company-wide intranet. Few people these days would purchase a desktop or notebook computer and *not* go "online" with it. Even fewer people would purchase a smartphone and *not* connect it to the mobile phone network to make calls, send messages, or "use data." It almost seems that it would take a conscious effort of will to deliberately stay offline in today's world.

In this chapter, we are going to attempt to understand how and why the world moved from having a handful of individual calculating machines to the highly interconnected pervasive computer networks of today. We will not yet look at the *technology* of networking—that will be covered in Chapter 11. Instead, it is important to consider the *concepts* of connectivity first. These concepts tend to be universal and are applied irrespective of the technology used. For example, many of the same concepts apply to Ethernet and IEEE 802.11 just as they do with WiMax and Bluetooth. Readers may even begin to see those concepts applied to other areas related to computer hardware or software, even to business operations.

10.1 Why Connect, How to Connect

When standalone computers were primarily only used in research laboratories as calculating machines, there would have been occasions when one research group wanted to share a set of data with another research group. Perhaps it was data collected as a result of an experiment or a calculation that they wished to verify.

In that era, the most obvious method of sending a "large" block of data (above about 1 Kibyte) would be to store it on punched cards, or later to tape or disc, that is, some form of physical secondary storage medium. They would then literally send the medium, by courier or postal service, to the recipient.

The recipient would load the storage media into a tape player, a disc unit, or punched card machine that would read it sequentially, transferring the data electronically down a wire into their computer. At some point an engineer would have looked at this arrangement and wondered whether they could simply use a very long wire to connect their computer to the remote computer and transfer the data directly down the wire.

The problem was that data signaling using high and low voltages to directly represent "0" and "1" does not work well over long distances and cannot convey data quickly, except over very short distances. What those engineers needed was a method of *encoding* the data to make it more robust to travel over long distances, and allow it to be communicated faster. Data coding, including compression and error correction, subsequently became a significant research field, leading to a number of very important discoveries that underpin electronic communications today.

10.1.1   One-to-One Communications

Once coded, data could be transferred reliably down a wire from one computer to another. The trade-off was that the transfer could either be "fast" or "far" but not both (although the limits of how far and how fast have both advanced rapidly over the years). This allowed one-to-one communications to take place, enabling one computer to connect to another using dedicated wiring. This very simple method of connectivity is shown in Figure 10.1 with the two computers and an electronic link between them. The same symbols and diagram type will be used to illustrate more complex connectivity as we develop it further.

There are no arrows shown on the link in Figure 10.1 (or over the next few pages) as we assume bidirectional links. These allow communication in either direction and are also known as *duplex*, but there are other possibilities as shown in the following table.

Type	Description
Simplex	Data is transferred in only one direction.
Half duplex	Data can be transferred in either direction, but not simultaneously.
Full duplex	Data can be transferred in both directions simultaneously.

Half-duplex systems require some kind of mechanism to decide which direction is currently transferring data (if not, they can get into a horrible mess). To use voice communications as an example, this is how CB (citizens' band) radio or a walkie-talkie works. With a conversation happening on a single channel, when someone speaks, they must end with the word "over." This indicates that they are handing over the channel to allow another person to speak next. People in a conversation speak in turn, switching whenever

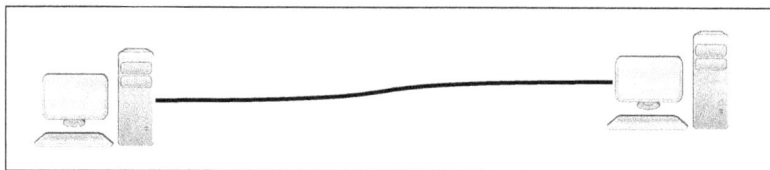

FIGURE 10.1   Diagram showing a one-to-one connection between two computers.

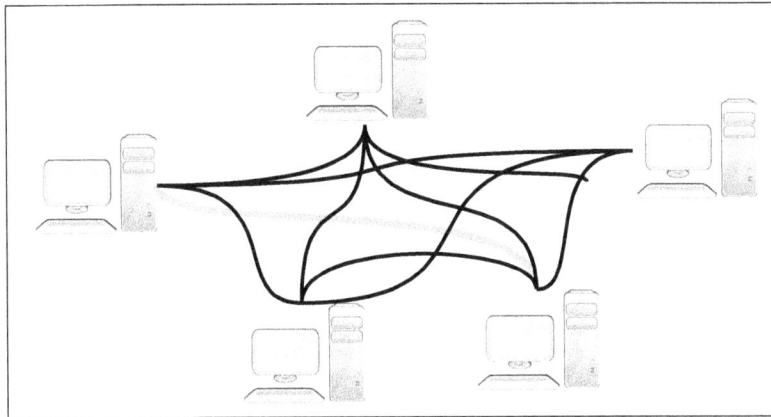

FIGURE 10.2   Many computers are interconnected, with direct links between each machine.

one hands "over" to the other. The final sentence ends with "over and out." This type of mechanism is exactly what computers need to use when they share a half-duplex link over a single wire (or channel), and there are many systems to do exactly that.

10.1.2   One-to-Many Communications

Computer engineers were of course not satisfied for long by one-to-one links and would have soon wanted to connect to other computers, necessitating additional long wires. In fact, there would need to be a dedicated wire between each computer being connected, requiring rather a lot of wires, with the number of wires growing significantly as the number of computers increases. This kind of arrangement is shown in Figure 10.2 in which five computers are linked to each other. To do this, each computer operator needs to maintain four separate links, and the diagram shows the profusion of (messy) wiring needed for even a small number of machines.

When data encoding techniques progressed further to enable sending signals over telephone wires using a device called a modem (modulator-demodulator), communications flexibility improved. This was thanks to now being able to route data links through exchanges, as shown in Figure 10.3.

The telephone network was circuit-switched, which meant that there was essentially a single wire from each computer to the nearest exchange, and then a bunch of wires between exchanges. To make a call (or a data connection), the wire from the caller to the local exchange would be connected to one of the wires between exchanges and then the active wire going into the remote exchange would be connected to the callee's wire. This kind of communications is known as circuit switched because there is always a dedicated electrical circuit from caller to callee. This dedicated circuit would remain fixed through the telephone call (or communications link).

Readers who have seen early telephone exchanges in a museum or on a documentary film may have noticed how the operator would physically plug and unplug wires from a switchboard. This was the process used to make connections; joining wires together to make a circuit from one caller to another. Since it was a manual operation,

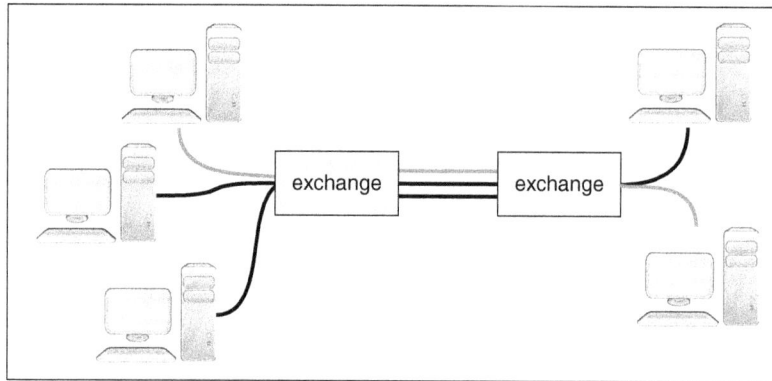

FIGURE 10.3   Computers from two locations communicate through exchanges, with each machine linked to an exchange and the exchanges linked together. An active call (the gray wire) occupies an entire physical wire in each segment.

early telephones had no dial or number pad; the user would simply ask the operator to connect them. Dials and, much later, buttons for the numbers 0 to 9, * and #, were conveniences that allowed the dialer to enter a sequence of numbers that would command an automated exchange to make the required connection without manual intervention.

Long-distance telephone calls would actually consist of several segments that were joined—a wire from the caller to the local exchange, a wire from that exchange to another local exchange, and then a wire from there to the recipient. Three wires, but by being connected at the exchanges, they appeared to be a single link for the duration of a call. A similar situation occurs with modern computer networks. Although they connect hundreds or even thousands of computers, in reality they consist of a large number of smaller one-to-one links. The main difference is that they are able to share the wires.

10.1.3   Packet Switching

While it was possible to communicate over telephone wires, this was neither convenient nor efficient for anything except the smallest and simplest messages. Data speeds were highly limited, despite some very complex coding techniques being developed, but links at that stage were still essentially one computer making a "call" to another computer over a dedicated wire that was fixed for the duration of the call. This was still a one-to-one link, except that users now had the ability to choose which "one" to communicate with.

Wires are expensive—not just because of the price of the copper used but also the cost of laying, maintaining, and repairing wires that are connected to almost every home within each developed country. There is therefore an understandable wish to use the wires more efficiently, and so various ways were found to share wires.

Probably the most important sharing mechanism—and the one that is used in the vast majority of network and communication systems today—is packet switching. A block of data to be communicated is first encapsulated into small packets. These are conveyed across links from the sender to receiver. The receiver then reassembles the block of data.

The basic idea, shown in Figure 10.4, is simple, but the arrangement has many hidden complexities that have led to many ingenious ideas. The important point is that one wire

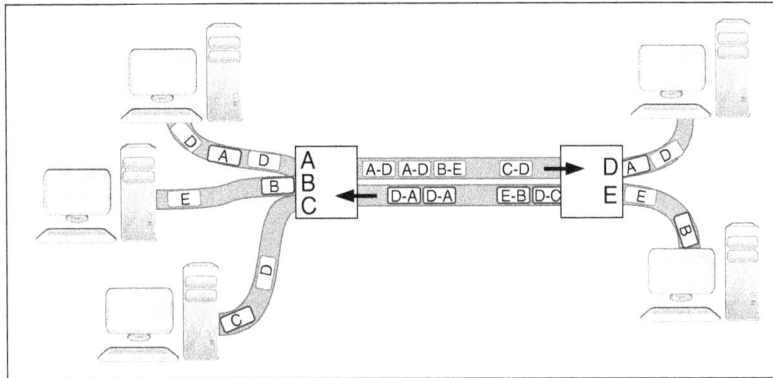

FIGURE 10.4 Packet switching allows multiple computers to share the same link, each forming and transmitting small packets of data that are routed to the correct destination.

can be used to send packets from different senders to different receivers. While the packets may in reality take turns to get transmitted, because this happens so fast, it appears to the user that communication is continuous and fully duplex.

Almost every digital communication system we use today is packetized. This includes mobile phone networks, WiFi, WiMax, Ethernet, cable modem systems, and digital television and digital radio standards as well as most of the landline telephone network. Chapter 11 will discuss the most common standard used for data packets, and that is called TCP/IP (transmission control protocol/Internet protocol), but many other methods exist for niche applications.

10.1.4 Simple Communications Topologies

A communication *topology* refers to the way in which different computers in a network are connected together. We also often refer to *network architectures*, which are discussed further in Section 10.5. The topology or architecture of a system can have a very large effect on its performance, and this has become a major part of network system design. As an example, consider two alternative methods of connecting five computers together in Figure 10.5. Each of these uses the same number of wires, but there are some major differences:

- *Choice of route*—if the ring network needs to send a packet from computer 1 to 3, it can send the packet via computer 2 (i.e., the link is $1 - 2 - 3$), but there is a second route possible, namely $1 - 5 - 4 - 3$. The second route has more "hops," but it means that if the network breaks at any point, it is still possible for packets to get from any computer to any other. This is not true of the star-connected network. In that topology there is no choice of route, and so no alternative if part of the network fails.

- *Packet latency/number of hops*—with just five computers, a fully functional ring network can route a packet from any computer to any other in no more than two hops, which is the same as the star network. However, as the network size

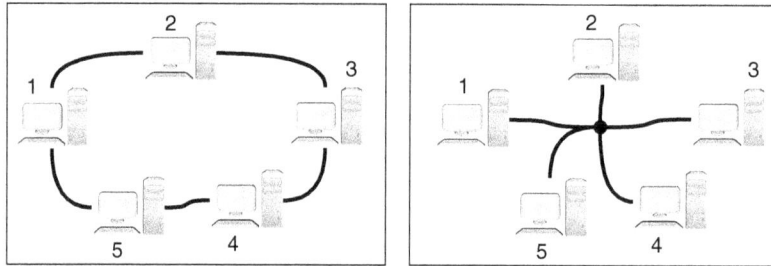

FIGURE 10.5 Five computers linked together in a ring topology (left) and in a star connection (right). In each case there is only a single wire per computer.

grows, the star network can still route a packet in just two hops, whereas the ring network cannot. For example, in a network with 100 computers, packets in the ring network going from computer 1 to 50 would need 50 hops. The problem with multiple hops is that they take time. Each hop, where a packet goes through a computer or routing node, takes a similar amount of time as it does for an electrical signal to travel half way around the world. Fifty hops are thus going to make for a slow communications link.

- *Routing* is the process where a packet from one network is transferred to another network. In Figure 10.5, the star network does routing at the center of the star, but individual computers do not need to route any packets except their own (assuming they are only sent correct packets). By contrast, the ring network needs to do packet routing at each computer; computers accept their own packets but pass on other packets. If computer 4 was sending a large amount of traffic to computer 2, this could overwhelm computer 3.

- *Bandwidth* is usually measured in bits per second (or Gbits/s in today's networks), and is a characteristic of all packet handling components in a network. The links (physical wires plus transceivers) will have a rated maximum bandwidth, as will each network interface on each computer. In the star topology network, the central node will also have a limited bandwidth, and it is this that becomes a major bottleneck in a large network; every packet needs to pass through the same point, and so this determines the maximum speed or latency of every piece of data in that network.

- *The router* at the center of the star network would have to be another computer in reality, because packet handling is a task that is under software control in most networks. In the figure shown, that computer would require five network interfaces, which is a significant number (although still easily achievable in practice). However, as the network grows, the number of network interfaces must increase proportionally and would quite rapidly exceed the physical and processing capabilities of most computers. The ring network only requires two network interfaces for each computer, no matter how large the ring becomes; however, with a big network, each computer would spend a substantial amount of time handling packets for *other* computers—something that never happens in a star network.

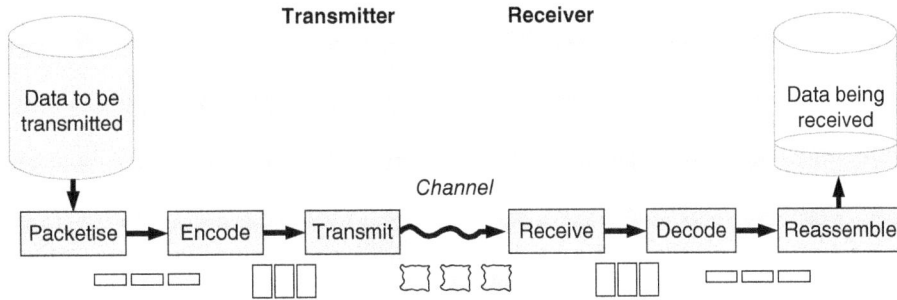

FIGURE 10.6 A generic transmitter-receiver pair sending data from the system on the left to be received by the system on the right over a channel, which could be a wired or wireless link.

This short discussion has highlighted some of the issues that would occur for two very different network topologies. The star network guarantees fewer hops than the ring network as the number of computers increases (and hence should communicate faster, on average); however, it is less tolerant to failure and relies upon a single central computer called a router that needs to perform complex packet handling and is a bottleneck to performance. Each computer in the ring network has twice as many network interfaces as for the star, but it only has to perform simple packet handling. It is tolerant to a single fault or failure but it becomes inefficient as its size grows.

So which is best—the star or the ring? The answer is that neither is used in modern network deployments, apart from in some specialized systems. However, both are frequently combined as part of other topologies, which inherit a number of the same advantages and disadvantages. Section 10.5 considers this further.

10.2 System Requirements

A wide variety of computer communication methods exist, most of which have developed unique solutions to various problems of transferring information between two computers. Some of these solutions are specific to the type of topology or media they were developed for, and may depend upon the exact networking technology being used (the subject of the following chapter). But there are a number of almost universal processes that occur no matter what network is being used. We will explore these individually, with reference to a generic communications situation in Figure 10.6. In this arrangement, the *channel* between transmitter and receiver experiences occasional errors, which is almost always the case in practice, and so communication needs to be managed with this in mind.

10.2.1 Packetization

Packetization is the process of splitting a larger block of data into smaller packets that are usually of equal size. It is done at the transmitter to prepare data to be transmitted. Also known as *segmentation*, this is matched by a reverse process at the receiver called *reassembly*.

Dividing data into smaller packets is very important because of errors that might occur between receiver and transmitter. Error rates are usually quoted in terms of the probability of a single bit being received in error. This might range from 10^{-2} (which is 1 bit in every hundred being in error, or a probability of 0.01) across a very "noisy" wireless link, but it could be as low as 10^{-12} (i.e., 1 in 10^{12} or a probability of 0.000000000001) for a high-quality wired connection.

In any computer executable or file of data, if just a single bit is erroneous, the entire file is corrupted. For a situation when a file is being transferred from one place to another, the entire file would have to be resent if just one error occurs unless we can identify that error and manage it. With large files, resending could be very troublesome and slow, and so packetization is used to divide a file (or block of data) into multiple smaller packets. Each packet has a simple error check code that allows it to be checked when it is received, to reveal whether any bits within the packet are erroneous. If they are, just the corrupted packets need to be retransmitted rather than the entire file.

In general, the higher the error rate of the "channel," the smaller the packets should be, because they are more likely to be received in error. In this way, wireless communication systems tend to have smaller packet sizes than wired systems (because the channels are "noisier," they have higher error rates).

The consequence of packetization in this way is that additional information needs to be added to every packet—an error check code, as well as a sequence number (the sequence number is needed so that the receiver can ask for which packet needs to be resent if one is received in error). This additional information occupies bandwidth. It adds to the number of bits being sent, but it is not user data; that is, it is not part of the file being transmitted. Since a fixed amount of information (bits) is added to each packet, it is more efficient to use fewer packets when sending a file, if possible. The following table summarizes the trade-off between overheads and error resilience:

Packet size	Error behavior	Overhead
Small packets	Less data to retransmit if errors are detected, and so better for noisier channels	Greater overhead because there is more additional data (e.g., sequence numbers and error check codes) needed to be sent
Large packets	More data to retransmit on error	Less overhead because there are fewer packets overall, so a more efficient choice for reliable communications channels

We will revisit the process of handling errors in Section 10.2.5. At this point we simply use error behavior to explain why packetization is so important.

In fact, it has another essential role and that is in real-time communications. If a computer could only transmit entire files at a time, streaming video or audio would not be possible, neither would voice or video calling. In each of these applications, a frame of video (or a slice of audio) is encoded and quickly transmitted, so that the receiving program can accept and display them. Neither application would be at all convenient if the entire video had to be saved to a file, then transmitted in full, and then received in full, before being replayed.

10.2.2   Encoding and Decoding

After packetization, the data to be transmitted exists as packets of binary numbers. These might range in size from 64 bits up to 2048 bytes or more, but no matter what data is being transmitted, it consists of a sequence of binary digits.

Computers understand the 1s and 0s of digital information, and this kind of data is very easy to work with inside a computer, but the physical world outside a computer is analog. Data is conveyed through this world by changes in physical properties such as voltage, current, or the arrangement of different frequency components.

Encoding is the method of transforming a packet of binary digits into "symbols" that are suitable for transmission over the particular medium being used. Many encoding methods employ one symbol to represent one bit of data, but methods exist that can transform 1024, 2048, or even more bits into a single symbol. Sometimes, an entire packet of digital information becomes just a single symbol to be transmitted (although this would be quite unusual, it normally becomes a sequence of symbols).

Encoding may be adaptive depending on distance, bitrate, and control arrangements as well as changing with the type of media. This is why a low-strength WiFi link generally works slower than a high-strength link, and why rural communities typically have slower Internet connections than urban communities.

Decoding is the opposite of encoding and happens at a receiver. It means transforming a received symbol back into binary data.

10.2.3   Transmission

The sequence of symbols to be transmitted are output one at a time onto the transmission media—either wired or wirelessly. In many systems, including Ethernet and WiFi, the transmitter has the responsibility to ensure that no other system is transmitting over the same channel (or wire) at the same time. To do this, it will sense the state of the channel before transmitting, and often for a short while after transmission begins (to catch situations when two devices start to transmit simultaneously). Later in Section 11.3, we look in more detail how this works for Ethernet, including how the transmitter pauses or "backs off" its transmission for a short period if a conflict occurs.

The transmitter is the device that bridges the divide between the digital world of the computer and the analog outside world. It needs to physically transform the symbols being transmitted into the voltage, current, or frequency modulations that are to be conveyed across the transmission medium.

10.2.4   Receiving

A receiver is the opposite of the transmitter. Receiving data means recognizing symbols from the transmission media and essentially bringing them into the computer in a digital form. Many receivers also have the responsibility to sense the channel and will often measure the link quality, something that can be used by communications software to adjust operations.

The receiver will not always receive exactly what was transmitted. Noise and errors in the transmission channel mean that some transmitted symbols may not be received at all, and some may be received in error and occasionally additional spurious signals are received. There might be interference from other transmitters, and sometimes packets will be overlapped, shifted slightly in time or single bits are missed out (causing all subsequent bits to be in the wrong position, and thus corrupted).

Symbol errors are very common in wireless networks, but less so in wired networks. One particular type of error that is relatively common in a wired network is when someone unplugs the network cable—meaning that the usual states of a wired network are either "almost perfect" or "getting nothing." Wireless systems, by contrast, frequently experience error rates right across the range.

Wireless communications is a large and significant research area in its own right—significantly bigger than computer architecture, for example. Many interesting inventions and advances have been made over the years to improve digital communications—not just for computer networking but also for voice communications such as mobile telephony, which is entirely digital in the modern world.

10.2.5   Error Handling

Error detection schemes allow a certain number of bit errors within a packet to be detected. An algorithm at the transmission end forms a checksum from the content of each packet and appends it to the transmitted packet. This is illustrated in Figure 10.7. At the receiver, the same algorithm forms a second checksum from the received packet data. This second checksum is compared to the received first checksum, and if both are identical, the received packet is assumed to be correct. Any single bit error within the received packet would cause the second checksum to differ from the first one.

By adding additional information to a packet in this way, it is possible to detect, and even correct, bit errors in a received packet. The method shown in Figure 10.7 is quite simple; the cyclic redundancy check (CRC) is low complexity and very fast to perform

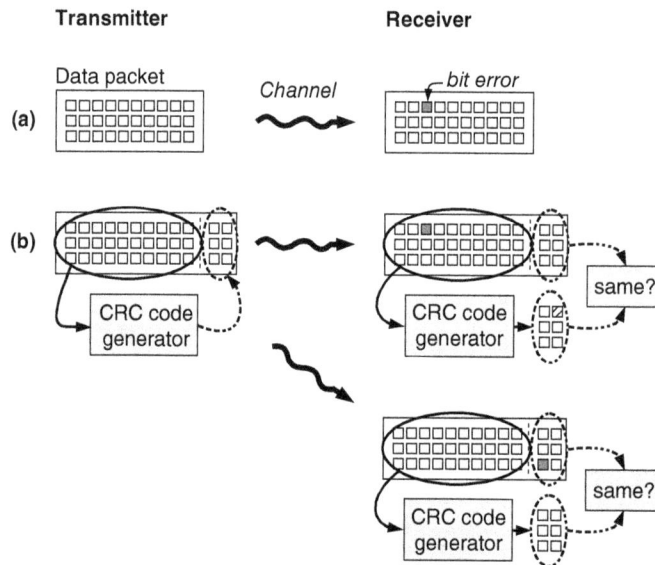

FIGURE 10.7   Packets transmitted over a noisy link may experience bit errors. (*a*) Without additional information, the receiver cannot detect the presence of a bit error. (*b*) Adding cyclic redundancy check (CRC) bits to the packet allows the receiver to detect an error in the transmitted bits, occurring either in the data or in the check bits themselves.

in hardware. Unfortunately it is also not sufficient protection for links that are noisy. If there was a bit error within the packet and also a bit error in the data check bits (i.e., both of the errors in Figure 10.7b occurred simultaneously), the receiver would run the CRC algorithm on the erroneously packet and get a checksum that could match the erroneously received checksum. The system would therefore not detect any error.

We briefly discussed error correcting methods for satellite memory in Section 7.10. The principle is the same for communications links. In general, we can use more complex (slower, using more energy) algorithms and/or append extra information to each packet, to detect significant errors. So while a simple CRC can detect a single bit error, a more powerful CRC or appending more check bits could detect if two bits happened to be in error, and an even more powerful CRC or even more check bits would be needed to detect three bit errors and so on.

The cost, in both complexity and additional bandwidth, of making a simple check for bit errors in a packet is small enough that almost all communication systems will do at least this much error detection. Some methods do much more.

So what happens when an error is detected? There are really three main system-level responses. The first is to *ignore* the error (which happens a lot more often that readers might expect; see Box 10.1), second is to ask for the erroneous data to be *resent*, and third is to find some way to *correct* the error. Since both of the active responses are important, we will discuss correction and resending in a little more detail before we move on to connection management.

10.2.5.1 *Error Correcting Codes*

Error correcting codes (ECCs) use powerful algorithms and additional check bits to enable correction of bit errors and are usually robust to multiple errors. In these systems, the receiver is able to not only detect errors but also correct them. As with the CRC, a more powerful algorithm (i.e., slower, consuming more energy) and even more check bits can be made to correct further errors.

How much ECC to add to data packets is an open-ended question. System designers can opt for solutions that range from doing no error checking to adding powerful codes that can correct enormous numbers of errors. But the price is paid in terms of computational and bandwidth efficiency. More powerful codes require a lot of computation, and hence tend to be slow and inefficient. They also require additional check bits, which means that a greater proportion of the information being transmitted is overhead; that is, it does not contribute to the "user data" that is being communicated. In reality, the system designer must make a pragmatic decision based on factors such as

- The expected range of error rates.
- How much bandwidth is available to transmit additional error coding bits.
- The data rate (highly complex algorithms may not be feasible in high data rate systems).
- Having sufficient computational resources at the transmitter and receiver to perform the error coding and correction. Some systems are asymmetrical in that they require more of the processing to be done at the transmit end or more to be done at the receive end.
- How important is the data; does it really need to be error free?

Box 10.1   Everyday Errors

Every communication channel that we make use of in daily life is subject to errors. Some are highly error prone whereas others experience only occasional errors. Most of the time, our computers handle the errors occurring in wireless and wired networking without us becoming aware of them—aside perhaps from a slowdown in connection speed. However, many of us will also have seen or heard the consequences of errors, perhaps without being aware of it. These noticeable errors occur mainly in uncorrected communications channels. In Chapter 11, for example, we will see how erroneous user datagram protocol (UDP) messages are usually left uncorrected and are not resent. UDP is commonly used for streaming or real-time communications such as video, audio, and voice.

In an audio call, or when listening to streaming music, errors are sometimes heard as "bloops," "blips," or "clicks" in the sound. Some news broadcasts feature reporters presenting from remote areas that can occasionally feature such artefacts. For video and images too, uncorrected errors are noticeable. In the example below, a picture of Mount Cook in New Zealand has been reproduced without error, as well as after a single error in two different positions in the images to the right. Each error corrupted three 32-bit values out of a 152-Kibyte JPEG image (a 0.001% error). Because of the JPEG coding used, the consequence of the error is more significant, causing a swathe of color and brightness changes in the central image, and a shift or offset for part of the right-hand image.

The large amount of damage caused by a small error is due to the sequential decoding of compressed JPEG images in blocks that start in the top-left corner, then proceed left to right in lines down the image until the bottom right-hand corner is reached. Sequential decoding means that an error in one block can propagate to cause errors in the subsequent blocks that are decoded. In the middle image, the error propagates right to the end of the image. Digital video, such as that used in DVD and BluRay systems, are also decoded sequentially within frames. However, for moving images, the sequential decoding does not always stop at the current frame but can propagate to subsequent frames. This explains why errors on a DVD sometimes cause the next few seconds of video to be corrupted, turn blue, pixelate into large blocks, and become jerky or shimmer like silver.

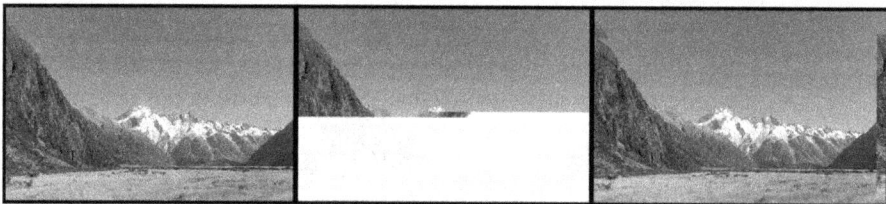

A Creative Commons CC0 image of the beautiful Mount Cook in the South Island of New Zealand, obtained freely, with thanks, from https://pixabay.com. The corrupted versions were created by the author using a simple hex editor.

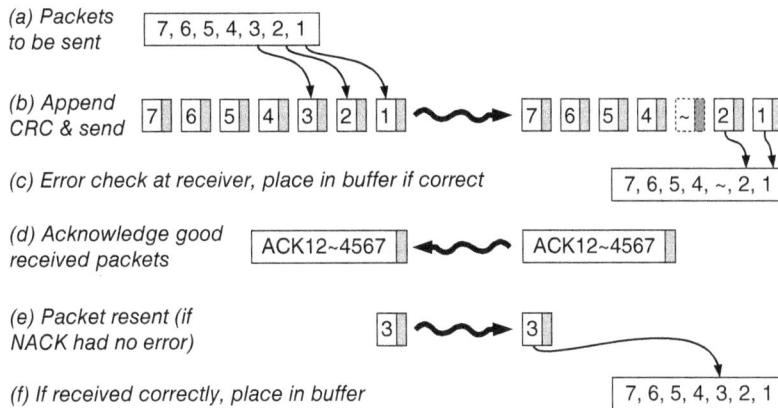

(a) Packets to be sent   `7, 6, 5, 4, 3, 2, 1`

(b) Append CRC & send

(c) Error check at receiver, place in buffer if correct

(d) Acknowledge good received packets   `ACK12~4567`   `ACK12~4567`

(e) Packet resent (if NACK had no error)

(f) If received correctly, place in buffer   `7, 6, 5, 4, 3, 2, 1`

FIGURE 10.8   An acknowledgment protocol where the transmitter is notified when packets are received correctly, and the transmitter has the responsibility to resend packets that are corrupted or not received. The alternative negative acknowledgment protocol is shown in Figure 10.9.

We will revisit some of these issues in our comparison between UDP and TCP in Chapter 11.

10.2.5.2   ACK/NACK Protocols

Where packets are short, bidirectional communication costs are low, and error rates are not high, it is often better to design a system where erroneous packets are resent on error, rather than corrected. This requires error detection, which is much simpler, faster, and more bandwidth efficient (under the circumstances mentioned) than error correction.

A resending protocol would not be appropriate where link costs are high or very slow (e.g., communicating with a remote satellite), or when communication is asymmetric or shared (e.g., broadcast services). It would also not work with high error rates, because the number of resent packets also being corrupted and thus requiring sending a third time means that most of the system bandwidth could end up being consumed by packets being sent, resent, re-resent, re-re-resent, and so on.

There are many types of resend protocols, but the two main classes are positive and negative acknowledgment schemes, abbreviated to ACK and NACK, respectively.

In an ACK protocol, illustrated in Figure 10.8, the transmitter has responsibility for keeping track of what has been transmitted correctly to the receiver. By adding CRC or other error check bits before transmission, the protocol requires the receiver to send an acknowledgment back to the transmitter when packets have been received without error. The transmitter, having a record of transmitted packets, can very easily compare what packets have been received with what packets *should* have been received. After waiting a certain time (in case a transmitted packet has been delayed), any packets missing at the receiver are retransmitted.

The NACK protocol, illustrated in Figure 10.9, puts more onus upon the receiver. In this case, the receiver needs to keep track of packets that have been received correctly, and must eventually inform the transmitter if one is missing by sending a NACK. The transmitter then resends the requested packet.

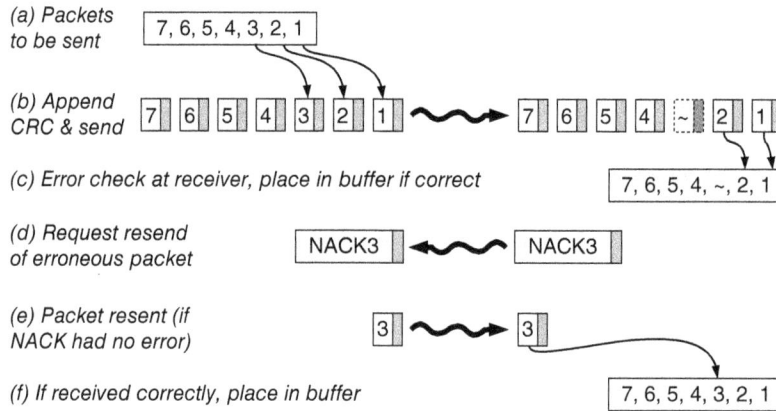

(a) Packets to be sent 7, 6, 5, 4, 3, 2, 1

(b) Append CRC & send 7 6 5 4 3 2 1 → 7 6 5 4 ~ 2 1

(c) Error check at receiver, place in buffer if correct 7, 6, 5, 4, ~, 2, 1

(d) Request resend of erroneous packet NACK3 ← NACK3

(e) Packet resent (if NACK had no error) 3 → 3

(f) If received correctly, place in buffer 7, 6, 5, 4, 3, 2, 1

FIGURE 10.9 A negative acknowledgment protocol where the transmitter is notified when packets are received incorrectly or are not received. The transmitter must resend the requested packets. The alternative positive acknowledgment protocol is shown in Figure 10.8.

Both methods require packets to have error check codes appended to them; both also require some way of identifying packets. In practice, most communication systems use a sequence number. They send packets that are numbered sequentially, 0, 1, 2, 3, 4, and so on. This is necessary for the NACK scheme, because otherwise it will not know whether a packet is missing (and is used in ACK protocols to identify which packets were received correctly). But it is also necessary for handling packets that are received out of order; practical networking systems like the Internet very commonly have situations where packets are received out of sequence (as we will see in Chapter 11).

Some of the characteristics of ACK and NACK schemes are shown in the following table:

Characteristic	ACK	NACK
Transmitted data	Packet of user data, sequence number and error check codes	Packet of user data, sequence number and error check codes
Error codes	Inserted at transmitter, read at receiver	Inserted at transmitter, read at receiver
On packet error or missing packet	Receiver does nothing	Receiver sends a NACK
On receiving a good packet	Receiver sends ACK	Receiver sends nothing
Resend	When transmitter realizes a packet has not been received	When receiver asks for it

To recap on the ACK and NACK schemes, both require error checking, both require sequence numbers to identify packets, and both require a list that keeps track of correctly received packets. For the ACK scheme, the list is at the transmitter, for the NACK scheme it is at the receiver. In most networking systems, there is also both a transmit and a receive buffer to hold packets in sequence. At the transmitter it holds packets ready to be sent, until they each have an ACK or have no NACK. At the receiver it holds packets until a block of data has been received, with all packets in sequence.

Assuming that more packets are received correctly than are corrupted, we might imagine that there could be more ACKs than NACKs; however, practical systems can use just 1 bit to represent each packet (i.e., an acknowledgment packet of 256 bits can show the status of the past 256 received packets, such as using a 1 to represent a successfully received packet and a 0 for one not yet received). Using methods like this is not (usually) necessary to send a single ACK or NACK for each received or not-received packet. Instead, a single ACK/NACK contains enough information to tell the transmitter about the status of multiple previously sent packets. This is highly important for ensuring efficiency.

Before we move on, one final potential catch with each of those schemes is that the *reverse channel* (information sent back from the receiver to the transmitter) will be just as error-prone as the forward channel was. So whatever type of ACK or NACK scheme is used, it needs to be robust enough to cope with situations when the ACK or NACK message is itself lost or corrupted.

Practical systems might send multiple ACK or NACK replies and have a number of timeout limits that cause the system to try again after a certain time, retry several times (but not forever), and eventually give up on a packet that does not work. We all see this kind of behavior when trying to load a webpage from a website that is not functioning correctly (i.e., one that is "down"). A Web browser will try to load the page, keep trying, and eventually will give up and display an error message instead. The time it takes to try and load the page is based partly on the networking system internal timeouts.

10.2.6   Connection Management

The NACK and ACK methods rely upon some form of overall control. This control system uses information such as ACK, NACK, sequence numbers, and so on to decide which packet gets transmitted at which time. But it does a lot more than that in many network systems.

10.2.6.1   Arbitration

Arbitration means deciding who or what has higher priority between multiple alternatives. In networking it most commonly refers to the situation where a link can be used by multiple devices, such as the one-to-one simplex link in Section 10.1.1. The link could be a wire or a wireless channel, but the important fact is that when it can only be used by one device at a time but multiple devices are able to transmit, *something* needs to arbitrate; it needs to decide which device transmits at which time. Arbitration or resource control is quite a complex topic that we will not discuss further in this book, but just note that in the simplex link case, arbitration often means "who talks next" while in the case of errors it controls how packets are resent. In the NACK and ACK examples mentioned previously, the decision was handled by the receiver and transmitter, respectively. In more complex cases with multiple transmitters and multiple receivers, such decisions could either be made by one central authority or by a distributed decision making process. The latter is the method used in many practical networking systems like Ethernet, WiFi, and so on because distributed decision making is almost always faster and is more robust to error, although it relies on systems cooperating.

A useful analogy is vehicular traffic within a city. Most drivers have control of when to begin their journey, how fast to drive (within limits), and which route to take. Our traffic systems impose controls at certain points, including traffic lights and speed limits, but otherwise rely upon drivers to cooperate on the road and provide them the freedom within which to do so. An alternative method would be for drivers to submit their journey plans to city traffic controllers in advance of any journey. The controllers would then provide them with directions for their route, specify speed, starting time, and driving behavior.

If everyone followed their city-mandated traffic plan, in theory there would be minimal journey disruption and maximum efficiency. However, the system would be incredibly inflexible. An accident, a late start, an unexpected event, could cause chaos. It would also require a method for drivers to submit a traffic plan, and for them to receive (and follow) their journey directions, which might need to be updated during a journey. The entire idea is impractical because the control overhead is too great. It is the same in most networks; in practice it is usually preferable to specify a few very robust rules for cooperation rather than detailed but inflexible rules for total control.

10.2.6.2   Control Packets

Knowing that networks do require some form of control and rules, we need to look at how these are implemented. Considering the situation of a transmitter and a receiver, there is already a mechanism for communication, and that is through sending packets, which usually has to be supported in both directions. The most obvious way of implementing control communication in such a system is to make some of the packets into "control" packets and some into "data" packets, and that is precisely what happens. Control packets tell the system how to work while data packets convey user data across the link. In reality, links that packetize data almost always have different packet types, with some being used exclusively to control the link.

Control packets might convey ACK or NACK information but also advise about network bandwidth or congestion. If alternative network paths exist between the transmitter and receiver (i.e., over multiple hops), they could also tell computers (nodes) on the network how to route certain packets.

In modern networks they perform quality-control functions as well as provide monitoring capabilities, including finding errors, blockages, or bottlenecks and identifying how a network is structured, such as which computer is located where.

With both Ethernet and WiFi, this kind of control happens continually and automatically. Most of the time the user is completely unaware that it is happening.

10.2.6.3   Addressing

The final topic in this section is addressing. Up to now we have mainly talked about one-to-one links, but we know that many alternative topologies exist that connect multiple computers together. For example, if we consider the ring topology network of Figure 10.5, there needs to be a way of sending a packet from Computer 1 to Computer 3 via Computer 2. Computer 2 needs to know which packets to route to computer 3 and which packets to not route.

The solution is that each computer has a unique address on the network. Any message destined for a particular computer is *addressed* to that computer. When computers receive a packet of information, they first look at its destination address. They should then either ignore, or onward route, packets that are addressed elsewhere.

In Chapter 11, when we discuss practical wired and wireless networking, we will discover hardware and IP addresses. The hardware address, unique for each computer interface on a particular network, is designed with Ethernet packets in mind. Each packet is addressed to a particular interface or is addressed to a range of addresses (which could include all addresses, with a broadcast packet).

10.3 Scalability, Efficiency, and Reuse

We started this chapter by considering the need to send data from one computer to another and have progressively built up a number of requirements as we proceeded, that would be needed to make this kind of communications more flexible and generic.

While it is still possible today (electrically speaking) to string a fixed wire between two remote computers sending pure binary, the vast majority of practical communications between remote computers uses packetized and encoded communications, predominantly based around the Internet Protocol.

As we have seen, packetization is flexible, convenient, and error-tolerant. However, there are overheads associated with packetization, which include the fact that we probably need to include additional information (bits) to handle error conditions, indicate control or data packets, and hold the address of the recipient of the packet. A full list of information that could be appended to a packet of data would be extensive, but the following is very common:

- Address of the intended recipient
- Address of sender
- Type of packet
- User data (sometimes called a payload) for a data packet
- Control information, for a control packet
- Sequence number
- Error check (or correction) bits

Other common information includes a timestamp (or time-to-live), details about routing and network conditions, a data length field, and ideally perhaps some form of security information to prevent spoofing.

All these items of information are important for generic and flexible networking topologies and methods, but of course they add an overhead; for every kilobyte of user information transmitted, a few bytes (or maybe more) of additional information must be sent. This reduces the usable bandwidth of the link. When designing bespoke networking solutions for a fixed high-performance link (like a wire linking two supercomputing nodes), simplifications could be made in the interests of efficiency, at the expense of lost flexibility. No other devices would be able to connect to the "network" but it could be very fast.

10.4   OSI Layers

The International Organization for Standardization (ISO) ratified the open systems interconnection (OSI) model (also known as a reference system) in the late 1970s as a way of classifying and managing computer connections. Yes, this really is quite confusing; it is the OSI from the ISO.

The model has seven layers, as shown in Figure 10.10. From the bottom up, these encompass everything from bit-level signaling to the application that makes use of that signaling (e.g., spanning the intermediate steps from voltage transitions on an Ethernet cable up to an Internet banking system). It is not always easy to remember the names, so many of us use a mnemonic such as "**P**lease **D**o **N**ot **T**hrow **S**ausage **P**izza **A**way" (from the bottom up, this spells out the initials of the layer names).

The idea is that each layer communicates only with the layers immediately above and below them on the stack, and that such communications are strictly defined. Thus developers of one layer need only be concerned with the communications to neighboring layers. These subdivisions allow for more regularity in communications, and in theory higher reliability.

The function of each layer is described in turn, starting from the lower physical layer:

Physical—handles the electrical connectivity of a unit to a communications medium. For example, the wires, timings, and voltages within a data bus, or the frequency and phase signals of a wireless connection to convey digital information (bits). The physical layer is responsible for ensuring that the unit can "talk to" and "listen to" the medium of transmission. It is also responsible for establishing a connection to the medium, to participate in schemes allowing that medium to be shared (where appropriate) as well as translating to and from logical bits and hardware-specific physical symbols.

Data link—bits are combined into blocks for transmission, and the flow is managed (which includes arranging for delivery to the right destination). The data link layer (DLL) imposes a point-to-point or (multi)point-to-multipoint structure onto the physical communications handled by the physical layer. In many systems it is required to handle errors that occur in the physical layer so that it presents an error-free frame interface to the network layer.

Network—information blocks are routed across a network that consists of individual data links, and possibly mixes different types of physical media. The idea is to communicate end-to-end from sender to receiver, relying on lower layers to handle whatever type of signaling is used (or types, in the case of multiple hops). Communicating

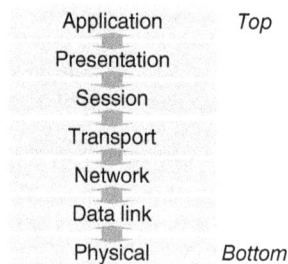

FIGURE 10.10 The open systems interconnection seven layer model of network communications.

Application	*Top*
Presentation	
Session	
Transport	
Network	
Data link	
Physical	*Bottom*

over one hop or many does not concern the network layer, nor does the type of technology used for communications.

Transport—ensures that all data arrives at the receiver and is sorted into the right order, before being passed to the correct application.

Session, presentation, application—these three software layers contribute different aspects of the way application programs make use of the network to communicate with each other.

These seven layers are something like a computer scientist's conception of how networking *should* work. However, the actual networks that we all rely upon every day were designed by computer engineers. Engineers are very pragmatic, and so they created a system that works very well in practice, but it only requires five layers. Presumably this was not approved of by the OSI inventors. We will meet those five TCP/IP layers in Section 11.2.

10.5   Topology and Architecture

Earlier chapters in this book explored the internal architecture of a computer, or a CPU. For example, Chapter 4 outlined how various functional units can be connected using different kinds of bus architecture. We saw how instructions moved items of data from memory to a functional unit for processing, then perhaps to a second functional unit, and eventually moved the result back to be stored in memory.

With this view of a computer, imagine now that one of the functional units is moved to outside the computer so that it becomes a peripheral. The concept is essentially the same, except that the data transfer to and from a peripheral is usually slower than to and from an internal functional unit (internal buses almost always run faster than external buses, because shorter is better—at least when it comes to computer buses).

Now imagine that the peripheral containing the functional unit is moved even further away. The interface between the main computer and the remote functional unit would need to stretch and change, but the concept of sending data to the unit, and then receiving transformed data, remains.

In some ways we can view a network from a functional or service perspective as a set of units that send information to other units to be transformed into something else, which is sent elsewhere.

10.5.1   Hierarchical Network

Most real-world networks these days have a hierarchical topology, such as the example in Figure 10.11. This includes the vast majority of company intranets (closed networks that exist completely within an organization), where the machine at the top would often be the company gateway to the outside Internet. As is usual these days, the network consists of machines that are linked together, with several servers linking to two or three different machines (in a school or university, such links could connect several hundred wired or wireless devices, not just three).

Computers at the top of the hierarchy tend to be servers for Web, mail, printing, and other applications, whereas those along the bottom tend to be users' machines. This means that most communication is between the lower layer and the upper layer

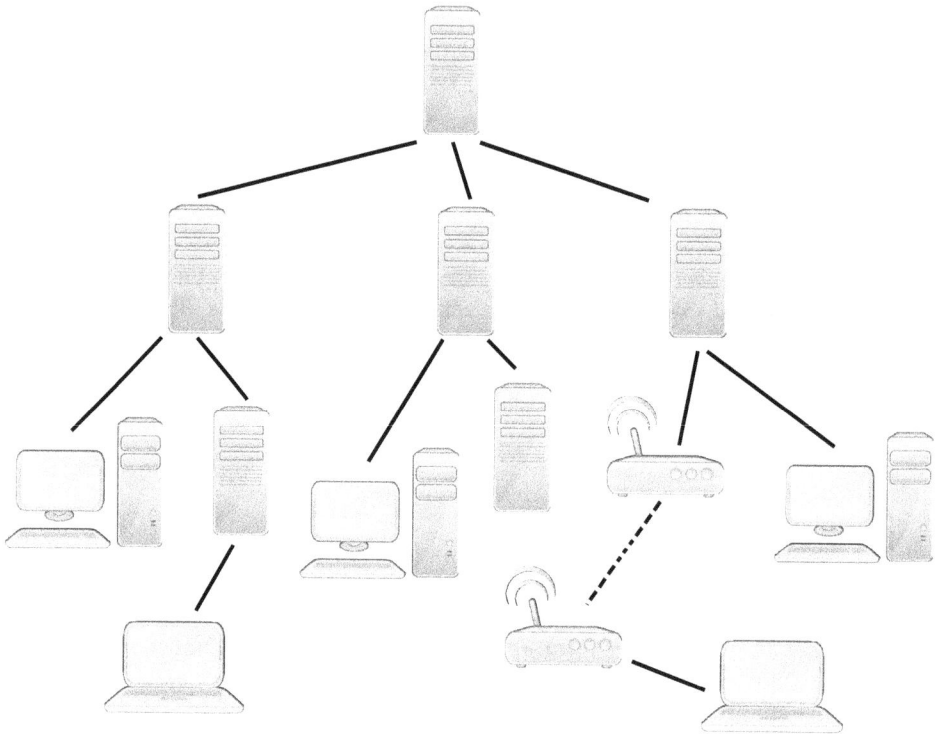

FIGURE 10.11   An example of a hierarchical network, showing different layers of computers.

(a relatively efficient link) rather than between two machines on the lower layer (which would be much less efficient, possibly involving twice as many hops as simple top-to-bottom communications).

10.5.2   Client-Server Architecture

A client-server architecture is an arrangement in which some computers are designated as servers, which exist to provide a function and are connected to a larger network such as the Internet (or an intranet). The structure exploits the ability of networks to support a data connection from one computer to any other. When one computer on the network requires a particular service, for example printing a document, it connects to a print server and passes the "print job" to the print server. Any other machine could, simultaneously, connect and upload print jobs. Individual print jobs would then be printed by the server, in order. The example of Figure 10.12 shows how four clients can use the Internet (or an intranet) to access different servers for specific tasks.

10.5.3   Peer-to-Peer Architecture

Not exactly a topology, a peer-to-peer system is any arrangement where two or more computers share tasks and information equally. It differs from a client-server architecture in that one machine is not designated to provide a service for the other.

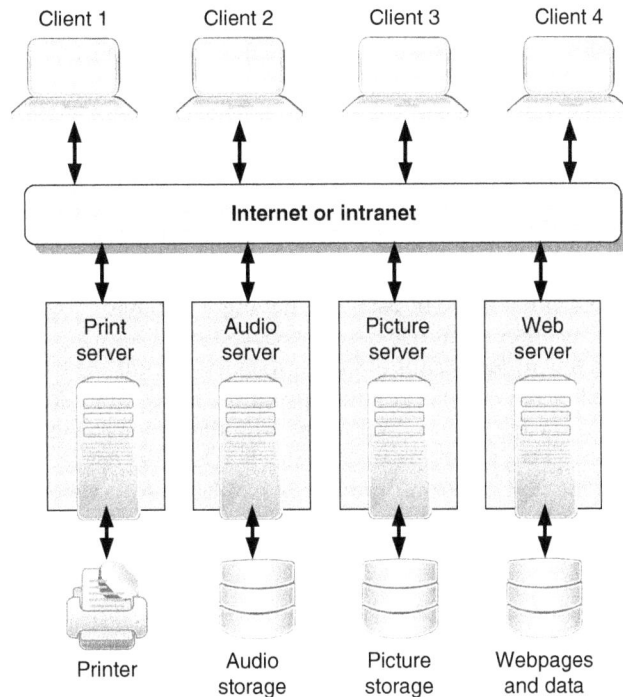

FIGURE 10.12   An example of a client-server arrangement.

Peer-to-peer systems can be more efficient than other systems because they do not involve any third-party machines. With many users, servers can become bottlenecks (or targets for denial-of-service attacks), whereas peer-to-peer links do not need to share resources apart from the basic functionality of the network itself.

10.5.4   Ad Hoc Connection

The networks we have encountered up to now have been, for the most part, carefully drawn and designed. This is the aim of a network architect, or a system planner—to have a nice neat diagram representing their network. Unfortunately the actual situation in many companies and most universities tends to be more cluttered and less regular. The opposite extreme to a completely planned network is a completely ad hoc network. This is one in which any device can connect into the network at any time and in any position. The beauty of the TCP/IP-based Internet that we use is that it can support both planned and ad hoc connectivity equally well.

Wireless ad hoc networks are especially interesting in that they can form a scattered group of devices (we call them nodes) with wireless networking capability that connects to the other nodes nearby. In such systems, a message passing from one node to another distant node could traverse an infinite number of possible paths along the way. Various network management algorithms are used to steer messages appropriately in order to meet a system objective (such as fastest message delivery, shortest physical distance, lowest energy path, and so on). Researchers have developed many special routing algorithms for ad hoc networks that have different properties, and this is a very active current research area.

10.5.5   Mobility and Handoff

The final topic in this section is a very brief discussion about mobile networking. For most readers, mobile networking will mean connecting to the Internet from a handheld device while walking or in a vehicle (hopefully as a passenger, not as a driver). Mobile networks are cell based, with each cell having a notional hexagonal footprint around a cell tower. Such hexagons are densely packed in developed urban areas, so that just walking for a few minutes could transition between multiple neighboring cells.

When a call or data connection is in progress, the communication is conducted with the nearest cell tower (technically, the one with the strongest signal, and that is often the closest one physically). Moving from that cell to another triggers a process called handoff, in which the first cell tower gracefully hands the responsibility for communication to the second cell tower. Modern mobile voice and data communications systems are all packet based, so what happens is that the handoff occurs between two adjacent transmission packets. Depending upon the actual technology being used, this may involve a frequency switch or a timeslot switch. In some systems a new link is made before the old one is broken (a so-called soft handoff), whereas in hard handoff systems, the old link is terminated simultaneously with the new one starting.

From what we know about networking systems already, if packets have source and destination addresses, as long as the wired network connecting the cell towers together is linked, packets received at either cell tower can "find their way" to the destination (this is called routing). Having sequence numbers also means that if the destination receives multiple versions of the same packet, it simply needs to ignore the duplicates.

10.6   Summary

This chapter has introduced the concept and requirements behind connecting one computer to another remote computer. From this very basic foundation, we considered how it would be possible to connect multiple computers together, which involves defining a topology. We then discussed the concept of transmitting data in packets.

Once data is packetized, communication becomes far more interesting because packets almost take on a life of their own, with addressing, routing, error checking, error correction, retransmission, acknowledgment, and even a time-to-live. Acknowledgment and negative acknowledgment schemes were two different ways of handling errors, and they went hand-in-hand with link control and management.

In an attempt to bring order to the field, the OSI model (developed first by the ISO), divided the process of communication between computers into seven interconnected stages or layers. However, we learned that five layers are all that we tend to use in real systems. This reality, the IP, TCP/IP, WiFi, and the vast World Wide Web, will now be explored as we embark on our survey of networking in Chapter 11.

10.7   Problems

10.1   Identify which of the following types of communication are simplex, half duplex, or full duplex:

- Talking excitedly one to one with a friend.
- Writing letters to, and receiving letters from, a pen-friend who lives in a different country.
- Watching a broadcast television program.

10.2   Given that the United States and the United Kingdom each have networked computers numbering in the hundred millions and most are operational for many hours each day, explain why transatlantic communications can no longer be circuit-switched.

10.3   The size of data packets being sent between two computers can be changed to suit communication conditions. Explain why packet size is reduced when the link error rate increases.

10.4   In a packet-switched network where several computers share a long-distance link (similar to the diagram in Figure 10.4), specify four pieces of information that each packet should contain, apart from the user data that is being transmitted.

10.5   Highlight the difference between *error detection* and *error correction* when referring to packetized network communications.

10.6   If ECCs are so powerful, identify two reasons why ECC data is not routinely appended to every packet transmitted across all networks.

10.7   Identify three types of information that might be conveyed by control packets in a busy network.

10.8   In the OSI model, what is the name of the layer that connects to the electrical wiring in a networked device?

10.9   In a typical household connecting to the Internet over cable modem, can you guess the name of the device which sits at the top of the local network hierarchy?

10.10   In an office network, with one computer permanently connected to the office printer, would a peer-to-peer architecture be best for managing the printer, or a client-server arrangement?

10.11   A friend watches YouTube videos on his smartphone as he walks from home to work and back every day—a distance of a couple of miles. Although he sets his smartphone to remain connected to the Internet continuously, he often experiences interruptions to the videos at a certain point on his journey. He notes that the same problem occurs at roughly the same place when walking in both directions. Identify what aspect of his Internet connectivity is causing the drop out.

10.12   What is the name of the process by which a packet, which has a destination address specified, is directed by network nodes down the correct network path needed to reach its intended destination.

10.13   Give two reasons why many packetized data transmission systems include a sequence number in the header information for each packet sent.

10.14   Would a star or a ring network topology be the best choice for a packet data transmission system in which reliability is key, ensuring that packets reach their destination is more important than efficiency or than how long it takes them to arrive.

10.15   Explain why it can take longer to download a file over a 400-Mbit/s WiFi link when the access point is 100 m away than over a 22-Mbit/s WiFi link with an access point in the same room. Both WiFi devices are connected to the same backbone network, and no other users are connected to either system.

CHAPTER 11

Networking

The idea of connecting computers together, at first one to one and later one to many, was explored in Chapter 10, which then went on to discuss some of the related issues, difficulties, and opportunities arising from that connectivity. Two major concepts that arose were the need to packetize data for transmission over a channel and the use of layered models for network communications. The famous seven-layered OSI model (Please Do Not Throw Sausage Pizza Away) was introduced but we mentioned that the Internet model actually makes use of only five layers.

In this chapter, we will be concentrating on the Internet Protocol (IP), examining those five layers and the concept of encapsulation, before building our understanding upward. In fact, we will build quite a long way upward in this chapter: Starting with physical packets, we will discuss transport mechanisms, the idea of sessions, addressing, various network protocols, application protocols, domains, the World Wide Web, and email. We will also touch very briefly on the important issue of network security.

The topic of computer networking itself is very much layered, in that the discussion of any aspect of a network depends upon the characteristics of lower as well as higher layers. This means that it is important to progress step by step through the topics. A misunderstanding of one layer might affect understanding of other layers.

Finally, we should clarify one point. In this book (as in many others discussing networking), "higher" refers to being closer to the application and the user, whereas "lower" refers to being closer to the actual hardware (voltages, signals, frequencies, and so on). So when we progress from lower to higher we mean moving from the physical communications mechanism to the application, or usefulness, that it has in daily life.

11.1 The Internet

The Internet is a frequent topic of conversation among many people, but it is useful to spend a moment defining it. In essence it is the name given to one particular connection protocol. The job of this protocol is to transfer information over one or many physical links, to allow one computer to communicate with another. Those links could be single or short hops, but are often multiple hops, and can operate over many different types of physical link technology.

The Internet allows data and resources to be shared between different computers and forms the underlying basis of communications between different places, people, and computers. Many of the computer applications that we use daily will communicate over the Internet: Web browsers accessing websites, Skype, WhatsApp, Facebook, SnapChat, Instagram, YouTube—all are services that rely on the Internet to communicate.

The Internet has four excellent main characteristics which explain its popularity:

1. *Flexibility*—it can operate over many different types of link, whether fast or slow, high or low bandwidth, error prone, crowded, or changing.

2. *Reliability*—even in the face of network errors and rapid changes, it can continue to operate reliably.

3. *Redundancy*—this is often a good thing in computing (except when it refers to your job), because it means that there are different ways of accomplishing a task. For example, two, three, or a hundred paths for getting a packet from computer A to computer B. The redundancy of the Internet is one of the key reasons why it can be so reliable in the face of individual node or link failure.

4. It is *standard*—meaning that different hardware from different vendors can work together to form a network, using different types of physical link, and conveying different types of data.

All of these characteristics mean that when we use a smartphone over a wireless link to access a server across the other side of the world, we do not need to concern ourselves with the type of networks and hardware used, the quality of the link, or the exact path of the link. It also means that when we physically plug a computer into a network somewhere, or connect to a new WiFi access point, the connection (usually) just works.

11.1.1   Internet History

Today's Internet largely owes its existence to research funded by the U.S. Defense Advanced Research Projects Agency (DARPA—but now called ARPA), in the 1970s. This research was for military purposes, and concentrated on creating a network with exceptional reliability and redundancy, that could continue operating even after multiple system failures. The network was called ARPANET.

In fact, there were other networks developed at the same time—including the Cambridge Ring (Cambridge University network) which probably became the world's first wide area network (WAN), but these had no standard way for different computers to communicate over different physical links. The alternative networks were also significantly less reliable and flexible than the Internet, and tended to require substantial knowledge and skill when used.

The first networks, including the Internet, were very much related to UNIX-based operating systems, only adopted by other operating systems such as Microsoft Windows, much later. Even today, UNIX-based computers tend to embrace networking more efficiently than Microsoft Windows–based systems and so most back-end servers, the network operating infrastructure itself, website servers, domain name servers, and so on are almost all UNIX based. The choice of UNIX-like systems by network specialists worldwide is primarily for reasons of reliability and security, but is also due to the strongly ingrained networking capabilities of UNIX.

11.1.2   Internet Governance

In its early years the nascent DARPA-hosted Internet extended gradually to reach various universities, first in the United States and then worldwide, and later to commercial and entertainment industries, before effectively taking over the world. Its basic design has actually changed very little since the DARPA days, which is one reason why the world is running out of IP addresses; at least for IP protocol version 4.

Today, Internet development is overseen by an independent organization known as the Internet Architecture Board (IAB)[1] and this umbrella organization oversees a number of smaller but vitally important groupings which include

- Internet Engineering Task Force (IETF)[2]
- Internet Research Task Force (IRTF)[3]

These are technical organizations that manage the evolution of the Internet and its future advances; no one organization or country runs the Internet. It may come as no surprise to readers to discover that they are managed by committees these days.

Despite today's governance by committee, many of the standards that make up the Internet and the World Wide Web began as someone's great idea, or were dreamed up in the lab rather than the boardroom. The first step on the path from an idea by a pioneering individual to being adopted worldwide was being issued as a request for comment (RFC). Box 11.1 explains more and highlights just a few.

Box 11.1   RFCs

Many of the so-called Internet "standards" that we rely upon today were defined by (relatively) ordinary computer engineers. They documented their ideas, experiments or findings, and passed them around to the community for comment, discussion, and refinement.

If or when those ideas seemed to be good, people started using them, and the best became de facto standards over time. Soon, these posted ideas began to be known as request for comments (RFCs). Each RFC was identified by a number for convenience. There are currently more than 8000 of them, and that number is growing all the time. Some are very technical, some are practical, some are records of ongoing conversations and a few are rather strange indeed.

Most RFCs cover a single idea or protocol, and all are publicly available, for anyone to view, at http://www.rfc-editor.org (or for anyone to propose their own). Some of the more famous RFCs include the following:

RFC 1 on "Host Software" was issued in April 1969, making it as old as the author of this book.

RFC 20 defines the American standard code for information interchange (ASCII).

RFC 527 is a poem entitled "ARPAWOCKY."

RFC 822 was the first one to define the format of an email message (superseded by **RFC 5322**).

---

[1] IAB: http://www.iab.org

[2] IETF: http://www.ietf.org

[3] IRTF: http://www.irtf.org

RFC 968 discusses debugging techniques and the problems that arise when bringing a new network into operation, but is written as a poem!

RFC 1149 is a description of a highly unusual way of transmitting data packets—it is very much worth a look.

RFC 1939 describes the post office protocol version 3 (POP3) for email retrieval, while **RFC 3501** describes an alternative method called the internet mail access protocol (IMAP). Both are in common use today.

RFC 2068 is where the hypertext transfer protocol (HTTP) can be found.

Despite this list looking like a historical index, RFCs are alive and well. They are still being issued, and the system continues to be in active use. As a method of proposing good ideas and perhaps seeing them adopted by others, it is unsurpassed.

11.2   TCP/IP and the IP Layer Model

The Internet layer model is shown in Figure 11.1 alongside the larger OSI model, introduced earlier in Section 10.4. Just as with the OSI model, the idea of the IP model is that each layer interfaces only with the layers immediately above and below, and is responsible for providing a particular aspect of the communications service. The IP model is occasionally depicted as having just four layers—in which case the physical and network layers combine to form a single *Network Interface* layer. But we will use the five-layer depiction shown, and explore the roles and responsibilities of each layer below:

- *Physical*—the lowest layer, this ensures data bits get communicated between nodes across a transmission medium. Like in the OSI physical layer, this handles the electrical connectivity, signaling, timing, media sharing, and translation between logical bits and symbols. In networking terms, the hardware device which handles the physical layer is often imaginatively called a "PHY."

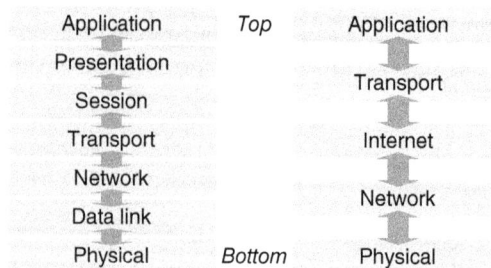

FIGURE 11.1   Comparison of the OSI seven-layer model (left) with the five-layer IP model (right).

- *Network*—this layer collects bits into larger units (packets) for transmission and hides the differences between different physical media from upper layers. It has the responsibility for media access control (MAC) as well as logical link control (LLC). Usually the MAC interfaces to the PHY of the lower layer, while the LLC interfaces upward to the higher layer. The MAC layer frames up data to be transmitted, verifies the frames of received data, and also provides services such as arbitration and flow control when the transmission media is shared by multiple units. The LLC, by contrast, handles errors and flow control with the higher layers. In some computers a single hardware device called a "MACPHY" combines the physical layer with the MAC sublayer, leaving the LLC to be performed in software.

- *Internet*—this layer uses the IP to ensure end-to-end transmission of packets (but does not guarantee that they arrive or are acknowledged).

- *Transport*—handles data packets that are to be transmitted to (or received from) another device. Packets that are transmitted from the transport layer are reassembled by the transport layer at the receiving device. Various protocols can be used for the transport (see Section 11.5), depending upon the type of information being sent and the desired characteristics of the communications.

- *Application*—may use higher layer protocols such as file sharing with network file system (NFS) or NetBIOS, file transfer protocol (FTP), simple mail transfer protocol (SMTP) post office protocol version 3 (POP3), and so on.

We will examine a few of these layers a little more closely, starting from the bottom and moving upward. In this way, we will see how a global structure like the World Wide Web can spread all around the world (and indeed off-planet, thanks to NASA), by using a foundation constructed from small but reliable and flexible link-building blocks. But first we need to explain one more important technique that is used in practical networking.

11.2.1   Encapsulation

Encapsulation, in its widest sense, describes the situation when an object is totally enclosed within another object. It is an important concept in some programming languages where it hides internal details from external interference. It is equally important in networking, not to hide information but to separate out the roles and responsibilities of different layers in a network.

Figure 11.2 shows the five-layer IP stack, giving the layer names, and showing an example of the data that is passed between layers. When passed layer by layer from top down (i.e., being transmitted), a block of data is successively encapsulated in a sequence of objects, each of which have a particular responsibility to implement the functionality of the corresponding layer. We will consider how data is transmitted by this stack.

11.2.1.1   Application Layer

Beginning at the top, imagine a file of data that needs to be transmitted from an application on one computer to an application on another. This data needs to be divided up into a number of blocks or packets (usually of fixed size—packetization was discussed in Section 10.2.1), each of which will be transmitted independently and then reassembled

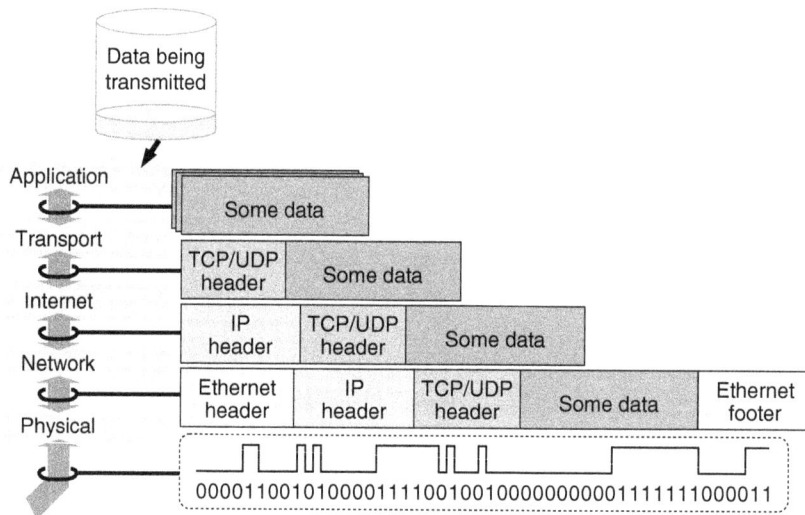

FIGURE 11.2   The principle of encapsulation enables data to be enclosed in a structure which is itself enclosed in another structure and itself enclosed in a further structure—each with different characteristics and responsibilities.

into a file at the receiving end. Each block of data is passed from the application layer down to the transport layer in turn.

11.2.1.2   Transport Layer

The transport layer forms each block of data into a packet (such as TCP or UDP, explained later in Section 11.5, that have different transmission characteristics, particularly in how they handle errors). Forming a TCP or UDP packet involves prepending various items of necessary information to the start of the data block in the form of a header. Once a packet has been formed, the entire thing is passed down to the Internet layer.

11.2.1.3   Internet Layer

Each packet from the transport layer is bundled up into an IP datagram—again by prepending various items of information to the start of the packet. We will look at exactly what information is added at each layer later, but for now, this datagram (i.e., a bundle of IP header and TCP or UDP packet) is passed to the network layer to prepare it for transmission over the physical connection. The Internet layer does not "know" what physical connection is being used, it just "trusts" that the network layer will get that IP datagram to where it is supposed to go.

11.2.1.4   Network and Physical Layers

IP datagrams are then prepared for transmission over the physical network. In Figure 11.2, the network type is Ethernet, which is commonly used for wired computer networking. Ethernet transmits data that is formed into frames, beginning with a header and ending with a footer. The network layer creates the header and footer to add to

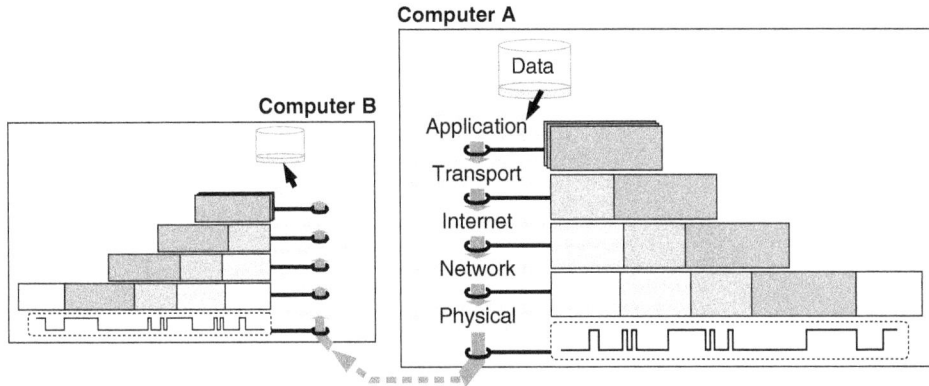

FIGURE 11.3 Diagram showing data from one computer transmitted to a second computer, highlighting the layer-wise encapsulation performed on the transmitting computer (right) and the corresponding de-encapsulation on the receiving computer (left).

the start and end of each IP datagram, and then passes to the physical layer to actually transmit it. The binary Ethernet data that eventually gets transmitted down the Ethernet wire (or wires) is shown at the bottom of the figure.

11.2.1.5   The Importance of Encapsulation

Encapsulation is also important in ensuring network independence, which means almost any type of lower layer is able to transmit packets (and data) from higher layers, providing the data is appropriately encapsulated and passed into the layer in the expected format. For example, TCP/IP can be encapsulated in IP datagrams and sent over Ethernet as shown, but also over WiFi (IEEE 802.11) links, in which case the encapsulation would be similar to Ethernet (which is actually more properly known as IEEE 802.3) but with different hardware and thus a different physical layer.

Apart from conveying TCP/IP as shown, Ethernet frames can also encapsulate a huge variety of different types of data, including speech (for telecommunications or voice conferencing), video, and even the control data for ATA hard disc interfaces. The beauty of encapsulation is that the structure of the Ethernet frame does not change when different data is conveyed, nor does the way it is processed by the network. Neither does a data packet change when it is conveyed over Ethernet, WiFi, USB, or other kinds of link.

One consequence of this is that, once a layer in the IP stack is working properly, it can be used with any lower layer protocol or type, as long as that lower layer "respects" the handling of the data (i.e., does what it is supposed to do).

What *is* it supposed to do? Well, that is best summed up by Figure 11.3, which shows data from one computer split into blocks with are gradually encapsulated into Ethernet frames by the physical layer. These are sent, at the bottom of the stack, to the physical layer in another computer. In the other computer, they move up through the stack, gradually stripping away the packet, frame and datagram information to provide the raw data blocks to the application layer on that other computer.

An example of this, specifically for Ethernet, will be discussed further in Section 11.3.2.

Computer A

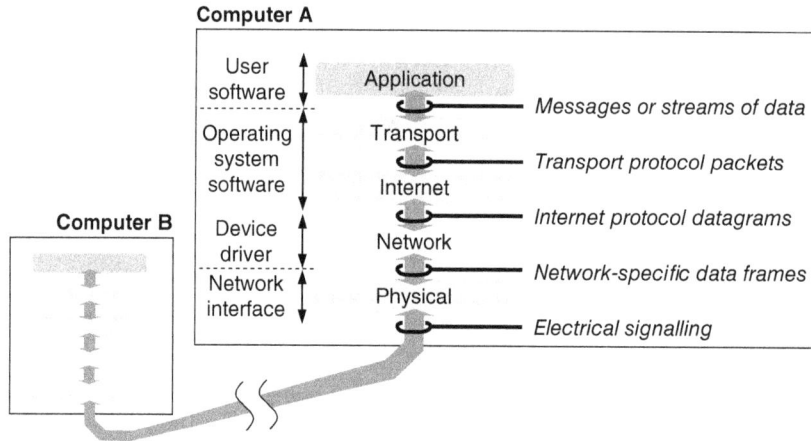

FIGURE 11.4 A diagram showing how computer A connects using TCP/IP to computer B, highlighting the types of data objects being passed from one layer to the next and indicating where each layer would typically be implemented.

11.2.1.6 Layer Interfaces

Looking again at this process, Figure 11.4 highlights the objects being passed from one layer to another, and also gives an indication of where the layers are implemented. In the figure, computer A connects using TCP/IP to computer B. The application layer in both computers is part of the software being used by a particular user, such as a Web browser or an email program. The transport, Internet, and network layers are typically implemented within the operating system while the physical layer normally resides in networking interface hardware (called a network interface card, or NIC, in older textbooks—since it used to be implemented on a physical plug-in circuit board on desktop PCs).

While the diagram shows data being passed up or down layer by layer, it also specifies *what* is being passed between layers. If the communication link is established and transmitting data from A to B without error, then the output from any layer in computer A (going down to the layer below) will be identical to the input to the same layer in computer B (being passed up from the layer below) a few moments later.

For example, the network-specific data frames generated by the network layer in computer A normally travel through the physical layer and on to computer B's physical layer. But if those frames were able to bypass the physical layer and be passed straight into the computer B's network layer, this layer would recognize those frames and be able to decode them straight away. The data within those frames would then be passed straight up into the Internet layer in computer B. Similarly, transport protocol packets from the transport layer in computer A would be recognized by the transport layer in computer B—no matter whether they went down the stack in A, across the link, and up the stack in B, or whether they were magically passed directly to the transport layer of computer B. Figure 11.5 illustrates how this process allows IP data to be conveyed within Ethernet data frames, which are discussed more fully in the following section.

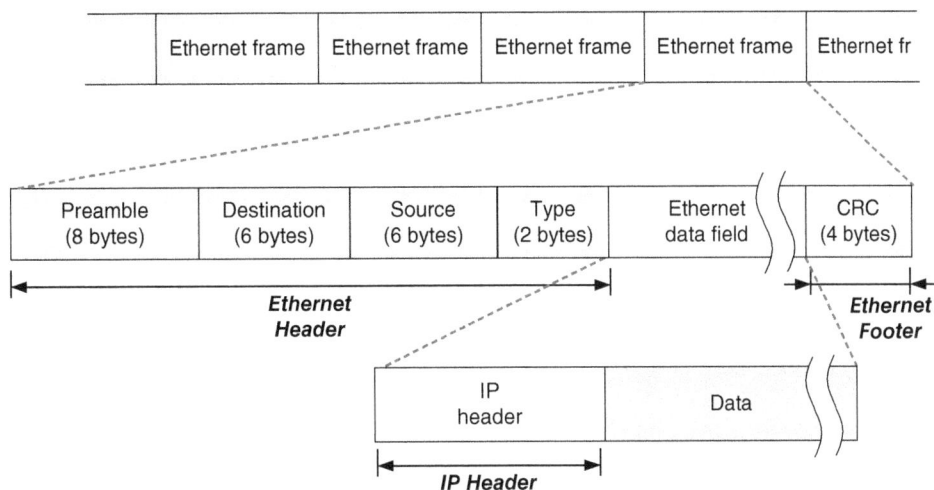

FIGURE 11.5 Examining data from an Ethernet connection (top), comprising individual data frames (middle) which are conveying individual Internet datagrams (bottom).

Viewed in this context the prime responsibility of every layer can be summed up simply. It is to ensure that the objects passed to it (data/packet/frame or whatever) are conveyed to the corresponding layer on the receiving computer.

11.3 Ethernet Overview

The most common hardware formats currently used for networking are probably wired Ethernet and WiFi. Both of these are defined by IEEE standards (802.2 and 802.11, respectively), and they share a number of similarities. Leaving WiFi for later (see Section 11.7.1), we will examine the pioneering Ethernet protocol.

Ethernet actually began as a networking standard that used coaxial cables for transmission (which look like old-fashioned wired television aerial cables). Eventually it became more commonly implemented over twisted pair cable with a phone-style connector (called an RJ45 connector), or over optical fiber, both of which are both more reliable in practice than the coaxial links. Another commonly used wired networking technology called *token ring* shared many of the concepts of Ethernet, but is now largely obsolete. Today, Ethernet is used not only for connecting fixed or desktop computers in offices, but also for providing telephone connections in many offices.

Wired Ethernet runs at different speeds depending on the type of cable used:

Coaxial up to 10 Mb/s (megabits per second, usually written as Mbps or Mbits/s)

CAT5 twisted pair operates up to 100 Mb/s

CAT6 twisted pair operates up to 1 Gb/s, also known as gigabit Ethernet

Optical fiber can exceed 100 Gb/s

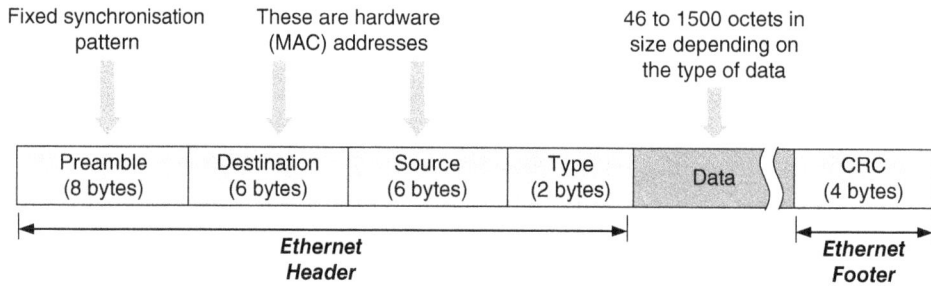

Figure 11.6 Internal details of an Ethernet frame, showing the header and footer that enclose a data payload that usually carries an IP datagram.

11.3.1 Ethernet Data Format

Data is conveyed over Ethernet in frames which can carry between 46 and 1500 bytes of data.[4]

Figure 11.6 displays the Ethernet frame format, showing the data payload preceded by an Ethernet header and followed by an Ethernet footer. The network layer would be passed an IP datagram by the Internet layer and told to transfer this from interface A to interface B (with the interface addresses being hardware or MAC addresses, see later). As far as Ethernet is concerned, it is sending a block of data, which just happens to be an IP datagram—since Ethernet can send other types of data. The network layer then forms an Ethernet frame by first prepending three items of information to the beginning of the data block, then computing a checksum which it adds to the tail of the data block. The three items of information are the destination address, source address, and data type. When this is transmitted over the physical interface, it will be preceded by a preamble of fixed data that is used to wake up any device connected to the Ethernet cable and inform it that a frame is about to follow.

The checksum is called a cyclic redundancy check (CRC), and is a 32-bit number made up from a computation of the main elements in the frame; the destination and source addresses, the data type, and, most importantly, the data payload itself. The purpose of the CRC is to allow the receiver to very quickly and easily detect whether a received frame contains errors.

The way that this works is that the receiver does the same computation on a received Ethernet frame; it computes a CRC from the received data payload combined with the received destination and source addresses and type. The receiver then compares its calculated CRC with the one that was inserted after the packet by the transmitter. If there is any difference between the two, this indicates an error.

In operation, the Ethernet frame is like an electronic "postman" which carries data from a sender to a receiver. The sender specifies the addressee (who it is being sent to), and gives its own address (so the addressee knows who sent it). It also indicates what is inside the letter (the type). The main difference between a letter and the Ethernet frame

---

[4] Note that in networking terminology, "bytes" are often called "octets" for historical reasons, but we will generally use the more common computer terminology of "bytes."

| Preamble (8 bytes) | Destination (6 bytes) | Source (6 bytes) | Type (2 bytes) | Ethernet Frame Payload | CRC (4 bytes) |

Ethernet Frame

| IP header | IP Datagram Payload |

IP Datagram

| UDP header | UDP Payload |

UDP Packet

| Data |

Application data packet

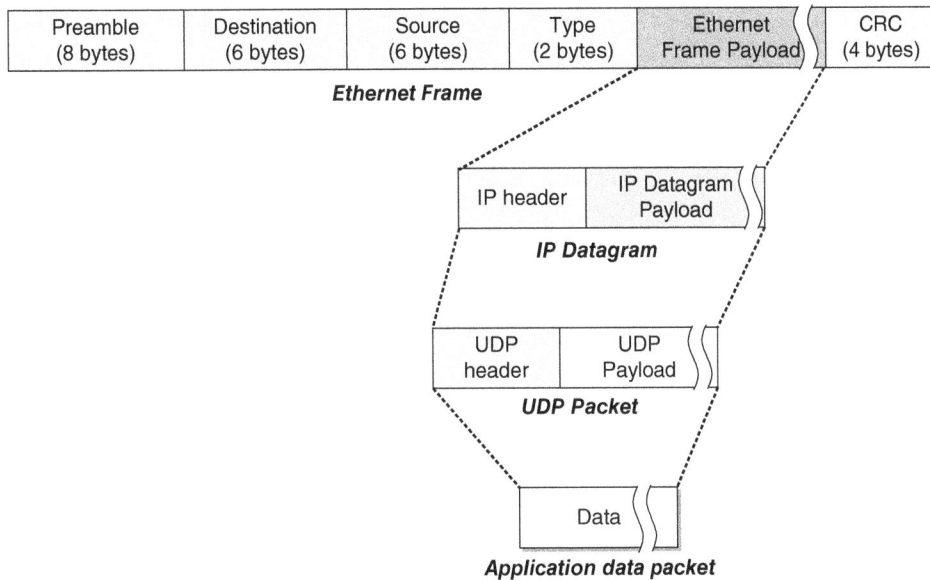

FIGURE 11.7 Illustration of how an Ethernet frame can carry an IP datagram which itself carries a TCP packet of application data.

is that every interface connected to that Ethernet network receives the same frame—but those interfaces who are not the addressee should ignore it.

And just like a letter, sometimes the frame will get lost, delayed, damaged en route, or data will be missing. Evil postman might read the contents, or even change them.

Interestingly, for testing purposes, it is possible to place a network interface into a so-called promiscuous mode which allows it to read *every* packet, not just those addressed to it.

11.3.2   Ethernet Encapsulation

Encapsulation is used to convey data within packets, datagrams, and frames so that TCP/IP can be carried over Ethernet. This was briefly covered previously in Section 11.2.1.5 when we discussed the layers in the TCP/IP model. The concept is explained in more detail in Figure 11.7 where an Ethernet frame is shown carrying a payload that contains an IP datagram. That IP datagram itself is carrying a UDP packet, which in turn is conveying a block of data. All of this encapsulation is designed for just one purpose—to send that block of data from one application to another application.

A very important point to highlight is that the Ethernet frame conveys its contents from one hardware interface to another hardware interface on the same network. These are usually two machines plugged into the same Ethernet or two devices with WiFi interfaces. But what happens when we need to send a packet of data to a machine on another network? The way that works is to make use of a gateway interface. Sending an IP datagram to another network means that it is conveyed over Ethernet on one network to the gateway interface. This process is illustrated in Figure 11.8 where an application

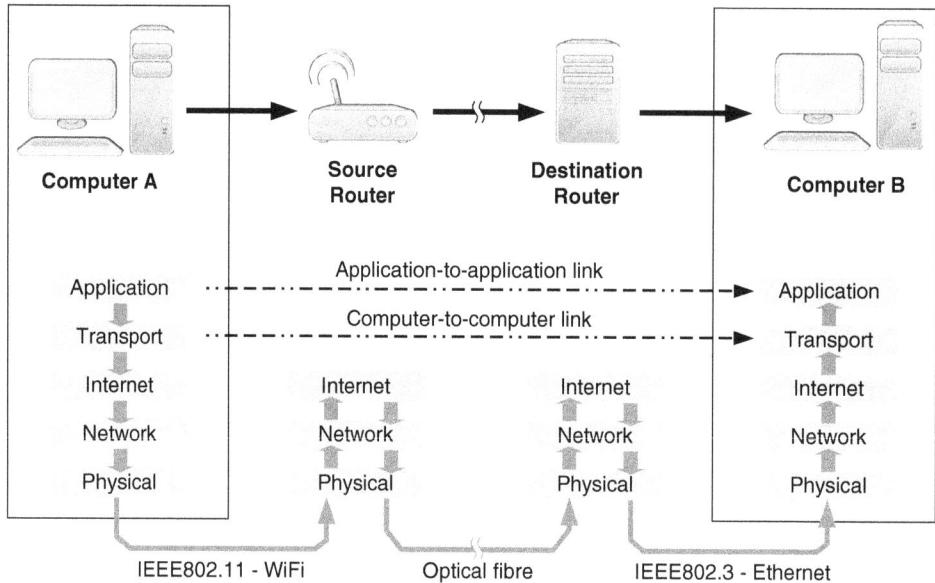

FIGURE 11.8   Illustration of an application on one computer sending data to an application on a second computer that resides on a different network, through two routers acting as gateway interfaces.

on computer A is sending data to computer B on a different network. The two networks shown are connected between two gateway machines (routers). In this case, when a WiFi data frame from computer A (similar to an Ethernet frame, but wireless) is received at the gateway router, the IP datagram is de-encapsulated from the WiFi frame, and then passed to the Internet layer. This will direct the datagram to a different network interface that connects to the destination router. This second interface re-encapsulates the IP datagram and sends it over the next link. This second link probably makes use of different technology to the local network and so it will be encapsulated in data frames appropriate to that network. When this data frame is received at the destination router, it is again de-encapsulated and the process repeated. In Figure 11.8 the final link is Ethernet. So we see the principle of encapsulation working to wrap data in a transport packet, then an IP datagram and then using a variety of data frames, the IP datagram finally reaches computer B where it is unpacked layer by layer to deliver the data to the application. Interestingly, to the application on both computers, the link process is invisible—they just communicate as if they were running on the same computer (and indeed, they could do so). Similarly for the transport layer, which does not "notice" the individual hops involved in getting packets from computer to computer.

A final note about this process is that (as usual) there are many ways to accomplish the same thing, and there is a lot of detail involved in networking that would need an entire textbook to explain. However, it is useful at this point to briefly explain the difference between a network *router*, a network *switch*, and a network *hub*. With reference to Figure 11.8, the routers shown each connect two networks at the Internet level. They examine IP datagrams, and learn the best path between source and destination routers.

An Ethernet switch, by contrast, would not involve the Internet layer. Switches operate (typically) at the network layer to learn the hardware (network-dependent) addresses of the devices connected to that network. Switches connect multiple subnetworks called network segments together; commonly 16, 32, 64, or more subnetworks in an office or school Ethernet system. Switches intelligently pass Ethernet frames from one subnetwork to another by examining the hardware addresses of the frames and matching those to the machines it knows are connected to each segment.

An Ethernet hub operates at an even lower layer than a switch. A hub is a purely physical layer device having multiple Ethernet ports. It copies every packet received at one port, to all other ports. Technically, all machines connected to that hub are on the same network segment (unlike the switch which connects different network segments together). Since they tend to be much less efficient than switches when traffic is heavy, hubs are increasingly rare.

11.3.3   Ethernet Carrier Sense

Ethernet, WiFi and many other networks are inherently shared bus systems. This means that all connected machines need to share access to the same transmission media. Even when there is a direct physical cable to each computer, as in CAT5 Ethernet, all machines connected to that network segment (including those communicating through a hub) must share the communications channel.

Ethernet frames are physically broadcast, so that all connected devices can "see" every frame. In normal operation, all interfaces will receive each frame, which they decode just enough to view the destination MAC address. If this is addressed to them, or addressed to all machines (e.g., a broadcast packet), then they will begin to process the packet appropriately. Switches will additionally handle packets that they are responsible for, such as those known to be addressed to a machine on another network segment.

The hardware delivery (physical layer) is called *best effort*. It attempts to get the payload datagrams delivered but does not guarantee anything. The hardware layer does not check to see whether frames were received by any machine, or whether they were received correctly. Higher layers in the TCP/IP model are responsible for such checks, or for guaranteed transmission.

No single station on the network "controls" the Ethernet link, which raises a possible problem if two machines must transmit data at the same time—who goes first? To solve this issue, the designers of Ethernet specified that all machines (we will call them stations) connected the shared link (we will call it a channel) must use something called a carrier sense multiple access with collision detect (CSMA/CD) strategy for transmitting data:

- *Carrier sense* (CS)—listens to see if another station is using the channel, and if so waits for the channel to be idle (i.e., waits for a gap). Once the channel is idle, it will start to transmit data. For Ethernet (Figure 11.7) the first part to be transmitted is the preamble, which begins with a fixed sequence of 56 alternating ones and zeros (`10101010...10`).

- *Multiple access* (MA)—means that other stations might also be waiting for idle periods on the same channel.

- *Collision detection* (CD)—if the channel is busy and there are many stations, it is very likely that two stations will start to transmit at the same time, or nearly the same time. So stations "listen" to the channel while they start to transmit. A station will notice that its transmission is in collision if it does not "hear" its own preamble correctly on the channel. This is usually because another station is also starting to transmit. So once a station notices this, it sends a jam signal to ensure that other stations also notice the collision. It then stops transmitting and waits for a random period of time (called a *back-off* time). After that it tries again. Meanwhile the other transmitting station also stopped and waited, but most likely for a different length of time.

CSMA/CD works very well in practice, except when networks become crowded, with many users and heavy traffic. In those circumstances, stations tend to spend more time waiting in back-off time than they do transmitting. It is therefore common to divide networks into segments, with stations sharing the physical channel within their own network segments, but not sharing the channel with machines on other segments. As we saw in Section 11.3.2, networks can be divided into different segments using switches.

Most variants of WiFi also make use of some kind of CSMA/CD to manage the sharing of transmission channels (it is primarily a variant called CSMA/CA which changes collision *detection* to collision *avoidance*, but the general idea is the same). Although segmentation can easily solve the issue of efficiency in busy Ethernet systems, this is much harder to solve in wireless networks where the signal is not confined to a physical cable. It is for this reason that faster wireless networking standards use multiple subchannels of many different frequencies, as well as small low-power cell sizes in busy locations.

11.4 The Internet Layer

The Internet layer uses the Internet Protocol (IP) to communicate host to host, where the hosts could be on opposite sides of the world, next door, or equally they could be the same computer.

Hosts are distinguished by their different IP addresses. The original idea was that every host would have a unique IP address; however, the available addresses (using the original IP standard) were exhausted long ago, and so several mechanisms were developed to reuse IP addresses, or allow different machines having the same IP address to coexist.[5]

11.4.1 IP Address

At the time of writing, most of us still use IP version 4 (IPv4) to identify machines throughout our homes and offices, although IP version 6 (IPv6) is used increasingly. The IP address is a software value (not built into the hardware like the physical MAC address is), and it can be changed by the operating system and even remotely, using some protocols.

---

[5] Technically, it is the network interfaces that have IP addresses, and so a machine with multiple network interfaces might have multiple IP addresses.

In IPv4, addresses are 32 bits in size and are specified using four 8-bit integers in decimal (not hexadecimal) such as

```
192.168.12.9
```

Every *visible* computer on a network should have a unique IP address (although later we will note one relatively recent class of exceptions). However, private (or isolated) networks can use whatever addresses they like. Each integer is 8 bits and hence lies in the range 0 to 255, and so there are $2^{32} = 430001000000$ possible addresses. This might seem like plenty, but many addresses and ranges of address within that 10^{12} range of possible values are reserved, and others are used only to indicate special operations, and so unavailable for general use. The structure of an IPv4 address is explored further in Box 11.2.

Box 11.2 IPv4 Address Allocation

IP address allocation

IP address ranges have historically been allocated to different top-level organizations based on the first octet (i.e., the "AAA" in the IP address AAA.bbb.ccc.ddd). In the early days of the Internet, many large organizations moved quickly to reserve their own IP address spaces. For example, all IP addresses beginning the 017 were allocated to Apple, and all addresses beginning with 056 by the U.S. Postal Service.

Later, as the world supply of IPv4 addresses became exhausted, most of those early allocations were relinquished, and instead were allocated to one of 5 regional organizations (Regional Internet Registries), such as APNIC (Asia Pacific Network Information Centre), to manage on behalf of the countries in their region.

More about IP addresses

Let us examine an IP address further, for example, 192.168.1.234. The first numbers (on the left) specify the network whereas the last numbers (on the right) identify the host interface connected to that network. Usually the host information identifies which computer is connected. Since there are different sized networks, with different numbers of computers, the split between the network and the host interface parts can be almost anywhere in that string of numbers.

For the example given, 192.168.1 might be the network and 234 might specify the machine—allowing between 0 and 255 machines (although some numbers in that range, including both 0 and 255, are reserved). However, in a network that has more than 255 computers, we could split the range differently. For example, 192.168 could specify the network, while the machine is 1.234. This would allow there to be almost 65,000 different computers on the network. The split could be anywhere—so it might be a lot easier to visualize it in binary. Here is a split where the final byte of the address specifies the host interface (in bold text): 11000000 10101000 00000001 **11101010** and here is a split where the final two bytes specify the host interface; 11000000 10101000 **00000001 11101010**.

The split does not need to be at a byte boundary, so if we wanted we could reserve the final 10 bits to specify the host interface; 11000000 10101000 00000001 11101010. This means that the same IP address could be interpreted differently depending upon the network architecture that is being used.

If one computer (host) has more than one network interface connected to the Internet, each of those interfaces can have a different IP address. However, if both interfaces connect to the *same* network, the IP addresses *must* be different.

11.4.1.1 IP Version 6

Eventually the world will switch over to IP version 6 (IPv6), which uses 128-bit addresses. For convenience, these are usually specified in hexadecimal with 16-bit groupings, such as 210F:6154:0017:9FBB:0000:0000:0000:0000.

While this is quite a long number (and certainly too long to conveniently write in binary), IPv6 addresses tend to contain long strings of zeros. In practice, any long strings of zeros in the address can be replaced by the shorthand ::. The standard also allows leading zeros in each 16-bit section to be omitted (i.e., 00ab can be written as ab). So we would instead write the above IPv6 address as 210F:6154:17:9FBB::, which requires significantly less typing.

With a 128-bit address range, IPv6 can support a very large number of host addresses. In fact, there are as many as $2^{128} = 340000000000000000000000000000000000000$ different addresses (although it includes some special, or reserved, addresses). With such a large number, we are unlikely to be running out of IPv6 addresses any time soon!

11.4.2 Internet Packet Format

Just like the Ethernet frame that often carries the IP datagram (Figure 11.7), the IP datagram itself has a well-defined structure. Apart from its own payload (which will often be a TCP or UDP packet), it includes a header that contains several items of information.

The most important header fields are the source and destination IP address and the protocol type (i.e., what it is carrying). Ethernet frames also have a source and address field, but those are hardware (MAC) addresses because they operate at a physical or network layer. IP datagrams are used by the Internet layer to specify the host interface (i.e., machine) which sent the information, as well as the destination host interface that will receive it. Since this is from the Internet layer, the addresses are specified as IP addresses. We already know that visible IP addresses should be unique within a network, and so a sending machine can specify where to send information by giving the destination IP address. The receiving host interface will likewise only be able to "see" one host interface with that IP address, and thus knows where it was sent from.

On a small network, where all IP addresses are visible to all connected machines, IP datagrams are sent from one host and received by another on the same network. However, considering the global Internet, your home computer cannot know where all other computers are located—there are simply too many computers to be listed in one place—and so mechanisms exist for delivering packets to unknown destinations. This (and more besides) is the function of routing (see Section 11.4.3).

Before discussing routing of packets, we should note that the IP datagram header also carries information fields specifying the length of data being carried, time-to-live (TTL) information, some flags that warn about congestion and give hints on packet handling and a header checksum (used by the receiving machine to determine that the header of a received packet has not been corrupted).

11.4.3   Routing

Routing is the name of the process that makes sure IP datagrams get passed on to the right network and host that they are destined for. If the network part of the destination address is the same as the current network, then routing means delivering the datagram directly to the right host on that network. If the network part of the destination address differs from the current network, then the destination machine is on a different network. In that case, a routing table must be consulted to find out where to send the datagram.

In a small home network connected to the Internet through an Internet service provider (ISP), the routing table might just indicate something like "send all external traffic to the ISP, and leave it to them to handle."

In a bigger network, the routing table might specify which machine interface to send which datagrams to, depending upon the network part of the destination IP address. That machine interface should connect to the correct network (or should know a path to the correct network) as well as to the current one. The host machine that does the routing is called a router. It is connected to at least two networks and transfers, or routes, IP datagrams from one network to the other.

The general approach is that an IP datagram is delivered directly if we know how to do so from the routing table, otherwise, it is delivered to a default router. In the case of a local network connected to a bigger network, and when that bigger network connects to an even bigger one, the routing machines are often called gateway machines.

Every host, router, and gateway contain an internal routing table. This table usually contains multiple rows. Each row specifies either a single IP address or a range of IP addresses. The entries identify which network interface should be used to deliver IP datagrams to different addresses, and there is almost always a default route (which shows how to handle datagrams sent to addresses that are unknown).

When sending an IP datagram to a remote computer for the first time, the sending machine will probably not know where to find that computer. The destination IP address will not be found in the routing table, and so the datagram is forwarded to the default route—probably a gateway router, that passes the packet to the other network it is connected to. If that network does not contain the destination machine, then the process is repeated at the router, which sends it to the gateway router for that second network. The process repeats again, and again, until the IP datagram reaches a network or router that "knows" which host has that IP address. That routing process ensures that no machine needs to know all of the details of how to get to the destination, it just needs to know where to send packets when it does not recognize the address. This is rather like the chain of postal employees when delivering a letter internationally. The local post office does not recognize the destination address, so passes the letter to an area office, which passes it up to a regional office, then a national office. The letter is then passed to the correct country, where it traverses the path to regional, then area, and then local office, before it is given to a postal worker who knows exactly which house to deliver it to.

After a packet has been successfully routed from source to destination, the sending machine—and all machines in between—learn the route by caching the routing information. Next time a packet is sent to that address, these machines will simply repeat the same routing path (assuming no errors, delays, or congestion).

11.4.4 Unicasting and Multicasting

IP datagrams are usually sent to a single IP address, which means one host interface on one machine. These normal IP addresses are called unicast addresses.

Multicast addresses, by contrast, allow messages to be sent that are received by any "interested host" on that network. Datagrams with destination address starting with 224, 239 or anything in between, contain control messages and other specific or reserved messages. These multicast messages will normally be received and decoded by all host interfaces connected to the network. For example, IP datagrams addressed to 224.0.1.1 contain network time protocol (NTP) broadcast messages, which would allow any computer on the network to make use of a network-attached timer to synchronize their internal software clocks.

An address where the nonnetwork part is all binary ones (e.g., x.y.255.255 or x.y.z.255) is a broadcast address that goes to all hosts on that network.

11.4.5 Anycasting

Anycasting means that a single message is sent to a set of servers which could be geographically spread out. Although we had noted that visible IP addresses should be unique, anycasting is a relatively recent technique to break that rule.

With anycasting, the IP datagram is routed along the shortest path and delivered to exactly one member of the group that has the same IP address. The delivery will be to the closest member in network terms (which means the one with the lowest latency—the shortest packet delivery time).

This is achieved by giving multiple hosts the same IP address to provide multiple redundant instances of a service tied to one IP address. This is used most effectively for DNS servers, which we will consider in the following two sections.

11.4.6 Naming

Since it would be very inconvenient for humans to have to remember IP addresses when accessing websites, sending email, and connecting to servers, the Internet has a naming service, so that an IP address can optionally have one or more names associated with it. All hosts must have IP addresses, but not all hosts need to have names (and of course you cannot have a name without an IP address).

Like IP addresses, the names have to be unique across the visible Internet. For example, www.kent.ac.uk maps, at the time of writing, to the IP address 129.12.10.249. Entering http://129.12.10.249 as a URL (uniform resource locator) in a Web browser will display the University of Kent website, but most people would find it a lot easier to remember www.kent.ac.uk than the full IP address.

Part of the ease of remembering is that names are arranged in a hierarchical fashion. For example, we can probably guess that library.kent.ac.uk will direct us to the library website of the University of Kent in the United Kingdom. Likewise, library.canterbury.ac.nz directs us to the library website of the University of Canterbury in New Zealand.

Analyzing library.kent.ac.uk from the least to the most specific (right to left), as we might do with a postal address:

- uk—designates the country or top-level organization. Authorities within each country are responsible for allocating names. In the United Kingdom, this is an organization called Nominet.

- `ac`—designates a subspace, administered separately. In this case, it is the UK academic community.[6]
- `kent`—designates an organization who control their own namespace (in this case, it is the University of Kent that "owns" the namespace ending in `kent.ac.uk`).
- `library`—a name assigned by the organization (the University of Kent) to indicate a particular service or host interface.

Responsibility for handling the set of names (the namespace) is delegated to different countries and organizations, so a single global directory is NOT required, except to handle the country and organization names. Some very prominent noncountry top-level domain names (including `.com`, `.edu`, and `.org`) are actually the responsibility of the United States for historical reasons, and are effectively administered as if they were part of the `.us` country code (which not many people use in practice).

As mentioned, top-level domain (country) names are assigned globally, but inside a country, namespaces can be handled very differently. Anyone can use a name from or within any country top-level domain, as long as the country registrar allows them to register it. There is no technical requirement that the host interface, the computers, or the organization needs to reside within that country. That fact provided a major economic boost to countries like Tuvalu; Tuvalu registers all subdomains ending in their country code, `.tv`, most of which have been sold to global television stations rather than to Tuvaluan companies.

11.4.7 Domain Name Servers

Domain Name Servers (DNS) are machines that actually perform the domain naming service. They handle the mapping of an Internet name to an IP address. Top-level domain name servers handle the country code or top-level domain name mapping, but with billions of machines in many domains, it would be impossible for one machine to maintain lists of every host interface within their domains. For this reason, as well as for the reason of avoiding traffic bottlenecks or single points of failure, naming is done by a distributed set of machines.

Each machine maintains a database of entries such as this:

```
library.kent.ac.uk    129.12.10.251
```

The distributed database is implemented over a large number of DNS machines, each handling part of the namespace. However, almost all name servers are duplicated in case one fails.

DNS queries come to those machines using either UDP or TCP, and generally on port number 53 (we will learn about port numbers in Section 11.5.1). The software in a client (i.e., in your computer) that contacts the DNS to get name resolution, is called a resolver.

---

[6] There are a few roughly agreed subspace names, with some country-specific variation. For example, `.ac` for academic and `.edu` for educational are both used worldwide to specify universities, with different countries choosing one or the other. Similarly, `.com` and `.co` are roughly equivalent to indicate company domains, used in various countries.

In most operating systems, resolvers have a number of ways of resolving a name to an IP address. Methods include keeping a local list of fixed hostnames, keeping a set of cached hosts and their IP addresses (ones that have been requested previously will be stored in the cache), and the fixed IP address of a DNS—because obviously it would not work if a client tried to access a DNS using a hostname instead of the IP address (think about why this would not work).

The local cache or previously resolved addresses have a time limit which specifies the time validity of each entry in the cache. If the time has expired, typically after a few hours, the DNS forgets the name from the cache (i.e., it deletes that entry).

11.4.7.1   Name Lookup Example

Here is an example for the domain `mcloughlin.eu`. The scenario is that a home-based computer user located in the United States wishes to visit the website located at the European Union–based URL `mcloughlin.eu` for the first time. They enter the URL into their Web browser and press enter. Their computer should now send an HTTP (hypertext transfer protocol) message to the machine at `mcloughlin.eu` to request the webpage "located" there. However, to send any message, as we know from our discussion of the Internet layer, the sending computer first needs to know the IP address of the host interface (webserver) it is communicating with. What happens next is explored below, and illustrated in Figure 11.9:

1. The browser sends a request to the operating system (OS) to lookup the host IP address (in Linux this might be a call to the `gethostbyname()` function).

2. The built-in OS resolver acts based on the local resolving methods. Usually it keeps a cache of hostname/IP pairs, which include recently accessed machines as well as the content of fixed machine in its local `hosts` file. Since this website has not been visited before from that machine, the resolver will not find the IP address locally.

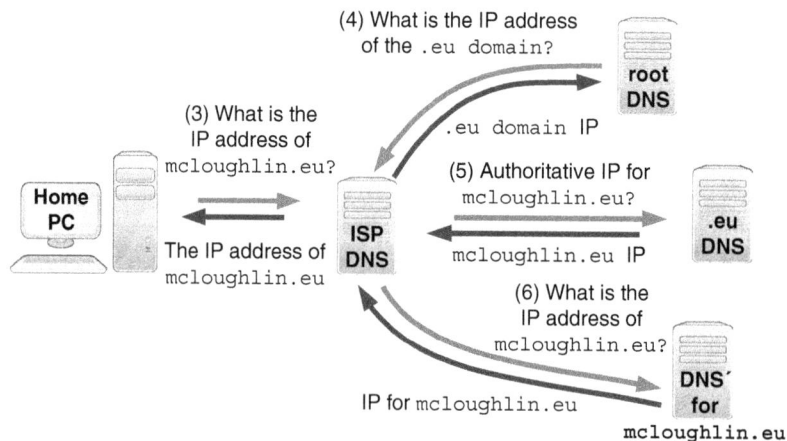

FIGURE 11.9   Sequence of messages between a user machine, their ISP and various DNS servers to resolve the domain `mcloughlin.eu`, with step numbers corresponding to the sequence in Section 11.4.7.1.

3. The resolver passes a request to their local DNS, which will be a machine maintained by their Internet service provider (ISP). The ISPs DNS server will probably not recognize the name (given that it seems to be a very unpopular website).

4. The ISPs DNS server passes the request to a root DNS, to find out where the top-level .eu domain names are stored. The root DNS returns an IP address of the .eu domain DNS.

5. The ISPs DNS server then sends a request to that machine, asking which is the authoritative domain server for mcloughlin.eu. The reply will contain an IP address of yet another DNS, but one that definitely recognizes that domain name.

6. A query is them made to the authoritative DNS server to find the actual IP address of the mcloughlin.eu server. The reply will identify which host interface, among all machines on the Internet, is the server for that URL. This reply is finally passed back to the original requesting computer.

7. The home computer in the United States then forms an HTTP message which it sends to mcloughlin.eu to port 80 (the default HTTP port) requesting the webpage available from that URL.

11.4.7.2 Reverse DNS Lookup

It also necessary sometimes to do a reverse lookup, when a host needs to find out the name associated with a particular IP address. To do this, the host operating system composes and then broadcasts the query IP address encoded as a special name. This name includes the IP address with the sequence of numbers reversed and ending with the words "in-addr.arpa." For example, to query the IP address 129.12.10.249 it will send the request 249.10.12.129.in-addr.arpa.

If that IP address is indeed associated with a name, then the nearest DNS server will reply with the name ("kent.ac.uk"). The DNS server can do this because its stored records generally hold several different types of information in their tables about each IP address, such as

Record	Function
A	Maps a name to an address.
PTR	Maps an address to a name.
CNAME	Aliases a name to another one.
SOA	Start of authority, this indicates which name server is ultimately responsible for the information held in the record. It also specifies the record validity time.
TXT	Various additional pieces of information can optionally be included here.
MX	Specifies which host will handle mail for that domain. There are usually at least two MX records to provide a backup in case one mail server goes down.

The top-level domains are specified in root DNS servers, of which there used to be 13, from different hardware manufacturers and running different operating systems, spread around the world. Later, these were extended to incorporate many machines, by using anycasting (see Section 11.4.5) to map the 13 IP addresses to 13 groups of DNS machines.

This allowed there to be many machines serving queries, each sharing a single IP address. Although it appeared to an observer that only 13 servers exist, there are actually many.[7]

Today, Google operates two public DNS services at the IPv4 addresses 8.8.8.8 and 8.8.4.4 (in IPv6, they are at 2001:4860:4860::8888 and 2001:4860:4860::8844, respectively).

11.5   The Transport Layer

The transport layer uses different kinds of protocol to transfer various types of data over networks. There are many protocols available at the transport layer, but software developers mostly make use of either:

- *Transmission control protocol* (TCP)—which guarantees end-to-end full-duplex[8] acknowledged transmission over sessions[9] which ensure that data packets are delivered to the Application layer in the correct sequence. TCP is reliable but less efficient than UDP in error-free conditions, yet can be prone to delays when errors or congestion occur.

- *User datagram protocol* (UDP)—is a best-effort, sessionless, unacknowledged, streaming transmission. There is no guarantee that packets will arrive, and no guarantee that they will be passed up to the application layer in the correct sequence. However, UDP is much faster to communicate a block of data than TCP, much faster to stream, and significantly more traffic efficient when the network is congested.

Given the choice of transport mechanisms, software developers would make a decision as to which protocol they want, based on assessing the network, data type, and importance of messages. We will explore both in turn, after first describing the concept of port numbers.

11.5.1   Port Number

Whatever transport protocol is being used, the task is to communicate from one application to another application (usually, but not necessarily, on different computers). There may be many applications running on each computer, which themselves will be communicating, probably with several other computers.

So with multiple packets traveling to and from a set of different computers, there needs to be a way of specifying which application a packet belongs to. This is done by using a 16-bit port number: Every TCP and UDP packet specifies both the source and the destination ports, just as the IP datagram specifies both the source and destination IP addresses.

Applications running on a computer can request the OS to allow them to "listen" to packets arriving on a particular port, which means they will receive any messages (UDP)

---

[7] For an up to date map of root servers, see http://www.root-servers.org.

[8] Full duplex: data flows in both directions.

[9] Session: meaning that a logical link is established before transmission begins, maintained during the transmission period, and then closed down when transmission has finished.

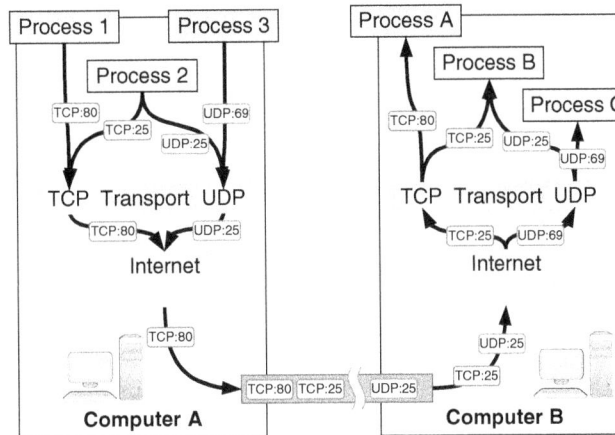

FIGURE 11.10 A network diagram illustrating a set of applications on one computer communicating over the Internet to a set of applications on another computer. The various TCP and UDP packets being sent are passed to the correct applications based on their port numbers.

or connections (TCP) sent to that port. Applications can also use the OS to send data to a specified port on a computer specified by its IP address. An illustration of this process is given in Figure 11.10 where several applications on two computers are able to communicate messages through different ports but share a single network interconnection.

11.5.2 User Datagram Protocol

The user datagram protocol (UDP) provides a best-effort transmission of packets from source to destination. Best effort means that the network (from Internet layer down) will attempt to convey and deliver the packets, but the UDP protocol itself does not guarantee transmission, and does not check to see whether packets have been received correctly.

At an application level, the programmer may wish to add such functionality—communicating from sender to receiver and checking whether the communication was successful—but this is not done automatically by UDP. TCP, by contrast, does check for errors and can retransmit missing packets to ensure delivery of an entire message.

UDP also lacks the concept of an "entire message," and as such is *connectionless*. This means that each packet is transmitted independently, and is not considered to be part of a large block being communicated. UDP packets are sent from one machine to another machine by specifying the receiver IP address and port number.

We will contrast TCP and UDP further in Section 11.5.4.

11.5.3 Transmission Control Protocol

The transmission control protocol (TCP) uses the concept of a connection from sender to receiver to transfer a stream of octets (bytes). The data is transported in pieces, but this is invisible to the application. To the application, a block of data (whether large or small) simply gets transferred from the sending machine to the receiving machine.

TCP is inherently reliable unlike UDP or even IP. This does not mean that TCP is better, it simply means that the responsibility to check whether transmission succeeded (and if not to do something about it) is built into TCP. It means that the application may not need to implement acknowledgments or checking.

To ensure transmission reliability, TCP has two main mechanisms of checking for missing or erroneous packets, and then resending packets that were missing or received in error.

A TCP link begins by forming a connection between a pair of endpoints, specified by port numbers as well as IP addresses. Once established, the TCP connection is bidirectional. The TCP protocol allows the same port (and IP address) to be used by multiple connections simultaneously (e.g., several email clients connecting to one email server).

When transferring a block of data, this is split into consecutive segments which are then encapsulated into IP datagrams and transmitted as a stream to the receiver.

Segments (we loosely refer to them as TCP packets) consist of a header and a data payload. The header specifies a source port, destination port, sequence number, and quite a lot of other pieces of information, including a checksum. The data payload contains a piece of the stream of data, and successive segments will be used to convey the entire stream in chunks. For a particular segment, the data payload might be the next chunk of the data being conveyed, or it might be a retransmitted chunk that was either not received at the receiver, or was received but contained errors (which can be discovered thanks to the checksum). The sequence number helps to maintain correct order, and allows the receiver to inform the transmitter which segments need to be retransmitted, in the case that some segments are missing or received erroneously.

Any packet-based communications system that retransmits packets that were not received, or received with error, needs a way for the receiver to inform the transmitter which packets to resend. There are two methods in general, called ACK (acknowledge) and NACK (negative acknowledge) schemes, which we met in Section 10.2.5.2. In ACK schemes, the receiver specifically tells the transmitter which packets have been received correctly. The transmitter then must decide to resend any packets that were not received. NACK schemes, meanwhile, rely on the receiver to inform the transmitter exactly which packets to resend. TCP uses an ACK mechanism (specifically, it is a block-based selective ACK mechanism), and this involves the receiver acknowledging each block of received packets to the transmitter.

When a packet is lost, there will be a delay until the receiver finishes receiving a block of data packets, then a delay while it composes and sends a block acknowledgment to the transmitter. Another delay is required for the transmitter to receive and decode the acknowledgment, and then to compose and resend the lost packet. All of this means that reliability comes at a cost—in both the time taken to ensure a packet is delivered, and also in the additional network traffic involved when resending a packet.

Figure 11.11 illustrates the TCP header, showing source and destination port numbers, as well as sequence number, ACK information, and checksum. The other fields shown are not relevant to the current explanation of the protocol.

11.5.4 UDP versus TCP

Thanks to their characteristics described in Sections 11.5.2 and 11.5.3, the two most common transport protocols have distinct uses in a number of applications. To recap those characteristics: When streaming live video or audio frames, where time delay and traffic

FIGURE 11.11
Internal structure of a TCP packet; the diagram is 32 bits wide, with the first word at the top, and data payload shown at the bottom.

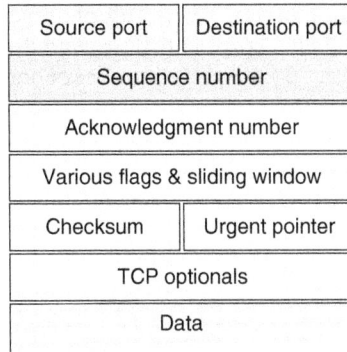

Source port	Destination port
Sequence number	
Acknowledgment number	
Various flags & sliding window	
Checksum	Urgent pointer
TCP optionals	
Data	

efficiency are of utmost importance (and certainly more important than a couple of milliseconds of missing data), UDP would tend to be the method of choice. This includes video conferencing and Internet telephony. When each packet counts, and it is more important to receive data fully and in the correct order than receiving them quickly, then TCP would be used. This means that something like financial records or bank transaction data would most likely be sent using TCP packets.

TCP	UDP
Reliable	Best effort
End-to-end connection	Connectionless
Takes time to establish a link	No link—simply send packets
Packets form a stream of data	Each packet is independent
Packets delivered to application in the correct order	Packets delivered to application as they are received (i.e., some packets may be out of sequence)

File transfers over protocols such as network filing system (NFS) can be set up to use either UDP (quicker) or TCP (more reliable), but TCP is the default. Email these days generally makes use of TCP, as does most Web traffic.

11.6 Other Messages

11.6.1 Address Resolution Protocol

As we have seen, when an application wishes to communicate with another application, it specifies an IP address to the OS, that it wishes to contact (or it specifies a name which is then resolved to an IP address as described in Section 11.4.6). It then provides data to the transport layer, which in turn provides TCP or UDP packets to the Internet layer. The Internet layer encapsulates the data in an IP datagram which specifies the source and destination IP addresses. This is passed to the network and then physical layers. Assuming an Ethernet or WiFi connection, the IP datagram is finally encapsulated into an Ethernet or WiFi frame for transmission over the physical medium. This frame specifies the physical MAC address of the computer that will receive the frame. We know that

IP datagrams will progress hop by hop over multiple links from source to destination. This was illustrated graphically in Figure 11.8. It showed that the route from source computer A to destination computer B followed a number of hops, and each hop is a separate physical link. From our Ethernet discussion we know that the physical hops communicate using MAC addresses and not IP addresses, and so how does an IP address get mapped to a MAC address? This is done by the address resolution protocol (ARP).

For example, computer A wishes to communicate with computer B, and knows its IP address. The routing table in computer A tells the computer whether computer B is on the same physical network, or whether it is external. If on the same network, computer A tries to communicate directly with it. If external, computer A sends the packet first to the gateway router.

In each case the OS sends out a special packet (an ARP message), at the same "level" as an IP datagram (i.e., it gets packed directly into a physical frame). This ARP message is small and efficient and simply specifies (a) the IP address for which a MAC address is being requested, (b) the sender's IP address, and (c) the sender's MAC address. The requested IP will either be that of computer B (if it is on the same network), or that of the gateway router (if computer B is external).

The ARP message is broadcast to all hosts on the local network. It will be ignored by all machines except the host that has the specified IP address, which responds by sending back its physical MAC address.

ARP results (IP/physical address mappings) are cached (i.e., remembered for later use) to avoid continual ARP queries. A timeout is applied so that the information has a limited lifetime before a fresh ARP query needs to be made. This is needed when computers move from one network to another—an increasingly common occurrence in the era of mobile computing.

ARP is handled entirely inside the Internet layer, and is part of its task of keeping both the local network and the Internet as a whole running smoothly.

11.6.2   Control Messages

The Internet layers on different hosts also need to communicate with each other to ask for and report status. This includes information on issues such as error conditions and network congestion. These messages use the Internet control message protocol (ICMP). With ICMP, these messages are wrapped inside IP datagrams (which means they are handled like UDP packets).

One of the most common control messages is ICMP echo, which is more commonly known as "ping." It is an "are you alive" message that is sent from one host to another, and is very useful for diagnosing network connectivity problems, as well as monitoring whether a given host has crashed or failed.

11.7   Wireless Connectivity

At the time of writing, an increasing number of people are connecting to the Internet from mobile devices, and that means wireless connectivity.

Wireless is also particularly important for embedded systems, as these are increasingly required to achieve mobile connectivity. The following subsections survey some of the main standards related to wireless communications—with the notable exception of mobile phone and data standards based on Global System for Mobile

Communications (GSM), General Packet Radio Service (GPRS), High Speed Packet Access (HSPA), and so on, which are beyond the scope of this book.

11.7.1 WiFi

The most famous of wireless network standards, WiFi was first approved by IEEE in late 1999 as part of the IEEE 802.11b standardization effort. Shortly after that, the IEEE 802.11a standard was ratified, using a new encoding scheme—orthogonal frequency division multiplexing (OFDM) to enable higher data rates as well as wireless channel availability. Since then, IEEE 802.11n was introduced, making use of multiple input, multiple output (MIMO) technology to increase peak data rates up to an advertised maximum of 600 Mbps. It is said to guarantee a minimum throughput of 100 Mbps (after subtracting protocol management features like preambles, inter-frame spacing, acknowledgment, and other overheads).

IEEE 802.11a is generally much faster than IEEE 802.11b, with a 54-Mbps maximum data rate in the 5 GHz frequency range, compared to IEEE 802.11b's 11 Mbps rate at 2.45 GHz. IEEE 802.11n is faster again—always assuming hardware is used with multiple antennas, and the environment supports multipath transmission (crowded urban areas and offices tend to do so, wide open spaces like desert communities and parkland may not do so).

The peak data rate of a standard like IEEE 802.11g sounds very promising, supposedly delivering 54 Mbps data rates. However, nearly half of its available bandwidth is consumed by transmission overheads. Putting this another way, the user will probably never be fortunate enough to experience data transfers as fast as the peak advertised rate, and much less than that as distance increases (i.e., the signal weakens) or interference increases. A WiFi device typically has a maximum range of 50 to 100 m.

Over the past decade, IEEE 802.11b, a, g, and n have been joined by a plethora of IEEE 802.11 substandards. Each have distinct characteristics, offering different strengths and application areas. As with Ethernet, IEEE 802.11 interfaces have a built-in hardware MAC address. This can be used to identify devices connecting to a network (and is, in theory, an unchangeable attribute of each interface).

Standard	Data rate
IEEE802.11b	up to 11 Mbps
IEEE802.11g	up to 54 Mbps
IEEE802.11af	up to 569 Mbps
IEEE802.11ay	up to 100 Gbps (100,000 Mbps)

11.7.2 WiMax

IEEE 802.16, also known as Worldwide Interoperability for Microwave Access (WiMAX), is a wireless broadband technology, supporting point to multi-point (PMP) wireless access. IEEE 802.16 was published in 2002 as a fixed-wireless standard based on line-of-sight (LOS) technology. It aimed to provide high-speed backbone interconnection service to enterprises that were operating in locations where it was infeasible to run a physical fiber or copper infrastructure.

IEEE 802.16 was originally targeted at business users and operates on licensed bands in the 10 to 66 GHz range over 20, 25, or 28 MHz channel widths (i.e., users have to purchase the frequency bands that they use). It requires line-of-sight (LOS) between base station and user. Data rates of up to 134 Mbps can be achieved, but limited to a range of 2 to 5 km around the base station. There are several other variants of this technology, having different operating parameters, frequency bands, data rates, ranges, and so on.

11.7.3   Bluetooth

Bluetooth,[10] originally developed by Ericsson is now an international standard short-range communications technology intended to replace the cables connecting portable and/or fixed devices, while also maintaining high levels of security. Bluetooth-enabled devices connect and communicate wirelessly through short-range, ad-hoc networks known as piconets or personal area networks (PANs).

Each device can simultaneously communicate with up to seven other devices within a single piconet, using a packet-based protocol, and in addition each device can belong to several piconets simultaneously. Networks can be established dynamically and automatically as Bluetooth-enabled devices enter and leave radio proximity. Version 2.0 + Enhanced Data Rate (EDR), adopted in November 2004 had a data rate of 3 Mbps, but the latest version 5 (June 2016) significantly increased that to 50 Mbps. Version 5 also included a long-range variant, reputed to be capable of operating over distances of 200 m.

Bluetooth technology operates in the unlicensed industrial, scientific, and medical (ISM) band at 2.4 to 2.485 GHz and comes in several classes based on transmitter power (and consequently range):

- Class 4 radios have a range of around 50 cm, with maximum transmitter power of just 500 μW.

- Class 3 radios have a range of around 1 m, with maximum transmitter power being 1 mW.

- Class 2 radios, usually found in mobile devices, have a range of around 10 m, and maximum transmitter power of 2.5 mW.

- Class 1 radios, used mostly in industry, have a range of 100 m, and maximum transmitter power of 100 mW.

Each Bluetooth device has two parameters that are involved in practically all aspects of Bluetooth communications. The first one is a unique 48-bit address assigned to each Bluetooth radio at manufacture time and which supposedly cannot be modified. The second parameter is a free running 28-bit clock that ticks once every 312.5 μs, which corresponds to half the residence time in a frequency when the radio hops at the nominal rate of 1600 or 800 hops/s.

11.7.4   ZigBee

Formally known as the IEEE 802.15.4 wireless personal area network (PAN) standard, ZigBee, ratified in 2004 was targeted mainly for embedded applications. ZigBee actually builds on top of IEEE 802.15.4 with mesh networking, security, and applications

---

[10]Bluetooth is named after a mythical Norse 10th-century king who united the far-flung Scandinavian tribes into a unified kingdom.

control—making it a practical and usable standard. The focus of network applications under ZigBee includes the aim of low power consumption, high density of nodes per network, low cost, and simple implementation.

Three devices type were specified in the original standard, namely: Network Coordinator, Full Function Device (FFD), and Reduced Function Device (RFD). Only the FFD defines the full ZigBee functionality and can become a network coordinator. The RFD has limited resources and does not allow some advanced functions (such as routing) since it is a low-cost endpoint solution. Each ZigBee network has a designated FFD that is a network coordinator.

The coordinator acts as the administrator and takes care of organization of the network, which tends toward a master-slave configuration. ZigBee supports up to 65,535 separate networks with topologies that include star, peer-to-peer, and mesh. When a ZigBee device is powered down (meaning that all circuitry switched off apart from its 32 kHz clock), it can wake up and transmit a packet in 15 ms—this exceptionally low latency can allow systems to wake, transmit, sleep on a duty cycle that consumes extremely low levels of power.

The defined channels for ZigBee are numbered 0 (868 MHz), 1 to 10 (915 MHz) and 11 to 26 (2.4 GHz). Maximum data rates allowed for each of these frequency bands are fixed at 250 kbps (at 2405 to 2480 MHz worldwide), 40 kbps (at 902 to 928 MHz in the Americas), and 20 kbps (at 868.3 MHz in Europe). As usual, the rates quoted are theoretical raw data rates rather than achievable ones, which will be reduced due to both interference and protocol overheads. ZigBee is a packetized standard, with 127 byte packets including header and 16-bit checksum, and a data payload up to 104 bytes in length. The maximum output power of the radios is generally 1 mW giving a range up to 75 m.

ZigBee supports symmetrical encryption and authentication, and has mechanisms for key handling and frame protection. In terms of protocol stack size when connected to a controlling CPU, ZigBee requires about 32 KiB, but can define a limited variant down to about 4 KiB (which is an extremely small amount of code for a standard that is as capable as ZigBee).

11.7.5 Near-Field Communications

Near-field communication (NFC) is one of the most commonly used wirelesses networking standards today. It provides dedicated short-range connectivity, and is the underlying technology for most smart card interfaces, including public transport cards, "tap and go" payment cards, electronic passports, Apple Pay, Android Pay, and so on. The original NFC was jointly developed by Sony and NXP and provides intuitive, simple, and safe communication between electronic devices over distances up to about 4 cm. It was approved as an ISO standard in 2003 and now has very wide industry following.

NFC operates at 13.56 MHz with a data rate up to 424 kbps and is compatible with some other contactless approaches, such as ISO 14443A and ISO 14443B (used with Sony's FeliCa technology). Like NFC, both operate in the 13.56 MHz frequency range. Several other NFC variants exist which may not be fully standard compatible, but operate with enhanced data rates as well as in different frequency ranges.

An NFC communications interface can operate in different modes which determine whether a device generates a radio frequency (RF) field of its own, or whether a device harvests its power from an RF field generated by another device. If the device generates its own field it is called an *active* device, otherwise it is called a *passive* device.

11.8   Network Scales

Networks span an impressive range of scales, from a few mm up to tens of km. While there are a bewildering array of acronyms to cover the types of networks, the following are in most common use:

Standard	Data rate
BAN	Body area network
PAN	Personal area network
CAN	Controller area network
LAN	Local area network
WLAN	Wireless local area network
WAN	Wide area network
MAN	Metropolitan area network

The Internet is a type of WAN, and some references refer to global area networks (GANs), but usually in terms of very wide-area wireless connectivity. Not included in the list are storage area networks (SANs) which are used for local storage, and interplanetary networks (which do not currently exist, to the best of the author's knowledge).

11.9   Summary

This chapter primarily concentrated on networking by means of the IP. We examined the five layers of the TCP/IP protocol stack, noting where they differ from the seven layers in the OSI model of Chapter 10. We introduced the important concept of encapsulation, and then built upward from that foundation. Starting with physical packets, we discussed transport mechanisms, sessions, addressing and routing, several network and application protocols, domains, the World Wide Web, and email. From bits on a wire, we built upward to the global Internet, domain name servers, and typical applications. We ended the chapter with a summary of common wireless networking standards— although we note that the world of computing contains far more de facto or even formalized standards than we could possibly capture fully in a single textbook.

11.10   Problems

11.1   Identify three of the networking features of the Internet that have led to its widespread adoption and popularity.

11.2   When data is sent from an application using TCP/IP, what will the transport layer encapsulate that data into?

11.3   When the information from the transport layer in question 11.2 is encapsulated by the layer below it, what will it be encapsulated in?

11.4   When receiving network information, which TCP/IP layer will receive Ethernet frames as its input?

11.5   Which three layers of the TCP/IP protocol stack are typically implemented within an operating system?

11.6   Name four pieces of information that are added to payload data when it is encapsulated in an Ethernet frame.

11.7   Name three pieces of information that form part of a TCP header.

11.8   In the early 1990s, the Computer Science department at the University of Cambridge created one of the first webcams, specifically to monitor the status of their coffee pot (namely, whether it was full or empty). If the webcam were replaced by a simple sensor that measured the weight of the coffee pot every millisecond, and streamed the weight values to a webserver for display on a webpage, identify whether UDP or TCP would be the protocol of choice to communicate the streaming weight values.

11.9   Is there any circumstance where two computers with the same IP address should be connected to the same network (i.e., so that they are both "visible" to each other)?

11.10   When scanning network traffic, a packet is spotted with the content `101.2.1.155.in-addr.arpa`. Which type of computer is likely to respond to this packet, and what information will be provided in the response?

11.11   When an application transmits TCP/IP data to a given port, X on a remote server, are the packets always sent from port X on the transmitting machine, or can they be sent from a different port to the one they will be received on?

11.12   If the IP address for `www.kent.ac.uk` is `129.12.10.249`, then is the university that has IP address `129.67.242.155` likely to be located in Canada, in Singapore, or in the United Kingdom.

11.13   In networking diagnostics, what is a typical function of an ICMP echo packet, commonly known as a "ping"?

11.14   Arrange the following type of networks in size order, from smallest to largest; PAN, CAN, WLAN, MAN, GAN.

CHAPTER 12

The Future

T his chapter focuses more on the next steps in the continuing evolution of computers and their architecture than in what has gone before—and as noted many times in previous chapters, that future is likely to be embedded. However, with the ever-increasing performance targets set by Moore's law, manufacturers have turned strongly toward parallel computation. This includes not only the large-scale server farms of companies such as Google and Amazon, but also parallel computation in mobile and portable devices. These trends toward parallelism are likely to continue; so perhaps we are looking at a future of embedded parallelism. Other emerging topics like ambient intelligence and pervasive computing (computers all around us, cooperating to provide services), quantum computers, biological computers, and so on are worth thinking about.

In trying to chart a future that differs from mainstream computing, some of this chapter is definitely with us already: it is included here to indicate growing importance and widening impact in the future. Some of our future computing techniques are those which have been tried and forgotten before, but are now being revisited. Others, such as quantum computers, sound more at home in a science fiction novel than a computer architecture textbook. In fact, we will take the science fiction aspects even further—and we note that there are many instances where science fiction became future reality.

Whatever the future holds, computing is very likely to remain a largely evolutionary discipline, enlivened by the occasional small leap forward, and disruptive breakthroughs that have characterized the past few decades of progress. Despite the majority computers of today (i.e., tiny embedded systems inside a myriad of everyday devices) not looking much like the laptops of the 2000s, the beige boxes of the 1980s, or the mainframes of the 1970s, we have seen that much of their CPU technology is very familiar, albeit highly miniaturized. Likewise, the operating systems of choice today for both the biggest and the smallest computer systems are UNIX based. The newer systems may be highly advanced, streamlined, secure, and extremely capable, but looking "under the hood" of an Android smartphone will reveal almost the same directory structure as AT&T UNIX of the 1970s, similar system files to IBM's AIX of the 1980s, and the same device driver interface as the original Linux of the 1990s. Meanwhile the connectivity of today (which is primarily wireless) may not much resemble the cable connections and dial-up links of yesteryear, but it relies on the same standards. A computer engineer from the 1990s who examines the IP datagrams, or even data frames, sent over a modern WiFi network will see traffic that they are already familiar with.

None of this is to say that today's technology is old, it is simply that as technology moves forward, developers retain the best of what they have and replace whatever they can improve. It is therefore more a testament to the excellent design of those technologies mentioned. It is also a reassurance that an understanding of the state of computer systems today—as encapsulated in this book—will be relevant for years to come.

12.1 Single-Bit Architectures

In Section 4.2.2, we designed an ALU from a combination of separate 1-bit ALUs. This approach is quite common (for instance, the ARM core historically used it), and is often called bit-slicing. In effect, each bit is dealt with separately in parallel because the bus to the ALU is parallel. Alternatively, the ALU could accept bits in a serial fashion, process them, and output its result serially. In fact, serial CPUs have been developed over the years that do all processing with a bit-serial architecture.

Processing in serial fashion means higher on-chip clock speed, but fewer on-chip bus connections, so the interface hardware is simplified. The processing hardware is also much simpler since it generally only needs to process a single bit at a time. However, the rest of the CPU internals are not always simplified because a serial controller is required to route all the serial operands around the CPU—and this means some complex timing circuitry in practice. However, bit serial approaches have one big advantage in that the same CPU and ALU hardware can cope with different word lengths, just by changing the timing of the operations. We will see an example below.

For some serial operations, processing can occur as the serial bits are being fed into the ALU and calculations proceed bit by bit. However, some operators require all bits to be fed in before processing can begin (which is obviously slower, and requires storage for the entire word). It is therefore important to carefully design bit-serial computations to maximize efficiency.

12.1.1 Bit-Serial Addition

As an example consider the addition of two bit-serial numbers. These are presented to the adder with the least-significant bit (LSB) first, and added bitwise, with the carry from a particular add being fed back ready for the subsequent add.

The example shown in Figure 12.1 has 4 bits of two 16-bit words already added and thus four answer bits have been produced. The logic within this adder need not be complex, in fact it may be somewhat similar to the block diagram of Figure 12.2.

FIGURE 12.1 An example of two serial streams of binary digits being added together bitwise with the carry in for each bitwise addition being the fed back carry output from the previous bit calculation.

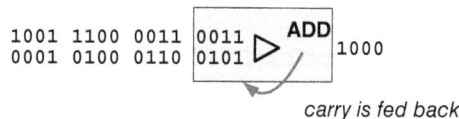

```
1001 1100 0011 0011    ADD
0001 0100 0110 0101  ▷      1000
```

carry is fed back

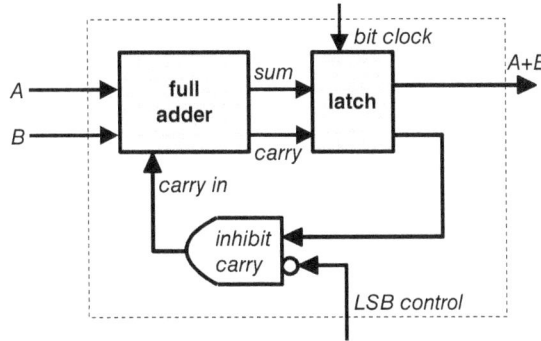

FIGURE 12.2   The full adder circuitry required by a bit-serial adder unit.

Within Figure 12.2, the two-bit streams are presented at A and B. These are added with the carry feedback to produce a sum and a carry. The LSB control signal is used to inhibit any carry feedback, so that there is no carry going in to the least-significant bit position (as expected). The latch delays the output of the adder to be synchronous to a bit clock, and delays the carry to be ready for the next bits to be added.

With this kind of scheme there is no need for any gap between one sequence of bits presented to the inputs and the next, as long as the LSB control signal prevents any inappropriate carry from being transferred from one sum to the next. The beauty of this is that literally *any* word length numbers can be added with exactly the same hardware, as long as the timing of the LSB control signal demarcates between input words, as shown in Figure 12.3.

An accumulator is equally simple to implement—this is left as an exercise for the reader.

12.1.2   Bit-Serial Subtraction

Consider the adder of Section 12.1.1, and note that any carry is naturally propagated from beginning to end. Since a subtraction may involve a *borrow* instead of a carry, there could be an analogous process. However, it is likely that a simple trick can make the process easier.

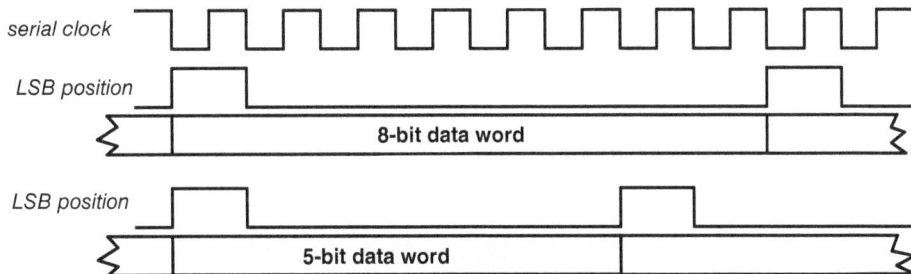

FIGURE 12.3   Timing waveforms relating bit clocks, the LSB position flag and the data words.

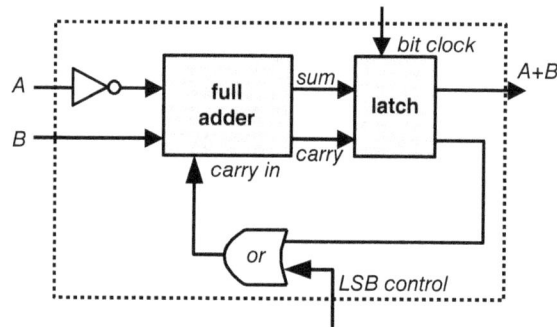

FIGURE 12.4 The subtraction circuitry required by a bit-serial adder unit. The main difference between this and the adder of Figure 12.2 is the inverter (NOT gate) inserted at input A, and the change to the LSB control gate at the bottom.

Remember from Chapter 2 that changing the sign of a two's complement number is relatively easy (although not quite as easy as it is for a sign-magnitude number). To change sign, it is necessary to simply swap all 1s with 0s and 0s with 1s then add a 1 to the least significant bit position. Then we use the fact that B-A is equivalent to B+(−A), so we simply need to perform an addition with one operand negated.

Inverting all the bits in the bit-serial input is as easy as placing a NOT gate at the input signal. Adding a 1 to the least significant bit position is also simple, it just needs the first carry to be set rather than clear (i.e., the LSB control signal needs to cause the carry to be set instead of cleared). The logic required to perform this kind of bit-serial subtraction is shown in the diagram of Figure 12.4. Comparing this and the hardware in Figure 12.2 it should be clear that it would be trivial to create a bit-serial ALU that uses an external control signal to switch between SUB and AND functions.

12.1.3 Bit-Serial Logic and Processing

Considering that arbitrary length words can be added or subtracted with this hardware that consists mainly of a single full adder and a latch, it should be evident that bit-serial logic is extremely hardware efficient. It can also be clocked at fast speeds (since there is little logic propagation delay between clocked latches), but with the disadvantage that however fast the clock, each arithmetic operation on length n-bit words will require $(n+1)$ clock cycles to complete.

Bit serial logic has found a home in many FPGA (field programmable gate array) logic designs due to its hardware efficiency which would allow many streams of processing to occur simultaneously. It also matches the hardware architecture of an FPGA, where the logic can be implemented in a single cell of the device. Such a cell may be called a logic element (LE), logic cell (LC), or similar name from different manufacturers, but the important point being that each cell contains a look-up table with the ability to perform as a full adder, allied with a flip-flop (which can implement a latch), or sometimes two of each. Interconnects between such cells are slower, with the speed being inversely proportional to the geographical distance between cells.

In general, for parallel-bit computation (i.e., the "normal" way of implementing an ALU), the interconnect delay can become crippling in an FPGA as the width of numbers being added increases substantially. This, along with carry propagate issues means that once numbers of above 256 bits are added together, it becomes extremely difficult to do so at high computation speeds. Moreover, current generation mid-range devices may not even contain sufficient routing interconnects to implement such functionality (which incidentally is not an obscure requirement, but rather a common function in many encryption algorithms). Under these circumstances, the single cell required for a bit-serial implementation becomes highly attractive.

12.2   More-Parallel Machines

With smaller and smaller CPUs or cores becoming available, and the easy convenience of interconnection by means such as Ethernet, it has become easy to cluster computers together with the aim of having them co-operate. Writing efficient software for such a system is another question, but from a hardware point of view, simply wiring together several off-the-shelf PCs with fast links constitutes a cluster computer.

Previously, in Section 5.8, we had outlined some of the many levels of parallelism that could be found within computers and had met the distinction between loosely and tightly coupled systems in that chapter. Here we will concentrate on the biggest level of parallelism that was listed, namely machine parallelism.

For computational problems with many loosely coupled tasks (such as groups of code functions that perform difficult and complex processing, but do not communicate between each other with a relatively low bandwidth), parallel execution of each function into separate tasks can speed up completion time. On the other hand, a system with tasks that communicate between each other either very frequently or with high bandwidth, may not run faster with parallel execution, due to bottlenecks in communications between CPUs. However, there are sufficient tasks which *can* be parallelized to have driven forward the parallel processing agenda over the past decade or so.

In large-scale parallel processing systems, tasks typically execute on physically separate CPUs, and this is what we will consider: groups of separate CPUs or perhaps PCs, rack or blade servers. The argument could even be extended to clusters-of-clusters, but that is outside the scope of this book, and will best be left to textbooks devoted to advanced parallel and distributed computing.

12.2.1   Clusters of Small CPUs

A decade ago it was almost unthinkable to mention the phrases "parallel computing" and "embedded systems" in the same sentence. Parallel systems were big—either clusters or mainframes ("heavy metal"), embedded systems were operated by single, usually simple, low-power CPUs. Today, however, the reality is quite different. Almost all mobile computing devices, whether smartphones, tablets, or wearables, employ a multicore architecture for their main CPU. Many such systems make use of a multicore GPU and employ a cluster of more specialized CPU cores for other tasks such as GPS navigation and power management.

Despite the improved internal computing power of mobile devices, we have seen in Chapter 5, the increasing tendency to offload complex processing. Offloading means sending complex tasks to be handed in the cloud or on server farms—which are themselves large parallel processing clusters.

The increasing availability, bandwidth, and connection speed of wireless Internet connectivity means that less needs to be done on a mobile device itself. The exception is where very fast latency is required (i.e., immediate response to something), or where the processing needs to be done in a location that has poor connectivity.

The second situation was one faced by the author in the early 2002, when he was asked to design the computers for a new research satellite. The need for high levels of autonomy, coupled with wireless links that only operated reliably twice a day (with one link being very slow at just tens of kilobits per second, and the other much faster but only usable for tens of seconds at a time while it drained the battery) meant powerful on-board processing was required. Unfortunately powerful CPUs tend to be unreliable in space, so the only alternative was to build a cluster of simpler CPUs. His design, a parallel cluster of ARM processors running Linux in what is called a Beowulf cluster configuration, was launched on the X-Sat satellite in 2009 to become the first parallel cluster computer in space.

12.2.1.1   The Parallel Processing Unit

The parallel processing unit (PPU) was designed to provide high-reliability computer services in a micro-satellite (i.e., a satellite weighing between 10 and 100 kg). The satellite was designed to capture images from a 500-km-high orbit, process these on board, and then downlink them to the ground.

We know from Section 7.10 that the space environment contains cosmic radiation that makes electronics unreliable, and so most satellite designers choose to design using radiation-hardened or radiation-tolerant CPUs. Unfortunately, due to the manufacturing, testing, and qualification processes involved, these devices tend to be very expensive, difficult to procure, not at all power efficient and rather slow. Most microsatellites contain 8086-era processors, and few exceed operating speeds of 10 MIPS. Even so, satellite computer designers are a conservative bunch (albeit with good reason—few would want to risk wasting a million dollar launch due to a poor design of on-board computer), and typically derate processors so that they operate at half of the maximum clock speed specified by the manufacturer.

With such feeble on-board computers, it is no wonder that satellites do not tend to perform processing on board. Most simply capture information and then download this to ground-based computers for processing. This is not the forum to argue the advantages and disadvantages of such approaches, but only to note the small but growing movement toward improving the capability of satellite on-board computers using commercial off-the-shelf (COTS) CPUs.

The PPU uses this approach. It is designed around Intel StrongARM devices arbitrated through four radiation-tolerant field programmable gate arrays (FPGAs). Initial designs, still shown in older research literature, used two FPGAs, but in the final construction four devices were used for arbitration. Since any COTS device is unlikely to survive long in space, there are in 20 separate CPUs provided in the PPU, with the system designed to accommodate the expected gradual failure of CPUs over time. Using published radiation tolerance information, the PPU was designed so that, while it would have 20 functioning CPUs at launch, at the end of its design lifetime of 3 years, most CPUs would have failed, but sufficient would still be alive for the PPU to maintain its mission objectives. One of the important features that allows individual devices to fail without compromising the rest of the system is by having individual power supplies to each

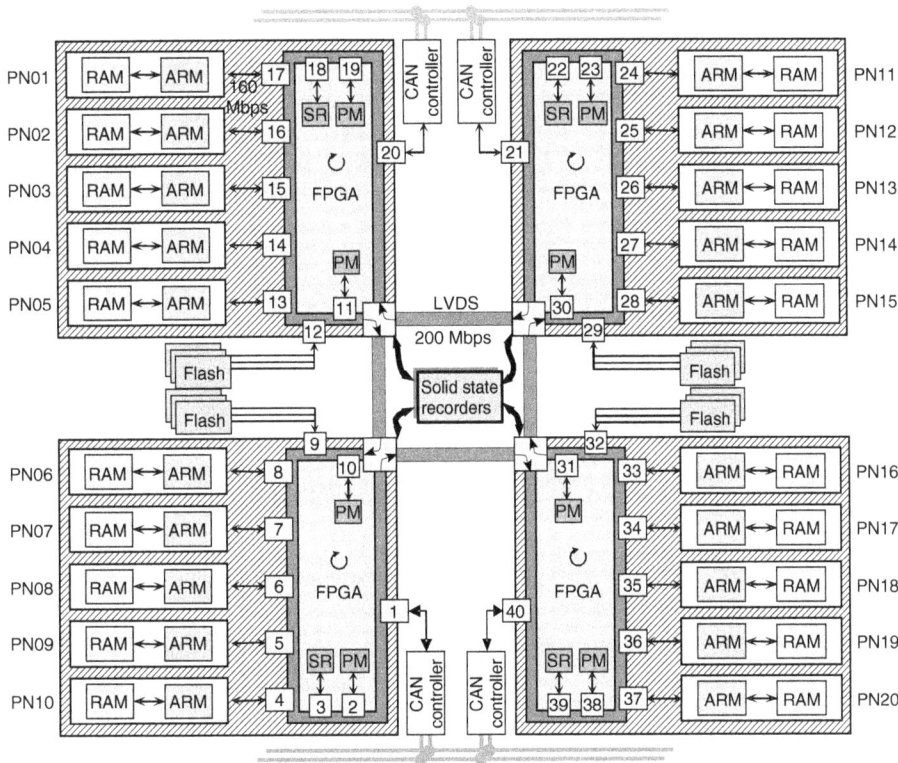

Figure 12.5 A block diagram of the parallel processing unit (PPU), showing 20 processing nodes (PNs), each containing an SA1110 CPU and 64 Mbytes of local memory connected to a local Actel AX1000 FPGA using dedicated buses. Four FPGAs each host 5 PNs and connect to one solid-state recorder and one controller area network (CAN) bus. The FPGAs interlink using dual bidirectional low-voltage differential signaling (LVDS) connections. A time-slotted global back plane bus conveys data between PNs, external links, internal configurable processing modules (PM), and internal status registers (SR).

of the 20 CPUs. By making these switchable, faulty devices can even be powered down remotely if necessary.

The PPU is shown in block diagram form in Figure 12.5. It can be seen that the four FPGAs (Actel AX1000 devices) each accommodate five processing nodes (PNs). Each PN connects to its FPGA over a dedicated parallel bus which will be explained a little later. Within the FPGAs, the author invented a communications system that he named the "time-slotted global back plane" (TGB). The TGB bus, operating like a token-ring system, sends messages and data between node addresses. Each PN has its own dedicated TGB node (and address), as do external connections, internal configurable processing modules (PM), and status register (SR). The external connections are to solid-state recorder, a large array of flash memory storage, and to a controller area network (CAN) bus, arbitrated by C515C controllers. The CAN bus conveys control information to and from the PPU from the satellite main computer. The two FPGAs connect together using

LVDS (see Section 6.3.2), over which the TGB normally traverses, and LVDS is also used for fast data connections to the solid-state recorders (and incidentally also to the camera module and high-speed data download radio also on the satellite).

Operating code for the PNs is stored in flash memory, and there are three identical copies of the code connected to each FPGA in a triple redundant fashion (see Section 7.10). The entire design showcases the concept of "reliability through redundancy" and was built from the bottom up with reliability in mind. Consider some of the reliability features of the design:

- Replicated PNs—with so many PNs, failure of a few can be tolerated so that the system will continue working.

- Individual buses—if the PNs shared a common bus, then it is quite possible for a cosmic-ray-induced error to cause one of the addresses or data bus driver circuits to fail, causing it to become stuck high or stuck low. The effect of this on a shared bus would be to prevent any of the connected devices from communicating properly. Thus, there is an individual parallel bus between each PN and the FPGA. When a PN "dies" this does not affect other PNs.

- Distributed memory—similarly, a failure in shared memory would affect all connected processors, and so this system does not rely upon any shared memory except for that in the solid-state recorder.

- Triple redundant operating code—three blocks of flash memory per FPGA allow the FPGA to perform bitwise majority voting on every word of operating code.

- Two links between FPGAs—if one of the bidirectional LVDS links fails, the other remains operational.

- Multiple links to the solid-state recorder—similarly if one LVDS link fails, the other remains operational.

- Multiple CAN bus links—two links for each FPGA again provides redundancy in case one fails.

- TGB bus nodes—these are very simple fault-tolerant units which track whether the device they connect to remains operational. Irrespective of their device failure, they do not prevent onward communications on the TGB.

- TGB data packets—are parity protected in source, destination address and data fields.

- TGB bus circuit—normally the TGB circulates around 32 nodes, distributed equally between the FPGAs. In the case of individual node failure, the bus remains unaffected; however, in the case of a broken link between FPGAs, the TGB buses on each side detect that break, "heal" the cut, and continue unaffected within their respective areas.

- Four FPGAs—in case one fails, the PPU remains. Since the radiation-tolerant FPGAs are more reliable in space than the SA1110 processors, fewer are required to ensure operation over the design lifetime.

Although the PPU is fault tolerant, it is also a traditional parallel processor. Each PN can operate independently, and can communicate (by TGB) with the nodes around it. There is a mechanism within the computer to allow the physical node

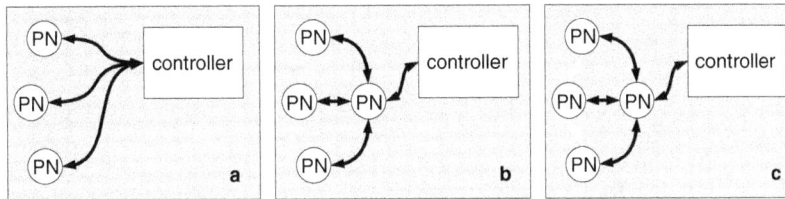

FIGURE 12.6 The remapping of PNs within the PPU and the establishment of links between the PNs can result in several alternative interconnection strategies to be used. In this case (*a*) shows three PNs operating independently, perhaps as a three-way majority voter arbitrated by an external controller. Diagram (*b*) shows that the majority voting process itself has been off-loaded from the controller onto a PN which has in turn called upon three other PNs to cooperate. Diagram (*c*) then shows four PNs fully interconnected, with one responsible for interactions with the controller.

numbers (0, 1, 2, up to 31) to be remapped into various types of logical connection, including any of those we will encounter in Section 12.2.3.

In fact, an example of this remapping can be seen in Figure 12.6. The node which "launches" any PN can restrict the connectivity of that PN to just itself, or to other PNs, leading to a very flexible set of operating arrangements.

At start-of-life, when all resources are operating correctly, the PPU has a respectable specification for an embedded computer (especially one that was designed over a decade ago), of 4000 MIPS, consuming 6 Watts of electrical power in an 1800-cm^2 package (about the same size as a small notebook computer). A typical microsatellite on-board computer will be 200 times slower, two or three times as big and consume a similar amount of power, and furthermore costs around 10 times as much—although cost is rarely the primary consideration during satellite design.

Although there are several further interesting design features of the PPU, including an unusual 17-bit parallel data bus arrangement for optimal data transfer speed, it is the parallelism that is the focus of this section. With that in mind, consider Figure 12.7 in which the speedup has been plotted for an image processing job shared across several processors. Speedup, defined in Section 5.8.2, indicates how well a system is able to parallelize its computation. Perfect speedup (shown as a diagonal line in Figure 12.7) means that a job will run n times faster on n processors than it does on one. The example algorithm running on the PPU does not achieve perfect speedup, but does very clearly benefit from increased parallelism.

12.2.2 Parallel and Cluster Processing Considerations

Parallel processing system design issues might include

- How many processors are required?
- How should they be interconnected?
- What capabilities should each processor have?
- Should the system be homogeneous or heterogeneous (i.e., should all CPUs be the same, or should there be a mixture)?

If n processors are in use, then the completion time, as we have seen, may not equal an nth of the time taken by a single processor—even in a homogeneous system. The actual

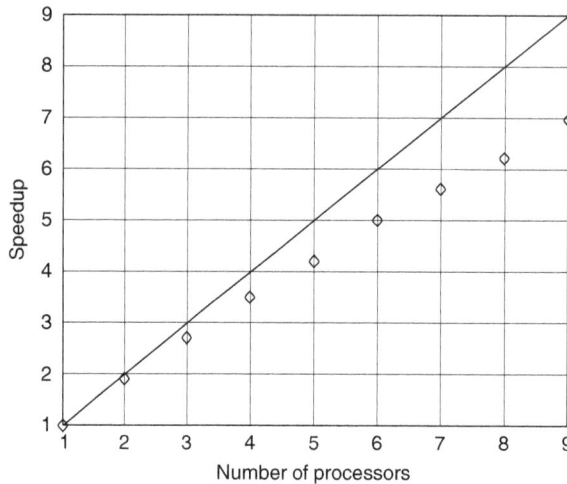

FIGURE 12.7 The degree of speedup achieved within the PPU by sharing an image processing job among up to nine PNs, with perfect speedup indicated by the diagonal line. Clearly the PPU does not achieve perfect speedup, but does evidently benefit from parallel processing. This result was obtained by PPU coinventor Dr. Timo Bretschneider and his students for processing involving unsupervised image classification tasks.

time may be more, or even less than an nth of the time taken by a single processor. All depends on the original *single-thread* implementation and the parallel implementation. Some calculation problems can be divided easily into a number of subtasks where the data transfer between these subtasks is small. Given this then each subtask could be allocated to a different processing unit. In other systems this process may not be so simple.

With subtasks of unequal complexity, the system could benefit from being heterogeneous—consisting of processors of different capabilities, and the interlinking of processors could even follow the requirements of the calculation to be solved, that is, heterogeneous interconnection is also possible. However, the control of such a system becomes more complex—especially if the dividing up of tasks is to be accomplished dynamically, and given different types of processors which are themselves being dynamically chosen.

12.2.3 Interconnection Strategies

Consider a more general system with identical (homogeneous) processors, which we shall refer to as *nodes*. If these nodes are linked in a regular fashion, two main system design issues are the type of interconnection used and the number or arrangement of interconnections.

Interconnection type will define the bandwidth of data which can travel over the link in Mbits/s or Gbits/s, as well as the latency of messages passed (i.e., the time taken for a message issued at one end of the link, to travel to the receiver at the other end.) Example types of interconnection are Ethernet, ATM (asynchronous transfer mode), optical interconnects, InfiniBand, and so on. These vary widely by both bandwidth and cost.

In addition, there are two paradigms of distributed parallel processing systems, with many variants in between—*shared memory* and *message passing*. Message passing

FIGURE 12.8   Six
different parallel
interconnection
arrangements
illustrated.

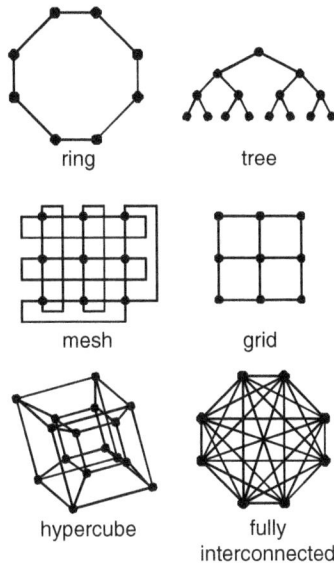

uses structured methods to communicate between nodes, such as message passing interface (MPI), and is suited well to loosely coupled tasks that require medium-to-low-bandwidth data interconnects. Shared memory is useful when separate processors operate on the same source data, or need to communicate with medium-to-high bandwidth. Shared memory systems were considered in the discussion of the MESI cache coherence protocol in Section 4.4.7. Shared memory parallel computing is similar in concept, but on a much larger scale.

In general, the number of interconnections that each node possesses limits the number of other nodes that it can be connected with. At one extreme is the possibility of a node being fully connected to all other nodes. On the premise that connection between processors is relatively slow, a fully linked system minimizes data transfer duration since each transfer is one hop. At the other extreme is a ring structure where each node connects to two others. These, and several other common interconnect strategies are shown in Figure 12.8.

Ring: Each element needs to support two connections. It is scalable with no changes in elements. There are potentially many data hops between target nodes. This is such a regular arrangement that it is easy to setup, manage, and visualize.

Tree: Each element (apart from the top and bottom layers) needs to support three connections. It is easily scalable and software data paths are simplified, but may require many data hops for some transfers when the number of nodes becomes large.

Mesh: Each element needs to support four connections. It is easily scalable, but some nodes require many hops to reach other nodes. The *grid* is similar but differs by not providing wrap-around edge connections (i.e., left-right and top-bottom). Both of these are so regular that they are easy to setup and run in practice, and easy to scale.

Hypercube: Each element needs to support only four connections in a tetrahedral fashion, while data path hops are also minimized. In many cases this is the architecture of choice: sometimes simply because it sounds so high-tech in company press releases—but is a pragmatic trade-off between number of links and latency (hops).

Fully interconnected: Each element needs to support a connection to every other element, making this very difficult to scale because adding one node means increasing the number of connections by the number of existing nodes. The large number of connections means there needs to be a large number of ports at each node as well as extensive physical interconnections in between. All of this indicates that the arrangement is better avoided for large-scale installations; however, it has the lowest latency of all methods because there is only one hop between a node and any other.

Of course there are many possibilities of hybrid schemes—for example, a ring of meshes where each "node" around the ring is itself a set of machines connected as a mesh, or perhaps a more commonly found example is a grid of hypercubes. These are cluster computers arranged in a grid, where each vertex within the grid comprises a multi-processor hypercube.

Grain size describes the level of parallelism. In the most fine-grained machines, actual machine instructions are issued in parallel (such as vector machines or VLIW processors) while coarse-grained machines can run large software routines, or separate items of software, in parallel. This relates back to the discussion in Section 5.8 on the different forms of parallelism.

Using an abstraction method such as MPI, course-grained parallel algorithms can execute in different program instances, and it does not matter whether these are run all on one CPU, or across multiple CPUs. Similarly, it does not matter whether these CPUs reside in a single box, in several boxes in a single data center, or in several geographical locations within a cloud or grid computing system.

Coarse-grained machines tend to be loosely coupled, whereas fine-grained machines tend to be more tightly coupled. The amount of data transfer between elements specifies the speed of the data connection between them, and the number of hops the data must traverse has both bandwidth and latency considerations (i.e., if inter-processor data must traverse two hops then each hop must be capable of twice the bandwidth). Data transfer requirements also have a bearing on memory architecture such as whether each processing element should use local memory or shared memory as mentioned previously. Local memory machines may have distributed memory or may simply use multiple copies of cached memory. Some examples of extremely large-scale systems are shown in Box 12.1.

Box 12.1   Some Examples of Large-Scale Cluster Computers

Looking at the current Top500 list (www.top500.org) of the world's fastest supercomputers, we see an interesting variety of computer systems represented.

Sunway TaihuLight is the current fastest supercomputer at the time of writing. It lives in the National Supercomputing Center of China in Wuxi, China, and consists of 10,649,600 Sunway SW26010 multicore processor cores, each clocked at 1.45 GHz. The processor racks are powered by a high-efficiency direct current (DC) power supply, and employ a water cooling system to maintain their temperature. Rather than making use of much more common Intel devices, the processors in this system have been custom designed by the Shanghai High-Performance Integrated Circuit Design Center. The software appears to be a customized version of the Linux operating system called Sunway RaiseOS. Peak performance is 125,436 TFlop/s, but the machine consumes more than 15 MW of electricity, about the same as a medium-sized Western European town. The system efficiency is 6.05 GFlops/W.

The fastest current European system, **Piz Daint**, based in the Swiss National Supercomputing Centre, Lugano, Switzerland, is a very different machine. It is a hybrid Cray XC50/XC40, making use of several generations of Intel Xeon cores each operating at between 2.1 and 2.3 GHz and each having at least 64 Gbytes of local RAM. Twelve of these are accompanied by NVIDIA Tesla P100 GPUs equipped with 16 Gbytes of RAM. The total of 206,720 cores is about 50 times fewer than TaihuLight, but achieves a very respectable peak performance of 15,988 TFlop/s (an eighth of TaihuLight). However, where Piz Daint excels is in its power consumption of about 1.3 MW, which equates to 7.45 GFlops/W, making this one of the most efficient large cluster computers in the world today. As with all of the fastest machines it runs a Linux-based operating system.

The Green500 list is a different ordering of the Top500 data. This list is in order of power efficiency. We have already noted the second most efficient system above, but consider the current third-placed system:

Shoubu, the ZettaScaler-1.6 system at the Advanced Center for Computing and Communication, RIKEN in Japan is an interesting heterogeneous machine. It contains 1,313,280 processor cores, made up from 2.3 GHz Xeon E5 processors with PEZY-SC accelerators. The PEZY-SC devices are themselves a very interesting heterogeneous architecture of two 0.7 GHz ARM9 cores combined with 1024 RISC Processor Elements. These are arranged in a geographical hierarchical fashion that spans the scale from "villages" of four cores, through "cities" of four villages up to "prefectures" of 16 cities. This arrangement is highly power-efficient, allowing PEZY-SC accelerated systems to occupy a number of positions on the Green500 list. Shoubu itself consumes less than 0.15 kW of electrical power and has a peak performance of 1533 TFlop/s. This is about a tenth of Piz Daint, at about a tenth of the electrical power, giving it only slightly increased efficiency despite it being an older architecture. Several manufacturers are currently working on modern ARM or Cortex architecture supercomputer accelerators (based on ARMv8), which will probably enable ARM architecture devices to make further inroads into both Green500 and Top500. In terms of operating system, Shoubu also runs Linux.

Conspicuously absent from these lists are server farms operated by companies like Google and Amazon which would, in all likelihood, top any list of the world's most powerful computers. However, little is publicly divulged by Google, Amazon, or their major commercial competitors.

12.3   Asynchronous Processors

All common CPUs are synchronous in operation, meaning that they are clocked by one or more global clocks (and domains) such as a processor clock, memory clock, system clock, instruction clock, bus clock, and so on.

Within a particular clock domain—being the physical area on-chip that contains elements acted upon by the same clock—flip flops, and units built upon the basic flip flop, will be synchronous, operating together. The speed of clock is determined for a particular domain with an upper limit being set by the *slowest* individual element. Typically that

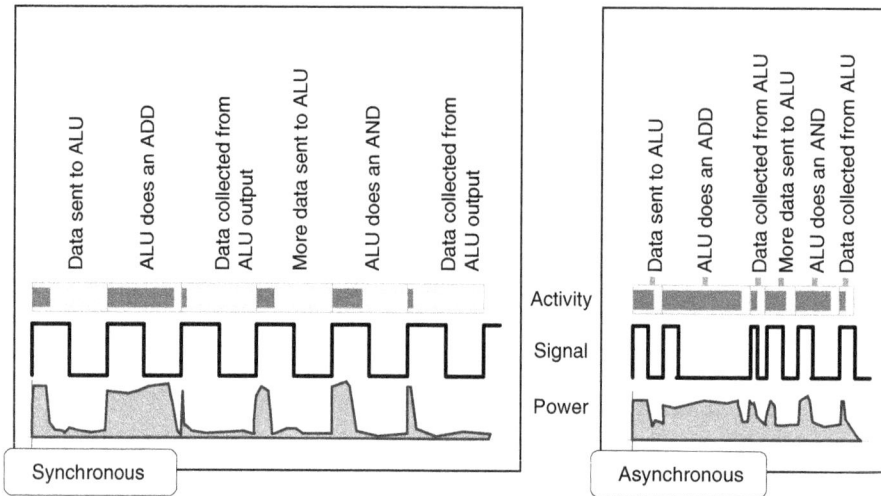

FIGURE 12.9 A comparison of the way a sequence of two operations would execute on synchronous and asynchronous RISC processors.

means that many individual elements could operate faster, but are held back by the slowest one.

For example, an ALU takes its input from two holding registers, and one clock cycle later, latches the result into an output register. If performing an ADD, the operation may be completed only just in time—perhaps the result is ready only 0.01 clock cycles early. However, if the operation was something simpler, such as an AND which has no carry propagation, then the operation may have been ready far earlier—perhaps 0.9 clock cycles early. So depending upon what the ALU is doing, it is either almost fully occupied, or is sitting waiting for its result to be collected. Irrespective, the fixed processor clock which controls it will be set to be slightly slower than the slowest operation.

An analysis of ALU operation would probably reveal that for a substantial amount of time, the unit lies idle. This indicates low usage efficiency. We have already met several techniques to overcome these efficiency limitations, including allowing parallel operation (i.e., several events occur simultaneously rather than sequentially), and pipelining (remember that pipelining breaks up the individual steps into smaller, faster steps, which then overlap with each other during execution. The idea is that, since each individual element is now faster, the overall clock speed can increase).

One very unusual technique is to allow asynchronous operation. An asynchronous processor allows each operation to perform at full speed without wasted parts of a clock cycle. In fact, there may be no need to have a clock at all since each individual element operates at maximum speed, informing the control hardware when the operation is complete. Figure 12.9 provides an example in which an ALU performs an ADD followed by an AND in which the previous ALU output (from the ADD) is an operand. Examining the diagram from top to bottom, the sub-steps in the tasks are shown at the top, followed by a bar showing the activity in the processor, then the synchronous clock and finally a power plot. Time moves from left to right in this diagram.

For the synchronous case, the activity bar shows large swathes of idle time when the ALU has finished operation and is waiting for the next clock cycle before proceeding.

The power plot oscillates rapidly between periods of high and low power (which is not good from either a power supply design perspective or from a CPU temperature fluctuation perspective). The asynchronous CPU, by contrast, packs each operation much closer together, with the clock signal operating just as fast as each sub-operation requires. The power is more smooth, although the total energy used (the shaded area under the power plot) is identical in both cases. Obviously the asynchronous device completes the overall task faster because it spends less time idle.

Advantages of the synchronous approach:

- Simpler to design, and far more design experience of this approach
- Almost, if not all, CPU design tools assume a clocked design
- Synchronous design eliminates race conditions
- Delays are predictable

Disadvantages of the synchronous approach:

- At higher speeds, clock skew or jitter can become a problem
- Almost all latches and gates switch in time with the clock (whether or not there is any new data to process), and as CMOS power usage is mostly caused by switching, this leads to relatively high power dissipation
- Less than theoretical maximum performance (wasted parts of cycles due to operation at the speed of the slowest element)
- Large areas of silicon are devoted to clock generation and distribution

Moving to an asynchronous approach makes sense, although each individual asynchronous element still has to be interfaced with its neighbors—and this requires some form of synchronization. Such synchronization may be relatively simple, but being replicated across an IC for many elements will result in extra logic.

In theory an asynchronous processor should operate with lower power, and at higher speed, than an equivalent synchronous processor. However, the designer has to pay very careful attention to the possibility of race conditions occurring. Avoiding these may actually make the asynchronous processor slightly larger than the synchronous processor would be.

One example of an asynchronous processor (in fact it is possibly the world's only commercial asynchronous architecture) is the AMULET. This was designed by Manchester University in the UK, based on the very popular ARM processor. In the design of the AMULET, certain problems had to be overcome. We will consider some of these and the approach the designers used to solve them in the sections below.

12.3.1   Data Flow Control

If there is no reference clock within the processor, then how can data flow from one unit to the next be controlled? AMULET uses a technique known as request-acknowledge handshaking, which follows the following sequence of events for a unidirectional parallel bus:

1. Sender drives its data onto the bus
2. Sender issues a request event

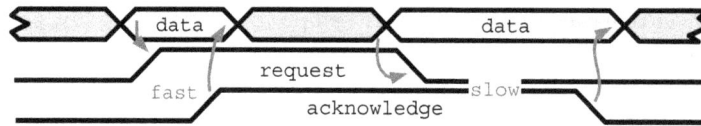

FIGURE 12.10   Request-acknowledge bus transactions for asynchronous bus communications.

3. When ready, receiver reads the data from the bus

4. Receiver then issues an acknowledge event

5. Sender can then remove data from the bus

The request and acknowledge signals are two separate wires that run alongside a standard bus. The diagram shown in Figure 12.10 illustrates these edge-sensitive (transition encoded) signals in use.

Use of request-acknowledge signaling allows each element to be self-timed. The AMULET pipeline elements each operate at different (optimal) speeds depending on the actual instructions being executed. In other words, a unit performing a simple operation will complete very quickly, whereas one performing a more complex operation will take longer. At worst-case, the pipeline is the same speed as if it was synchronously clocked (which means it is clocked as slow as the slowest element—so it is performing a continuous sequence of the slowest instructions). However, in any real-world application, the pipeline would operate faster than the fully synchronous version.

12.3.2   Avoiding Pipeline Hazards

If instructions complete at different, possibly unknown, times, how can RAW hazards be avoided within the pipeline?

Since the processor shares a load-store architecture with its parent, the ARM, almost all CPU operands are register to register. So a method was needed to prevent a register from being overwritten by a result if its content is required as input to an intermediate operation.

The solution is a method of register-locking based on a register-lock FIFO. When an instruction is issued that needs to write to a particular register, it places a lock in the FIFO, and then clears this when the result is written. When a register is read, the FIFO is examined to look for locks associated with that register. If a lock exists, the register read is paused until that register clears.

An example is given in Figure 12.11 which shows the first eight-register lock FIFOs, and the locks corresponding to pipeline position being entered from the top and flowing downward in step with the corresponding flow through the pipeline. In the program that is running, the result of the first instruction goes to r1 (and the lock is then removed). The result of the second instruction goes to r3 (and that lock is then cleared). The result of the third instruction goes to r8 (and that lock is then cleared). At the current time, writes to r1, r3, or r8 will pause until the instructions currently in the pipeline are cleared.

While register locking solves potential hazards, it has been shown to result in frequent pipeline stalls, so more recent AMULET processor developments utilized register forwarding techniques that have been adapted for asynchronous use.

FIGURE 12.11 The register locking hardware in the AMULET processor helps to prevent pipeline hazards.

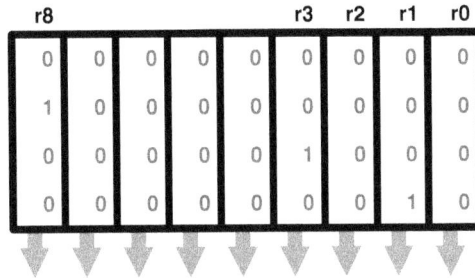

12.4   Alternative Number Format Systems

Looking at the world of computers and CPUs today, we see a profusion of binary logic devices, where a logic 1 is represented by a high-voltage level, and a logic 0 by a low-voltage level. A binary word consists of several such bits—typically, 4, 8, 16, 24, 32, or 64. Whether two's complement or unsigned numbers are used, the weighting of each bit in the binary word represents a power of 2. Floating point extends the concept further to represent mantissa and exponent separately, but these are also binary words, with weightings based on powers of 2.

Analog computers were a refreshing alternative, but these were phased out in the 1980s as digital computers became faster, cheaper, smaller, and more accurate. It seems that the world has converged on digital binary systems for their computing needs.

However, there are some research alternatives with niche applications, that may well become mainstream in time. These will be explored in the following sections.

12.4.1   Multiple-Valued Logic

Faster and faster clock speeds have been the rule for several years in the computing industry, but there are limits. True, these limits shift outward regularly, but they do constrain progress. As we have seen in Chapter 6 as well as earlier in this chapter, if clock rate is constrained, more data can be shifted down a bus by having more wires in parallel. Or by clocking data on both edges of a clock (known as double data rate or DDR).

There is another alternative though, and this is to allow each wire to convey more information simultaneously. Normal binary signaling uses high and low values to represent 1 and 0. For example, in a 5-V system, a binary 1 may be any voltage above 2.5 V while binary 0 is any voltage below that.

Multi-valued logic allows intermediate voltage values to convey more information. Extending the 5-volt, example above to four-level multi-valued logic, every wire can convey 2 bits of information simultaneously, instead of a single bit:

    00   0.0V
    01   1.7V
    10   3.3V
    11   5.0V

This is rather different from the CMOS voltage levels commonly used in electronics, and requires more complicated driver and detection circuitry, but doubles the amount of data that can be represented on a single wire. It does, however, reduce the noise immunity of the data when compared to a system utilizing two (binary) voltage levels. A CPU that uses such logic would need to carefully consider analog as well as digital design issues. With reduced noise immunity it would also be less tolerant to noise and interference spikes, but the trade-off is that it could convey more data faster.

Although the author does not know of any current commercial CPUs utilizing multi-valued logic, the technique has found a niche application in memory storage. Vendors of flash memory are continually under pressure to deliver "larger" devices—meaning devices that can store more bits of memory in a given volume. Manufacturers have typically relied on reduced silicon feature sizes to allow smaller transistors, which can be packed more densely into an IC. However, Intel introduced a more radical design over a decade ago, which allowed 2 bits or 4 bits of data to be stored in a single-transistor cell, using a multiple-valued logic approach. The commercial NOR-flash devices resulting from this research are marketed as Intel StrataFlash® and have found widespread adoption in smartphones, music players, and other advanced embedded systems where size is a primary constraint.

Note that there are diminishing gains as the number of multiple values increases—moving from 1 to 2 bits means halving the voltage threshold (but doubling the amount of data that can be represented). Moving from 2 to 3 bits means halving the threshold again, but only increasing data representation by 50%. These decreasing returns, coupled with increased noise sensitivity, tend to limit the technique in practice to a maximum of 2 or 3 bits per cell/transistor/wire.

One final point here is the effect of cosmic ray irradiation as mentioned briefly in Section 7.10, where the occurrence of single event upsets (SEUs) was discussed. A cosmic ray impinging on a silicon gate induces a change in stored charge, which manifests itself as a voltage fluctuation. Multiple-valued logic devices, as we have discussed, exhibit reduced immunity to voltage noise. Charge storage devices, such as flash memory, have similarly reduced immunity to charge fluctuation. All of this implies that multi-valued logic and memory devices are best avoided in systems destined for use in high-altitude locations: such as on aircraft, in electronic climbing equipment and in consumer electronics destined for Mexico City or Lhasa.

12.4.2 Signed Digit Number Representation

Signed digit (SD) representation is an extension of binary representation such that there is redundancy in the stored number (i.e., there is more than one way to represent each number). The redundancy comes about by introducing the possibility of a sign for each digit in a binary word. By sign, this means that each bit position in a SD word can hold a "1" a "0" or a "−1." The actual bit position weightings are the same as for any standard binary number format.

Consider just the first, second, and third bit positions in binary. Reading "backward," from right to left, these have weightings of 1, 2, and 4 in any binary system. Standard binary can use these digits to store zero or one units (1), zero or one 2s and zero or one 4s. This means that a binary number 001b has value 1 in decimal (written 1d) and a binary number 101b has value $1 + 4 = 5$d. But in SD each bit position could also hold a "−1," so we might see a SD binary number like 10-1b which has value $-1 + 4 = 3$d. Another example would be -1-10b, which has value $-2 - 4 = -6$d. The bit weightings are illustrated in Figure 12.12 in which binary, decimal, and SD binary representations are compared.

FIGURE 12.12   Unlike standard binary (top) or decimal representations (middle), signed digit allows individual digits to be negative as well as positive. Signed digit binary (bottom) allows each separate digit to be 0, $+1$, or -1.

Another unusual aspect of SD representations is that there is a degree of freedom associated with the representation because there are typically multiple ways of writing a single integer (assuming it is not too big to be represented with the chosen number of bits). This contrasts to standard binary in which there is only ever one combination of binary digits that represents a particular value: Every binary number has a unique value, and every decimal number has a unique value in binary. The flexibility of SD representation is illustrated in Figure 12.13 for the decimal value 23.

Knowing what we have learned about binary in Chapters 2 and 3, we can immediately see that there are problems with using signed digit—one being that an ALU or adder which handles SD arithmetic needs to be able to cope with negative digits in any position. Unsigned binary, by contrast, has no negative digits at all, and even two's complement binary only has negative weights in the most significant bit position. But in both cases, a bit is either a 0 or a 1, so to support SD, any ALU needs to be able to cope with digits of -1 as well. It turns out that while this is definitely a complication, and it does need to be taken into consideration when designing an ALU, it does not require much additional hardware, and therefore does not significantly reduce efficiency.

Further SD binary examples are as shown in the table below, which lists some of the many alternative ways that the number equivalent to decimal value 3 could be written:

SD vector	Value	Density
(0 0 0 0 1 1)	$2+1=3$	2
(0 0 0 1 0 −1)	$4-1=3$	2
(0 0 1 −1 0 −1)	$8-4-1=3$	3
(0 1 −1 −1 0 −1)	$16-8-4-1=3$	4
(1 −1 −1 −1 0 −1)	$32-16-8-4-1=3$	5
(0 0 1 −1 −1 1)	$8-4-2+1=3$	4
(0 1 −1 −1 −1 1)	$16-8-4-2+1=3$	5
(1 −1 −1 −1 −1 1)	$32-16-8-4-2+1=3$	6

We shall see later that choosing an alternative with more zero digits, will require fewer operations (gates) when implementing adders and, particularly, multipliers. We define the *density* of a signed digit number as the total number of nonzero digits used to represent that number. A lower density is better because it would result in a faster partial product multiply.

	128	64	32	16	8	4	2	1	each box can contain
Binary	0	0	0	1	0	1	1	0	1 or 0
Signed digit binary	0	0	0	1	1	0	0	−1	1, 0, −1
	0	0	1	0	−1	0	0	−1	1, 0, −1

FIGURE 12.13   Standard binary has just a single way of representing a value such as 23d, whereas signed digit typically has multiple ways of representing the same value. This example shows the decimal number 23 represented in binary (top) and in two different signed digit binary representations below.

Any radix-2 binary number (i.e., standard binary) can be converted to SD representation using the following algorithm:

Let $a_{-1}, a_{-2}, \ldots a_b$ denote a binary number, and the desired SD representation be $c_{-1}, c_{-2}, \ldots c_b$. Each bit in the SD representation may be determined through:

$$c_{-1} = a_{-i-1} - a_{-1}, \quad \text{where} \quad i = b, b-1, \ldots, 1$$

where $a_{-b-1} = 0$.

In order to better exploit the redundancy involved in this representation when it is used in a logic device such as an FPGA or similar system, it is better to ensure that there are as few nonzero digits as possible within the number representation, and this is achieved through the employment of a *minimal signed digit vector*. This is the one (or possibly many) SD representation among the alternatives that has the lowest density.

All the examples in the table shown earlier represented the same number (3), with the first or second entries in the table having the minimum density of 2 and are thus the minimum signed digit vectors for decimal value 3d. Note the second row (0 0 0 1 0 −1), which is a minimum signed digit vector and additionally there is a zero digit between the two nonzero digits. It is, in fact, possible to prove that for every number a SD alternative exists where there are no nonzero digits next to each other. Sometimes there is more than one alternative where this is the case. These numbers are called canonical.

Hence canonical signed digit (CSD) numbers are minimum density signed digit vectors that are guaranteed to have at least one zero between any two nonzero digits.

Apart from the reduction in hardware that results from having many zeros in a calculation, there is another excellent reason why CSD numbers are attractive. This relates back to the parallel adder of Section 2.4.2, where the maximum speed at which additions can occur is limited by the propagation of a carry bit upward. Of course there is the carry look-ahead or prediction technique, but this requires large amounts of logic when the number of bits in the operand words becomes large. However, if we can guarantee that, for a nonzero digit, the next most significant digit is always a zero, there can never be any upward carry propagation from that point.

This ensures that using CSD to perform an addition is always fast since there is no delay caused by carry propagation.

One method of generating such a number (given in the excellent *Computer Arithmetic: Principles, Architecture and Design* by Kai Hwang, published in 1979) is as follows:

We start with an $(n+1)$ digit binary number denoted by vector $B = b_n b_{n-1}...b_1 b_0$, where $b_n = 0$ and each element $b_i \in \{0, 1\}$ for $0 \leq i \leq n-1$. From this we want to find the $(n+1)$ digit canonical signed digit (CSD) vector $D = d_n d_{n-1}...d_1 d_0$ with $d_n = 0$ and $d_i \in \{1, 0, -1\}$. Within their own representation formats, both B and D should represent the same value.

Remember that in terms of determining the value of a number (and in fact any signed digit vector including SD, CSD, and so on), the normal rules of binary apply in relation to the weighting value of each bit position:

$$\alpha = \sum_{i=0}^{n} b_i \times 2^i = \sum_{i=0}^{n} d_i \times 2^i$$

The heuristic described below is a simple but logical method of obtaining the CSD representation of a binary number, based on the method by Hwang:

Step 1	Start with the least significant bit in B and set the index $i = 0$ and initial carry $c_0 = 0$.
Step 2	Take two adjacent bits from B, b_{i+1} and b_i and the carry c_i and use these to generate the next carry c_{i+1}. The carry is generated in the same way as for full addition: thus, $c_{i+1} = 1$ iff[1] there are two or three 1s among $\{b_{i+1}, b+i, c_i\}$.
Step 3	Calculate the current digit in the CSD word from $d_i = b_i + c_i - 2c_{i+1}$.
Step 4	Increment i and go to step 2. Terminate when $i = n$.

Notice that before the calculation, the most significant bit of the original binary number is fixed at 0 (and thus the number of bit positions is effectively lengthened by 1 bit). Thus the CSD representation may have one extra digit over and above binary. See Box 12.2 for some further example CSD numbers.

Box 12.2   Example CSD Number

As an example, consider an 8-bit binary number:

(0   1   0   1   0   1   1   1) which has a value in decimal of 87d.

Applying Hwang's heuristic, the CSD representation becomes

(0   1   0 −1   0 −1   0   0 −1) with a value of $128 - 32 - 8 - 1 = 87$.

In this example, since it is canonical, there are no adjacent nonzero digits in the resulting number, and the density of the resulting CSD number is four.

---

[1] "iff" means "if and only if."

12.5 Optical Computation

Advanced researchers have turned to some novel technologies to try and improve CPU performance. This section presents two interesting ideas based on optical processing. Optical signals are interesting because they propagate at the speed of light—in other words they are the fastest signals that are thought to be possible. For a world in which electron propagation speed through silicon is the defining limitation for computer performance, extremely fast optical signaling is an understandably attractive proposition.

Any digital computer needs to rely on the existence of a switch. Because of this, optical switching technologies have received significant amounts of research effort over the past two decades or so. Miniature all-optical switches are still elusive laboratory creations for the most part, but integrated optics is a branch of optical technology that attempts to build optical circuitry on silicon using fabrication technology similar to electronic ICs (and sometimes mixed with electronics on the same substrate). Current examples of integrated optical devices include multiplexers, filters, and laser diode transmitters.

Although all-optical computers are the major research goal, hybrid electro-optic systems have found several applications in recent years inside computers. The driving factor behind these devices is that all optical signals travel at the speed of light, but also that several signals can coexist in the same physical location without interfering (i.e., crossed beams of light do not have the same problems as crossed electrical wires). Optical interference is also much easier to control and limit than electrical interference.

12.5.1 The Electro-Optical Full Adder

Remember the carry-propagate delay in the full adder of Section 2.4.2? The problem was that the output is not available until the carry has propagated from the least to the most significant bit. This upward propagation delay then became the major limiting factor on adder speed.

The electro-optical full adder works on the principle of making the carry operate at the speed of light. The carry circuitry shown in Figure 12.14 aims to achieve that. The important things to note are that bits x and y are input as *electrical* signals, whereas the carry in and carry out are optical (light beams). The boxes with diagonal crosses in them are electronically controlled optical switches—either blocking or not blocking the light beams below. Light paths travel from left to right, at some points being split into two, and in two places beams are joined together (which is a logical OR).

$$c_{OUT} = x.y + x.c_{IN} + y.c_{IN}$$

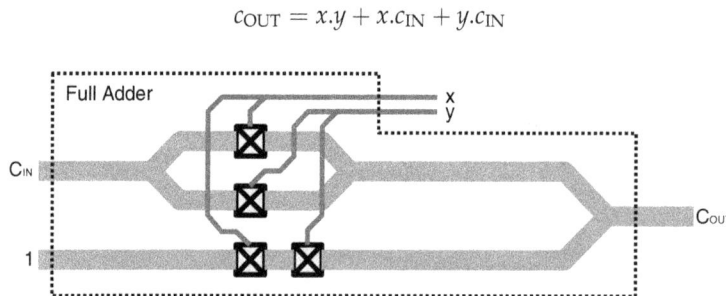

FIGURE 12.14 The electro-optic full adder combines electronic switches and light paths to create a very fast adder not limited by the propagation speed through layers of logic gates.

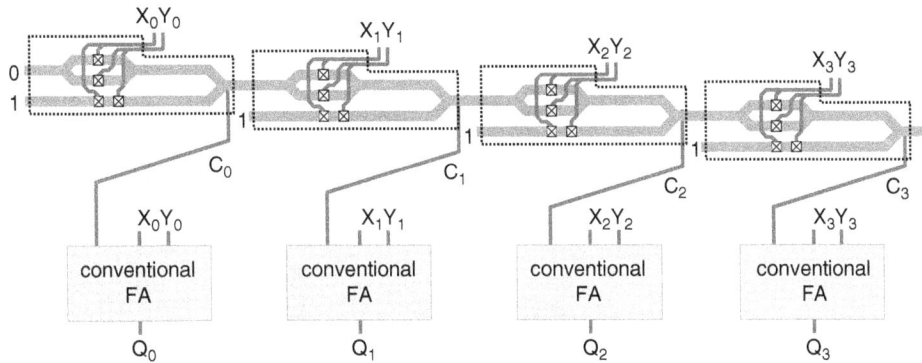

FIGURE 12.15 An example of the electro-optic adder of Figure 12.14 being used in a 4-bit electro-optic hybrid parallel adder that has no electrical propagation delay.

For such a structure arranged as a parallel adder, C_{in} would be fed from the C_{out} beam of the next less significant bit. There are two switch elements per bit, and these switch as soon as the input bits are present. In other words, all switches, for all bit calculations happen simultaneously. The optical carry signal propagates at the speed of light through the entire structure. Other circuitry (not shown) is used to calculate the output result for each bit position (which depends upon the input bits and the C_{in}, which has just been determined).

Compare this speed-of-light propagation delay to the propagation delay of a standard n-bit full adder, which is n times the delay of a single add element (which itself is the propagation delay of several AND and OR gates). This just illustrates one technique out of many possible optically assisted elements that comprise current research topics in computer architecture, and could well be found inside future high-performance CPUs.

An example of using the electro-optic full adder device as a full 4-bit adder is shown in Figure 12.15. Four devices are arrayed to compute just the carry signals of a 4-bit addition. Conventional adders are used to perform the actual addition of input bits with the carry signals (computed at the speed of light) coming from the electro-optic adders. Other combinations are possible, but this illustrates the main principle behind the electro-optic adder to virtually eliminate propagation delay—which would mean that a huge 4096-bit adder constructed using this technology could compute a result just as fast as an 8-bit conventional adder.

12.5.2 The Electro-Optic Backplane

As buses become wider (64 bits or more for both data and address), a larger number of signals have to be connected between modules or blocks within any computer that comprised plug-in modules or cards. At the same time, computers and modules are also becoming faster in terms of clock speed, thereby both causing more, and becoming more susceptible to, electromagnetic interference (EMI). This complicates the job of designing buses, such that 12 or 16 layer printed circuit boards (PCBs) are not uncommon for embedded computer designs. In particular, it can be difficult to design PCBs when large buses need to cross each other. Worst of all (from an EMI perspective) are long parallel

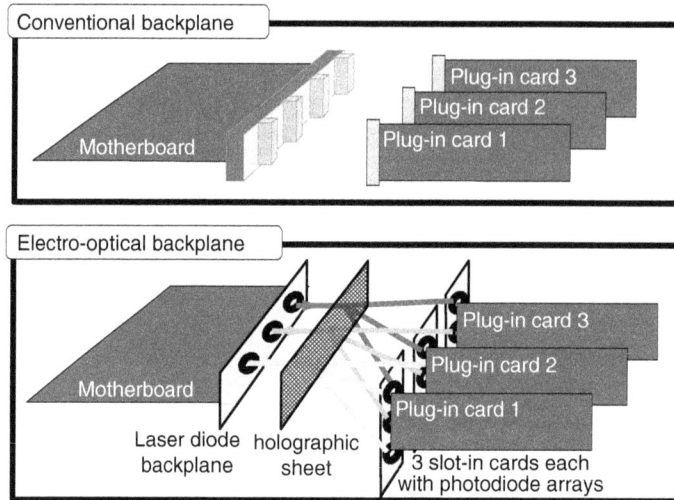

FIGURE 12.16　The electro-optic backplane (bottom) uses a holographic sheet to split optical signal beams in free space from laser diode (or LED) transmitters to multiple receiver arrays which are located on the edge of physically separate plug-in cards (motherboard and cards are all electrical printed circuit boards). With a conventional backplane (top), each card would be fitted with an electrical edge connector that requires physical and mechanical connection between each item.

runs of buses which are side by side. Both of these aspects are found around the large connectors needed to connect parallel buses from one PCB to another.

While relentless miniaturization of devices and subsystems has substantially reduced the need to have plug-in cards of any description (as has the invention of fast and reliable serial bus technologies), there are many cases where they are still required—from ultrahigh performance computers, to avionics systems (computers on aircraft), and to telecommunication systems. The latter two areas tend to require plug-in cards to enable fast replacement of damaged or degraded subsystems.

Mechanical plug-in cards rely on electrical connectivity through edge connectors that can degrade with repeated insertion, with dirt or corrosion, or mechanical misalignment. They are thus problematic from mechanical, electrical, reliability, and interference perspectives, as well as placing a limitation on performance (signal speed). One solution to these problems involves electro-optical technology. In this case, the big advantage of optics is the ability for beams of light to intersect or run close together without causing mutual interference: different light beams can occupy the same physical space at the same time without difficulty. These advantages inherent in optical interconnects have been demonstrated in optical backplanes that use individual laser diodes for every signal output, and individual photodiodes for every signal input. Transmission hologram sheets can be used to route signals to receiver arrays as shown in the diagram in Figure 12.16.

Optical backplanes have no maximum clocking speed: They are limited only by the laser diode modulation and photodiode bandwidth, which can be very fast—at least in the GHz range. They also allow hot-insertion (multiple cards can be unslotted, and slotted in while the system is running without causing the backplane signals to change).

By contrast, fast electrical buses require termination which varies with the load so that fast buses generally cannot support hot-insertion. Above all, with no electrical contacts to degrade, optical backplanes can have very high reliability.

However, very careful alignment of slot-in cards is required (so that the signal beams hit only the correct photodiode), and they are a lot more expensive than passive (electrical) backplane connections. We also assume that the beams propagate through free space (so a spider walking along the edge of a card could cause a problem), although it is entirely possible that the same techniques could be used in other optically transparent media, such as silicates, which can be used to safely encase circuits.

12.6   Science Fiction or Future Reality?

This section is unlikely to be required by the core syllabus of any Computer Systems, Engineering, Science or Architecture course anywhere in the world. It probably cannot even be examined—except in hindsight. However, science fiction ideas are fun, unusual and innovative. Consider Section 12.6 to be your reward for reaching this point after working through the entire book up to now: a glimpse of some of the wild and wonderful ideas taking shape at (or beyond) the fringes of the computing research arena.

Sometimes researchers do look to science fiction for inspiration (as well as for a little light relaxation from time to time). Often, science fiction is the logical extension of current trends, but occasionally it yields some ideas that appear remarkably strange for their time that actually come true. Consider a few examples:

The iPad   A. C. Clarke's "2001: A Space Odyssey" film of 1969 directed by Stanley Kubrick shows a touch screen computer displaying video.

The Internet   Mark Twain wrote an article for the Times of London in 1904 describing a "telelectroscope," which predicted the Internet connectivity and features (although it was probably more along the lines of Google Earth that he was thinking about).

Calculator   Jules Verne in 1983 predicted the existence of calculators, high-speed trains, a global communications network (as well as, of course the future moon landings).

CCTV   Key to George Orwell's famous book "1984," written in 1949. Big brother really *is* watching you.

Tricorder   The 1966 Star Trek (original series) featured a handheld data analyzer that appeared to contain a small screen. Medical and engineering variants of this amazing device were available. While several researchers have claimed to have built a modern tricorder, the Tricorder X-Prize, first awarded in 2014, promoted the development of a "mobile solution that can diagnose patients better than or equal to a panel of board certified physicians."

Looking at the examples above, we can see that computers, video technology, and mobile phones were all predicted in fiction long before they became a reality. Given such examples, perhaps it is a good idea to look at what today's Science Fiction is predicting about tomorrow.

12.6.1 Distributed Computing

As mentioned in the preface, the world of computing is becoming more and more embedded. It is interesting that this trend coincides with another—the shift toward wireless connectivity. As technology progresses, the logical convergence point is a future in which humans may be outnumbered several thousand to one by miniaturized processing units that are interconnected by wireless networks. Even though most of these processors will be nominally dedicated to particular functions (such as a microwave processor, or a telephone processor, or a central heating/air conditioning processor), it is quite likely that at any given time many of them will not be required for their nominal functions.

Simply allowing idle processors to communicate and cooperate would make available an aggregate computing power several orders of magnitude greater than that available in a desktop PC. Given new forms of human-computer interfacing, and advances in software, we may each expect to be able to interact with our own dedicated computer personality. This personality would be hosted on a constantly changing set of basic processors, but would present a consistent interface to the user. It would be a distributed computer program, remotely accessible, and existing as our personal assistants.

Does this sound like science fiction? Much of the basic technology exists today. Furthermore, with recent products like Amazon's Alexa, Samsung's Bixby, Google Home, and Apple HomePod coming on the market, all of which use speech for their human-computer interface, this prediction is now looking an increasingly likely future.

12.6.2 Wetware

This is perhaps moving much further into the realm of science fiction, but if current advances in genetics and bio-computation continue, we could start to see viable computation performed on biological machines during the next decade or so. This is not so far-fetched when we consider that the most complex and capable computer available today is located inside our heads. Advances in medical analysis techniques are constantly unveiling more and more detail on the operation of human and mammalian brains, and our understanding of these systems continues to improve.

Now consider that *hardware* refers to the electronic systems inside a computer, *firmware* is the low-level reprogrammable behavior build into a computer and *software* is the program that runs on the hardware, then what is *wetware*?

Wetware is the augmentation of the human brain with additional biological based or biologically interfaced capabilities, and is firmly in the realm of science fiction at present. However, direct interfacing with the brain and nervous system has been performed for decades using electrical sensors and stimulators. Examples include vision systems for the blind (since the 1970s), and cochlear implants for the hearing-impaired (since the 1990s). It does not require a large stretch of the imagination to envisage the interfacing of computer units to the brain in a kind or "co-processor" arrangement.

Human ingenuity, despite decades of progress, has not yet invented a computer that can approach the computational abilities of the human brain, except in the single area of fast computation of predetermined tasks, or in fixed and narrow tasks such as playing go and chess. In almost all areas of competition, and certainly in flexibility, the human brain wins. Since we have the example of such amazing calculating machines in nature, researchers have long tried to emulate or copy these capabilities—both by reproducing the principles of the human brain in silicon, but also in biological processing.

Researchers have demonstrated the use of biological and/or chemical building blocks for performing computation. Something like an artificial biological computer, where processing is performed on artificial biological neurons, perhaps based on the structure of the human brain, is not beyond the realm of present science. A biological transistor (controlled switch) has already been demonstrated in the laboratory, so we know that binary logic—the foundation of modern computing—is possible in biological systems. However, it is likely that novel structures would be better suited to biological computation rather than simply copying the methods used with silicon into a biological construct.

Personally the author is quite happy with his brain the way it is, but a natural progression into the future could see artificial neural aids being developed for people with disabilities, including learning disabilities or memory loss. Such augmentation could take many forms, but might include a higher-bandwidth computer interface for realistic gaming, a memory-storage and recall unit, and access to senses beyond the five natural ones of sight, sound, smell, touch, and taste. The possibilities are endless once the basic problems of making a brain-compatible interface are solved, but the real advance will be brain augmentation with an artificial all-biological computer (which at least means we will not have to carry a battery around with us).

12.7   Summary

In this chapter, we have tried to plumb the depths of the future of computers. We began with some fairly safe bets—single-bit architectures, parallel and asynchronous systems (safe in that each of these is an established technique, albeit most are largely confined to one or two niche applications). Parallel processing is clearly on the agenda for the future of most of the computing industry: with dual, quad, and eight-core processors being widely available, it does not take a large leap of imagination to see this trend toward greater parallelism continuing. Massively parallel computing is also a safe bet since most of us enjoy the benefits that such computation brings to the world, used by the likes of Google, Amazon, and others to provide services that are increasingly integrated into our daily lives.

This chapter also provided an overview of alternative number formats, encompassing another class of techniques having significant penetration in niche computing areas, but also with potential for impacting the future of mainstream computing. Beyond this, we considered electro-optical hybrids which, despite having been technically feasible for over two decades, have yet to make any major impact in the computing world.

Finally, science fiction. Let us be honest: science fiction was the path that led many of the author generations into the study of science and engineering in the first place. Whether it was the sonic screwdriver and TARDIS of Dr. Who, the phasor arrays and transporters of the Starship Enterprise or the 'droids and lightsabers in Star Wars, most scientists and engineers have been impacted strongly by technologically inspired fiction. Let us try and maintain that "cool" technology factor, and in turn inspire the generations that will follow us.

And now, having performed the written equivalent of a "core memory dump" by committing decades of work experience, education, and research to paper, it is the turn of the author to pass on the baton. As he sits back and relaxes after a year of book writing, he is looking forward to the future of embedded, pervasive, parallel, and advanced computing systems that you, the reader, are ready to create. In that, may the force be with you.

While you can and certainly will make great *evolutionary* progress in computing systems over the next few years, do not forget to tap occasionally into the imaginative genius of science fiction to inspire leaps of progress that are truly *revolutionary* too.

Let the future begin.

APPENDIX **A**

Standard Memory Size Notation

Most people are taught the SI (System International) scheme of units at school, in which the prefix of the unit denotes the power of 10. For example, a millimeter is 10^{-3} meter, a centimeter is 10^{-2} meter, and a kilometer is 10^3 meters. This is very simple. Here are some of the more useful prefixes:

Prefix name	Prefix letter	Multiplier
exa	E	10^{18}
peta	P	10^{15}
tera	T	10^{12}
giga	G	10^9
mega	M	10^6
kilo	k	10^3
milli	m	10^{-3}
micro	μ	10^{-6}
nano	n	10^{-9}
pico	p	10^{-12}

However, when it comes to counting computer memory sizes, which are constructed in powers of two, 2, 4, 8, 16, 32, 64, and so on, the SI units are inconvenient and confusing.

The reason is that it turns out that 2^{10}, being equal to 1024, is too close to 1000 and so the value of 2^{10} had come to be referred to as a "kilo." Thus in popular usage 1 kbyte was actually 1024 bytes, which is not the correct SI definition of a "kilo." While this discrepancy may be fine for everyday usage, there are many occasions when it is necessary to be more precise, and where the non-SI usage can become confusing.

Thus the International Electrotechnical Commission (IEC) introduced a new and nonambiguous set of terms for the storage of computer data, similar to, but distinct from, the SI units. In the range of sizes useful for computers, these prefixes are as follows:

Prefix name	Prefix letter	Multiplier
exbi	Ei	2^{60}
pebi	Pi	2^{50}
tebi	Ti	2^{40}
gibi	Gi	2^{30}
mebi	Mi	2^{20}
kibi	Ki	2^{10}

Thus, a computer hard disc having a capacity of 1 Tibyte (tebibyte) actually contains 1,099,511,627,776 bytes, which is almost 10 percent more than a hard disc having a 1-Tbyte (terabyte) capacity (1,000,000,000,000 bytes).

Throughout this book, the IEC units, which have also been ratified by the IEEE and others, will be adopted wherever appropriate.

Examples

128 Kibytes
128 KiB
128 kibibytes
Means $128 \times 2^{10} = 131{,}072$ bytes

20 Mibytes
20 MiB
20 mebibytes
Means $20 \times 2^{20} = 20{,}971{,}520$ bytes

500 Pibytes
500 PiB
500 pebibytes
Means $500 \times 2^{50} = 562.96 \times 10^{15}$ bytes

Standard Logic Gates

The standard gates used in digital logic are shown below, along with their truth tables (the sequence of outputs, **X**, given input bits **A** and **B**). The exclusive-OR gate is abbreviated as either "EOR" or, more commonly, "XOR."

Gate	Diagram	A	B	X
AND		0	0	0
		0	1	0
		1	0	0
		1	1	1
OR		0	0	0
		0	1	1
		1	0	1
		1	1	1
EOR / XOR		0	0	0
		0	1	1
		1	0	1
		1	1	0
NOT		0		1
		1		0

The "bubble" in the inverter (NOT gate) can be combined with any other logic gates. As an example, the following NAND and NOR gates are the negated versions of AND and OR, respectively (you can see the "bubble" on their outputs).

Gate	Diagram	A	B	X
NAND		0	0	1
		0	1	1
		1	0	1
		1	1	0
NOR		0	0	1
		0	1	0
		1	0	0
		1	1	0

Index

Note: Page numbers followed by *f*, *t*, or *n* represent figures, tables, or footnotes, respectively.

A

AAPCS. *See* ARM Architecture Procedure Call Standard
ABI. *See* Application Binary Interface
Absolute address, 93
Access efficiency, for cache memory, 155
ACK protocol, 449–451
  TCP and, 484
Acorn Computers Ltd.:
  Archimedes of, 82
  BBC and, 82–83
  RiscPC by, 9–10, 10*f*, 82
Actel, 366
Ad hoc:
  networking, 457–458
  scheduling, 272
ADCs. *See* Analog-to-digital converters
ADD, 134
  addressing modes and, 100–101
  ARM and, 84
  with asynchronous processors, 506
  control unit and, 74, 77
  dual-bus architecture and, 130
  RPN and, 101
Addition, 32–34, 32*f*, 38
  ALU for, 21
  bit-serial, 494–495, 494*f*, 495*f*
  of IEEE754, 59–62
Address handling, 204–207, 207*f*
Address resolution protocol (ARP), 485–486
Address translation cache (ATC), 142
Addresses, 93. *See also* Internet Protocol; Media access controller
  for DRAM, 318–321, 318*f*, 320*f*
  I/O, PCI and, 265
  modes of, 97–101
  for networking, 452–453
  offset for, 90, 187
  virtual, 21–22
ADSP21xx, 20, 71, 84, 95, 125*n*2
  addressing modes and, 99
  buses and, 124–127, 126*f*
  cache memory and, 155
  DAGs for, 204–207, 207*f*
  registers and, 99
  VLIW of, 249
  ZOL for, 202–206
Advanced interrupt controller (AIC), 280
Advanced microcontroller bus architecture (AMBA), 255, 261
Advanced RISC Machine, 82
Advanced Technology Attachment (ATA), 329
AHB. *See* ARM host bus
AIC. *See* Advanced interrupt controller
Altera Nios II, 366
ALU. *See* Arithmetic logic unit
Amazon Echo, 11, 11*f*
AMBA. *See* Advanced microcontroller bus architecture
Ambient intelligence, 236
AMD, 14
  performance of, 115
  Phenom, 294, 294*f*
  3DNow! of, 160, 164
Amdahl's law, 234
American Standard Code for Information Interchange (ASCII), 105, 106*t*
AMULET, 507–508, 509*f*
Analog Devices. *See* ADSP21xx
Analog-to-digital converters (ADCs):
  RE of, 346
  of Samsung S3C2410, 292

525

Analytical Difference Engine, 1, 1*f*, 4
AND, 523
    ALU for, 21, 134
    ARM and, 85, 91
    control unit and, 77, 79
    CPUs pipelines and, 189
    soft core processors and, 362
Android:
    embedded systems for, 395
    shell, 384
    smartphones with, 393
Anycasting, 478
APB. *See* ARM peripheral bus
APCS. *See* ARM procedure call standard
API. *See* Application programming interface
Apple:
    Firewire, 268
    iMac, 10, 11*f*
    iOS, embedded systems for, 395
    iPhone, 11
    iPod, 306
    MacOS, 393, 394
        dual-booting of, 420
Application Binary Interface (ABI), 376*n*2
Application layer, 465–466
Application programming interface (API),
        400*n*3
Application-specific isolated circuit
        (ASIC), 15
    DMC and, 358
    RE for, 346, 350, 356–357
Arbitration, for networking, 451–452
Arithmetic, 31–37. *See also* Addition;
        Subtraction
    with fractional numbers, 47–48
    for IEEE754, 59–62
Arithmetic logic unit (ALU), 21, 31–37, 70,
        133*f*, 135*f*
    addressing modes and, 99–100
    for asynchronous processors, 506–507
    branch prediction for, 209
    BTB for, 224
    buses and, 124, 124*f*, 125, 125*f*, 126*f*,
        129–132, 129*f*, 130*f*, 132*f*
    clock speed of, 113
    clocks and, 298–300
    control unit and, 74, 77
    design of, 133–134
    fetch and, 209*n*5
    functionality of, 113, 132–133

hardware acceleration and, 201
    MMX and, 164
    offset and, 224
    programming for, 371
    propagation delays, 134–136
    registers of, 21, 298
    relative branching and, 187–188
    RPN and, 101
    superscalar architectures and, 194, 194*f*
    Tomasulo's algorithm for, 240
    VLIW and, 246–250
ARM, 5, 14, 71
    addresses and, 93
    addressing modes and, 97–101
    AMBA, 261
    bit width of, 113–114
    boot vector of, 371
    booting on, 405
    branching and, 92–93, 374
    buses and, 123, 124, 255
    cache memory and, 145
    condition codes in, 91–92
    CPI for, 114
    CPUs and, 82–86
    CPUs pipelines and, 178
    cross-compiling, 379–380
    DAG for, 204
    data processing by, 109
    data types and, 103
    DCP of, 231–232, 232*f*
    design of, 83
    DMA and, 257
    embedded systems coprocessors and, 165
    emulation by, 380
    in fourth-generation computers, 9
    FPU on, 160
    Huffman coding and, 95–97
    IBM and, 83
    IFU and, 88–95
    immediate constant and, 93–95
    instruction sets and, 87–89, 88*f*, 91
    interrupts of, 273, 275, 276, 277–278, 279,
        280
    in iPhone, 11
    JTAG for, 333, 334*f*, 335
    LR of, 375
    memory of, 372–373, 373*f*
        for embedded systems CPUs, 323–324,
        325*f*
    microcode and, 81

ARM (*Cont.*):
mobile multiprocessing and, 238
MOV and, 93–94
multiplication and, 37
NOT and, 85*n*5
registers and, 99, 124
relative branching and, 187
RISC and, 82
Samsung S3C2410 for, 291–292
soft core processors for, 359
SWI and, 281–282
VFP and, 166
ARM Architecture Procedure Call Standard
(AAPCS), 376*n*2
ARM host bus (AHB), 255
ARM peripheral bus (APB), 261
ARM procedure call standard (APCS), 376,
376*n*2
ARP. *See* Address resolution protocol
ARPANET, 462
ASCII. *See* American Standard Code for
Information Interchange
ASIC. *See* Application-specific isolated
circuit
ASR. *See* Automatic speech recognition
Assembly language, 20*f*, 74, 373, 376
interpreted languages and, 382
Asynchronous processors, 505–508, 508*f*,
509*f*
data flows in, 507–508
pipelines and, 508
ATA. *See* Advanced Technology Attachment
Atanasoff-Berry machine, 4
ATC. *See* Address translation cache
ATMEL, 260, 311
Automatic speech recognition (ASR), 237
Automobiles, 69–70
Availability:
of OS, 394
of soft core processors, 361
awk, 387–388

B

Babbage, Charles, 1, 1*f*, 4
Baby, 5
Backup system, 431–433
Backward compatibility:
of CPUs, 79, 79*n*4
of Intel Pentium, 163
of third-generation computers, 9

Ball grid array (BGA), 329
JTAG for, 335
RE for, 348
Bandwidth, 442
Barcelona Supercomputer Center, 12, 13*f*
Barriers, of FS, 423
Bash. *See* Bourne again shell
BASIC. *See* Beginner's All-purpose Symbolic
Instruction Set
Basic blocks, 225–227, 226*f*
Basic input output stream (BIOS), 20–21
HDD and, 419
ISA and, 263
programming for, 371
BBC. *See* British Broadcasting Corporation
BCD. *See* Binary-coded decimal
BDTi, 115–116
Beginner's All-purpose Symbolic Instruction
Set (BASIC), 82, 115
Bell Labs, 4–5
BEQ, 374
BGA. *See* Ball grid array
BGT. *See* Branch if condition flags greater
than 0
BGTD, 189
Big endian, 22–25
Big.LITTLE, 238–239, 239*n*13
Bill of materials (BOM), for RE, 347–348
Binary-coded decimal (BCD), 29
Binary logic, 7
Binary16, 50
Binary32 floating point, 50
Binary64 floating point, 50
Binary128, 50
Binary256, 50
BIOS. *See* Basic input output stream
BIST. *See* Built-in self-test
Bit-level parallelism, 228
Bit-serial addition, 494–495, 494*f*, 495*f*
Bit-serial logic and processing, 496–497
Bit-serial subtraction, 495–496, 496*f*
Bit widths, of CPUs, 113–114
BL. *See* Branch and link
Black box, 103*n*10, 107
Bluetooth, 237, 437, 488, 488*n*10
BLVDS. *See* Bus LVDS
BODMAS, 101*n*9
Bogomips, 114
BOM. *See* Bill of materials
Boot vector, 371

Booth's method, for multiplication, 42–44, 42t
Booting, 374
    flash memory and, 325, 406–407, 407f, 408f
    from HDD, 405–406, 408–409, 420–421
    JTAG for, 335
    kernel for, for embedded systems CPUs, 325–326
    from NV, 325–327
    OS and, 405–410, 407f, 408f
    from SSD, 405–406, 408–409
Bootloader, 405
    memory and, 279–280
Boundary scan data (BSD), 335
Boundary scan data logic (BSDL), 335
Bourne again shell (bash), 384
Branch and link (BL), 375
Branch if condition flags greater than 0 (BGT), 183–184, 186, 374
Branch target buffer (BTB), 224–227, 224f, 225f
Branching:
    ARM and, 92–93, 374
    interrupts, 276–278
    for pipelines, 183–185
    prediction of, 185, 208–227, 213f, 215f, 218f, 220f–226f
    relative, 187–188
British Broadcasting Corporation (BBC), 92–93
Brownout detectors, 341–343, 342n6
BSD. See Boundary scan data
BSDL. See Boundary scan data logic
BTB. See Branch target buffer
Buffers:
    BTB and, 224–227, 224f, 225f
    control unit and, 74, 74f
    MOB, 294
    ROB, 294
    TB, 142
    tristate, 21, 75, 75n3, 77
Built-in self-test (BIST), 331–333, 331f, 332f
Burst mode, for EDO DRAM, 321
Bus LVDS (BLVDS), 266
Buses. See also specific types
    ALU and, 21, 129–132, 129f, 130f, 132f
    architecture of, 123–132, 124f–126f, 128f–131f
    ARM and, 255
    in computer organization, 17–18

control signals of, 256–257
control unit and, 74–75, 74f, 77
DMA and, 257–258
DSP and, 125, 127–128, 128f
dual architecture for, 129–130, 129f
for embedded systems CPUs, 295
on ICs, 257
    pin swapping of, 352–353
instruction sets and, 193
registers and, 22
Busy wait loop, 306
BYPASS, 334
Byron, Ada, 4
Bytes, 107

C

C (programming language), 380
    compilers, 377, 377f, 378f
    data types for, 103, 105
    for embedded systems CPUs, 366
    internal data and, 108
    interrupts and, 276
    Linux and, 143
    MMU and, 140
    registers and, 327
    trapping errors in, 143–144
    ZOL and, 203–204
C++ (computer language), 104, 367, 380
    for embedded systems CPUs, 366
C (carry) flag, 36–37
    for pipelines, 182
C-shell (csh), 384
Cache memory, 144–158, 151f, 156f, 157f
    access efficiency for, 155
    coherency of, 156–158
    for data storage, 108
    direct, 146–148
    for embedded systems CPUs, 297
    full-associative, 149
    locality principles for, 149–150
    performance of, 154–156
    replacement algorithms for, 150–154
    set-associative, 148–149
    spatial locality for, 149–150
    speed of, 116
    temporal locality for, 149–150
Calling process, for programs, 375–376
Cambridge University, 5, 80, 92
    WAN at, 462
CAN. See Controller area network

Canonical signed digit (CSD), 512–513
Carrier sense (CS), 473–474
Carry predictors, 34, 34*f*
Catastrophic errors, 401
CCR. *See* Condition code register
CD. *See* Collision detection
CDB. *See* Common data bus
CDC6000, 7
Cell processor, 234–235
Central processing units (CPUs), 15, 21,
    21*n*1. *See also specific components*
  ARM and, 82–86
  as asynchronous processors, 505–508, 508*f*,
    509*f*
  in automobiles, 69–70
  backward compatibility of, 79, 79*n*4
  basics, 69–119
  bit widths of, 113–114
  clock speed of, 113
  in computer organization, 17, 20
  control unit and, 74–79, 74*f*–78*f*
  data handling by, 102–112
  for embedded systems, 291–368
    brownout detectors for, 341–343
    clocks for, 298–306, 299*t*, 300*f*
    DRAM for, 315–321, 316*f*–318*f*, 320*f*
    error handling for, 336–343, 337*f*, 339*f*,
      342*f*, 343*f*
    flash memory for, 298, 307, 309–313, 313*f*
    hardware software codesign for,
      363–365, 363*n*11, 364*f*, 366*f*
    I/O for, 295, 329–330
    memory for, 295, 306–321, 323–328, 324*f*
    memory pages for, 321–323
    MMU for, 321–323
    off-the-shelf, 365–367
    OS for, 323
    RAM for, 314–321
    RE for, 343–358, 345*f*, 346*f*, 349*t*, 356*t*,
      357*t*
    required functionality for, 295–298
    reset supervisors for, 341–343, 342*f*, 343*f*
    ROM for, 298, 307–313, 308*f*
    as soft core processors, 358–363
    SRAM for, 314–315, 315*f*
    test and verification of, 328–336,
      329*f*–332*f*, 334*f*
    volatile memory for, 327–328
    WDTs for, 340–341
  externals of, 255–285

in fourth-generation computers, 9
  functionality of, 113
  hardware acceleration and, 201–208, 203*f*,
    207*f*
  IFU and, 88–95
  instruction handling by, 86–102
  internals of, 123–167
  interrupts and, 269–270, 273–282
  IPC for, 198–201
  memory hierarchy and, 71–72, 72*f*
  memory in, 114
  microcode and, 79–81, 80*f*
  multiplication and, 37
  OS and, 393, 411–413
  performance of, 114–118, 173–250
  pipelines of, 174–192, 175*f*, 177*f*, 178*f*, 180*f*,
    184*f*, 190*t*, 191*f*, 192*f*
  power for, 301–306, 302*f*, 305*f*
  program storage and, 70–71, 71*n*1
  program transfer and, 73–74, 73*n*2
  registers and, 22
  resource manager for, 396
  RISC and, 21
  speedups for, 174
  superscalar architectures for, 194–198,
    194*f*, 195*t*, 198*f*
  in third-generation computers, 8
  Tomasulo's algorithm for, 239–246, 241*f*,
    243*t*, 244*f*, 245*t*
  top-down view of, 113–118
  VM and, 396–397
  in von Neumann architecture, 19
CF. *See* Compact flash
Churchill, Winston, 3
CISC. *See* Complex instruction set computer
Client-server architecture:
  for networking, 456, 457*f*
  for OS, 402, 404–405, 405*f*
Clock cycles, 22
  with asynchronous processors, 506–507
  CPUs pipelines and, 175
  of RISC and CISC, 81–82, 81*f*
  SDRAM and, 321
  superscalar architectures and, 194
Clock speed:
  in computer organization, 18
  of CPUs, 113
  for DRAM, 316
  propagation delay and, 303
  speedups for, 174

Clocks:
    for asynchronous processors, 505–506
    for embedded systems CPU, 298–306,
        299t, 300f
    power and, 301–306, 302f, 305f
    for SoC, 298–306, 299t, 300f
Cloud computing, 235, 237
    OS for, 395
Cluster computers, 235
    OS for, 395
CMOS. *See* Complementary metal-oxide
    semiconductor
Co-simulation, for hardware software
    codesign, 365
Co-synthesis, for hardware software
    codesign, 365
COBOL, 380
Coherency, of cache memory, 156–158
Collision detection (CD), 473–474
Colossus, 3–4, 4f, 7
Column address strobe (nCAS), 318, 318f
Common data bus (CDB), 241–242, 241f,
    245t
Compact flash (CF), 268, 312
Compile time, pipelines and, 185–186
Compilers, 49, 86
    linkers and, 377–379, 378f, 379f
    for programs, 377–381, 377f–379f
    VLIW and, 250
    WAW and, 328
Complementary metal-oxide semiconductor
    (CMOS), 301, 302f
    for asynchronous processors, 507
    current flow in, 303
    multiple-valued logic for, 509
Complex instruction set computer (CISC),
    21, 71
    addressing modes and, 98, 100
    CPI for, 114
    embedded systems coprocessors and, 166
    instruction sets for, 193
    IPC for, 198
    vs. RISC, 81–83, 81f
    soft core processors and, 360
    speedups for, 174
Complex numbers, 112, 112n13
Compressed instruction sets, 95–97
Computer generations, 5–11
    first, 6–7
    second, 7–8

third, 8–9
    fourth, 9–10
    fifth, 10–11
Computer organization, 17–21, 20f
Concurrency, OS and, 411–413
Condition code register (CCR), 374, 375
Condition codes, in ARM, 91–92
Conditional flags, for pipelines,
    181–183
Console, UNIX, 383
Control packets, 452
Control signals, of buses, 256–257
Control unit, 74–79, 74f–78f
Controller area network (CAN), 70, 260
    PPU for, 499
Coprocessors, 21, 87, 158–159
    for embedded systems, 165–166
Cortex devices, 88
Counter registers, 214
CP/M, in fourth-generation computers, 9
CPI. *See* Cycles per instruction
CPU time, 397–398, 398f
CPUs. *See* Central processing units
CRC. *See* Cyclic redundancy check
Cross-compiling, 379–380
CS. *See* Carrier sense
CSD. *See* Canonical signed digit
Csh. *See* C-shell
Current flow, 302, 303–304
Cycle-by-cycle timing, 75–76, 75f
Cycles per instruction (CPI), 114–115
Cyclic redundancy check (CRC), 446–447,
    446f
    for booting, 406

D

DACs. *See* Digital-to-analog converters
DAGs. *See* Data address generators
DARPA. *See* Defense Advanced Research
    Projects Agency
Data abort, 143
Data address generators (DAGs), 204–207,
    207f
Data dependency hazard, for pipelines,
    180–182
Data flows, 107
    in asynchronous processors, 507–508
    power and, 306
    in UNIX, 385–387, 385f–387f

Data handling:
   by CPUs, 103–112
   for internal data, 108–109
   processing in, 109–112, 110*f*
   storage, 107–108
Data link layer (DLL), 454
Data memory address (DMA), 126–127
Data output register, of serial ports, 328
Data types, 103–107, 104*t*
   ASCII for, 105, 106*t*
   for Ethernet, 470–471
Data width, 25
DCP. *See* Dual core platform
DDR. *See* Double data rate
Deadline monotonic scheduling, 272
Debugging:
   for embedded systems CPUs, 295
   JTAG and, 333
   OS and, 392
DEC. *See* Digital Equipment Corporation
Decoding, 445
Defense Advanced Research Projects
      Agency (DARPA), 462
Delay lines, 306
Delay-locked loop (DLL), 295, 301
Delayed allocation, 422–423
Delayed branch, 188
Denormalized numbers, 53–54, 56–57
Device control, 403
Device manufacture test, 330
Dhrystone, 116
Digital Equipment Corporation (DEC), 7–8
   Alpha 21264, 223
   ARM and, 85
   parity memory for, 336
   StrongARM, 20, 85, 145
      PPU for, 498
   VAX, 275
Digital filters, for fractions, 46, 46*n*5
Digital signal processing (DSP), 29–30
   buses and, 125, 127–128, 128*f*
   cache memory and, 155
   CPUs pipelines and, 178
   fractions and, 46
   speeds of, 116
   superscalar architectures and, 194
   VLIW and, 248, 249
   ZOL and, 202
Digital-to-analog converters (DACs):
   DRAM and, 321
   RE of, 346

Dijkstra, Edsger, 413
DIP. *See* Dual in-line package
Direct cache, 146–148
Direct memory access (DMA), 108
   buses and, 257–258
   PC/104 standard and, 263
   wireless and, 284
Directories, of FS, 428–430, 430*f*, 431*f*
Directory look-aside table (DLT), 142
Disc image, for booting, 406
Distributed computing, 518
Distributed control, 76
Division, 44–46, 46*f*
   of fractional numbers, 48–49
   of IEEE754, 62–63
DLL. *See* Data link layer; Delay-locked loop
DLT. *See* Directory look-aside table
DMA. *See* Data memory address; Direct
      memory access
Domain Name Servers (DNS), 479–482,
      480*f*
Dongarra, Jack, 14*f*
DOS. *See* Microsoft Disk Operating System
Double data rate (DDR), 321
Double precision, 50
DRAM. *See* Dynamic RAM
DSP. *See* Digital signal processing
Dual-booting, 420–421
Dual-bus architecture, 129–130, 129*f*
Dual core platform (DCP), 231–232, 232*f*
Dual in-line package (DIP), 12, 262
Duplex, 438
Dynamic pipelines, 177–178, 178*f*
Dynamic priority, 416
Dynamic RAM (DRAM), 80
   addresses for, 318–321, 318*f*, 320*f*
   for embedded systems CPUs, 315–321,
      316*f*–318*f*, 320*f*
   fetch by, 117
   RE of, 346

E

EABI. *See* Embedded ABI
Earliest deadline first scheduling, 272
EBR. *See* Extended boot record
ECCs. *See* Error correcting codes
EDAC. *See* Error detection and correction
EDO. *See* Extended data out
EDR. *See* Enhanced Data Rate
EDSAC. *See* Electronic Delay Storage
      Automatic Calculator

EDVAC. *See* Electronic Discrete Variable Automatic Computer
EEPROM. *See* Electrically erasable PROM
EFLOP. *See* ExaFLOP
EIA. *See* Electronic Industries Alliance
EIA232, 267
EIDE. *See* Enhanced IDE
EISA. *See* Extended ISA
Electrical conductivity, RE of, 349–350
Electrically erasable PROM (EEPROM), 308–309, 309*f*
  RE of, prevention of, 356
Electro-optic backplane, 515–517, 516*f*
Electromagnetic interference (EMI), 285
  current flow and, 304
  electro-optic backplane and, 515
Electronic Delay Storage Automatic Calculator (EDSAC), 5
Electronic Discrete Variable Automatic Computer (EDVAC), 5
Electronic Industries Alliance (EIA), 267
Electronic Numerical Integrator and Computer (ENIAC), 3, 6, 7*f*
Electrotechnical Commission (IEC), 521–522
Embedded ABI (EABI), 376*n*2
Embedded cloud, 237
Embedded systems. *See also* Central processing units (CPUs), for embedded systems
  coprocessors for, 165–166
  data flows in, 107
  data types in, 104–105
  OS for, 395
  Tomasulo's algorithm for, 245–246
Embedded wireless connectivity, 282–285
EMI. *See* Electromagnetic interference
Emulation:
  OS and, 401
  of programs, 380
  in third-generation computers, 8
Encapsulation:
  for Ethernet, 471–473, 472*f*
  for TCP/IP, 465–469, 466*f*–468*f*
Enhanced Data Rate (EDR), 488
Enhanced IDE (EIDE), 263
ENIAC. *See* Electronic Numerical Integrator and Computer
Enigma code, 3, 7
EOR. *See* Exclusive-OR gate

EPIC. *See* Explicitly parallel instruction computing
Erasable programmable ROM (EPROM), 72, 308–309. *See also* Electrically erasable PROM
  JTAG and, 335
ERC32, 255
Error correcting codes (ECCs), 447–449, 449*f*
Error detection and correction (EDAC), 338, 339*f*
Error handling:
  for embedded systems CPUs, 336–343, 337*f*, 339*f*, 342*f*, 343*f*
  for networking, 446–451, 446*f*, 449*f*, 450*f*
  OS and, 401
  for wireless technology, 283
Ethernet, 284, 437, 445, 469–474, 469*f*
  CS for, 473–474
  data types for, 470–471
  encapsulation for, 471–473, 472*f*
Everywhere computing, 236
EX. *See* Execution unit
ExaFLOP (EFLOP), 14
Excess-*n*, 28–29
Exclusive-OR gate (EOR,XOR), 523
Execute-in-place (XIP), 372
  for Linux, 323
Execution unit (EX), 191
Explicitly parallel instruction computing (EPIC), 248
  IPC for, 199
Extended boot record (EBR), 421
Extended data out (EDO), 321
Extended ISA (EISA), 261, 262
  PCI and, 265
External fragmentation, of MMU, 140–142
External stimuli, 269
EXTEST, 334

F

Fast Fourier transform (FFT), 206
Fast interrupt request (FIQ), 277–278, 280
FDES. *See* Fetch, decode, execute, and store
Ferranti Mark 1, 5
Fetch, 91. *See also* Instruction fetch and decode
  ALU and, 209*n*5

Fetch (*Cont.*):
  control unit and, 74, 78
  by DRAM, 117
  DSP and, 128
  MMU and, 143
  PC and, 374
  superscalar architectures and, 194*f*, 197
Fetch, decode, execute, and store (FDES), 78,
    78*f*, 90*f*
  CPUs pipelines and, 175, 175*f*
FFD. *See* Full Function Device
FFT. *See* Fast Fourier transform
Field programmable gate array (FPGA), 42,
    42*n*4, 166, 234, 246
  embedded systems CPUs and, 323
  hardware software codesign of, 363–365,
    364*f*
  JTAG and, 336
  PPU for, 498–500
  RE for, 346, 346*f*, 350, 352
    prevention of, 356–357
  soft core processors and, 358–362
FIFO. *See* First-in first-out
File handling, in UNIX, 385–386
File system (FS):
  directories of, 428–430, 430*f*, 431*f*
  inodes of, 425–428, 428*f*
  JFS for, 424, 431, 432*f*
  OS and, 421–433
  superblocks of, 424–425, 426*f*
File transfer protocol (FTP), 465
Finite impulse response filter (FIR), 127
Finite-state machine (FSM), 74
FIQ. *See* Fast interrupt request
FIR. *See* Finite impulse response filter
Firewire, 268
First-in first-out (FIFO), 139, 151
  asynchronous processors and, 508
  interrupts and, 273
  temporal scope and, 271
First-level interrupt handler (FLIH), 403
Flash memory, 72
  booting and, 325, 406–407, 407*f*, 408*f*
  for data storage, 107
  for embedded systems CPUs, 298, 307,
    309–313, 313*f*
  JTAG and, 334–335
  for NAND, 309
  for NOR, 309
  RE of, 352

slowing of, 312–313
soft core processors and, 359
for storage, 418
FLIH. *See* First-level interrupt handler
Floating point emulator (FPE), 160–162
Floating point numbers, 49–58
  data processing of, 111–112
  IEEE754, 50–51, 58–64
  processing of, 58–64
  SSE and, 165
Floating point operations (FLOP), 13, 14*f*
Floating point unit (FPU), 21, 58, 159–162,
    161*f*
  on ARM, 160
  coprocessors for, 158
  data processing by, 112
  instruction sets and, 87
  MMX and, 164
  relative branching and, 187
  resource manager for, 396
  superscalar architectures and, 194,
    194*f*
  on VIA Isaiah, 294
  VLIW and, 249
FLOP. *See* Floating point operations
Floppy discs, 115
  backup system and, 433
Flowers, Tommy, 3
Flynn, Michael, 18, 19*f*
Flynn's taxonomy, 18
FORmula TRANslation (FORTRAN),
    112*n*13, 380
FPE. *See* Floating point emulator
FPGA. *See* Field programmable gate array
FPU. *See* Floating point unit
Fractional notation, 29–30
Fractional numbers:
  arithmetic with, 47–48
  division of, 48–49
  multiplication of, 48–49
  number formats for, 46–49
FreeBSD, 392, 393, 394
  triple-boot of, 420–421
FS. *See* File system
FSM. *See* Finite-state machine
FTP. *See* File transfer protocol
Full adder, 32, 32*f*
Full-associative cache, 149
Full Function Device (FFD), 489
Functional blocks, 18

G

GAN. *See* Global area network
GEC Plessey Semiconductors, 261
General Packet Radio Service (GPRS), 487
Generalized floating point, 49
GFLOPS, 115
Global area network (GAN), 490
Global branch predictor, 185, 217–219, 218*f*
Global positioning satellite (GPS), 238
    SEUs for, 336
Global System for Mobile Communications
    (GSM), 486–487
Glue logic, 362
GNU, 118
    UNIX and, 387
Google. *See also* Android
    ASR, 237
GPRS. *See* General Packet Radio Service
GPS. *See* Global positioning satellite
GPU:
    for parallel machines, 497
    resource manager for, 396
Grain size, for parallel machines, 504
Graphical user interface (GUI), 371
    directories for, 428
    OS and, 392
    UNIX, 383, 384, 385
grep, 387
Grid computing, 235, 237
Gselect branch predictor, 219–220, 220*f*
Gshare branch predictor, 221, 221*f*
GSM. *See* Global System for Mobile
    Communications
GUI. *See* Graphical user interface

H

HAL. *See* Hardware abstraction layer
Half-duplex, 438–439
Half-precision, 50
Hamming codes, 337, 338
    matrices for, 340
Handoff, for networking, 458
Hard deadlines, 270
Hard disc drive (HDD), 72
    backup system and, 433
    booting from, 405–406, 408–409, 420–421
    for data storage, 107
    FS and, 421–433
    MMU on, 136, 137*f*
    partitioning of, 419–420, 421*f*
    for storage, 417–420, 419*f*
    UNIX shell on, 384
Hard real-time system, 270
Hardware abstraction layer (HAL), 403
Hardware acceleration:
    address handling for, 204–207, 207*f*
    CPUs and, 201–208, 203*f*, 207*f*
    shadow registers for, 207–208
    ZOL for, 201–206, 203*f*
Hardware software codesign, for embedded
    systems CPUs, 363–365, 363*n*11, 364*f*,
    366*f*
Harvard architecture, 19
    DSP in, 128, 128*f*
HCLK, 259
HDD. *See* Hard disc drive
Hewlett-Packard, 101
Hierarchical networks, 455–456, 456*f*
High-level languages (HLL), 86
    VLIW and, 250
High Speed Packet Access (HSPA), 487
HLL. *See* High-level languages
Homogeneous architectures, 234
Host, for cross-compiling, 380
HSPA. *See* High Speed Packet Access
HTML. *See* Hypertext markup language
HTTP. *See* Hypertext transfer protocol
Hubs, 473
Huffman coding, 95–97
Hybrid branch predictors, 222–223, 222*f*,
    223*f*
Hyperblocks, 226
Hypercubes, 503
Hypertext markup language (HTML), 383,
    464
Hypertext transfer protocol (HTTP), 464
    DNS and, 480

I

I/O. *See* Input/output
IAB. *See* Internet Architecture Board
IBM:
    ARM and, 83
    Cell processor, 234–235
    FORTAN, 112*n*13
    ISA, 262
    JFS, 424
    MCA, 261
    PowerPC RISC processor, 234

IBM (*Cont.*):
 speculative execution by, 186
 System/360, 8–9, 9f, 80
ICE. *See* In-circuit emulator
ICMP. *See* Internet control message protocol
ICs. *See* Integrated circuits
IDE. *See* Integrated development
  environment; Integrated drive
  electronics
Idle tasks, 272
IEC. *See* Electrotechnical Commission
IEEE754, 21
 addition of, 59–62
 arithmetic for, 59–62
 denormalized numbers, 53–54, 56–57
 division of, 62–63
 floating point numbers, 50–51,
  58–64
 FPE and, 160–161
 FPU and, 159
 modes, 51–55
 multiplication of, 62–63
 normalized mode, 52–53, 55–56
 number ranges, 55–58
 rounding of, 63–64
 subtraction of, 59–62
IEEE802, 284, 437, 467, 487
IEEE1149 JTAG, 295
IEEE1284, 267
IEEE1394, 268
IETF. *See* Internet Engineering Task Force
IFU. *See* Instruction fetch and decode
IIC. *See* Inter-IC Communications
IIR. *See* Infinite impulse response filter
ILP. *See* Instruction-level parallelism
IMAP. *See* Internet mail access protocol
Immediate constant, ARM and, 93–95
In-circuit emulator (ICE), 333
Index:
 buses and, 126–127
 in registers, 109
Industrial, scientific, and medical band
  (ISM), 488
Industry standard architecture (ISA),
  262–263
 PCI and, 265
Infinite impulse response filter (IIR),
  127
Infinity, 54
Inodes, 425–428, 428f

Input/output (I/O), 73, 77. *See also* Basic
  input output stream
 addresses, PCI and, 265
 buses, 267
 for embedded systems CPUs, 295,
  329–330
 OS and, 392
 SATA and, 329
 on Texas Instruments MSP430, 295–297
Instruction cycle, 22
 interrupts and, 275
Instruction fetch and decode (IFU), 90–95
 branching and, 92–93
 immediate constant and, 93–95
Instruction handling:
 addressing modes and, 97–101
 by CPUs, 86–102
 Huffman coding and, 95–97
 IFU and, 90–95
 RPN and, 100–102
 stack processing and, 100–102, 102f
Instruction-level parallelism (ILP), 228, 246
Instruction queue (IQ), for Tomasulo's
  algorithm, 240
Instruction sets:
 ARM and, 87–89, 88f, 91
 for CISC, 193
 compressed, 95–97
 for CPUs, 86–90, 88f
 for pipelines, 188–189
 for RISC, 193
Instruction translation, 81
Instructions per cycle (IPC):
 for CPUs, 198–201
 in parallel machines, 233
Integrated circuits (ICs). *See also* Printed
  circuit boards
 ARM and, 83, 85
 buses on, 257
  pin swapping of, 352–353
 in computer organization, 17
 in CPUs, 76
 design and manufacture problems of,
  328–331, 330f
 in fourth-generation computers, 9
 in multicore machines, 234
 in parallel machines, 233
 propagation delay and, 303
 of Samsung S3C2410, 292
 in third-generation computers, 8

Integrated development environment (IDE), 382–383
Integrated drive electronics (IDE), 73
Intel, 14
  DIP, 262
  IXP425, 258
  MMX, 162–163
  Pentium, 163
  Pentium Pro, 145
  performance of, 115
  StrongARM, 20, 85, 145
    PPU for, 498
  transistors and, 12
  XScale, 145, 279
Intel 80486, FDIV bug in, 328
Inter-IC Communications (IIC), 260
Interfaces. *See also specific types*
  for networking, 268, 269f
  wireless technology and, 284
Internal data, 108–109
Internal fragmentation, of MMU, 140
International Organization for
    Standardization (ISO), 454
Internet, 461–464
  cables for, 469
  governance of, 462–463
  history of, 462
  RFC for, 463–464
Internet Architecture Board (IAB), 463
Internet control message protocol (ICMP), 486
Internet Engineering Task Force (IETF), 463
Internet layer, 465, 466, 474–482
Internet mail access protocol (IMAP), 464
Internet Protocol (IP), 282, 284. *See also* TCP/IP
  addresses, 474–476, 474n5
    ARP for, 485–486
    naming, 478
  layer, 464–469, 464f
  routing, 477
Internet Research Task Force (IRTF), 463
Internet service provider (ISP), 477
  DNS for, 481
Interpreted languages, 381–383, 382f
Interrupt request (IRQ), 280
  interrupts and, 276, 277–278
  PC/104 standard and, 263

Interrupt service routine (ISR), 208, 278
  interrupts and, 276
  OS and, 273
Interrupts:
  branching, 276–278
  CPU and, 269–270, 273–282
  FLIH for, 403
  redirection of, 278–279
  reentrant code for, 281
  sharing, 280–281, 281f
INTEST, 334
Inverse of CPI (IPC), 115
IP. *See* Internet Protocol
IPC. *See* Instructions per cycle; Inverse of CPI
IQ. *See* Instruction queue
IRQ. *See* Interrupt request
IRTF. *See* Internet Research Task Force
ISA. *See* Industry standard architecture
ISM. *See* Industrial, scientific, and medical band
ISO. *See* International Organization for Standardization
ISP. *See* Internet service provider
ISR. *See* Interrupt service routine

J

Java, 103, 380
  embedded systems coprocessors and, 166
Javascript, 383
Jazelle, 166
JEDEC, 347
JFFS2. *See* Journaling flash filing system
JFS. *See* Journaled FS
JIS, 347
Joint Test Action Group (JTAG), 295, 333–336, 334f
  for booting, 335
Journal checksumming, 423
Journaled FS (JFS), 424, 431, 432f
Journaling flash filing system (JFFS2), 424
JTAG. *See* Joint Test Action Group

K

Kernel:
  for booting, 406
    for embedded systems CPUs, 325–326
  of OS, 402–403
Korn Shell (kshell), 384

L

Large-scale integrated devices (LSI), 347
Last-in first-out (LIFO), 101
Lattice semiconductor, 366
Layered OS, 402–404, 404*f*
LC. *See* Logic cell
LCD. *See* Liquid crystal display
LDM. *See* Load multiple registers
LDR. *See* Load register
LE. *See* Logic element
Least frequently used (LFU), 151
Least recently used (LRU), 139, 151
Least significant bit/byte (LSB), 23–25, 494,
    495, 495*f*
    for fractional notation, 29
    for subtraction, 35
LED:
    BCD for, 29
    power for, 306
    temporal scope and, 272
LFU. *See* Least frequently used
LIFO. *See* Last-in first-out
LINE, 146
Line-of-sight (LOS), for WiMax, 487–488
Link register (LR), 375
Linkers, compilers and, 377–379, 378*f*, 379*f*
Linux, 392, 393
    Beowulf, 235
    booting on, 406
    compiled languages, 377
    directories of, 430, 430*f*
    embedded systems for, 395
    FS of, 422
    kernel of, 406
    language and, 143
    as layered OS, 404*f*
    memory of, for embedded systems CPUs,
        323–324
    scheduling on, 416
    shell, 384
    triple-boot of, 420–421
    XIP for, 323
Liquid crystal display (LCD), 29, 292
LITTLE, 238–239, 239*n*13
Little endian, 22–25
LLC. *See* Logical link control
Load multiple registers (LDM), 193
Load register (LDR), 74
Load-store, 73
    addressing modes and, 98
    instruction sets and, 193

Load store unit (LSU), 242–245, 245*t*
Local branch predictor, 185, 214–217, 215*f*
Logic cell (LC), 496
Logic element (LE), 496
Logic gates, 523–524
Logic propagation delay, 174
Logical link control (LLC), 465
Look-up tables (LUTs), 142, 154, 366
LOS. *See* Line-of-sight
Low-level scheduler, 403
Low-voltage differential signaling (LVDS),
    261–262, 265–267, 266*f*
    PPU for, 500
Locality principles, for cache memory,
    149–150
LR. *See* Link register
LRU. *See* Least recently used
LSB. *See* Least significant bit/byte
LSI. *See* Large-scale integrated devices
LSU. *See* Load store unit
LUTs. *See* Look-up tables
LVDS. *See* Low-voltage differential signaling

M

MA. *See* Multiple access
MAC. *See* Media access controller;
        Multiply-accumulate controller
Machine code, 71
Machine parallelism, 228
MacOS, 393, 394
    dual-booting of, 420
Mailboxes, 270
Manchester University, 5, 507
Marconi Ltd., 261
MareNostrum 4, 12, 13*f*, 15
Mask-programmed gate array (MPGA),
    356
Mass storage, 417–418
Massachusetts Institute of Technology, 5
Master boot record (MBR), 420–421
MCA. *See* Microchannel architecture
MCM. *See* Multichip module
Mean time before failure (MTBF), 6
    for OS, 393
Media access controller (MAC), 268, 269*f*
    addresses:
        ARP for, 485–486
        for WiFi, 487
    DMA and, 258
    network layer and, 465
    superscalar architectures and, 194

Media-independent interface (MII), 268

MEMC, 83

Memory. *See also specific types*

  addressing modes and, 99

  of ARM, 372–373, 373*f*

    for embedded systems CPUs, 323–324, 325*f*

  for booting, 405

  bootloader and, 279–280

  bytes and, 107

  in CPUs, 114

  DSP and, 128

  for embedded systems CPUs, 295, 306–321, 323–328, 324*f*

  EMI and, 285

  hierarchy of, 71–72, 72*f*

  IPC and, 201

  of Linux, for embedded systems CPUs, 323–324

  maps:

    for Texas Instruments MSP430, 326–327

    WDTs for, 341

  OS and, 398–400, 399*f*

  pages of, 137

    for embedded systems CPUs, 321–323

  power and, 306

  silicon for, 307

  standard size notation for, 521–522

  variables in, 108

  VLIW and, 249

Memory management unit (MMU), 21–22, 72, 136–145, 137*f*, 138*f*, 141*f*, 142*f*

  advanced, 142

  cache memory and, 144

  for data storage, 108

  direct cache and, 146

  for embedded systems CPUs, 321–323

  external fragmentation of, 140–142

  internal fragmentation of, 140

  memory protection in, 143–144

  RAM and, 22, 137–140

  retirement algorithms for, 139–140

  of Samsung S3C2410, 292

  segmentation of, 140

  virtual memory and, 136

Memory reorder buffer (MOB), 294

Mesh, for parallel machines, 503

MESI protocol, 156–158, 157*f*

Message passing interface (MPI), 503

MFLOPS. *See* Millions of floating point operations per second

Microchannel architecture (MCA), 261

Microcode, 79–81, 80*f*

Microkernel, 405, 405*f*

Microsoft Disk Operating System (MS-DOS), 394

  directories of, 430, 431*f*

  in fourth-generation computers, 9

  partitioning of, 421

  prompt, 384

  reset supervisors for, 341

Microsoft Windows, 393

  directories of, 430, 431*f*

  multi-tasking on, 416

  triple-boot of, 420–421

  VM, 394

MII. *See* Media-independent interface

Millions of floating point operations per second (MFLOPS), 115

Millions of instructions per second (MIPS), 71, 114

  for embedded systems CPUs, 337

  RISC and, 82

MIMD. *See* Multiple instruction stream, multiple data

MIPS. *See* Millions of instructions per second

MISD. *See* Multiple instruction stream, single data

Mitsubishi, 248, 250

MMC. *See* Multimedia card

MMU. *See* Memory management unit

MMX. *See* Multimedia extensions

MOB. *See* Memory reorder buffer

Mobile multiprocessing, 238

Mobile networking, 458

Modeling, for hardware software codesign, 365

Modulo-2, 338*n*4, 339

Monolithic OS, 402

Moore, Gordon, 1*n*1

Moore's law, 1, 1*n*1, 233

MOS Technology Inc., 12

Most important first scheduling, 272

Most significant bit/byte (MSB), 23–25, 214

  for IEEE754 normalized mode, 56

  for sign extension, 31

  for sign magnitude, 26

  for subtraction, 36

  for two's complement, 27

Motorola 68000, 71, 204
MOV, 93–94
MP3, 118
    DCP for, 231
MPGA. *See* Mask-programmed gate array
MPI. *See* Message passing interface
MS-DOS. *See* Microsoft Disk Operating
    System
MSB. *See* Most significant bit/byte
MTBF. *See* Mean time before failure
MUL:
    superscalar architectures and, 194*f*, 196
    VLIW for, 246
Multiple-issue superscalars, 197
Multitasking:
    on Microsoft Windows, 416
    in third-generation computers, 8
Multicasting, 478
Multichip module (MCM), 234
Multicore machines, 234
Multifunction pipelines, 176–177, 177*f*
Multimedia card (MMC), 268
Multimedia extensions (MMX), 162–165,
    163*f*
    coprocessors for, 158–159
    embedded systems coprocessors and,
        165
    FPU and, 159–160
    SIMD and, 228
Multiple access (MA), 473–474
Multiple instruction stream, multiple data
    (MIMD), 18, 19*f*, 227
    interrupts and, 273
    SISD to, 230–232, 231*f*
Multiple instruction stream, single data
    (MISD), 18, 19*f*, 227
Multiple-valued logic, 509–510
Multiplication, 37–44
    of fractional numbers, 48–49
    of IEEE754, 62–63
    partial products for, 38–41, 39*f*, 41*f*
Multiply-accumulate controller
    (MAC), 87
    buses and, 125, 125*f*, 126*f*
Multiuser systems, 394
Mutual exclusion, 413

N

N (negative) flag, for pipelines, 182
NACK protocol, 449–451, 450*f*
    TCP and, 484

NaN. *See* Not a number
NAND, 418, 523
    CMOS for, 301, 302*f*
    flash memory for, 309
nCAS. *See* Column address strobe
NDA. *See* Nondisclosure agreement
Near-field communications (NFC), 489
NEON-advanced SIMD, 166
NetBIOS, 465
Network file system (NFS), 465
Network layer, 465, 466–567
Networking, 437–458, 461–490. *See also*
    Ethernet; Internet; Wireless
    ad hoc, 457–458
    addresses for, 452–453
    arbitration for, 451–452
    architectures, 441, 455–458
    client-server architecture for, 456, 457*f*
    control packets for, 452
    decoding for, 445
    ECCs in, 447–449, 449*f*
    encoding for, 445
    error handling in, 446–451, 446*f*, 449*f*, 450*f*
    handoff for, 458
    hierarchical, 455–456, 456*f*
    interfaces for, 268, 269*f*
    management of, 451–453
    mobile, 458
    for one-to-many communications,
        439–440, 439*f*, 440*f*
    for one-to-one communications, 438–439,
        438*f*
    OSI for, 454–455, 454*f*
    packet switching for, 440–441, 441*f*
    packetization for, 443–445
    peer-to-peer architecture for, 456–457
    receivers for, 445–446
    scales for, 453, 490
    system requirements for, 443–453
    topologies for, 441–443, 442*f*, 443*f*, 455–458
    transmitters for, 445
    UNIX for, 383–384
NFC. *See* Near-field communications
NFS. *See* Network file system
nGCS, 259–260
Nice, 416
No operation (NOP), 177, 186
    CPUs pipelines and, 183
    interrupts and, 276–277
    for serial ports, 328
    VLIW and, 249

Non-volatile memory (NV), 405, 417
  booting from, 325–327
  RE for, 350, 351
Nondisclosure agreement (NDA), 347
NOP. *See* No operation
NOR, 418, 523
  flash memory for, 309
Normalized mode, 52–53, 55–56
NOT, 523
  ARM and, 85*n*5
  CPUs pipelines and, 189
Not a number (NaN), 54
nRAS. *See* Row address strobe
NRE, 358
Number formats, 25–31
  alternatives for, 509–513
  for fractional numbers, 46–49
NV. *See* Non-volatile memory
nWAIT, 257, 259

O

Octuple precision, 50
OCXO. *See* Oven-controlled crystal oscillators
OFDM. *See* Orthogonal frequency division multiplexing
Offset:
  for addresses, 90, 187
  ALU and, 224
  in branching, 92–93
  for superblocks, 425
One's complement, 27
Open, 400
Open systems interconnections (OSI), 21, 261, 454–455, 454*f*
  IP layer, 464, 464*f*
Operating system (OS), 391–433. *See also* Real-time operating system; *specific operating systems*
  availability of, 394
  backup system of, 431–433
  BIOS and, 21
  booting and, 405–410, 407*f*, 408*f*
  client-server architecture for, 402, 404–405, 405*f*
  for cloud computing, 395
  for cluster computers, 395
  concurrency and, 411–413

CPU and, 393, 411–413
CPU time and, 397–398, 398*f*
directories of, 428–430, 430*f*, 431*f*
DNS for, 480
for embedded systems, 395
for embedded systems CPUs, 323
emulation and, 401
error handling and, 401
in fourth-generation computers, 9
FS and, 421–433
interrupts and, 279
ISR and, 273
kernel of, 402–403
layered, 402–404, 404*f*
memory and, 398–400, 399*f*
for multiuser systems, 394
mutual exclusion for, 413
processes on, 410–413, 412*f*
programming for, 371
reliability of, 393–394
resource manager for, 396
safety of, 394
scheduling and, 413–417
for single-user systems, 394
storage and, 400–401, 417–433, 419*f*
structure of, 402–405
SWI and, 281
synchronization of, 413
task-switching by, 397–398, 397*n*1
in third-generation computers, 8
types of, 395–396
for VM, 394, 396–397
WDTs for, 341
Optical computation, 514–517
OR, 134
  ALU for, 21
  control unit and, 79
  for subtraction, 35, 35*f*
Orthogonal frequency division multiplexing (OFDM), 487
OS. *See* Operating system
OSI. *See* Open systems interconnections
Oslo PDP-7, 8*f*
Oven-controlled crystal oscillators (OCXO), 301
Overflow (V) flag, for ALU, 134, 182

P

Packet latency, 441–442
Packet switching, 440–441, 441*f*

Packetization:
   in application layer, 465
   for Internet, 476–477
   for networking, 443–445
Pages, memory, 137
   for embedded systems CPUs, 321–323
PANs. *See* Personal area networks
Parallel adder, 32–34
Parallel buses, 258–260
Parallel machines, 227–239
   of future, 497–505, 499*f*, 501*f*–503*f*
   IPC for, 199–200
   for performance, 232–234
Parallel ports, 107
Parallel processing unit (PPU), 498–501, 499*f*,
   501*f*
Parity memory, 336
Partial products, for multiplication, 38–41,
   39*f*, 41*f*
Partitioning:
   for hardware software codesign, 365
   of HDD, 419–420, 421*f*
PATA, 263
PC. *See* Personal computer; Program counter
PC/104 standard, 263–265, 264*t*
PCB. *See* Process control block
PCBs. *See* Printed circuit boards
PCI. *See* Peripheral component interconnect
PCI express (PCIe), 265
   LVDS and, 266–267
PCMCIA. *See* Personal Computer Memory
   Card International Association
PDP-1, 7–8
Peer-to-peer architecture, for networking,
   456–457
Peripheral component interconnect (PCI),
   262, 265
Peripheral device buses, 267–268
Personal area networks (PANs), 488
Personal computer (PC), 12
Personal Computer Memory Card
   International Association (PCMCIA),
   268
Pervasive computing, 236–237
PFLOPS, 115
Phase-locked loops (PLLs), 294, 295, 300–301
Philips, 248, 250
PHP, 383
PHY. *See* Physical layer driver
Physical address, 21–22

Physical layer driver (PHY), 268, 464,
   466–567
PIC, 71
PIDs. *See* Process identifiers
Pipelines:
   asynchronous processors and, 508
   branching for, 183–185
   compile time and, 185–186
   conditional flags for, 181–183
   of CPUs, 174–192, 175*f*, 177*f*, 178*f*, 180*f*,
     184*f*, 190*t*, 191*f*, 192*f*
   data dependency hazard for, 180–182
   dynamic, 177–178, 178*f*
   instruction sets for, 188–189
   interrupts and, 275–276
   multifunction, 176–177, 177*f*
   in parallel machines, 233
   run time, 189–192, 191*f*, 192*f*
   speedups with, 176
Piz Daint, 505
PLDs. *See* Programmable logic devices
PLLs. *See* Phase-locked loops
PMA. *See* Program memory address
PMs. *See* Processing modules
PNs. *See* Processing nodes
POP:
   interrupts and, 276
   RPN and, 102
POP3. *See* Post office protocol version 3
Port number, for transport layer, 482–483,
   483*f*
POST. *See* Power on self-test
Post office protocol version 3 (POP3), 464,
   465
Power:
   clocks and, 301–306, 302*f*, 305*f*
   for CPUs, 301–306, 302*f*, 305*f*
   for DRAM, 316
   for embedded systems CPUs, 295
Power on self-test (POST), 325
Power processing element (PPE), 234–235,
   236*f*
PPE. *See* Power processing element
PPU. *See* Parallel processing unit
Prefetch abort, 143
Printed circuit boards (PCBs), 329
   for electro-optic backplane, 515–516
   RE of, 344–345, 345*f*, 348–349, 349*f*
Pro-Electron, 347
Probabilistic branching, 186

Process control block (PCB), 403, 349*t*
  current flow in, 303
Process identifiers (PIDs), 416
Process switching, 416–417
Processes, on OS, 410–413, 412*f*
Processing modules (PMs), 499
Processing nodes (PNs), 499, 501*f*
Program counter (PC), 80, 84
  branch prediction for, 209
  BTB for, 224
  fetch and, 374
  interrupts and, 276, 280
  registers and, 124*n*1
  relative branching and, 187
  ZOL and, 203–204
Program memory address (PMA), 126–127
Programmable logic devices (PLDs), RE for, 350
  prevention of, 356
Programs, 371–388
  calling process for, 375–376
  compilers for, 377–381, 377*f*–379*f*
  emulation of, 380
  IDE for, 382–383
  interpreted languages for, 381–383, 382*f*
  passive obfuscation of, 355, 356*t*
  RE of, 351–353
    prevention of, 355, 356*t*
  registers for, 375–376
  running, 372–376
  storage for, 70–71
  transfer of, 73–74, 73*n*2
  for UNIX, 383–388
  writing, 376–383
Propagation delay, 302, 303
Punched cards:
  program storage and, 71
  in second-generation computers, 7
PUSH:
  ARM and, 98
  RPN and, 101

Q

Q-format, 29, 46
  FPE and, 162
Quadruple precision, 50
Queues, 270

R

Radix, 29, 29*n*3

Random-access memory (RAM), 72, 73, 80.
    See also Dynamic RAM; Static RAM;
    Synchronous dynamic RAM
  cache memory and, 154
  for data storage, 108
  direct cache and, 147–148
  for embedded systems CPUs, 314–321
  flash memory and, 312–313, 313*f*
  interrupts and, 279
  MMU and, 22, 137–140
  programming and, 372
  soft core processors and, 359
  virtual memory and, 136
  in von Neumann architecture, 19
Rate monotonic scheduling, 272
RAW. *See* Read-after write
RE. *See* Reverse engineering
Read, 400–401
Read-after write (RAW), 181, 191, 196
Read-only memory (ROM), 80. *See also*
    Electrically erasable PROM; Erasable
    programmable ROM
  for embedded systems CPUs, 298,
    307–313, 308*f*
Real-time clock module (RTC), 297
Real-time issues, 269–273
  packetization for, 444
Real-time operating system (RTOS), 270
  for embedded systems CPUs, 367
  interrupts and, 279
  memory for, 323–324
  SWI and, 281
  temporal scope and, 271
Reassembly, 443
Receivers, for networking, 445–446
Redirections, in UNIX, 385–387
Reduced Function Device (RFD), 489
Reduced instruction set computer (RISC),
    21, 71, 100
  addressing modes and, 98
  CPI for, 114
  DMA and, 258
  embedded systems coprocessors and, 165,
    166
  instruction sets for, 193
  interrupts and, 275
  IPC for, 198–199
  soft core processors and, 360
  speedups for, 174
  VLIW for, 246, 247*f*, 249

Reed-Solomon encoding (RS), 337, 338
Reentrant code, for interrupts, 281
Refresh process, for DRAM, 315
Registers, 22, 22*n*2, 71*n*1, 72, 73, 73*n*2. *See also*
    specific types
    ADSP2181 and, 99
    of ALU, 21, 298
    ARM and, 99, 124
    branch prediction and, 214
    C (programming language) and, 327
    control unit and, 74–75, 74*f*, 77
    DAGs and, 204–205
    for data storage, 108
    index in, 109
    instruction sets and, 193
    MMX, 162, 163*f*, 164
    PC and, 124*n*1
    for programs, 375–376
    SSE and, 164
    for subtraction, 46
ReiserFS, 424
Relative address, 93
Relative branching, 187–188
Reliability, of OS, 393–394
Renice, 416
Reorder buffer (ROB), 294
Repeated addition, or multiplication, 38
Repeated subtraction, for division, 44–46
Replacement algorithms, for cache memory,
    150–154
Request for comment (RFC), 463–464
Reservation stations (RS), 240–241, 245*t*
Reservation tables, 175, 179, 183, 190*t*
    for Tomasulo's algorithm, 242, 243*f*, 245*f*
    trace tables and, 211, 211*n*7
Reset circuitry, for embedded systems CPUs,
    295
Reset supervisors, 341–343, 342*f*, 342*n*5, 343*f*
Resource manager, 396
Retirement algorithms, 139–140
Reverse engineering (RE):
    active mitigation of, 357–358
    BOM for, 347–348
    defined, 344*n*7
    of electrical conductivity, 349–350
    for embedded systems CPUs, 343–358,
        345*f*, 346*f*, 349*t*, 356*t*, 358*t*
    functionality for, 346
    physical structure analysis for, 346–347
    prevention of, 353–358

process of, 344–348
    of programs, 351–353
        prevention of, 355, 356*t*
    system architecture analysis for, 348
Reverse Polish notation (RPN), 100–102
RFC. *See* Request for comment
RFD. *See* Reduced Function Device
Rings, for parallel machines, 503
Ripple-carry:
    adder, 33
    IEEE754 normalized mode and, 56
RISC. *See* Reduced instruction set computer
ROB. *See* Reorder buffer
Robertson's method, 42–44
Rockwell 6502, 12
ROM. *See* Read-only memory
Root directory, 428–430, 430*f*
Round robin:
    cache memory, 152
    scheduling, 272
Rounding, of IEEE754, 63–64
Routers and routing, 442
    for Ethernet, 472
    IP, 477
Row address strobe (nRAS), 318, 318*f*
RPN. *See* Reverse Polish notation
RS. *See* Reed-Solomon encoding;
    Reservation stations
RS232, 267
RTC. *See* Real-time clock module
RTOS. *See* Real-time operating system
Run time, 189–192, 191*f*, 192*f*

S

Safety, of OS, 394
Samsung:
    Galaxy smartphones, 15*f*
    S3C2410, 291–292, 298
        JTAG for, 335
        soft core processors and, 359
SAN. *See* Storage area network
SATA. *See* Serial ATA
Scheduling:
    low-level scheduler, 403
    OS and, 413–417
    of tasks, 270–272, 271*f*
Scoreboarding, for WAW, 196
SD. *See* Secure digital; Signed digit
SDR. *See* Software-defined radio
SDRAM. *See* Synchronous dynamic RAM

Secure digital (SD, xD), 268, 312
Secure socket layer (SSL), 383
sed, 387
Segmentation:
    of MMU, 140
    packetization and, 443
Segmentation fault error, 143
Self-timed control, 76, 77–78
Semaphores, 270, 413
Separate the condition-setting instruction
        (SUBS), 184, 186, 189
Serial ATA (SATA), 262, 263, 329
Serial buses, 18
Serial flash, 309, 310f
Serial Peripheral Interconnect (SPI), 260
Serial ports:
    bytes and, 107
    data output register of, 328
    NOP for, 328
    UART, 192
Server farms, 228
Set-associative cache, 148–149
SEUs. See Single event upsets
Shadow registers, 207–208
Shift-add method, for multiplication, 42
Shift register, 214
Shoubu, 505
SI. See System International
Sign extension, 30–31
Sign magnitude, 26
Signed digit (SD), 510–513, 511f, 512f
Silicon:
    for asynchronous processors, 507
    for memory, 307
SIMD. See Single instruction stream,
        multiple data
Simple mail transfer protocol (SMTP), 465
Simplification, 76
Sinclair, 115
    ZX Spectrum, 285
Sinclair, Clive, 83
Single-bit architectures, 494
Single-bus architectures, 130f, 131–132, 132f
Single-cycle execution, 193
Single event upsets (SEUs), 336
    multiple-valued logic and, 509
Single instruction stream, multiple data
        (SIMD), 18, 19f, 227
    hardware acceleration and, 201
    MMX and, 228

NEON, 166
    on VIA Isaiah, 294
Single instruction stream, single data (SISD),
        18, 19f, 227
    to MIMD, 230–232, 231f
Single precision, 50
Single T-bit branch prediction, 211–212
Single-user systems, 394
SISD. See Single instruction stream, single
        data
Smartphones, 15f
    with Android, 393
    ARM in, 15–16
    bit width of, 113–114
    mobile multiprocessing by, 238
SMP. See Symmetrical multi-processing
SMTP. See Simple mail transfer protocol
SoC. See System on chip
Soft core processors, 358–363
    availability of, 361
Soft deadlines, 270
Soft real-time system, 270
SoftBank, 5
Software. See Programs
Software-defined radio (SDR), 323
Software interrupt (SWI), 280, 281–282
Solid-state drive (SSD):
    booting from, 405–406, 408–409
    for data storage, 107
    for storage, 417–420
Sony:
    Cell processor, 234
    Playstation, 234
SP. See Stack pointer
SPARC:
    ERC32 and, 255
    RISC and, 82
Spatial locality, for cache memory, 149–150
SPE. See Synergistic processing elements
SPEC. See Standard Performance Evaluation
        Corporation
SPECfp, 116
SPECint, 116
Speculative execution, 183, 186–187, 210
Speech recognition, 237
SPI. See Serial Peripheral Interconnect
Spill code, 109
    for embedded systems CPUs, 328
SQUID. See Superconducting quantum
        interface device

SR. *See* Status register
SRAM. *See* Static RAM
SSD. *See* Solid-state drive
SSE. *See* Streaming SIMD extensions
SSEM, 5
SSL. *See* Secure socket layer
Stack pointer (SP), 375
Stacks, 375–376
  processing of, 100–102, 102*f*
    addressing modes and, 100–101
Stall time, 139
Standard error, 385
Standard input, 385
Standard interfaces, 260–268
Standard logic gates, 523–524
Standard output, 385
Standard Performance Evaluation
  Corporation (SPEC), 116
Star network, 441–443
Static RAM (SRAM), 80, 117
  DRAM and, 316–317
  DSP and, 128
  for embedded systems CPUs, 314–315,
    315*f*
  RE of, 346
    prevention of, 356
Status register (SR), 499
STM. *See* Store multiple registers
Storage:
  of data, 107–108
  delay lines for, 307
  flash memory for, 418
  OS and, 400–401, 417–433, 419*f*
  for programs, 70–71
  RE for, 350
Storage area network (SAN), 490
Store multiple registers (STM), 193
Streaming SIMD extensions (SSE), 164–165
  coprocessors for, 158–159
  embedded systems coprocessors and,
    165
  FPU and, 159
  instruction sets and, 87
SUB:
  ARM and, 85
  control unit and, 77, 79
  interrupts and, 276
Subnormal numbers, 51, 53–54
SUBS. *See* Separate the condition-setting
  instruction

Subtraction, 34–37
  OR for, 35, 35*f*
  ALU for, 21
  of IEEE754, 59–62
SUN Microsystems:
  PicoJava by, 174
  speedups by, 174
Sunway TaihuLight, 504
Superblocks, 226
  of FS, 424–425, 426*f*
Supercomputers, 12
Superconducting quantum interface device
  (SQUID), 358
SuperIO chip, 329, 330*n*3
Superscalar architectures:
  for CPUs, 194–198, 194*f*, 195*t*, 198*f*
  IPC for, 199
  multiple-issue, 197
  performance of, 197–198
  VLIW and, 249, 250
Supervisor, 273
  SWI and, 281
SWI. *See* Software interrupt
Switches, 473
Symbol errors, 446
Symmetrical multi-processing (SMP), 234
Synchronization, of OS, 413
Synchronous dynamic RAM (SDRAM), 80,
  321
  booting and, 325
  DSP and, 128
  for embedded systems CPUs, 298
  interrupts and, 278
  RE of, 346
  of Samsung S3C2410, 292
Synergistic processing elements (SPE),
  234–235, 236*f*
System/360, of IBM, 8–9, 9*f*
System control interfaces, 260
System data buses, 260–267, 264*t*, 266*f*
System International (SI), 521
System on chip (SoC), 292–293, 292*n*1
  clocks for, 298–306, 299*t*, 300*f*
  required functionality for, 295–298
  reset supervisors for, 342

T

TAG, 146, 149
Take/do not take bit (TDTB), 210, 299*t*, 210*n*6

Target, for cross-compiling, 380
Task parallelism, 228
Task-switching, 397–398, 397*n*1
Tasks:
  scheduling of, 270–272, 271*f*
  temporal scope for, 270–272, 271*f*
TB. *See* Translation buffer
TCK, 334
TCP. *See* Transmission control protocol
TCP/IP, 457, 464–469
  encapsulation for, 465–469, 466*f*–468*f*
TDI, 334
TDTB. *See* Take/do not take bit
Temporal locality, for cache memory, 149–150
Temporal scope, for tasks, 270–272, 271*f*
Texas Instruments:
  DSP code of, 81
  MSP430:
    I/O on, 295–297
    memory map for, 326–327
  TMS320, 84, 202
    interrupts and, 276
  VLIW of, 248
TGB. *See* Time-slotted global back plane
Throughput, 175
  VLIW and, 248
Thumb, 88
  ARM and, 96
Time-slotted global back plane (TGB), 499–500
Time-to-live (TTL), 476–477
TMR. *See* Triple module redundancy
Tomasulo's algorithm, 239–246, 241*f*, 243*t*, 244*f*, 245*t*
Toshiba, 234
Trace tables, 211, 211*n*7
Transistors:
  Bell Labs and, 4–5
  Intel and, 12
  Moore's law and, 1*n*1
  in second-generation computers, 7
  on VIA Isaiah, 294
Translation buffer (TB), 142
Transmission control protocol (TCP), 485*f*.
    See also TCP/IP
  DNS for, 479
  port number for, 482–483, 483*f*
  vs. UDP, 484–485
Transmitters, for networking, 445

Transport layer, 465, 466, 482–485, 483*f*
Trees, for parallel machines, 503
Triple-boot, 420–421
Triple module redundancy (TMR), 337, 337*f*
Tristate buffer, 21, 75, 75*n*3, 77
TTL. *See* Time-to-live
TWI. *See* Two-wire interface
Two-bit branch prediction, 212–214, 213*f*
Two-wire interface (TWI), 260
Two's complement, 27
TX-0, 5

U

UART. *See* Universal asynchronous receiver/transmitter
UBIFS. *See* Unsorted block image FS
uCLinux, 323
uCOS, 393
  VM, 394
UDP. *See* User datagram protocol
Unicasting, 478
Unicode, 105
Universal asynchronous receiver/transmitter (UART), 192, 297
Universal serial bus (USB):
  backup system and, 433
  peripheral device buses, 267–268
  UNIX, 383
  wireless technology and, 282
Universal synchronous/asynchronous receiver/transmitter (USART), 297
UNIX:
  backup system of, 433
  for cloud computing, 395
  CPU time for, 397–398
  data flows in, 385–387, 385*f*–387*f*
  embedded systems for, 395
  file handling in, 385–386
  FS of, 429
  Internet and, 462
  Linux and, 392
  for multiuser systems, 394
  parity memory for, 336
  programming for, 372
  programs for, 383–388
  redirections in, 385–387
  shell, 384–385, 396
  utility software for, 387–388
Unsigned binary, 26

Unsorted block image FS (UBIFS), 424
USART. *See* Universal
    synchronous/asynchronous
    receiver/transmitter
USB. *See* Universal serial bus
User datagram protocol (UDP), 448
    DNS for, 479
    port number for, 482–483, 483*f*
    vs. TCP, 484–485
User interface. *See also* Graphical user
    interface
    DCP for, 232
    in third-generation computers, 8
Utility software, for UNIX, 387–388

V

V (overflow) flag, 134, 182
Vacuum tubes, 5*n*2, 6
Variables, in memory, 108
VAX machines, 21
VBR. *See* Volume boot record
VDU. *See* Visual display unit
Vector floating point (VFP), 166
Vector parallelism, 228
Verification:
    of embedded systems CPUs, 328–336
    for hardware software codesign, 365
Very large-scale integration (VLSI), 9
Very long instruction word (VLIW), 246–250,
    247*f*, 248*f*
    IPC for, 199
VESA, 262, 263
VFP. *See* Vector floating point
VHSIC hardware description language
    (VHDL), 358, 358*n*8
    soft core processors and, 360
VIA Isaiah (Nano), 293–294, 293*f*, 298
VIDC, 93
Video RAM (VRAM), 321
Virtual address, 21–22
Virtual machine (VM), 394, 396–397
Virtual memory:
    cache memory and, 146
    MMU and, 136
    OS and, 399–400
    in third-generation computers, 8
Virtualization technology, 237
Visual display unit (VDU), 115
VLIW. *See* Very long instruction word

VLSI. *See* Very large-scale integration
VM. *See* Virtual machine
Volatile memory, 405, 417
    for embedded systems CPUs, 327–328
Volume boot record (VBR), 421
von Neumann architecture, 6, 19
    cache memory and, 145
VRAM. *See* Video RAM
VxWorks, 393
    as layered OS, 402–403
    VM, 394

W

WAN. *See* Wide area network
WAR. *See* Write-after-read
Watchdog timers (WDTs), 340–341
Watchpoint, 171*n*7
WAW. *See* Write-after-write
WDTs. *See* Watchdog timers
Wetware, 518–519
Whetstone, 116
Wide area network (WAN), 462, 490
WiFi, 467, 474, 487
Wilkes, Maurice, 80
WiMax. *See* Worldwide Interoperability for
    Microwave Access
Windows. *See* Microsoft Windows
Wireless, 282–285, 283*f*, 486–489
    interfaces and, 284
    symbol errors in, 446
Worldwide Interoperability for Microwave
    Access (WiMax), 437, 487–488
Write, 401
Write-after-read (WAR), 181, 191, 196, 244
Write-after-write (WAW), 181
    compilers and, 327
    scoreboarding for, 196
Write back, 147
Write deferred, 147
Write through, 147
Write-through with no write-allocate
    (WTNWA), 147

X

xD. *See* Secure digital
Xilinx Microblaze, 366
XIP. *See* Execute-in-place
XOR. *See* Exclusive-OR gate

Y

Yet another flash filing system (YAFFS), 424

Z

Z (zero) flag, for pipelines, 182, 183
Z shell (zsh), 384

Zero, IEEE754, 54, 55
Zero-overhead loops (ZOL), 201–206, 203*f*
ZigBee, 488–489
ZOL. *See* Zero-overhead loops
zsh. *See* Z shell
Zuse, Konrad, 4

www.ingramcontent.com/pod-product-compliance
Lightning Source LLC
Chambersburg PA
CBHW061925190326
41458CB00009B/2652